Methods in Cell Biology

VOLUME 41
Flow Cytometry
SECOND EDITION, PART A

Series Editors

Leslie Wilson
Department of Biological Sciences
University of California, Santa Barbara
Santa Barbara, California

Paul Matsudaira
Whitehead Institute for Biomedical Research and
Department of Biology
Massachusetts Institute of Technology
Cambridge, Massachusetts

Methods in Cell Biology

Prepared under the Auspices of the American Society for Cell Biology

VOLUME 41
Flow Cytometry
SECOND EDITION, PART A

Edited by

Zbigniew Darzynkiewicz
Cancer Research Institute
New York Medical College
Valhalla, New York

J. Paul Robinson
Department of Physiology and Pharmacology
School of Veterinary Medicine
Purdue University
West Lafayette, Indiana

Harry A. Crissman
Biomedical Sciences Division
Los Alamos National Laboratories
Los Alamos, New Mexico

ACADEMIC PRESS

San Diego   New York   Boston   London   Sydney   Tokyo   Toronto

Front cover photograph (paperback edition only): Multicolor PRINS. For more details see Color Plate 3 (Chapter 6) in Volume 42 of *Methods in Cell Biology.*

Academic Press, Inc.
A Division of Harcourt Brace & Company
525 B Street, Suite 1900, San Diego, California 92101-4495

United Kingdom Edition published by
Academic Press Limited
24-28 Oval Road, London NW1 7DX

International Standard Serial Number: 0091-679X

International Standard Book Number: 0-12-564142-7 (Hardcover)

International Standard Book Number: 0-12-203051-6 (Paperback)

PRINTED IN THE UNITED STATES OF AMERICA
94 95 96 97 98 99 EB 9 8 7 6 5 4 3 2 1

I dedicate these volumes to my friends, the late Zalmen A. Arlin and Joel Brander, who recently lost their battles with cancer. The courage they showed fighting this disease inspires us in the continuing efforts to eradicate it from the face of the earth, so that future generations may be spared the tragedies that affected them and their families. We are also inspired by the support offered by Ann Arlin, Julie Brander, and Michael Bolton, through the "This Close" Foundation for Cancer Research, Inc., established in memory of Zal and Joel.

This book could not have been prepared without the help of my dearest friend Irene Logsdon.

Zbigniew Darzynkiewicz

CONTENTS

4. Multiparameter Analysis of Leukocytes by Flow Cytometry

Carleton C. Stewart and Sigrid J. Stewart

5. Immunophenotyping Using Fixed Cells

George F. Babcock and Susan M. Dawes

6. Simultaneous DNA Content and Cell-Surface Immunofluorescence Analysis

Paul P. T. Brons, Piet E. J. Van Erp, and Arie H. M. Pennings

19. Immunochemical Quantitation of Bromodeoxyuridine: Application to Cell-Cycle Kinetics

Frank Dolbeare and Jules R. Selden

20. Application and Detection of IdUrd and CldUrd as Two Independent Cell-Cycle Markers

Jacob A. Aten, Jan Stap, Ron Hoebe, and Piet J. M. Bakker

21. Cell-Cycle Analysis Using Continuous Bromodeoxyuridine Labeling and Hoechst 33358–Ethidium Bromide Bivariate Flow Cytometry

Martin Poot, Holger Hoehn, Manfred Kubbies, Angelika Grossmann, Yuchyau Chen, and Peter S. Rabinovitch

Contents of Volume 42
Flow Cytometry, Second Edition, Part B

CONTRIBUTORS

Numbers in parentheses indicate the pages on which the authors' contributions begin.

Jacob A. Aten (317), Centre for Microscopial Research University, 1105 A2 Amsterdam, The Netherlands

George F. Babcock (81), Department of Surgery, University of Cincinnati Medical Center, and Shriners Hospital for Crippled Children–Burns Institute, Cincinnati, Ohio 45267

Piet J. M. Bakker (317), Centre for Microscopial Research University, 1105 A2 Amsterdam, The Netherlands

Kenneth D. Bauer (351), Department of Immunology, Genentech, Inc., South San Francisco, California 94080

Michael J. Boyer (135), Department of Medical Oncology, Royal Prince Alfred Hospital, Sydney, New South Wales 2050, Australia

Robert A. Bray (103), Department of Pathology, Emory University Hospital, Atlanta, Georgia 30322

Paul P. T. Brons (95), Department of Hemathology, University Hospital at Nijmegen, 6500 HB Nijmegen, The Netherlands

Wayne O. Carter (437), Department of Immunopharmacology, School of Veterinary Medicine, Purdue University, West Lafayette, Indiana 47907

Yuchyau Chen (327), Department of Pathology, School of Medicine, University of Washington, Seattle, Washington 98195

Ib Jarle Christensen (219), Finsen Laboratory, DK-2100 Copenhagen, Denmark

Harry A. Crissman (175, 195, 341), Cell Biology Group, Los Alamos National Laboratory, Los Alamos, New Mexico 87545

John D. Crissman (1), Department of Pathology, Harper Hospital, Detroit, Michigan 48201

Zbigniew Darzynkiewicz (15, 185, 401, 421, 527), Cancer Research Institute, New York Medical College, Valhalla, New York 10595

Susan M. Dawes (81), Shriners Hospital for Crippled Children–Burns Institute, Cincinnati, Ohio 45229

Caroline Dive (469), M.R.C. Clinical Oncology Unit, The Medical School, Cambridge CB2 2QH, United Kingdom

Frank Dolbeare (297), Biology and Biotechnology Program, Lawrence Livermore National Laboratory, Livermore, California 94550

Lynn G. Dressler (241), Lineberger Comprehensive Cancer Center, School of Medicine, University of North Carolina–Chapel Hill, North Carolina 27599

Ger J. van den Engh (509), Department of Molecular Biotechnology, University of Washington, Seattle, Washington 98915

Walter Giaretti (389), Laboratorio di Biofisica e Citrometria, Istituto Nazionale per la Ricerca sul Cancro, 16132 Genoa, Italy

Jianping Gong (15, 421), Cancer Research Institute, New York Medical College, Valhalla, New York 10595

Angelika Grossmann (327), Department of Comparative Medicine, School of Medicine, University of Washington, Seattle, Washington 98195

Robb Habbersett (341), Cell Biology Group, Los Alamos National Laboratory, Los Alamos, New Mexico 87545

David W. Hedley (135, 231), Department of Medicine and Pathology, Princess Margaret Hospital, Toronto, Ontario, Canada M4X 1K9

Hans Herweijer (509), Department of Medical Oncology, University Hospital, 3015 GD Rotterdam, The Netherlands

Greg T. Hirons (195), Cell Biology Group, Los Alamos National Laboratory, Los Alamos, New Mexico 87545

Ron Hoebe (317), Centre for Microscopial Research University, 1105 A2 Amsterdam, The Netherlands

Holger Hoehn (327), Department of Human Genetics, Biocenter Am Hubland, University of Würzburg, 97074 Würzburg, Germany

James W. Jacobberger (351), Cancer Center, Case Western Reserve University, School of Medicine, Cleveland, Ohio 44106

Richard R. Jonker (509), Department of Medical Oncology, University Hospital, 3015 GD Rotterdam, The Netherlands

Carl H. June (149), Department of Immunobiology, Naval Medical Research Institute, Bethesda, Maryland 20889

Wolfgang Kellermann (449), Institut für Anästhesiologie, Klinikum Grosshadern der Universität, D-81377 München, Germany

Sven Klingel (449), Arbeitsgruppe Zellbiochemie, Max-Planck-Institut für Biochemie, D-82152 Martinsried, Germany

Manfred Kubbies (327), Boehringer–Mannheim Research Center, D-8122 Penzberg, Germany

Judith Laffin (543), Department of Microbiology, Immunology and Molecular Genetics, Albany Medical College, Albany, New York 12208

Jørgen K. Larsen (377), Finsen Laboratory, Haematological/Oncological Center, Rigshospitalet, DK-2100 Copenhagen, Denmark

John M. Lehman (543), Department of Microbiology, Immunology and Molecular Genetics, Albany Medical College, Albany, New York 12208

Xun Li (15), Cancer Research Institute, New York Medical College, Valhalla, New York 10595

Padma Kumar Narayanan (437), Department of Immunopharmacology, School of Veterinary Medicine, Purdue University, West Lafayette, Indiana 47907

Kees Nooter (509), Department of Medical Oncology, University Hospital, 3015 GD Rotterdam, The Netherlands

Michael Nüsse (389), GSF-Forschungszentrum für Umwelt und Gesundheit, Institut für Biophysikalische Strahlenforschung, AG Durchflusszytometrie, D-85758 Oberschleissheim, Germany

Noboru Olshi (341), Cell Biology Group, Los Alamos National Laboratory, Los Alamos, New Mexico 87545

Friedrich J. Otto (211), Fachklinik Hornheide, D-48157 Münster, Germany

Arie H. M. Pennings (95), Department of Hemathology, University Hospital at Nijmegen, 6500 HB Nijmegen, The Netherlands

Martin Poot (327), Department of Human Genetics, Biocenter/Am Hubland, University of Würzburg, 97074 Würzburg, Germany

Peter S. Rabinovitch (149, 263, 327), Department of Pathology, School of Medicine, University of Washington, Seattle, Washington 98195

J. Paul Robinson (437, 461), Department of Physiology and Pharmacology, School of Veterinary Medicine, Purdue University, West Lafayette, Indiana 47907

Gregor Rothe (449), Arbeitsgruppe Zellbiochemie, Max-Planck-Institut für Biochemie, D-82152 Martinsried, Germany

Larry C. Seamer (241), University of New Mexico Cancer Center, Flow Cytometry Facility, School of Medicine, Albuquerque, New Mexico 87151

Jules R. Selden (297), Department of Safety Assessment, Merck Research Laboratories, West Point, Pennsylvania 19486

Howard M. Shapiro (121), Howard M. Shapiro, M.D., P.C., West Newton, Massachusetts 02165

Jan Stap (317), Centre for Microscopial Research University, 1105 A2 Amsterdam, The Netherlands

John A. Steinkamp (175), Cell Biology Group, Los Alamos National Laboratory, Los Alamos, New Mexico 87545

Carleton C. Stewart (39, 61), Laboratory of Flow Cytometry, Roswell Park Cancer Institute, Buffalo, New York 14263

Sigrid J. Stewart (39, 61), Laboratory of Flow Cytometry, Roswell Park Cancer Institute, Buffalo, New York 14263

Frank Traganos (185, 421), Cancer Research Institute, New York Medical College, Valhalla, New York 10595

John J. Turek (461), Department of Veterinary Anatomy, Purdue University, West Lafayette, Indiana 47907

Günter K. Valet (449), Arbeitsgruppe Zellbiochemie, Max-Planck-Institut für Biochemie, D-82152 Martinsried, Germany

Piet E. J. Van Erp (95), Department of Hemathology, University Hospital at Nijmegen, 6500 HB Nijmegen, The Netherlands

Lars L. Vindeløv (219), Department of Haematology, Rigshospitalet, DK-2100 Copenhagen, Denmark

Daniel W. Visscher (1), Department of Pathology, Harper Hospital, Detroit, Michigan 48201

James V. Watson (469), M.R.C. Clinical Oncology Unit, The Medical School, Cambridge CB2 2QH, United Kingdom

PREFACE TO THE SECOND EDITION

The first edition of this book appeared four years ago (*Methods in Cell Biology*, Vol. 33, *Flow Cytometry*, Z. Darzynkiewicz and H. A. Crissman, Eds., Academic Press, 1990). This was the first attempt to compile a wide variety of flow cytometric methods in the form of a manual designed to describe both the practical aspects and the theoretical foundations of the most widely used methods, as well as to introduce the reader to their basic applications. The book was an instant publishing success. It received laudatory reviews and has become widely used by researchers from various disciplines of biology and medicine. Judging by this success, there was a strong need for this type of publication. Indeed, flow cytometry has now become an indispensable tool for researchers working in the fields of virology, bacteriology, pharmacology, plant biology, biotechnology, toxicology, and environmental sciences. Most applications, however, are in the medical sciences, in particular immunology and oncology. It is now difficult to find a single issue of any biomedical journal without an article in which flow cytometry has been used as a principal methodology. This book on methods in flow cytometry is therefore addressed to a wide, multidisciplinary audience.

Flow cytometry continues to rapidly expand. Extensive progress in the development of new probes and methods, as well as new applications, has occurred during the past few years. Many of the old techniques have been modified, improved, and often adapted to new applications. Numerous new methods have been introduced and applied in a variety of fields. This dramatic progress in the methodology, which occurred recently, and the positive reception of the first edition, which became outdated so rapidly, were the stimuli that led us to undertake the task of preparing a second edition.

The second edition is double the size of the first one, consisting of two volumes. It has a combined total of 71 chapters, well over half of them new, describing techniques that had not been presented previously. Several different methods and strategies for analysis of the same cell component or function are often presented and compared in a single chapter. Also included in these volumes are selected chapters from the first edition. Their choice was based on the continuing popularity of the methods; chapters describing less frequently used techniques were removed. All these chapters are updated, many are extensively modified, and new applications are presented.

From the wide spectrum of chapters presented in these volumes it is difficult to choose those methods that should be highlighted because of their novelty, possible high demand, or wide applicabilities. Certainly those methods that offer new tools for molecular biology belong in this category; they are presented in chapters on fluorescence *in situ* hybridization (FISH), primed *in situ* labeling (PRINS), mRNA species detection, and molecular phenotyping. Detection of intracellular viruses and viral proteins and analysis of bacteria, yeasts, and plant cells are broadly described in greater detail than before in separate chapters. The chapter on cell viability presents and compares ten different methods for identifying dead cells and discriminating between apoptosis and necrosis, including a new method of DNA gel electrophoresis designed for the detection of degraded DNA in apoptotic cells. The chapter describing analysis of enzyme kinetics by flow cytometry is very complete. The subject of magnetic cell sorting is also described in great detail.

Numerous chapters which focus on the analysis of cell proliferation also should be underscored. The subjects of these chapters include univariate DNA content analysis (using a variety of techniques and fluorochromes applicable to cell cultures, fresh clinical samples, or paraffin blocks), the deconvolution of DNA content frequency histograms, multivariate (DNA vs protein or DNA vs RNA content) analysis, simple and complex assays of cell cycle kinetics utilizing BrdUrd and IdUdr incorporation, and studies of the cell cycle based on the expression of several proliferation-associated antigens, including the G_1- and G_2-cyclin proteins. Approaches to discriminating between cells having the same DNA content but at different positions in the cell cycle (e.g., noncycling G_0 vs cycling G_1, G_2 vs M, and G_2 of lower DNA ploidy vs G_1 of higher ploidy) are also presented.

Many of the methods described in these volumes will be used extensively in the fields of toxicology and pharmacology. Among these are the techniques designed for analysis of somatic mutants, formation of micronuclei, DNA repair replication, and cumulative DNA damage in sperm cells (DNA *in situ* denaturability). The latter is applicable as a biological dosimetry assay. A plethora of methods for analysis of different cell functions (functional assays) will also find application in toxicology and pharmacology.

The largest number of chapters is devoted to methods having clinical applications, either in medical research or in routine practice. Chapters dealing with lymphocyte phenotyping, reticulocyte and platelet analysis, analysis and sorting of hemopoietic stem cells, various aspects of drug resistance, DNA ploidy, and cell cycle measurements in tumors are very exhaustive. Diagnosis and disease progression assays in HIV-infected patients, as well as sorting of biohazardous specimens, new topics of current importance in the clinic, are also represented in this book.

Individual chapters are written by the researchers who either developed the described methods, contributed to their modification, or found new applications and have extensive experience in their use. Thus, the authors represent a "Who's Who" directory in the field of flow cytometry. This ensures that the essential details of each methodology are included and that readers may easily learn these techniques by following the authors' protocols. We would like to express our gratitude to all contributing authors for sharing their knowledge and experience.

The chapters are designed to be of practical value for anyone who intends to use them as a methods handbook. Yet, the theoretical bases of most of the techniques are presented in detail sufficient for teaching the principle underlying the described methodology. This may be of help to those researchers who want to modify the techniques, or to extend their applicability to other cell systems. Understanding the principles of the method is also essential for data evaluation and for recognition of artifacts. A separate section of most chapters is devoted to the applicability of the described method to different biological systems. Another section of most chapters covers the critical points of the procedure, possible pitfalls, and experience of the author(s) with different instruments. Appropriate controls, standards, instrument adjustments, and calibrations are the subjects of still another section of each chapter. Typical results, frequently illustrating different cell types, are presented and discussed in yet another section. The Materials and Methods section of each chapter is exhaustive, providing a detailed, step-by-step description of the procedure in a protocol or cookbooklike format. Such exhaustive treatment of the methodology is unique; there is no other publication on the subject of similar scope.

We hope that the second edition of *Flow Cytometry* will be even more successful than the first. The explosive growth of this methodology guarantees that soon there will be the need to compile new procedures for a third edition.

Zbigniew Darzynkiewicz
J. Paul Robinson
Harry A. Crissman

PREFACE TO THE FIRST EDITION

Progress in cell biology has been closely associated with the development of quantitative analytical methods applicable to individual cells or cell organelles. Three distinctive phases characterize this development. The first started with the introduction of microspectrophotometry, microfluorometry, and micro-interferometry. These methods provided a means to quantitate various cell constituents such as DNA, RNA, or protein. Their application initiated the modern era in cell biology, based on quantitative—rather than qualitative, visual—cell analysis. The second phase began with the birth of autoradiography. Applications of autoradiography were widespread and this technology greatly contributed to better understanding of many functions of the cell. Especially rewarding were studies on cell reproduction; data obtained with the use of autoradiography were essential in establishing the concept of the cell cycle and generated a plethora of information about the proliferation of both normal and tumor cells.

The introduction of flow cytometry initiated the third phase of progress in methods development. The history of flow cytometry is short, with most advances occurring over the past 15 years. Flow cytometry (and, associated with it electronic cell sorting) offers several advantages over the two earlier methodologies. The first is the rapidity of the measurements. Several hundred, or even thousands of cells can be measured per second, with high accuracy and reproducibility. Thus, large numbers of cells from a given population can be analyzed and rare cells or subpopulations detected. A multitude of probes have been developed that make it possible to measure a variety of cell constituents. Because different constituents can be measured simultaneously and the data are recorded by the computer in list mode fashion, subsequent bi- or multivariate analysis can provide information about quantitative relationships among constituents either in particular cells or between cell subpopulations. Still another advantage of flow cytometry stems from the capability for selective physical sorting of individual cells, cell nuclei, or chromosomes, based on differences in the variables measured. Because some of the staining methods preserve cell viability and/or cell membrane integrity, the reproductive and immunogenic capacity of the sorted cells can be investigated. Sorting of individual chromosomes has already provided the basis for development of chromosomal DNA libraries, which are now indispensable in molecular biology and cytogenetics.

Flow cytometry is a new methodology and is still under intense development, improvement, and continuing change. Most flow cytometers are quite complex and not yet user friendly. Some instruments fit particular applications better than others, and many proposed analytical applications have not been extensively tested on different cell types. Several methods are not yet routine and a certain degree of artistry and creativity is often required in adapting them to new biological material, to new applications, or even to different instrument designs. The methods published earlier often undergo modifications or improvements. New probes are frequently introduced.

This volume represents the first attempt to compile and present selected flow cytometric methods in the form of a manual designed to be of help to anyone interested in their practical applications. Methods having a wide immediate or potential application were selected, and the chapters are written by the authors who pioneered their development, or who modified earlier techniques and have extensive experience in their application. This ensures that the essential details are included and that readers may easily master these techniques in their laboratories by following the described procedures.

The selection of chapters also reflects the peculiarity of the early phase of method development referred to previously. The most popular applications of flow cytometry are in the fields of immunology and DNA content–cell cycle analysis. While the immunological applications are now quite routine, many laboratories still face problems with the DNA measurements, as is evident from the poor quality of the raw data (DNA frequency histograms) presented in many publications. We hope that the descriptions of several DNA methods in this volume, some of them individually tailored to specific dyes, flow cytometers, and material (e.g., fixed or unfixed cells or isolated cell nuclei from solid tumors), may help readers to select those methods that would be optimal for their laboratory setting and material. Of great importance is the standardization of the data, which is stressed in all chapters and is a subject of a separate chapter.

Some applications of flow cytometry included in this volume are not yet widely recognized but are of potential importance and are expected to become widespread in the near future. Among these are methods that deal with fluorescent labeling of plasma membrane for cell tracking, flow microsphere immunoassay, the cell cycle of bacteria, the analysis and sorting of plant cells, and flow cytometric exploration of organisms living in oceans, rivers, and lakes.

Individual chapters are designed to provide the maximum practical information needed to reproduce the methods described. The theoretical bases of the methods are briefly presented in the introduction of most chapters. A separate section of each chapter is devoted to applicability of the described method to different biological systems, and when possible, references are provided to articles that review the applications. Also discussed under separate subheads are the critical points of procedure, including the experience of the authors with different instruments, and the appropriate controls and standards. Typical results, often illustrating different cell types, are presented and discussed in the "Results" section. The "Materials and Methods" section of each chapter is the most extensive, giving a detailed description of the method in a cookbook format.

Flow cytometry and electronic sorting have already made a significant impact on research in various fields of cell and molecular biology and medicine. We hope that this volume will be of help to the many researchers who need flow cytometry in their studies, stimulate applications of this methodology to new areas, and promote progress in many disciplines of science.

Zbigniew Darzynkiewicz
Harry A. Crissman

CHAPTER 1

Dissociation of Intact Cells from Tumors and Normal Tissues

Daniel W. Visscher and John D. Crissman

Department of Pathology
Harper Hospital
Wayne State University
Detroit, Michigan 48201

I. Introduction

The great majority of flow cytometric (FCM) DNA analysis literature published to date has made use of nuclear suspensions—obtained by enzyme digestion either from formalin-fixed, paraffin-embedded tissue blocks (so-called Hedley method) or from unfixed tissue samples utilizing the trypsin-detergent technique (Vindeløv *et al.*, 1983). Nuclear extraction from paraffin blocks offers the considerable advantage of allowing retrospective analysis of archival tumor samples. However, it is generally associated with creation of nuclear debris and is poorly reproducible owing to variable fixation and tissue handling conditions at the time of initial pathologic evaluation. Although the Vindeløv protocol optimizes DNA histograms (i.e., low coefficients of variation with minimal debris), its use is limited exclusively to nuclear suspensions (Frierson, 1988).

Using two (or three)-color analysis, solid tumor cytometry has been adapted to evaluation of multiple features including nuclear DNA content (Feitz *et al.,* 1985). These include protooncogene expression and cell-lineage-specific analysis. Obviously, FCM provides an optimal means for comparing various phenotypic traits with ploidy and/or cell-cycle distribution. However, cytoplasmic or plasma membrane distribution of most epitopes mandates use of intact cell suspensions for most studies of this type. Intact, viable cells are also required if heterotransplantation or *in vitro* cell analyses (i.e., labeled substrate incorporation) are incorporated into FCM research protocols.

Dissociation protocols which yield intact cell suspensions are generally less technically demanding than enucleation methods. However, they require fresh, unfixed tissues which may not be readily available to all flow cytometry laboratories. Without tissue sections to guide sample selection in fresh tumors, neoplastic cell representation becomes an important factor. Equally critical is preservation of plasma membrane and cytoplasmic constituents from excessive traumatization.

Criteria for quality of cell dispersion are poorly defined in the cytometry literature. Moreover, controlled studies comparing dissociation techniques, especially for whole-cell methods, are difficult to find. In this chapter, disaggregation success is assessed by the following parameters: (1) overall yield (i.e., cells per gram of tissue), (2) proportion of detectable neoplastic events in DNA histogram, (3) histogram quality (coefficient of variation and baseline profile), and (4) morphologic appearance of disaggregated cells in cytologic preparations. Concepts of two-color, whole-cell analysis are also introduced in order to illustrate the effect of certain dissociation artifacts on DNA histograms.

II. Application

As previously noted, the fact that whole-cell dissociation must be performed on fresh, unfixed samples creates a set of problems not encountered with enucleation procedures. First, autolysis limits both cell and cytoplasmic recovery. Tissue samples therefore need to be partitioned as soon as possible from resected tumors. If immediate dissociation is not possible, samples should be held in a balanced media (such as Hanks' balanced salt solution) at refrigerator temperature. Although there are no published studies systematically examining the effects of autolysis, it has been the experience of our laboratory that histogram quality and cell integrity deteriorate perceptibly after approximately 6 hr of unfixed storage, becoming unacceptable in most cases after 24–36 hr. Also, since tumor aliquots are obtained prospectively, without microscopic assessment, this step mandates careful supervision by an experienced pathologist (i.e., for clinical patient specimens). Apart from identifying and sampling quantitatively sufficient neoplasm it is critical to account for the qualitative heterogeneity intrinsic to all solid tumors. One important source of heterogeneity is phenotypic variability within neoplastic populations per se (Kallioniemi, 1988). In colorectal carcinoma, for example, it is important to sample the area of deepest muscular layer invasion since the surface may consist in large part of benign adenoma. The deepest component also contains the most biologically aggressive, and thereby most analytically relevant, cells. Tumors such as prostatic adenocarcinoma present other problems in prospective sampling studies by virtue of their multifocality and poor gross demarcation from surrounding benign glandular tissue. For studies of this type we recommend use of morphologic controls (i.e., tissue sections from immediately adjacent areas) to confirm presence and type of neoplastic tissue.

Finally, the amount of fibrous stroma, necrosis, and host inflammatory infiltrate varies dramatically between different tumors of the same primary and especially in breast carcinomas. Clearly, this makes it difficult to rigidly define how much tissue is required in order to obtain an adequate cellular yield.

Whole-cell dissociation techniques are also fundamentally impacted by ultrastructural architectural features of the particular tumor type being examined. The most important characteristics affecting intact cell dissociability are: (1) number and type of adhesive cell junctions, (2) relative amount and nature of extracellular matrix, and (3) intrinsic cytoplasmic integrity of the neoplastic population. It should be noted that each of these features varies considerably, not only between tumor systems of different primary site or histologic classification, but within the spectrum of tumors from any given pathologic category (e.g., ductal adenocarcinoma of the breast). This inherent biologic variability represents an important source of dissociation inconsistencies, even in laboratories with rigorously uniform procedures.

Two common strategies which have been employed to disrupt adhesive contacts between epithelial cells and their extracellular matrices are mechanical force (shearing, mincing, aspiration) and enzymatic cleavage of stromal components (with collagenase) or intercellular linkages (with any number of proteases). Chelating agents, such is EDTA, have been successfully employed in some studies to achieve cell suspensions (Eade et al., 1981). Use of proteases, such as trypsin, risks damage to cell-surface molecules and this factor should be considered if membrane-associated proteins, such as ERBB-2, are being evaluated. Enzymatic digestions simultaneously employ a DNase—in order to prevent reaggregation by "sticky" free DNA strands released from damaged cells. Obviously enzymatic cell dispersion protocols require careful monitoring (time, temperature, enzyme concentration) to achieve technical reproducibility. They should employ only highly purified enzyme sources (Hefley et al., 1983). In general, tumors of glandular origin (colon carcinomas, breast carcinomas, prostate carcinomas), renal cell carcinoma, and transitional carcinoma may be dissociated either mechanically or enzymatically. Squamous carcinomas, in contrast, have well-developed cell attachments (desmosomes) which require enzymatic digestion in order to yield representative single-cell suspensions.

Table I summarizes six studies which have compared mechanical and enzymatic disaggregation, predominantly in neoplasms of glandular derivation (all nonsquamous). It should be noted these protocols are quite variable with respect to enzymes employed and their incubation conditions (time, temperature, and concentration). Trypsin or collagenase is utilized in most studies, although a variety of other enzymes (hyaluronidase, bacterial protease) have also been used for solid tumor dispersion (Pallavicini *et al.*, 1981; Engelholm *et al.*, 1985; Rong *et al.*, 1985). Corresponding methods of mechanical dissociation were also variable (scraping, mincing, or needle aspiration) and generally not described in great detail.

Authors who evaluated postdispersion viability uniformly reported this parameter was optimized in enzymatic protocols (Slocum *et al.*, 1981). This is expected, given the gross cytoplasmic trauma induced by shearing and tearing, versus the relatively selective, molecular level, cleavages produced by enzyme digestion. Although viability is critical to success of assays involving clonogenicity or active cell metabolism, it is not requisite for analysis for DNA content or antigen expression using immunocytochemical techniques.

With respect to histogram quality and overall cell yield neither approach was clearly superior. Five of six authors, however, reported enzymatic disaggregation techniques were associated with lower relative yields of DNA aneuploid (i.e., neoplastic) populations. The mechanisms responsible for this consistent finding are not clear and have yet to be systemically investigated. Therefore,

Table I
Summary of Technical Studies Comparing Mechanical vs Enzymatic Dissociation of Glandular Tumors

		Techniques			Method superior for		
Author	Tumor	Mechanical	Enzyme	Overall yield	Aneuploid recovery	Histogram quality	Viability
Bach *et al.* (1991)	Colon Breast	FNA[a] (*in vitro*)	Trypsin[a] and collagenase	NA[b]	Mechanical	Mechanical	NA
Costa *et al.* (1987)	Multiple	Mincing	Collagenase	Similar	Mechanical	NA	Enzyme
Chassevent *et al.* (1984)	Breast	Mincing	Collagenase	Mechanical	Mechanical	NA	Enzyme
Engleholm *et al.* (1985)	Breast	FNA (*in vivo*)	Trypsin	NA	Similar	Similar	Enzyme
McDivitt *et al.* (1984)	Breast	Scrape	Collagenase	NA	Mechanical	Similar	NA
Crissman *et al.* (1988)	Colon (mouse)	Mincing	Collagenase	Mechanical	Mechanical	Similar	NA

[a] FNA, fine needle aspiration.
[b] NA, not available.

there is little evidence in the published literature to suggest more complicated enzymatic protocols are necessary for dissociation of (nonsquamous) carcinomas in most flow cytometric studies.

Several other points made in these studies are worthy of brief mention. First, most authors employed either Ficoll separation or micropore filtering (pore sizes 30–200 μm) steps following disaggregation to remove residual macroscopic tissue aggregates. Although this is advisable from the standpoint of maintaining cytometer fluidic components, these procedures add little to improving histogram quality or neoplastic representation (McDivitt *et al.*, 1984). Second, as already noted, characteristics intrinsic to individual tumors per se were a major variable in dissociation efficiency—generally affecting all methods compared (Costa *et al.*, 1987) to a similar extent. Most authors reported a two- to fivefold cell yield variability (per gram of tissue) between different tumors of any given system. Finally, comparison of different mechanical dispersion protocols is rarely performed. Some investigations suggest needle aspiration (*in vitro* or *in vivo*) is a particularly effective manner of releasing neoplastic aneuploid populations (Greenebaum *et al.*, 1984; Remvikos *et al.*, 1988).

Thus, although enzymatic digestions add little to dissociation of glandular tumors, neoplasms having squamous differentiation require this approach in order break the well-developed linkages between cells of these tumors (Bijman *et al.*, 1985). Ensley *et al.* (1987a,b) have reported both overall yield and viability are enhanced by incubations using a mixture of collagenase and trypsin, presumably reflecting synergy due to combined lysis of extracellular matrix and cell adhesive proteins. Their procedure is dependent on both time and temperature of enzyme treatment, emphasizing the need for careful optimization of proteolytic cell dispersion. A noteworthy finding in this, and other, studies is that prolonged enzyme digestions are associated with lower viabilities due to excessive cellular damage (Allalunis-Turner and Siemann, 1986).

III. Methodology: Mechanical Cell Dispersion

A. Procedure

Tissue aliquots are held at 4°C until processed.

1. Working in a biological cabinet, remove tissue from *holding medium* and blot on gauze.

2. Using a surgical blade, trim fat and areas of necrosis, then weigh remaining tissue.

3. Place tissue in a Petri dish containing 5 to 10 ml of RPMI 1640 medium (Gibco, Grand Island, New York). Using forceps and a No. 22 surgical blade, gently scrape the tissue until the solution becomes turbid. Then bisect the specimen to provide more surface area and repeat.

4. Pour the solution through an 80-μm mesh metal sieve, transfer to a 15-ml conical centrifuge tube, and centrifuge at 250g for 10 min.

5. Remove supernatant and resuspend pellet in 1 ml each of RPMI medium and inactivated fetal calf serum. Using a hemacytometer, perform a viability and cell count.

6. Fix cells slowly by adding dropwise 6 ml of cold 70% ethanol while vortexing.

7. Cap tube and store at 4°C for 12 hr before staining.

Following mechanical dispersion, the following enzymatic incubation may be performed to fully dissociate neoplasms of squamous original/differentiation (beginning after step 3).

1. Prepare enzyme cocktail consisting of 2.5 mg/ml trypsin, 0.5 mg/ml collagenase (type II), and 20 μg/ml DNase I.
2. Place tissue in fresh tube containing 10 ml/1 g tissue of enzyme cocktail.
3. Incubate on rocker 37°C, 1 hr.
4. Sieve, then add 1–2 ml FBS to stop enzyme, and wash.

B. Reagents

1. RPMI 1640 medium (Gibco No. 320-1875AJ)
2. Holding medium: Add to 10 ml of RPMI 1640 tissue culture medium 0.1 ml of 100× concentrated penicillin/streptomycin solution (Sigma Chemical Co, St. Louis, MO. No. P0781). Store in 50-ml conical centrifuge tubes at 4°C for up to 30 days.
3. Trypan blue

POSSIBLE CARCINOGEN!

4. Trypsin, collagenase, and DNase (Worthington Biochemical, Freehold, NJ).

IV. Results

A. Overall Yield

In our laboratory the mean weight of fresh colorectal carcinoma samples submitted for mechanical dissociation is 571 mg, with a resulting mean cellular yield of 1.1 × 10^8 cells. The mean weight of fresh breast carcinoma samples is 457 mg, resulting in an average yield of 8.7 × 10^7 cells. In specimens weighing less than 100 mg, neoplastic cell yields become unreliable, particularly in breast carcinomas.

We have also attempted dissociation by scraping cut tumor surfaces with a scalpel, followed by rinsing the blade with saline (i.e., without submitting intact tumor aliquots). Repetitive (i.e., three to five) scrapes yield in the range of 1/2 million–2 million cells per exposed surface.

Variability in cell yields is considerable (three- to fivefold), particularly in breast carcinomas, reflecting intertumoral heterogeneity as documented in the previously cited literature. In general, breast tumors with prominent desmoplasia result in lower yields, thus necessitating submission of larger samples.

B. Proportion of Detectable Neoplastic Events

Our laboratory routinely employs a two-color, multiparametric technique in which suspensions are dual-labeled. Nuclear DNA is stained with propidium iodide (PI) and cytoplasm is labeled with cytokeratin (Ck)-FITC or with leukocyte common antigen (LCA)-FITC (Zarbo et al., 1989). Although neither cytoplasmic label is tumor-cell specific, this method allows assessment (and gating) of populations derived from epithelium and host inflammatory infiltrates, respectively. It is assumed that the preponderance of cytokeratin-positive cells in a carefully selected carcinoma represents neoplastic elements.

In addition to allowing dissection of DNA histograms based on cytoplasmic markers, we have employed this technique to systematically evaluate neoplastic cell representation in our human tumor specimens (Visscher et al., 1991). Cytokeratin-positive cells accounted for an average of 40% of the histogram events in breast carcinomas. Since only approximately 50% of cells in a given breast tumor are neoplastic, representation in our dissociated specimens simulates the complex host/tumor cellular composition of human cancers.

Evaluating percentage cytokeratin staining among DNA aneuploid events may be utilized to assess dissociation efficiency (i.e., percentage of intact neoplastic cells). The average aneuploid recovery in our human breast carcinomas was 48% of histogram events (range 10–75%). This is expected given the relative proportion of host-derived cells as noted above. Given the average cytokeratin-positive yield of 40% of total events, it may be concluded (assuming adequate staining and detection sensitivity) that, on average, 80% of neoplastic cells have sufficient remaining cytoplasm following dissociation for multiparametric analysis. Creation of some bare nuclei accounts in great part for the 10% discrepancy between biologic and cytometric neoplastic cell representation noted previously. Neoplastic bare nuclei may be identified in two-dimensional dot plots (i.e., DNA vs cytokeratin) of appropriately gated flow cytometric data (see Fig. 1).

C. Histogram Quality

Mean G0/G1 peak coefficient of variation (CV) for breast carcinomas in our laboratory is 3.8%. This is substantially lower than reported CVs in series dissociated from fixed, paraffin-embedded tissues (Crissman et al., 1988). Also, DNA histograms derived from fresh tissues lack the characteristic descending

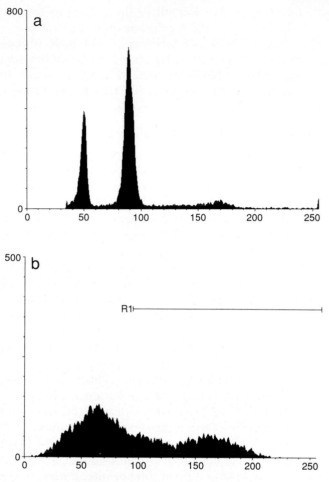

Fig. 1 Cytometric detection of bare nuclei vs intact cells following mechanical dissociation. (a) Ungated DNA histogram showing DNA aneuploid population (DNA index 1.8). (b) Fluorescence 1 (FITC) histogram of cytokeratin-labeled cells. The line indicates the position of computer gate (R1) designing Ck-positive events.

"debris slope" found in many histograms following paraffin enucleation procedures. We have not been able to identify DNA histogram artifacts associated with intact cell status. Finally, doublets or triplets are not identified in large numbers following whole-cell dissociation, presumably because cytoplasmic reassociation is less likely than nuclear DNA aggregation of enucleated suspensions. To a great extent, this reflects abnormal cell adhesion properties of neoplastic cells which are known to lack completely developed intercellular junctions.

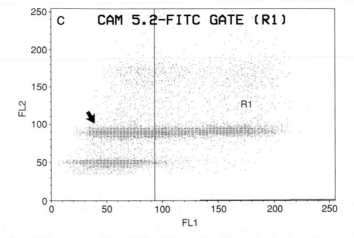

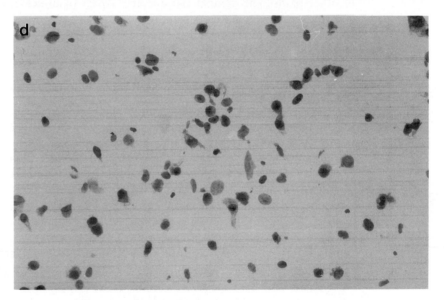

Fig. 1 (*Continued*)   (c) Two-dimensional dot plot [vertical axis, DNA(PI); horizontal axis, cytokeratin (FITC)] showing lack of cytokeratin staining in some DNA aneuploid events (arrow) due to cytoplasmic stripping during disaggregation procedure. Note presence of diploid range, cytokeratin-negative inflammatory/stromal cells. (d) Corresponding cytospin, showing intact tumor cells, bare nuclei, and host-derived inflammatory cells.

D. Morphologic Appearance of Suspensions

We strongly recommend the use of cytospin preparations as a quality control measure in whole-cell dissociation procedures. In addition to reflecting neoplastic vs host cell representation, Papanicolaou-stained cytospin pellets also may be used to assess degree of cellular trauma/cytoplasmic stripping (see Fig. 2). Our lab employs a cytocentrifuge manufactured by Shandon Southern Instruments (Sewickley, PA).

Immunostained cytospin preparations are an effective method for verifying both sensitivity and specificity of cytoplasmic dual-labeling procedures. We have demonstrated an excellent correlation between manually counted cytokeratin-positive cells in cytospins and corresponding automated flow cytometric counts (Visscher *et al.*, 1991). Although such comparisons are trivial in the case of a widely distributed antigen such as cytokeratin, they may have considerable value in protocols which quantitate expression of oncoproteins or other pathologic cellular parameters. Apart from serving as a morphologic control over dissociation and multiparameter FCM, unstained smears or cytospin preparations may be utilized for parallel investigations which require direct morphologic visualization.

V. Conclusion

Cell dissociation remains a major obstacle to the technical evolution of solid tumor flow cytometry. Unfortunately, intertumoral heterogeneity represents the most significant, and most difficult to control, variable in these procedures. We have accumulated evidence that cytoplasmic loss due to mechanical trauma is greater in more rapidly cycling (hence more clinically aggressive) tumors (Visscher *et al.*, 1991). Breast carcinomas having below-median cytokeratin-positive events are more frequently estrogen receptor negative, are of high nuclear grade, and have above-median SPF. Similarly, colorectal carcinomas typically have fewer cytokeratin-positive (and cytokeratin-positive aneuploid) events than breast carcinomas, indicative of their generally greater proliferative capacity. These considerations imply there is probably no single dispersion protocol which optimizes yield and efficiency for every human carcinoma, even from a given primary site.

It is likely, therefore, that studies employing whole-cell FCM analysis of human tumor suspensions will need to employ extensive quality control measures in order to ensure adequate representation of intact neoplastic cells. We recommend routine cytologic examination of cell suspensions in addition to use of a positive control system when cytoplasmic or plasma membrane epitopes are under analysis. Obviously, immunologic detection methods are subject to a number of potential interferences in addition to cytoplasmic trauma, especially fixation effects and inappropriate antibody concentrations. Similar considerations apply to cell metabolic studies. Thus, although whole-cell analysis opens

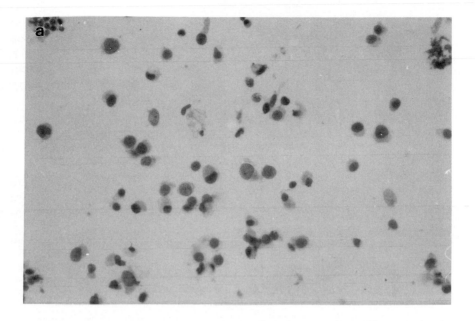

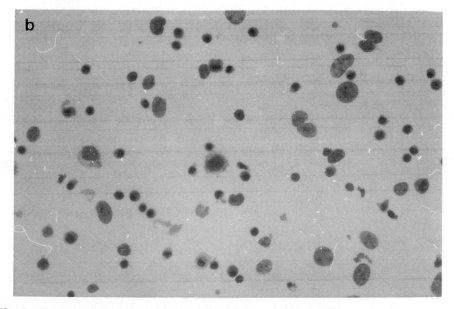

Fig. 2 Cytologic appearances of dissociated solid tumors. (a) Breast carcinoma with 68% cytokeratin-positive (Ck +) events by flow cytometry. Note intact cytoplasm surrounding dissociated single neoplastic cells. (b) Breast carcinoma, medullary subtype, showing inflammatory cells and tumor cell nuclei with stripped cytoplasm. Only 17% of events were Ck +, due to prominent inflammatory infiltrate (34% total events) as well as cytoplasmic loss. The latter is typically a problem in rapidly cycling populations, which have "fragile" cytoplasm.

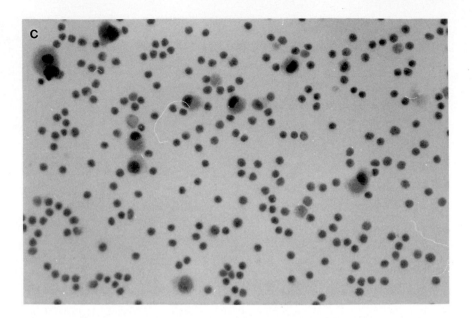

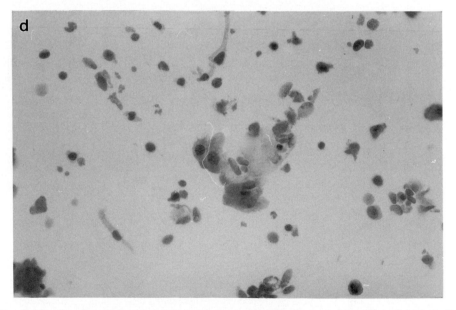

Fig. 2 (*Continued*)   (c) In this breast carcinoma, only 13% of events were Ck + ; however, the tissue was obtained from a regional lymph node metastasis. Note the presence of occasional intact epithelial cells and a background rich in lymphocytes. (d) Dissociated colorectal carcinoma, showing admixture of intact tumor cells, stripped nuclei, stromal and inflammatory cells, and fragments of debris.

many new doors in cytometric study of neoplasia, it creates a set of difficult, but not insurmountable, standardization problems.

References

Allalunis-Turner, M. J., and Siemann, D. W. (1986). *Cancer (Philadelphia)* **54**, 615–622.

Bach, B. A., Knape, W. A., Edinger, M. G., and Tubbs, R. R. (1991). *Ann. J. Clin. Pathol.* **96**, 615–628.

Bijman, J., Wagener, D. J., van Rennes, H., Wessels, J. M. C., and van den Brock, P. (1985). *Cytometry* **6**, 334–341.

Chassevent, A., Daver, A., Bertrand, G., Coic, H., Geslin, J., Bidabe, M.-C., George, P., and Larra, F. (1984). *Cytometry* **5**, 263–267.

Costa, A., Silvestrini, R., Del Bino, G., and Motta, R. (1987). *Cell Tissue Kinet.* **20**, 171–180.

Crissman, J. D., Zarbo, R. J., Niebylski, C. D., Corbett, T., Weaver, D. (1988). *Mod. Pathol.* **1**(3), 198–204.

Eade, O. E., Andre-Ukena, S. S., and Beeken, W. L. (1981). *Digestion* **21**, 25–32.

Engelholm, S. A., Spang-Thomsen, M., Brunner, N., Nohr, I., and Vindeløv, L. L. (1985). *Cancer (Philadelphia)* **51**, 93–98.

Ensley, J. F., Maciorowski, Z., Crissman, J., Pietraszkiewicz, Klemic, G., Kukuruga, M., Sapareto, S., Corbett, T., and Crissman, J. (1987a). *Cytometry* **8**, 479–487.

Ensley, J. F., Maciorowski, Z., Crissman, J., Pietraszkiewicz, Hassan, M., Kish, J., Al-Sarraf, M., Jacobs, J., Weaver, A., and Atkinson, D.(1987b). *Cytometry* **8**, 488–493.

Feitz, W. F. J., Beck, H. L. M., Smeets, A. W. G. B., Debrugne, F. M., Voojis, G. P., Herman, C. J., and Ramackers, F.C.S. (1985). *Cancer (Philadelphia)* **36**, 349–356.

Frierson, H. F., Jr. (1988). *Hum. Pathol.* **19**, 290–294.

Greenebaum, E., Koss, L. G., Sherman, A. B., and Elequin, F. (1984). *Am. J. Clin. Path.* **82**, 559–564.

Hefley, T. J., Stern, P. H., and Brand, J. S. (1983). *Exp. Cell Res.* **149**, 227–236.

Kallioniemi, O. P. (1988). *Cytometry* **9**, 164–169.

McDivitt, R. W., Stone, K. R., and Meyer, J. S. (1984). *Cancer Res.* **44**, 2628–2633.

Pallavicini, M. G., Folstad, L. J., and Dunbar, C. (1981). *Cytometry* **2**, 54–58.

Remvikos, Y., Magdelenat, H., and Zajdela, A. (1988). *Cancer (Philadelphia)* **61**, 1629–1634.

Rong, G. H., Grimm, E. A., and Sindelar, W. F. (1985). *J. Surg. Oncol.* **28**, 131–133.

Slocum, H. K., Pavelic, Z. P., Rustum, Y. M., Kanter, P. M., and Nowak, N. J. (1981). *Cancer Res.* **41**, 1428–1434.

Vindeløv, L. L., Christensen, I. J., and Nissen, N. I. (1983). *Cytometry* **3**, 323–327.

Visscher, D. W., Wykes, S., Zarbo, R. J., and Crissman, J. D. (1991). *Anal. Quant. Cytol. Histol.* **13**, 246–252.

Waymouth, C. (1982). *Cell Sep.: Methods Sel. Appl.* **1**, 1–29.

Whitehead, R. H., Brown, A., and Bhathal, S. (1987). *In Vitro Cell. Dev. Biol.* **23**, 436–442.

Zarbo, R. J., Visscher, D. W., and Crissman, J. D. (1989). *Anal. Quant. Cytol. Histol.* **11**, 391–402.

CHAPTER 2

Assays of Cell Viability: Discrimination of Cells Dying by Apoptosis

Zbigniew Darzynkiewicz, Xun Li, and Jianping Gong

Cancer Research Institute
New York Medical College
Valhalla, New York 10595

METHODS IN CELL BIOLOGY, VOL. 41

I. Introduction

A. The Modes of Cell Death: Apoptosis and Necrosis

Two distinct modes of cell death, apoptosis and necrosis, can be recognized based on differences in the morphological, biochemical, and molecular changes of the dying cell. The assays of cell viability presented in this chapter are discussed in light of their applicability to differentiate between these two mechanisms.

1. Apoptosis

The terms "apoptosis," "active cell death," "cell suicide," and "shrinkage necrosis" are being used, often interchangeably, to define a particular mode of cell death characterized by a specific pattern of changes in nucleus and cytoplasm (Wyllie, 1985, 1992; Arends *et al.*, 1990; Compton, 1992). Because this mode of cell death plays a role during the programmed cell death, as originally described in embryology, the term "programmed cell death" is also being used synonymously (albeit incorrectly, in the context of denoting the mode of cell death) with apoptosis.

The role of apoptosis in embryology, endocrinology, and immunology is the subject of several reviews (e.g., in Tomei and Cope, 1991). The wide interest in apoptosis in oncology, so apparent in recent years, stems from the observations that this mode of cell death is triggered by a variety of antitumor drugs, radiation, or hyperthermia and that the intrinsic propensity of tumor cells to respond by apoptosis is modulated by expression of several oncogenes such as *bcl*-2, c-*myc*, or tumor suppressor gene p53 and may be prognostic of treatment (Wyllie, 1992; Sachs and Lotem, 1993; Schwartzman and Cidlowski, 1993).

Extensive research is underway in many laboratories to understand the mechanism of apoptosis. Knowledge of the molecular events of this process may be the basis for new antitumor strategies. Apoptosis affecting CD4$^+$ lymphocytes of HIV-infected patients also appears to play a pivotal role in pathogenesis of AIDS (Meyaard *et al.*, 1992).

The most common feature of apoptosis is active participation of the cell in its self-annihilation. The cell mobilizes a cascade of events that lead to its disintegration and the formation of the "apoptotic bodies" which are subsequently engulfed by the neighboring cells without invoking inflammation (Wyllie, 1985, 1992; Arends *et al.*, 1990; Tomei and Cope, 1991; Compton, 1992). Increased cytoplasmic Ca^{2+} concentration, cell dehydration, increased lipid peroxidation, chromatin condensation originating at the nuclear periphery, activation of endonuclease which has preference to DNA at the internucleosomal (linker) sections, proteolysis, fragmentation of the nucleus, and fragmentation of the cell are the most characteristic events of apoptosis. On the other hand, even during advanced stages of apoptosis, the structural integrity and the transport function of the plasma membrane are preserved. Also preserved and functionally active are the mitochondria and lysosomes.

Thus, regardless of cell type, or the nature of event which triggers apoptosis, this mode of cell death has many features in common. Some of these features can be analyzed by image or flow cytometry, and several methods have been described to identify apoptotic cells (Darzynkiewicz *et al.*, 1992).

Mitotic death, also termed delayed reproductive death, shows some features of apoptosis and thus may represent delayed apoptosis (e.g., Chang and Little, 1992; Tounecti *et al.*, 1993); it occurs as a result of cell exposure to relatively low doses of drugs or radiation, which induce irreparable damage, but allow cells to complete at least one round of division.

2. Necrosis

Necrosis is an alternative to the apoptotic mechanism of cell death. Most often it is induced by an overdose of cytotoxic agents and is a cell response to a gross injury. However, certain cell types do respond even to pharmacological concentrations of some drugs or moderate doses of physical agents by necrosis rather than apoptosis and the reason for the difference in response is not entirely clear. While apoptosis requires active participation of the involved cell, often even in terms of initiation of the *de novo* protein synthesis, necrosis is a passive and degenerative process. *In vivo*, necrosis triggers the inflammatory response in the tissue, due to a release of cytoplasmic constituents to intercellular space, often resulting in scar formation. In contrast, remains of apoptotic cells are phagocytized not only by the "professional" macrophages, but also by other neighboring cells, without evoking any inflammatory reaction. The early event of necrosis is swelling of cell mitochondria, followed by rupture of the plasma membrane, and release of the cytoplasmic content (reviews in Tomei and Cope, 1991).

B. Types of Cell Viability Assays

1. Assays of Plasma Membrane Integrity

It has been recognized during the past several decades that one of the major features discriminating dead from live cells is loss of the transport function and physical integrity of the plasma membrane. A plethora of assays of cell viability has been developed based on this phenomenon. Thus, for example, because the intact membrane of live cells excludes a variety of charged dyes such as trypan blue or propidium iodide (PI), incubation with these dyes results in selective labeling of dead cells, while live cells show no, or minimal, dye uptake (Horan and Kappler, 1977). By virtue of its simplicity, the PI exclusion assay appears to be the most popular in flow cytometry. A short (~5 min) incubation with 10–20 μg/ml of this dye, in isotonic media, labels the cells that have impaired transport function of the plasma membrane. Such cells cannot exclude this charged dye, which, after crossing the plasma membrane, binds to DNA and dsRNA and fluoresces intensively.

An assay based on a similar principle makes use of the nonfluorescent substrate of esterases, fluorescein diacetate, which is taken up by live cells, and, upon hydrolysis, the product (fluorescein) is detected due to its strong green fluorescence (Hamori *et al.*, 1980). Incubation of cells in the presence of both PI and fluorescein diacetate labels live cells green (fluorescein) and dead cells red (PI). This is a convenient assay, widely used in flow cytometry. Another assay combines the use of PI and Hoechst 33342 (HO342). Both are described later in this chapter.

Ethidium monoazide, as PI, is also excluded from live cells but enters cells that have damaged plasma membrane (Riedy *et al.*, 1991). This dye, upon binding to nucleic acids, can be covalently attached to the latter in the photochemical reaction. Therefore, exposure of cells to this dye, followed by their illumination with white light, irreversibly labels dead cells. The cells can then be fixed and counterstained with another dye of different color: the distinction between the dead and live cells at the time of their exposure to the dye remains preserved after subsequent fixation, washings, and counterstaining. This method, which is often used to exclude dead cells during immunophenotyping, is described in Chapter 3 of this volume.

Integrity of the plasma membrane can also be probed by cell resistance to trypsin and DNase: while live cells remain intact during incubation with these enzymes, dead cells are nearly totally digested and removed from the analysis (Darzynkiewicz *et al.*, 1984a).

2. Function of Cell Organelles

There are several assays of cell viability based on the functional tests of cell organelles. Thus, for example, the cationic dye rhodamine 123 (Rh123) accumulates in mitochondria of live cells due to the mitochondrial transmembrane potential (Johnson *et al.,* 1980). Cell incubation with 1–10 μg/ml of Rh123 results in labeling of live cells while dead (necrotic) cells show no Rh123 uptake. As in the case of fluorescein diacetate and PI, the combination of Rh123 and PI labels live cells green (Rh 123) and dead cells red (Darzynkiewicz *et al.,* 1982).

Another functional assay involves lysosomes. Incubation of cells in the presence of 1–2 μg/ml of the metachromatic dye acridine orange (AO) results in uptake of this fluorochrome by lysosomes of live cells which luminesce in red wavelength. Dead (necrotic) cells exhibit weak green and no red fluorescence at that low an AO concentration. This assay is described in Chapter 12 of this volume, as a test of the lysosome active transport (proton pump).

As discussed above, cells which undergo apoptosis have a preserved plasma membrane and several cell functions remain little changed, relative to live cells. Therefore, this simple discrimination of live and dead cells by dye exclusion, or based on functional assays of some organelles, is inadequate to identify cells that die by apoptosis, especially at early stages of cell death. Other assays, therefore, have to be used in such situations.

3. Identification of Apoptotic Cells

Two distinct features of apoptotic cells, extensive DNA cleavage and preservation of the cell membrane, provide the basis for the development of most flow cytometric assays to discriminate between apoptosis and necrosis. DNA cleavage can be detected by extraction of the degraded, low MW DNA from the ethanol-prefixed or detergent-permeabilized cells and subsequent cell staining with DNA fluorochromes. Apoptotic cells then show reduced DNA stainability (content of high MW DNA) (Darzynkiewicz *et al.,* 1992). Alternatively, the numerous DNA strand breaks in apoptotic cells can be labeled with biotinylated or digoxigenin-conjugated nucleosides in the reaction employing exogenous terminal deoxynucleotidyl transferase (TdT) or DNA polymerase (nick translation) (Gorczyca *et al.,* 1992, 1993b,c).

Another marker of apoptosis is chromatin condensation, which can be assayed by DNA *in situ* sensitivity to denaturation (see Chapter 33 of this volume). Changes in plasma membrane permeability to HO342 resulting in differential staining of DNA with this dye can also discriminate between live and apoptotic cells (Dive *et al.,* 1992; Ormerod *et al.,* 1992, 1993).

Practical aspects of several cell viability assays, their specificity in terms of discrimination between apoptosis and necrosis, and applicability in different cell systems are described below.

II. Changes in Light Scatter during Cell Death

Cell death is accompanied by a change in its property to scatter light and thus light scatter measurement is one of the simplest assays of cell viability. At early stages, apoptosis is characterized by markedly reduced cell ability to scatter light in forward direction and either by an increase (Swat *et al.*, 1991) or no change in the 90° angle light scatter (Darzynkiewicz *et al.*, 1992). At later stages, both the forward and right angle light scatter signal are decreased. These changes are a reflection of chromatin condensation, nuclear fragmentation, cell shrinkage, and shedding of apoptotic bodies. The major advantage of the assay based on light scatter measurement is its simplicity and that it can be combined with analysis of the surface immunofluorescence, e.g., to identify the phenotype of the dying cell. It can also be combined with functional assays such as of the mitochondrial potential, lysosomal proton pump, exclusion of PI or plasma membrane permeability to such dyes as Hoechst 33342 (see further).

Reduced forward light scatter, however, is not a very specific marker of apoptosis. Mechanically broken cells, isolated cell nuclei, and necrotic cells also have low light scatter properties. Furthermore, a loss of cell-surface antigens may accompany apoptosis, especially at later stages of cell death, and this may complicate the bivariate light scatter/surface immunofluorescence analysis. This approach, therefore, requires several controls and should be accompanied by another, more specific assay, or at least by confirmation of apoptosis by microscopy.

Because different models of flow cytometers have somewhat different positions and characteristics of their light scatter detectors, the degree of light scatter changes in the same cell systems may vary from instrument to instrument and be responsible for the lack of reproducibility of this assay between laboratories.

III. Cell Sensitivity to Trypsin and DNase

A short exposure of live cells simultaneously to trypsin and DNase has little effect on their morphology, function, or viability. Conversely, cells with damaged plasma membranes are digested by these enzymes to such a degree that, for all practical purposes, they can be gated out during measurement, e.g., by raising the triggering threshold of the light scatter, DNA, or protein fluorescence signals (Darzynkiewicz *et al.*, 1984a).

A. Reagents

1. DNase I, 200 μg/ml in HBSS (with Mg^{2+} and Ca^{2+}). Aliquots of 100 μl of DNase in microfuge tubes may be stored at -40°C.
2. Trypsin, 0.5%, in HBSS. Aliquots of 10 μl of trypsin may be stored at -40°C in microfuge tubes.

B. Procedure

1. Centrifuge cells (e.g., dispersed mechanically from the solid tumor). Suspend the pellet (approximately 10^6 cells) in 0.2 ml of HBSS.

2. Add a 100 μl aliquot of DNase I, incubate 15 min at 37°C, then add 100 μl of trypsin, and incubate for an additional 30 min. Add 5 ml of HBSS with 10% calf serum or HBSS containing the soybean trypsin inhibitor, to inactivate trypsin. Centrifuge.

3. Rinse cells once in HBSS, suspend in HBSS, fix cells in suspension, or stain with a desired fluorochrome (e.g., after permeabilization with 0.1% Triton X-100 in the presence of 1% albumin) shortly after this procedure.

C. Results

This procedure removes all cells with damaged plasma membrane (dead cells) from the suspension and thus all remaining cells exclude PI. However, because of the use of trypsin, which digests the cell coat, epitopes of many cell-surface antigens of live cells may not be preserved. Incubation of so-treated cells in culture medium at 37°C for several hours restores most antigens.

Since elimination of cells by digestion with these enzymes is selective to cells with damaged plasma membrane (exclusion of trypsin and DNase), this method can be used to remove necrotic or very late apoptotic cells, as well as those cells which have been mechanically damaged (e.g., by sonication or syringing), but not early apoptotic cells. Also digested are isolated nuclei. Somewhat similar results can be obtained by separation of live cells by the Ficoll-Hypaque gradient centrifugation.

IV. Fluorescein Diacetate (FDA) Hydrolysis and PI Exclusion

FDA is a substrate for esterases, the ubiquitous enzymes that are present in all types of cells. It can penetrate into live cells. The product of hydrolysis, fluorescein, is highly fluorescent and charged. It becomes, therefore, entrapped in the cell. This assay, which combines counterstaining of dead cells with PI, has been widely used in classical cytochemistry and in flow cytometry (Hamori *et al.*, 1980).

A. Reagents

1. FDA, dissolved in acetone, 1 mg/ml (stock solution).
2. PI, dissolved in distilled water, 1 mg/ml, (stock solution).

B. Procedure

1. Suspend approximately 10^6 cells in 1 ml of HBSS.
2. Add 2 μl of the FDA stock solution.
3. Incubate cells at 37°C for 15 min.
4. Add 20 μl of the PI stock solution.
5. Incubate 5 min at room temperature, and analyze by flow cytometry.

Both dyes are excited in blue light (488-nm line of the argon ion laser). Measure green fluorescence at 530 ± 20 nm and red fluorescence at >620 nm. Live cells show green, dead cells and red fluorescence.

This method, which is based on analysis of the integrity of the plasma membrane, as other methods based on the same principle, discriminates between necrotic cells (or cells with damaged membrane) and live cells, but not between live and apoptotic cells.

V. Rh123 Uptake and PI Exclusion

Uptake of the cationic fluorochrome Rh123 is specific to functionally active mitochondria: the dye accumulates in the mitochondria due to the transmembrane potential of these organelles (Johnson *et al.*, 1980). Thus, live cells with intact plasma membrane and charged mitochondria concentrate this dye and exhibit strong green fluorescence. In contrast, dead cells fail to stain significantly with this dye. A dual-staining with Rh123 and PI was proposed as the cell viability test (Darzynkiewicz *et al.*, 1982). This is also an assay of the mitochondrial transmembrane potential.

A. Reagents

1. Rh123 (available from Molecular Probes), dissolved in distilled water, 1 mg/ml (stock solution).
2. PI, dissolved in distilled water, 1 mg/ml (stock solution).

B. Procedure

1. Add 5 μl of Rh123 stock solution to approximately 10^6 cells suspended in 1 ml of tissue culture medium (or HBSS). Incubate 5 min at 37°C.
2. Add 20 μl of the PI stock solution. Keep 5 min at room temperature. Analyze by flow cytometry.

Both dyes are excited with blue light (e.g., 488-nm laser line). Rh123 fluoresces green (530 ± 20 nm), while PI, as described above, in red.

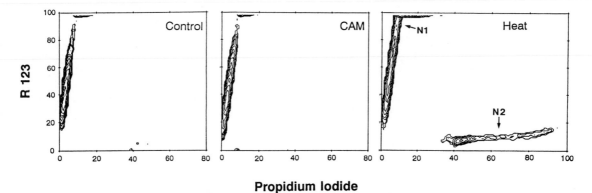

Propidium Iodide

Fig. 1   Stainability of live, apoptotic, and necrotic cells with Rh123 and PI. The mitochondrial probe Rh123 (green fluorescence) is taken up by live cells from control HL-60 cultures (Control); live cells exclude PI. Induction of apoptosis in approximately 40% cells by incubation with 0.15 μM camptothecin for 4 hr (CAM) neither changes Rh123 uptake nor affects exclusion of PI. Cell heating (45°C, 2 hr) causes necrosis (Heat). Early necrotic (N1) cells have elevated Rh123 uptake but exclude PI. Late necrotic cells (N2) have minimal Rh123 fluorescence and stain intensely with PI (reprinted with permission from Darzynkiewicz *et al.*, 1992).

C. Results

Live cells stain with Rh123 green and exclude PI. Dead cells, which have damaged mitochondria, have minimal green fluorescence but stain with PI (Fig. 1) (Darzynkiewicz *et al.*, 1982).

VI. PI Exclusion Followed by Counterstaining with Hoechst 33342

As mentioned, the charged dye PI is excluded by live cells. On the other hand, the bisbenzimidazole dye HO342 penetrates through the plasma membrane and can stain DNA in live cells. The method combining these dyes to discriminate between live and dead cells was introduced by Stöhr and Vogt-Schaden (1980) and modified by several authors (e.g., Wallen *et al.*, 1983; Pollack and Ciancio, 1990). In this method the cells are first exposed to PI and subsequently stained with HO342. Compared with live cells, HO342 fluorescence is suppressed in dead cells. The latter, however, stain more intensely with PI. This method, in modification of Pollack and Ciancio (1990), is described below:

A. Reagents

1. Stock solution of HO342: 1 mM HO342 in distilled water.
2. Staining solution of PI: 20 μg/ml of PI in PBS.

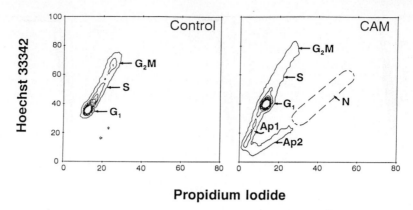

Fig. 2 Stainability of apoptotic and necrotic cells with PI and HO342 under conditions when HO342 is applied after cell exposure to PI and following their permeabilization. HL-60 cells, untreated (Control) and treated with 0.15 μM camptothecin (CAM) for 6 hr to induce apoptosis (Del Bino *et al.*, 1991) were incubated with 10 μg/ml of PI, then permeabilized with 25% ethanol and stained with HO342, as originally described by Pollack and Ciancio (1990). In this assay the cells with undamaged plasma membrane exclude PI and stain predominantly blue with HO342. The cells that cannot exclude PI have a more intense PI fluorescence and proportionally lower HO342 fluorescence. The live cells from control culture stain strongly with HO342, in proportion to their DNA content. The early apoptotic cells from the CAM-treated cultures show diminished HO342 fluorescence (Ap1). Late apoptotic cells (Ap2) cannot exclude PI and stain more intensely with PI. The position of necrotic cells (N), which also cannot exclude PI but have higher DNA content than Ap2, is indicated with a broken outline (reprinted with permission from Del Bino *et al.*, 1991).

3. Staining solution of HO342: Dilute the HO342 stock solution 1 : 4 in PBS (Ca^{2+} and Mg^{2+} free).

4. Fixative: 25% ethanol in PBS.

B. Staining Procedure

1. Centrifuge 3×10^5–10^6 cells, decant medium, and vortex the pellet.

2. Add 100 μl of PI staining solution, vortex, and keep on ice for 30 min.

3. Add 1.9 ml of fixative (reagent No. 4) and vortex.

4. Add 50 μl of HO342 staining solution, vortex, and keep on ice for at least 30 min. Samples are stable for up to 3 days in this solution, when kept at 0–4°C.

C. Instrumentation

Excitation of HO342 is in UV light, with maximum at 340 nm. PI can be excited with blue or green light, but it has also an absorption band in the UV light spectrum. Excitation of both PI and HO342, thus, can be achieved using, e.g., the 351-nm line of the argon ion laser or, in the case of illumination with a high pressure mercury lamp, using the UG1 filter.

Fluorescence of HO342 is in blue wavelength and a combination of optical filters and dichroic mirrors is required to obtain maximum transmission at 480 ± 20 nm. Fluorescence of PI is measured with long pass filter >620 nm.

D. Results

Figure 2 illustrates positions of live, apoptotic, and necrotic cells following cell staining with PI and HO342. As is evident, the early apoptotic cells (Ap1) have lowered stainability with HO342 compared to live cells from the control culture. The loss of HO342 stainability is likely a result of DNA degradation and extraction of low MW DNA from the cells following their permeabilization with ethanol. PI fluorescence of Ap cells is also low. The loss of plasma membrane function late in apoptosis results in increased stainability with PI (Ap2). Necrotic cells, or cells with destroyed plasma membrane function (e.g., by repeated freezing and thawing) have high PI and low HO342 fluorescence.

VII. Hoechst 33342 Active Uptake and PI Exclusion

The PI–HO342 method presented in Fig. 2, based on cell staining with PI followed by their exposure to HO342 after permeabilization (fixation) with 25% ethanol, is very much different from the method which utilizes the same dyes, but detects apoptotic cells based on their increased uptake of HO342, under supravital conditions in isotonic media, as described by Dive *et al.* (1992) and Ormerod *et al.* (1992, 1993). These authors observed that during a short exposure to HO342 apoptotic cells have stronger blue fluorescence, perhaps as a result of the increased permeability to this dye, compared to nonapoptotic cells. The combination of supravital HO342 uptake and PI exclusion, together with analysis of the cells' light scatter properties, provides an attractive assay of apoptosis.

A. Reagents

1. HO342, dissolved in distilled water 0.1 mg/ml, stock solution.
2. PI, dissolved in distilled water, 1 mg/ml, stock solution.

B. Procedure

1. Suspend approximately 10^6 cells in 1 ml of culture medium, with 10% serum. Add 10 μl of stock solution of HO342.

2. Incubate cells for 7 min.

3. Cool the sample on ice and centrifuge.

4. Resuspend cell pellet in 1 ml PBS. Add 5 μl of the PI stock solution. Analyze.

C. Results

As described above, excitation of HO342 is in UV light. Although optimally excited at green or blue wavelength, PI can be also excited in UV. Apoptotic cells have increased blue fluorescence of HO342, compared to live, nonapoptotic cells (Dive *et al.*, 1992; Ormerod *et al.*, 1993). The intensity of the cells' blue fluorescence changes with time of incubation with HO342, and the optimal time should be determined for best discrimination of apoptotic cells. The decrease in light scatter, combined with HO342 uptake, is also helpful to identify apoptotic cells. Necrotic cells are counterstained with PI.

VIII. Controlled Extraction of Low MW DNA from Apoptotic Cells

A. Degradation of DNA in Apoptotic Cells

One of the early events of apoptosis is the activation of endonuclease which nicks DNA preferentially at the internucleosomal (linker) sections (Arends *et al.*, 1990; Compton, 1992). Although exceptions have been reported (Collins *et al.*, 1992; Cohen *et al.*, 1992), DNA degradation is a very specific event of apoptosis and the electrophoretic pattern of DNA indicating the presence of DNA sections the size of mono- and oligonucleosomes is considered a trademark of apoptosis.

Fixation of cells in ethanol is inadequate to preserve the degraded, low MW DNA inside apoptotic cells: this portion of DNA leaks out during the subsequent rinse and staining procedure, and therefore less DNA in these cells stains with any DNA fluorochrome (Darzynkiewicz *et al.*, 1992). Thus, the appearance of cells with low DNA stainability, lower than that of G_1 cells ("sub-G_1 peaks", "A_0" cells) in cultures treated with various cytotoxic agents, has been considered to be a marker of cell death by apoptosis (Umansky *et al.*, 1981; Nicoletti *et al.*, 1991; Telford *et al.*, 1991).

Because the degree of DNA degradation varies depending on the stage of apoptosis, cell type, and often the nature of the apoptosis-inducing agent, the degree of extraction of low MW DNA during the staining procedure (and thus separation of apoptotic from live cells) is not always reproducible. It has been observed, however, that addition of phosphate-citric acid buffer at pH 7.8 to the rinsing fluid enhances extraction of the degraded DNA (Gorczyca *et al.*, 1993c; Gong *et al.*, 1993). This approach can be used to modulate the extent of DNA extraction from apoptotic cells to the desired level.

B. Reagents

1. Phosphate-citric acid buffer: Prepare solutions of 0.2 M Na$_2$HPO$_4$ and 0.1 M citric acid. Mix 192 ml of 0.2 M Na$_2$HPO$_4$ with 8 ml of 0.1 M citric acid; pH should be approximately 7.8.
2. *Hanks' buffered salt solution (HBSS).*
3. *Fixative:* 70% ethanol.

C. Procedure

1. Fix cells in suspension in 70% ethanol (i.e., add 1 ml of cells suspended in HBSS containing approximately 10^6 cells into 9 ml of 70% ethanol in tube) on ice. Cells in fixative (ethanol) can be stored at $-20°C$ for up to 1 week.

2. Centrifuge cells, decant ethanol, suspend cells in 10 ml of HBSS, and centrifuge.

3. Suspend cells in 1 ml of HBSS, into which you may add 0.2–1.0 ml of the phosphate-citric acid buffer (reagent No. 1). Add less (e.g., 0.2 ml or none) buffer if DNA degradation in Ap cells is extensive and DNA extraction effective, so the Ap cells are well separated from G$_1$ cells. Add more of the buffer (up to 1.0 ml) if DNA is not markedly degraded and there are problems with separating Ap cells (Ap cells overlap with G$_1$ cells). Incubate at room temperature for 5 min.

4. Centrifuge cells and add 1 ml of HBSS containing 20 μg/ml of PI and 5 Kunitz units of the DNase-free RNase A. (Boil RNase for 5 min before use if it is not DNase free.) Incubate cells for 30 min at room temperature. Use the laser excitation line at 488 nm or blue light (BG12 filter) and measure cell red fluorescence (>620 nm) and forward light scatter. Ap cells should have a diminished forward light scatter signal and decreased PI fluorescence compared to the cells in the main peak (G$_1$).

5. Cellular DNA may be stained with other fluorochromes instead of PI, and other cell constituents may be counterstained as well. The following is the procedure to simultaneously stain DNA and protein with DAPI and sulforhodamine 101, respectively:

• After point No. 3 of this procedure suspend cell pellet in 1 ml of a staining solution that contains:

Triton X-100, 0.1%

MgCl$_2$, 2 mM

NaCl, 0.1 M

Pipes buffer, 10 mM; final pH 6.8

DAPI, 1 μg/ml

Sulforhodamine 101, 20 μg/ml.

(*Nota Bene:* This solution is stable and can be stored, in a dark bottle, at 0–4°C for several weeks). Cells are analyzed while suspended in this solution.

• Use excitation with UV light (e.g., 351-nm argon ion line or UG1 filter for mercury lamp illumination). Measure the blue fluorescence of DAPI (460–500 nm) and red fluorescence of sulforhodamine 101 (>600 nm).

D. Results

Results are shown in Fig. 3. As is evident, apoptotic cells are well separated from live cells based on differences in DNA content. They also have diminished protein content. As mentioned, the degree of separation may be modified by varying the concentration of the phosphate-citric acid buffer (reagent No. 1).

This buffer can also be added to the staining medium (e.g., in a 1 : 10 to 1 : 5 proportion) when unfixed cells are stained with the DNA-specific dyes, following cell permeabilization with detergents, to extract low MW DNA from apoptotic cells.

Because the cell-cycle position of the nonapoptotic cells can be estimated (Fig. 2), this method offers the opportunity to investigate the cell-cycle specificity of apoptosis. Another advantage of this method is its simplicity and applicability to any DNA fluorochrome (Telford *et al.,* 1991) or instrument.

This method, however, is not very specific in terms of detection of apoptotic cells. In addition to apoptotic cells, the "sub G_1" peak can also represent mechanically damaged cells, cells with lower DNA content (e.g., in a sample containing cell populations with a different DNA index), or cells with different chromatin structure (e.g., cells undergoing erythroid differentiation) in which accessibility of DNA to the fluorochrome is diminished (Darzynkiewicz *et al.,* 1984b).

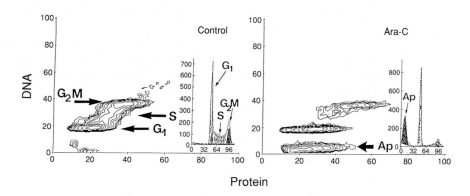

Fig. 3 Bivariate analysis of DNA and protein content distribution of control HL-60 cells and cells treated with 25 μM 1 β-arabinofuranosylcytosine (Ara-C) for 4 hr. DNA content frequency histogram of these cells is shown in the inset. The presence of apoptotic cells (Ap) in Ara-C-treated culture correlates with disappearance of S phase cells. A clear distinction between Ap and nonapoptotic cells was obtained by rinsing the ethanol-prefixed cells with phosphate-citric acid buffer (pH 7.8), as described in the procedure.

It should be stressed that this method can be used only on cells that are fixed in ethanol or permeabilized with detergents. Cell fixation in formaldehyde or glutaraldehyde crosslinks low MW DNA to other constituents and precludes its extraction. Only very late apoptotic cells, in which DNA degradation is significantly advanced, show diminished DNA content following fixation with the crosslinking agents.

IX. Sensitivity of DNA *in Situ* to Denaturation

Chromatin condensation is one of the early events of apoptosis. It has been observed that sensitivity of DNA *in situ* to denaturation is increased during chromatin condensation, e.g., as during mitosis or in G_0 (Darzynkiewicz *et al.*, 1987), and that chromatin condensation in apoptotic cells is also accompanied by increased DNA denaturability (Hotz *et al.*, 1992). This method, and its application to identify apoptotic cells, are described in Chapter 33 of this volume.

X. Detection of DNA Strand Breaks in Apoptotic Cells

As mentioned, extensive DNA cleavage is a characteristic event of apoptosis. The presence of DNA strand breaks in apoptotic cells can be detected by labeling the 3′OH termini in DNA breaks with biotin- or digoxigenin-conjugated nucleotides, in the enzymatic reaction catalyzed by exogenous TdT (Gorczyca *et al.*, 1992, 1993a,b), or DNA polymerase (Gold *et al.*, 1993). The method of detection of apoptotic cells by labeling DNA strand breaks with biotinylated dUTP (b-dUTP) by TdT is as follows:

A. Reagents

1. Fixatives: 1% methanol-free formaldehyde (paraformaldehyde; available from Polysciences, Inc., Warrington, PA) in PBS, pH 7.4; 70% ethanol.
2. Reaction buffer for TdT (5× concentrated):
    Potassium (or sodium) cacodylate, 1 *M*
    Tris–HCl, 125 mM, pH 6.6 at 4°C
    Bovine serum albumin (BSA), 1.25 mg/ml
    Cobalt chloride, 10 m*M*.
   (The complete buffer is available from Boehringer-Mannheim Biochemicals, Indianapolis, IN; Cat. No. 220 582.)

3. TdT in storage buffer
    Potassium cacodylate, 0.2 *M*
    EDTA, 1 m*M*
    2-Mercaptoethanol, 4 m*M*
    Glycerol, 50% (v/v), pH 6.6 at 4°C
    TdT, 25 units per 1µl.
    (Available from Boehringer-Mannheim; Cat. No. as above.)
4. Biotin-16-dUTP, 50 nmol in 50 µl. (Available from Boehringer-Mannheim; Cat. No. 1093 070.)
5. Saline-citrate buffer. Dilute 20x concentrated saline-sodium citrate buffer (SSC) fivefold with distilled water to obtain 4x concentrated solution (0.6 *M* NaCl, 0.06 M Na citrate), and add:
    Fluoresceinated avidin, 2.5 µg/ml (final concentration)
    Triton X-100, 0.1% (v/v)
    Dry milk, 5% (w/v) final concentration.
6. Rinsing buffer. Dissolve in HBSS:
    Triton X-100, 0.1% (v/v)
    BSA, 0.5%.
7. PI buffer. Dissolve in HBSS:
    PI, 5 µg/ml
    0.1% DNase-free RNase A.

B. Procedure

1. Fix cells in suspension in 1% formaldehyde for 15 min on ice.
2. Sediment cells, resuspend pellet in 5 ml HBSS, and centrifuge.
3. Resuspend cells in 70% ethanol on ice. The cells can be stored in ethanol, at −20°C for up to 3 weeks.
4. Rinse cells in HBSS and resuspend the pellet (less than 10^6 cells) in 1.5 ml microfuge tube in 50 µl of the solution which contains:
    Reaction buffer (reagent No. 2), 10 µl
    Biotin-16-dUTP (reagent No. 4), 1 µl (1 µg)
    TdT (reagent No. 3), 0.2 µl (5 units)
    Distilled H_2O, 38.8 µl.

Proportions of the reagents are scaled up proportionally to the number of samples examined at a given time, and the solution is then subdivided into 50-µl aliquots per sample.

5. Incubate cells in this solution for 30 min at 37°C, then add 1.3 ml of the rinsing buffer (reagent No. 6), centrifuge, and resuspend the pellet in 100 µl

of the saline-citrate buffer containing fluoresceinated avidin (reagent No. 5). Incubate at room temperature for 30 min.

6. Add 1.3 ml of the rinsing buffer and centrifuge. Rinse again with rinsing buffer. Resuspend the cell pellet in 1 ml of PI solution (reagent No. 7). Incubate 30 min at room temperature. Measure cell green (fluorescinated avidin) and red (PI) fluorescence after illumination with blue light (488 nm).

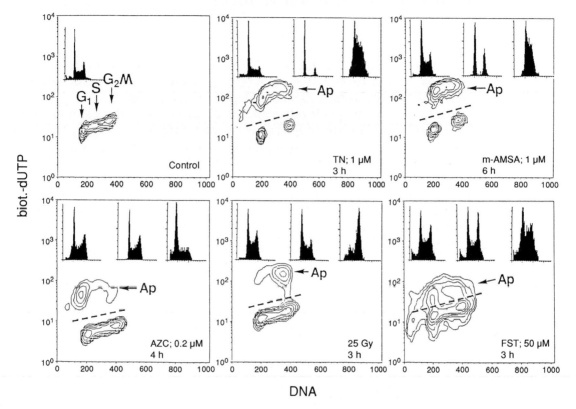

Fig. 4 Detection of the apoptosis-associated DNA strand breaks in HL-60 cells treated with various drugs and γ radiation. DNA strand breaks in apoptotic cells (Ap) are labeled with biotinylated dUTP (b-dUTP) in the reaction catalyzed by terminal deoxynucleotidyl transferase (TdT). Simultaneous counterstaining of DNA allows correlation of the presence of DNA strand breaks with cell position in the cell cycle. By gating analysis of the b-dUTP-labeled and -unlabeled cell populations, the cell-cycle distribution of nonapoptotic and apoptotic cells, respectively, can be estimated. In contrast to the staining of DNA of the ethanol-fixed cells (Fig. 3), only a minor amount of DNA is extracted from apoptotic cells in this method. Note the S phase specificity of apoptosis after treatment with teniposide (TN) and amsacrine (m-AMSA), G_1 specificity after treatment with azacytidine (AZC), and G_2M specificity after γ radiation (25 Gy). Apoptosis induced by Fostriecin (FST) is not cell-cycle specific. Insets show the DNA content frequency histograms of the total (left), nonapoptotic (middle), and apoptotic (right) cell populations; the frequency scales are arbitrary (scaled to maximum frequency). Dashes, the gating thresholds (reprinted with permission from Gorczyca *et al.*, 1993c).

C. Commercially Available Kits for Labeling DNA Strand Breaks

ONCOR, Inc. (Gaithersburg, MD), provides a kit (Apopt Tag kit, Cat. No. S7100-kit) specifically designed to label DNA strand breaks in apoptotic cells, which utilizes digoxigenin-conjugated dUTP and TdT. Description of the method is included with the kit. The results obtained with use of this kit compare favorably with the data shown in Fig. 4 (X. Li, M. James, F. Traganos, and Z. Darzynkiewicz, submitted for publication).

D. Results

Simultaneous detection of DNA strand breaks and analysis of DNA content makes it possible to identify the cell-cycle position of both cells in apoptotic population as well as the cell with undegraded DNA (Fig. 4).

Identification of apoptotic cells based on the detection of DNA strand breaks is the most specific assay of apoptosis, by flow cytometry. Namely, the cells with DNA strand breaks which accompany necrosis, or the cells with primary breaks induced by X-irradiation, up to a dose of 25 Gy, have markedly lower incorporation of b-dUTP compared to that of apoptotic cells (Gorczyca *et al.*, 1993a). This method is applicable to clinical material, to measure apoptosis of blast cells in peripheral blood or bone marrow induced during chemotherapy of leukemias (Gorczyca *et al.*, 1993a) or in solid tumors, from needle biopsy specimens (Gorczyca *et al.*, 1994).

XI. A Selective Procedure for DNA Extraction from Apoptotic Cells Applicable to Gel Electrophoresis and Flow Cytometry

A. Recovery of the Degraded DNA from Apoptotic Cells and Its Analysis by Gel Electrophoresis

As mentioned earlier in this chapter, one of the most characteristic features of apoptosis is activation of an endonuclease which has preference to the linker DNA sections; products of this enzyme are discontinuous DNA sections of a size equivalent to mono- and oligonucleosomose, and they form a typical "ladder" during gel electrophoresis (Arends *et al.*, 1990). Such an electrophoretic pattern is considered to be a hallmark of apoptosis, and although exceptions have been observed (e.g., Cohen *et al.*, 1992), analysis of DNA size on agarose gels is a widely used procedure to reveal apoptosis. Because such analysis should often be used to confirm the data by flow cytometry, a simple procedure of DNA electrophoresis in agarose gels is included in this chapter.

One of the methods to identify apoptotic cells, presented in this chapter, is based on cell fixation in 70% ethanol followed by cell hydration and extraction of low MW DNA with phosphate-citric acid buffer at pH 7.8 (see Section VIII,

A). This method can be combined with the analysis of DNA extracted with the buffer by gel electrophoresis (Gong *et al.*, 1994), as described below.

B. Reagents

1. Phosphate-citric acid buffer: Mix 192 ml of 0.2 M Na_2HPO_4 with 8 ml of 0.1 *M* citric acid; the final pH is 7.8.
2. Nonidet NP-40: Dissolve 0.25 ml of Nonidet NP-40 in 100 ml of distilled water.
3. RNase A: Dissolve 1 mg of DNase free RNase A in 1 ml of distilled water.
4. Proteinase K: Dissolve 1 mg of proteinase K in 1 ml of distilled water.
5. Loading buffer: Dissolve 0.25 g of bromophenol blue and 0.25 g of xylene cyanol FF in 70 ml of distilled water. Add 30 ml of glycerol (dyes are available, e.g., from Bio-Rad Laboratories, Richmond, CA).
6. DNA molecular weight standards: Use the standards that provide DNA between 100 to 1000 bp size (e.g., from Integrated Separation Systems, Natick, MA).
7. Electrophoresis buffer (TBE, 10 × concentrated): Dissolve 54 g Tris base and 27.5 g boric acid in 980 ml of distilled water; add 20 ml of 0.5 *M* EDTA (pH 8.0).
8. Agarose gel (0.8%): Dissolve 1.6 g of agarose in 200 ml of TBE.
9. Ethidium bromide (EB): Stock solution, dissolve 1 mg of EB in 1 ml of distilled water; for staining gels (working solutions), add 100 μl of the stock solution to 200 ml TBE.

C. Procedure

1. Fix 10^6–10^7 cells in suspension in 10 ml of 70% ethanol on ice. Cells may be subjected to DNA extraction and analysis after 4 hr fixation in ethanol, or stored in fixative, preferably at $-20°C$, for up to several weeks.

2. Centrifuge cells at 1000*g* for 5 min. *Thoroughly remove ethanol*. Resuspend cell pellet in 40 μl of phosphate-citric acid buffer (reagent No. 1) and transfer to 0.5-ml volume Eppendorf tubes. Keep at room temperature for at least 30 min, occasionally shaking.

3. Centrifuge at 1500*g* for 5 min. Transfer the supernatant to new Eppendorf tubes and concentrate by vacuum e.g., in SpeedVac concentrator (Savant Instruments, Inc.; Farmingdale, NY) for 15 min.

4. Add 3 μl of 0.25% Nonidet NP-40 and 3 μl of RNase A solution. Close the tube to prevent evaporation and incubate at 37°C for 30 min.

5. Add 3 μl of proteinase K solution and incubate for an additional 30 min at 37°C.

6. Add 12 μl of the loading buffer, transfer entire contents of the tube to 0.8% agarose horizontal gel.

7. Load a sample of DNA standards on the MW standard lane of the gel.

8. Run electrophoresis at 2 V/cm for 16–20 hr.

9. To visualize the bands, stain the gel with 5 μg/ml of ethidium bromide and illuminate with UV light.

10. Cells remaining in the pellet (after step 3 of this procedure) can be counterstained with PI (in the presence of RNase A) or DAPI, as described in Section VII, for analysis by flow cytometry.

Further details of the procedure of gel preparation and electrophoresis can be found in Sambrook *et al.* (1989).

D. Results

This simple procedure allows one to analyze the MW of DNA extracted from apoptotic cells by gel electrophoresis and to detect apoptotic cells, from the very same preparations, by flow cytometry (Gong *et al.*, 1994). Figure 5 illustrates both DNA analysis by gel electrophoresis and DNA content of the cells from which DNA was extracted, measured by flow cytometry.

E. Advantages of the Procedure: Comparison with Other Methods of DNA Extraction

The procedure of DNA analysis presented above offers several advantages over other methods used to extract DNA from apoptotic cells (e.g., Arends *et al.*, 1990; Compton, 1992):

1. The cells may be fixed in ethanol and stored indefinitely, a feature which is attractive for use with clinical material, which may be collected at different times, transported, and analyzed later.

2. Fixation in ethanol inactivates endogenous enzymes, preventing autolysis after sample collection.

3. Ethanol also inactivates HIV and other viral or bacterial pathogens, making the sample safer to handle.

4. The method is rapid and uses less toxic reagents (no phenol, chloroform, etc.).

5. The low MW DNA is selectively extracted and is not mixed with high MW DNA in the sample. Bands of uniform intensity, therefore, can be more easily obtained on gel electrophoretograms, when equal amounts of DNA (e.g., estimated by spectrophotometry; absorption at 260 nm) are applied per well;

6. The ratio of high to low MW DNA (extractable and nonextractable from the ethanol-fixed cells) may serve as an index of apoptosis in cell populations. High MW DNA can be extracted by standard biochemical procedures.

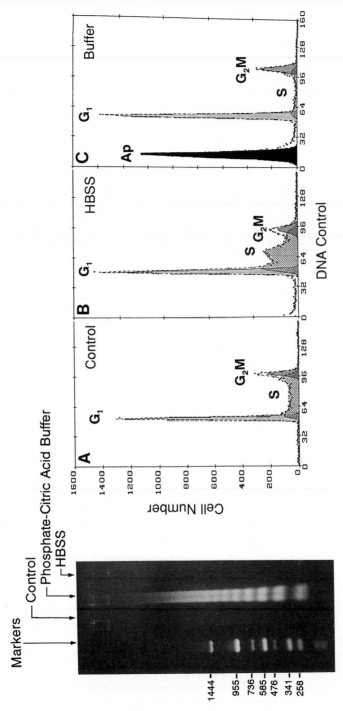

Fig. 5 DNA gel electrophoresis and corresponding DNA content frequency distribution histograms of HL-60 cells, untreated (Control, panel A) and treated with 0.15 μM camptothecin for 3 hr (B and C). As shown by us before (Del Bino *et al.*, 1991) such treatment induces apoptosis of S phase cells. The cells shown in panel B were rinsed with Hanks' buffered salt solution (HBSS) while those in panels A and C were rinsed with phosphate citric acid (PC) buffer. DNA extracted with HBSS and PC were analyzed by gel electrophoresis as shown in the respective lanes. Markers: MW markers, as indicated (base pairs).

7. The cells remaining after extraction of low MW DNA can be subjected to flow cytometry and, thus, the percentage of apoptotic cells can be estimated in literally the same samples from which DNA was used for electrophoresis.

XII. Comparison of the Methods: Confirmation of the Apoptotic Mode of Cell Death

Each of the presented methods has its advantages and limitations. Some of them have already been discussed. In general, the methods based on analysis of integrity and transport function of the plasma membrane (exclusion of PI, FDA hydrolysis, trypsin, and DNase sensitivity), although rather simple and inexpensive to use, fail to identify apoptotic cells. They can be used to identify necrotic cells, mechanically damaged cells, or very late stages of apoptosis.

Identification of apoptotic cells is more complex. None of the methods described in the literature, when used alone, can provide total assurance of detection of apoptosis. The most specific appears to be the method based on the detection of DNA strand breaks by labeling them with biotin- or digoxigenin-conjugated nucleotides. The number of DNA strand breaks in apoptotic cells is so large that the degree of cell labeling in this assay appears to be an adequate discriminator between apoptotic and necrotic cells. This assay also allows one to discriminate between apoptotic cells and cells with primary DNA strand breaks caused by high doses of radiation (Gorczyca *et al.*, 1992). However, one has to keep in mind that there may be situations when apoptosis is induced in the absence of DNA degradation (e.g., Cohen *et al.*, 1992) and, vice versa, extensive DNA degradation even selective to internucleosomal DNA may accompany necrosis (Collins *et al.*, 1992).

A higher degree of assurance of identification of apoptosis can be obtained by using more than one viability assay on the same sample. It is advisable, therefore, to simultaneously probe DNA cleavage (e.g., by labeling DNA strand breaks) and integrity of the plasma membrane (e.g., PI exclusion). Preservation of the plasma membrane integrity and extensive DNA breakage is a more specific marker of apoptosis than DNA strand breakage alone. Likewise, the status of mitochondria, probed by Rh123, which remains little changed in apoptotic cells, may be used as an additional marker of the mode of cell death.

In situations when apoptosis is not paralleled by DNA degradation (Collins *et al.*, 1992; Cohen *et al.*, 1992), analysis of DNA sensitivity to denaturation (see chapter 33 of this volume) appears to be the method of choice. The increased supravital cell stainability with HO342 combined with a decrease in forward light scatter may also be applied in such cases.

It should be stressed, however, that regardless of the flow cytometric method(s) used to identify apoptosis, this mechanism of cell death should be confirmed by inspection of cells under light or electron microscopy. Morphological changes during apoptosis have a very specific pattern and analysis of cell morphology should be the deciding factor in situations when there is any ambiguity regarding the mechanism of cell death.

Acknowledgments

Supported by NCI Grant RO1 CA28704, the "This Close" Foundation, the Carl Inserra Fund, the Chemotherapy Foundation, and the Dr. I Fund Foundation. I thank Drs. Frank Traganos for his comments and suggestions.

References

Arends, M. J., Morris, R. G., and Wyllie, A. H. (1990). *Am. J. Pathol.* **136,** 593–608.

Chang, W. P., and Little, J. B. (1992). *Int. J. Radiat. Biol.* **60,** 483–496.

Cohen, G. M., Sun, X.-M., Snowden, R. T., Dinsdale, D., and Skilleter, D. N. (1992). *Biochem. J.* **286,** 331–334.

Collins, R. J., Harmon, B. V., Gobe, G. C., and Kerr, J. F. R. (1992). *Int. J. Radiat. Biol.* **61,** 451–453.

Compton, M. M. (1992). *Cancer Metastasis Rev.* **11,** 105–119.

Darzynkiewicz, Z., Traganos, F., Staiano-Coico, L., Kapuscinski, J., and Melamed, M. R. (1982). *Cancer Res.* **42,** 799–806.

Darzynkiewicz, Z., Williamson, B., Carswell, E. A., and Old, L. J. (1984a). *Cancer Res.* **44,** 83–90.

Darzynkiewicz, Z., Taganos, F., Kapuscinski, J., Staiano-Coico, L., and Melamed, M. R. (1984b). *Cytometry* **5,** 355–363.

Darzynkiewicz, Z., Traganos, F., Carter, S., and Higgins, P. J. (1987). *Exp. Cell Res.* **172,** 168–179.

Darzynkiewicz, Z., Bruno, S., Del Bino, G., Gorczyca, W., Hotz, M. A., Lassota, P., and Traganos, F. (1992). *Cytometry* **13,** 795–508.

Del Bino, G., Lassota, P., and Darzynkiewicz, Z. (1991). *Exp. Cell Res.* **193,** 27–35.

Dive, C., Gregory, C. D., Phipps, D. J., Evans, D. L., Milner, A. E., and Wyllie, A. H. (1992). *Biochim. Biophys. Acta* **1133,** 275–282.

Gold, R., Schmied, M., Rothe, G., Zischler, H., Breitschopt, H., Wekerle, H., and Lassman, H. (1993). *J. Histochem. Cytochem.* **41,** 1023–1030.

Gong, J., Li, X., and Darzynkiewicz, Z.(1993). *J. Cell. Physiol.* **157,** 263–270.

Gong, J., Traganos, F., and Darzynkiewicz, Z. (1994). *Anal. Biochem.* **218,** 314–319.

Gorczyca, W., Bruno, S., Darzynkiewicz, R. J., Gong, J., and Darzynkiewicz, Z. (1992). *Int. J. Oncol.* **1,** 639–648.

Gorczyca, W., Bigman, K., Mittelman, A., Ahmed, T., Gong, J., Melamed, M. R., and Darzynkiewicz, Z. (1993a). *Leukemia* **7,** 659–670.

Gorczyca, W., Gong, J., and Darzynkiewicz, Z. (1993b). *Cancer Res.* **52,** 1945–1951.

Gorczyca, W., Gong, J., Ardelt, B., Traganos, F., and Darzynkiewicz, Z. (1993c). *Cancer Res.* **53,** 3186–3192.

Gorczyca, W., Tuziak, T., Kram, A., Melamed, M. R., and Darzynkiewicz, Z. (1994). *Cytometry* **15,** 169–175.

Hamori, E., Arndt-Jovin, D. J., Grimwade, B. G., and Jovin, T. M. (1980). *Cytometry* **1,** 132–135.

Horan, P. K., and Kappler, J. W. (1977). *J. Immunol. Methods* **18,** 309–316.

Hotz, M. A., Traganos, F., and Darzynkiewicz, Z. (1992). *Exp. Cell Res.* **201,** 184–191.

Johnson, L. U., Walsh, M. L., and Chen, L. B. (1980). *Proc. Natl. Acad. Sci. U.S.A.* **77,** 990–994.

Meyaard, L., Otto, S. A., Jonker, R. R., Mijnster, M. J., Keet, R. P. M., and Miedema, F. (1992). *Science* **257,** 217–219.

Nicoletti, I., Migliorati, G., Pagliacci, M.C., Grignani, F., and Riccardi, C. (1991). *J. Immunol. Methods* **139,** 271–279.

Ormerod, M. G., Collins, M. K. L., Rodriguez-Tarduchy, G., and Robertson, D. (1992). *J. Immunol. Methods* **153,** 57–65.

Ormerod, M. G., Sun, X.-M., Snowden, R. D., Davies, R., Fearnhead, H., and Cohen, G. M. (1993). *Cytometry* **14,** 595–602.

Pollack, A., and Ciancio, G. (1990). *In* "Methods in Cell Biology" (Z. Darzynkiewicz and H. A. Crissman, eds.), Vol. 33, pp. 19–24. Academic Press, San Diego.

Riedy, M. C., Muirhead, K. A., Jensen, C. B., and Stewart, C. C. (1991). *Cytometry* **12,** 133–139.

Sachs, L., and Lotem, J. (1993). *Blood* **82,** 15–21.

Sambrook, J., Fritsch, E. F., and Maniatis, T. (1989). "Molecular Cloning: A Laboratory Manual," 2nd ed. Cold Spring Harbor Lab., Cold Spring Harbor, NY.

Schwartzman, R. A., and Cidlowski, J. A. (1993). *Endocr. Rev.* **14,** 133–155.

Stöhr, M., and Vogt-Schaden, M. (1980). *In* "Flow Cytometry IV" (O. D. Laerum, T. Lindmo, and E. Thorud, eds.), pp. 96–99. Universitetforslaget, Bergen.

Swat, W., Ignatowicz, L., and Kisielow, P. (1991). *J. Immunol. Methods* **137,** 79–87.

Telford, W. G., King, L. E., and Fraker, P. J. (1991). *Cell Proliferation* **24,** 447–459.

Tomei, L. D., and Cope, F. O., eds. (1991). "Apoptosis: The Molecular Basis of Cell Death." Cold Spring Harbor Lab., Plainview, NY.

Tounecti, O., Pron, G., Belehradek, J., and Mir, L. M. (1993). *Cancer Res.* **53,** 5462–5469.

Umansky, S. R., Korol', B. R., and Nelipovich, P. A. (1981). *Biochim. Biophys. Acta* **655,** 281–290.

Wallen, C. A., Higashikubu, R., and Roti Roti, J. L. (1983). *Cell Tissue Kinet.* **16,** 357–365.

Wyllie, A. H. (1985). *Anticancer Res.* **5,** 131–142.

Wyllie, A. H. (1992). *Cancer Metastasis Rev.* **11,** 95–103.

CHAPTER 3

Cell Preparation for the Identification of Leukocytes

Carleton C. Stewart and Sigrid J. Stewart

Laboratory of Flow Cytometry
Roswell Park Cancer Institute
Buffalo, New York 14263

I. Introduction

Immunophenotyping is the term applied to the identification of cells using antibodies to antigens expressed by these cells. These antigens are actually functional membrane proteins involved in cell communication, adhesion, or metabolism. Antibodies bind to three-dimensional molecular structures on these antigens called epitopes and each antigen contains several hundred different epitopes. A monoclonal antibody is specific for a single epitope, while polyclonal antibodies are actually the natural pool of several hundred monoclonal antibodies produced within the animal, each one binding to its unique epitope. Immunophenotyping using flow cytometry has become the method of choice in identifying and sorting cells within complex populations. Applications of this technology have occurred in both basic research and clinical laboratories. The National Committee for Clinical Laboratory Standards has prepared guidelines for flow cytometry that describe in detail the current recommendations for processing clinical samples (Landay, 1992).

Advances in flow cytometry instrumentation design and the availability of new fluorochromes and staining strategies have led to methods for immunophenotyping cells with two or more antibodies simultaneously (Bender *et al.*, 1991; Riedy *et al.*, 1991; Sneed *et al.*, 1989; Stewart *et al.*, 1988, 1989; Stewart, 1990a,b; 1992). With this progress has come the realization that the use of a single antibody is insufficient for identification of a unique cell population. Instead cells within one population may have proteins in common with cells of another population. By evaluating the unique repertoire of proteins using several antibodies together each coupled with a different fluorochrome the repertoire for any given cell population can be determined.

There are many problems and pitfalls that can be encountered when labeling cells with antibodies. These problems need to be recognized and addressed when they occur and solutions devised, if the data obtained are to be correctly interpreted. These problems, their probable solutions, and several strategies for staining cells with antibodies labeled with different colored fluorochromes are the focus of this chapter. Some of them are improvements over those presented previously (Stewart, 1990, 1992).

II. Antibodies

A brief review of the characteristics of antibodies germane to their use in immunophenotyping is appropriate. Both mono- and polyclonal antibodies are used to determine the immunophenotype of cells. Monoclonal antibodies are derived *in vitro* from hybrids of mouse or rat origin, while polyclonal antibodies are prepared *in vivo* by challenging the immune system of a suitable animal with the antigen and isolating the antibodies produced from the blood.

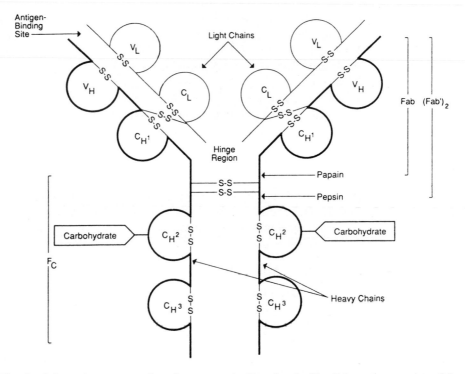

Fig. 1 Schematic representation of a monomeric Ab molecule. The Fab portion consists of the light chain and a fragment of the heavy chain. These two chains are held together by disulfide bonds. For any given Ab molecule there is a region of variable amino acid sequence (V_1) on the light chain and the heavy chain (V_{II}) that produces the epitope-binding site. Papain digestion produces two Fab fragments for each Ab monomer. There are also regions where the amino acid sequence is constant for Ab molecules of similar isotype and subclass. There is one constant region for light chains (C1) and three for heavy chains ($C_{II}1$, $C_{II}2$, and $C_{II}3$). The heavy chains are held together by disulfide bonds. If pepsin is used to digest the Ig, a fragment [F(ab')$_2$] is produced containing two epitope-binding sites. The Fc fragment is that portion of the Ab posterior to the hinge region. From "Pierce-Immunotechnology Catalogue and Handbook," Section C (1989).

A. Antibody Structure

The basic unit of all antibodies comprises a heavy and light chain, the former defining the antibody isotype. As depicted in Fig. 1, the minimum functional antibody molecule consists of two heavy and two light chains linked together with disulfide bonds. The part of the molecule which contains only heavy chains is known as the Fc[1] portion, whereas the part which contains both heavy and

[1] Abbreviations used: Ab, antibody; B, biotin; EMA, ethidium monazide; F, fluorescein; F(ab')$_2$, antigen binding portion of antibody molecules; Fc, the fragment (F) of the antibody molecule that exhibits a nonvariable or constant (c) amino acid sequence within a given isotype and subclass; FcR, receptor on cells for the Fc portion of the antibody; FGAM, fluoresceinated goat anti-mouse Ig; Ig, immunoglobulin; mAb, monoclonal antibody; MCF, mean cellular fluorescence; MNC, mononuclear cells; PBS, phosphate-buffered saline; PE, phycoerythrin; PECY5, phycoerythrin–CY5 tandem complex; PEGAM, phycoerythrinated goat anti-mouse Ig; PETR, phycoerythrin–Texas red tandem complex; TC-Ab, third-color antibody; TR, Texas red.

light chains is the F (ab')$_2$ portion. The Fc receptor binding domain and the complement binding and activation domain (Fig. 1) are in the Fc portion. Variations in the composition of the heavy chains lead to "isotypes" and there are "subclasses" within some isotypes. These variations are of practical importance because they define the repertoire of cells to which the antibody will bind. Table I shows the immunoglobulins and several of their properties which are due to amino acid and structural variations in the Fc portion. In the F(ab')$_2$ portion, the light chains are of two types: κ and λ, and they are associated with the heavy chains by disulfide bonding; the designation "ab" is "antigen binding" and the numeral 2 refers to the fact that there are two antibody binding sites in the basic functional subunit. These two can be chemically separated to produce an F(ab) but this reagent is not generally available. It is noteworthy that murine and rat antibodies almost always have κ light chains and that most *in vitro* hybridoma-produced monoclonal antibodies are of the γ (IgG) isotype.

B. Primary and Secondary Antibody

For immunophenotyping using a single color, it is customary to first stain cells with a primary antibody, usually a monoclonal antibody, directed to the epitope of interest. A secondary polyclonal antibody which is specific to epitopes on the primary antibody (the primary antibody therefore acting as an antigen in this reaction) and which is covalently coupled with fluorescein (or some other fluorochrome) is then used to bind to the primary antibody, effectively coloring the cell.

Table I
Human Immunoglobulins

Isotype	Subclass	Serum conc (mg/ml)	Complement fixing	Protein A reaction	NK cells	B cells	Monocyte	Neutrophil	Eosinophil	Basophil
IgG	—	—	—	—	III	II	I,II	II,III	II	II,III
	IgG1	9	Y	Y	—	—	—	—	—	—
	IgG2	3	Y	Y	—	—	—	—	—	—
	IgG3	1	Y	N	—	—	—	—	—	—
	IgG4	0.5	N	Y	—	—	—	—	—	—
IgM		1.5	Y	N	—	—	—	—	—	—
IgA	IgA1	3	N	N	—	—	II,III	—	—	—
	IgA2	0.5	N	N	—	—	—	II,III	—	—
IgD		0.3	N	N	—	—	—	—	—	—
IgE		NIL	N	N	—	—	—	—	II	I

Modified and expanded from Roitt *et al.* (1989).

This method has the advantage that a single second antibody can be used for staining many different primary antibodies. Unconjugated antibodies are also less expensive than conjugated primary antibodies; the latter may also be unavailable. Using a second antibody also provides an amplification of fluorescence that may be useful for resolving cells that have a low frequency of epitopes.

Whenever possible the F(ab')$_2$ fragment of an affinity-purified second antibody should be used. It is of course implicit in this scheme that the second antibody does not react with the cellular epitope. Second antibodies are available with the fluorochromes, fluorescein (F), phycoerythrin (PE), and the tandem complexes phycoerythrin–Texas red (PETR) or phycoerythrin–CY5 (PECY5) conjugated to them.

A typical second antibody might be labeled "FITC conjugated goat F(ab')$_2$ anti-mouse IgG antibody (γ- and light-chain specific, purified by affinity chromatography)." This label contains considerable information in abbreviated form. "Goat anti-mouse IgG" means the antibody was prepared by immunizing goats with mouse IgG. "Purified by affinity chromatography" means the goat serum was passed over an affinity column (usually Sepharose beads) to which mouse IgG was bound and, after elution from the column, the F(ab')$_2$ fragments were prepared and conjugated with fluorescein isothiocyanate ("FITC conjugated").

Since this goat polyclonal second antibody was prepared using a mouse IgG affinity column, this preparation contains antibodies that are specific for the heavy chain of mouse IgG. Because all isotypes found in serum, i.e., IgG, IgM, and IgA, have light chains, they will also bind this second antibody because it is also "light-chain specific." Therefore, this polyclonal second reagent is *not* specific for mouse IgG at all. In order for a second antibody to be specific for IgG it must have no light-chain activity. If the antibody were only heavy-chain specific the label would read "γ specific."

Any polyclonal second antibody contains all the isotypes found in the serum of the animal used to produce it (e.g., IgM, IgG, IgA). A simple way to determine if a second antibody to murine IgG has light-chain activity is to stain murine spleen cells with it, none should be positive. If there are positive cells, the reagent is defective and should not be used in applications where heavy-chain specificity is required.

C. Blocking

Most immunophenotyping experience has been derived using hematopoietic cells in general and lymphocytes in particular. New problems arise, however, when the staining methods derived for lymphocytes are applied to other types of cells. These problems include increased autofluorescence and Fc and nonspecific binding. As summarized in Color Plate 1, all blood leukocytes can be identified using an antibody panel to Fc receptors because each lineage expresses a unique repertoire. Fc receptor expression also creates a problem that may compromise the interpretation of results because antibodies bind very specifically to these receptors. This potential problem can be reduced by blocking the receptors with IgG prior to staining with a primary antibody.

In the hypothetical mixture of cells shown in Fig. 2, some may have both the desired epitope and Fc receptors, FcR (A), some cells may have only FcR (B), other cells may have only epitopes (C), and some cells may have neither (D). Only the cells with the desired epitope (A and C) should be identified by the mouse monoclonal antibody (mAb) and the second antibody (FGAM). When the mAb is added to the cells it can bind to the desired epitope via the $F(ab')_2$ portion (A and C) or to the cell via the Fc portion (A and B). If the sample contains an equal portion of each cell type, 3/4 of the cells present will bind the mAb. Since the second antibody is a fluoresceinated $F(ab')_2$ goat anti-mouse IgG, it binds to the murine mAb [but not to any FcR(B)]. Since 1/3 of the cells have only FcR, these cells are inappropriately labeled and counted as epitope positive, thereby overestimating the percentage of epitope positive cells by 1/4.

To account for this problem, the common practice is to use an "isotype" control. This control consists of the same cells, incubated with a myeloma protein having no epitope specificity, but having the same isotype and subclass as the primary antibody. As shown in Fig. 2, this myeloma protein is presumed to bind to cells having FcR (E and F). The second antibody is added, as before, to stain cells that have bound the isotype protein. The percentage of positive cells revealed by the isotype protein is then subtracted from the percentage obtained using the primary antibody. This procedure leads to an underestimate of the epitope-positive cells because 1/4 of the cells express both FcR and epitopes and this group is subtracted from the total. Thus, isotype controls may actually lead to an erroneous conclusion.

The autofluorescence control should always be analyzed. It contains no antibodies and is otherwise processed the same way as the other samples. This control provides a baseline to determine the minimum fluorescence above which positive cells are identified. Ideally, the isotype control and autofluorescence control would give identical results and the extent that they differ is indicative of a problem.

The Fc and nonspecific binding of antibodies to cells can be reduced by treating the cells with normal immunoglobulin (blocking). For the example shown in Fig. 3, a murine monoclonal antibody and FGAM were used. Because the second antibody is derived from goat, the blocking immunoglobulin is also derived from the goat. The cells are incubated for 10 min with an excess of goat IgG where the IgG binds the FcR (A and B). Without washing the cells, the primary murine mAb is added wherein it binds only to its epitopes (E and G). Because the labeled second antibody was made against murine IgG in the goat, it binds to the desired mAb but not the blocking goat IgG. To be sure that the block is effective, the "isotype control" antibody can be used in place of the primary antibody. Even if it did bind to unblocked cells, it should not bind to the blocked ones. The general rule is

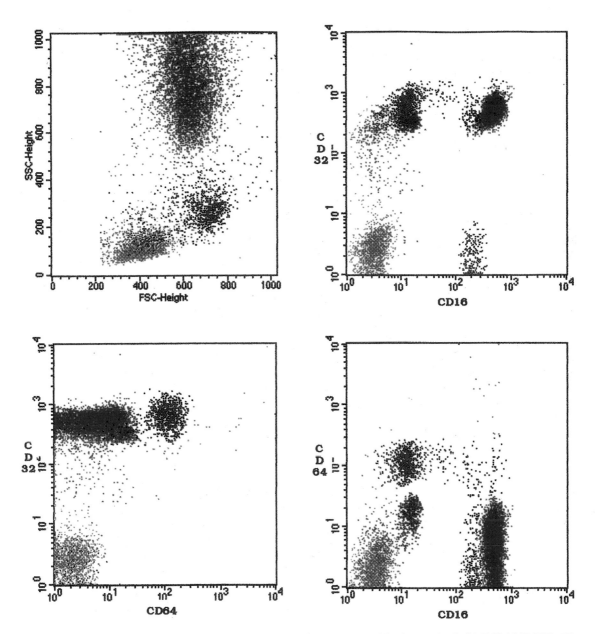

Color Plate 1 (Chapter 3) Fc receptor expression on blood leukocytes. Human blood was stained with F-CD16 (FcRIII), PE-CD32 (FcRII) and TC-CD64 (FcRI). B cells (orange, 43%) express only CD32, NK cells (cyan, 3.4%) express only CD16, and T cells (yellow) express no Fc receptors. Monocytes (green, 7.8%) express both CD32 and CD64 but their position along the CD16 axis is due to autofluorescence and not because they are expressing CD16. Eosinophils (red, 4.8%) are characterized by high-side scatter and express about the same level of CD32, but considerably less CD64 as monocytes. They are also autofluorescent and not CD16-positive. Both neutrophils (pink, 58%) and basophils (black, 1%) express CD16 and CD32, but not CD64; the fluorescence heterogeneity along the CD64 axis is due to autofluorescence. Basophils can be resolved from neutrophils because they exhibit low-side scatter caused by granules that have a similar refractive index to their cytosol. Finally, immature myeloid cells (blue, 1.6%) are homogeneous in CD32 and CD64 expression but heterogeneous in CD16 expression. Note how they form a continuum that merges into each myeloid subset and NK cells.

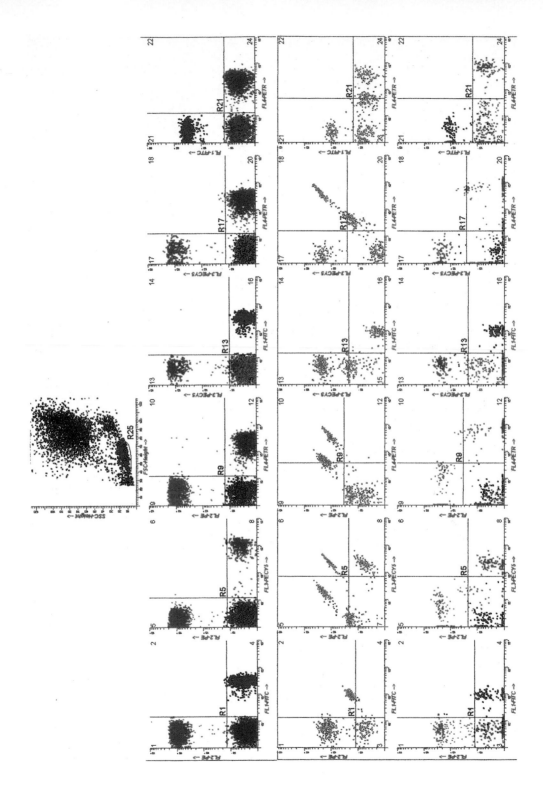

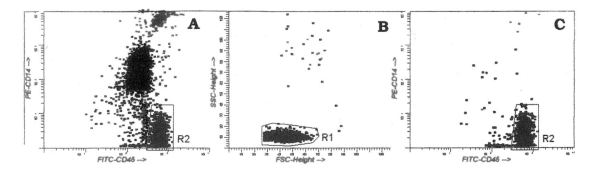

Color Plate 3 (Chapter 4) Determining a lymphogate. Blood is stained with CD45 and CD14 and the data are acquired. The bivariate distributions CD45 vs CD14, FSC vs SSC, and CD45 vs CD14 are displayed. As shown in A, region R2 is drawn circumscribing the CD45-bright, CD14-negative lymphocytes (cyan). Care must be taken to include B cells that may express slightly less CD45 than do T and NK cells and to exclude eosinophils (red). Next, the FSC vs SSC for cells in R2 is displayed (B) and the lymphogate R1 is drawn. Cells outside R1 are lymphocytes that will be excluded from the analysis and, in this example, some eosinophils exhibiting high SSC (green) were present in region R2. The proportion of lymphocytes inside R1 (97%) should be greater than 85% if gated cells (CDC 1992). Finally, the CD45 vs CD14 bivariate distribution of the cells in R1 is displayed (C) to determine the purity of the lymphogate. It is recommended (CDC) that more than 90% of cells are in region R2 (97%) when gated on region R1.

Color Plate 2 (Chapter 4) Compensation for up to four-color immunophenotyping. Human blood leukocytes were stained with B-CD45 in four separate tubes followed by either F-Av, PE-Av, PETR-Av, or PECY5-Av. After the erythrocytes were lysed and fixed, the cells were pooled and the data acquired using a lymphocyte gate. The optical filtration (that may be a part of the manufacturer's configuration) is shown in Table II. The detection channels were FITC for FL1, PE for FL2, PETR for FL4, and PECY (or PCP) for FL3. The following instrument compensation networks are required: FL1-%FL2, FL2-%FL1, FL3-%FL2, FL2-%FL3, FL3-%FL4, FL2-%FL4, and FL4-%FL2. Other compensation possibilities exist, but they are not necessary for the combination of fluorochromes used. Although cells were pooled for generation of this figure, they should be used separately for performing the actual compensation for the generation of the first compensation reference file. The six possible combinations of bivariate views for the parameters measured are displayed. In the top row, cells have been electronically compensated. In the middle row, no compensation has been applied. Because of the increased dimensionality of the data each population is clearly resolved in one or more views and while the pattern created may be different than what one is used to seeing, no compensation is required to visualize the clusters of single-labeled cells. In the bottom row, compensation has been accomplished in software (Verity Software House, Topshum, ME).

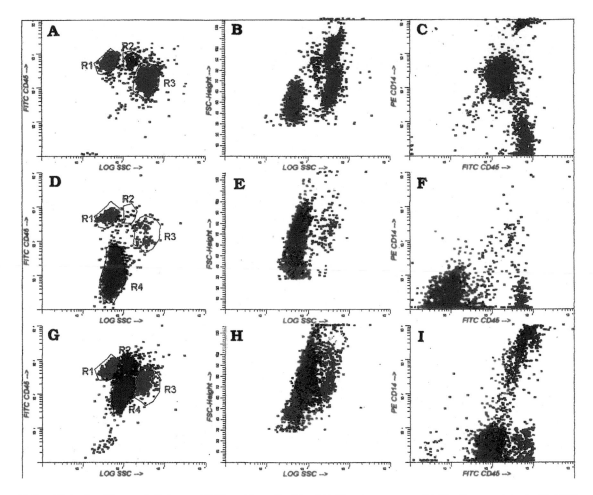

Color Plate 4   (Chapter 4) Gating on CD45-positive cells. Blood (A,B) or bone marrow (C–F) cells were stained with FITC CD45 PECD14 and the data acquired ungated. In A, lymphocytes were resolved in the bivariate display log SSC vs CD45 (R1) and then their FSC vs SSC (B) pattern for the cells was displayed. This method is also ideal for resolving leukemias. As shown in C and D, the AML cells are clearly resolved in D. The HLADr and CD14 expression as well as the FSC vs SSC characteristics can also be easily defined.

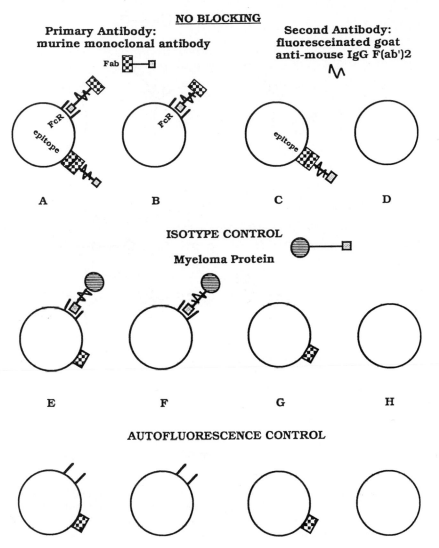

Fig. 2 Indirect immunofluorescence staining. The primary Ab is shown as a stick symbol that binds to cells with epitopes shown by the patterned rectangle. The Fc protein binds to its receptor, shown by the square. The second Ab binds to the primary Ab (A–D). The stick symbol for the isotype control (a myeloma protein) has the same shape for the FcR but a circle for the $F(ab')_2$ portion, for which there are no epitopes on the cells (E–H). And cells that have not been stained serve as an autofluorescence control!

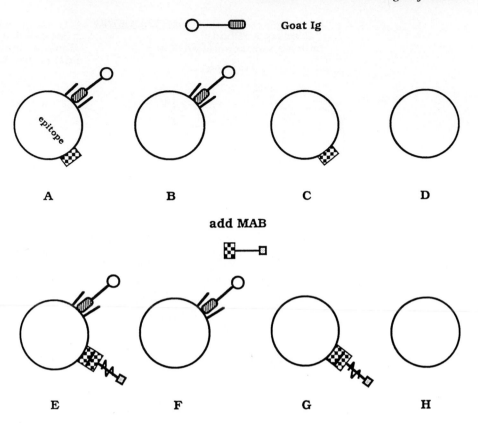

Fig. 3 Blocking Fc receptors. Cells are incubated for 10 min with goat Ig, shown as a stick symbol with its epitope-binding site a circle. The murine mAb is then added (shown as a patterned rectangle), and the cells are incubated for 15 min. After being washed, the pelleted cells are labeled with FGAM (shown as a jagged line), incubated 15 min, and again washed.

FOR INDIRECT STAINING, BLOCK WITH THE IMMUNOGLOBULIN FRACTION FROM THE SAME SPECIES IN WHICH THE SECOND ANTIBODY WAS MADE.

It is best to use the IgG fraction rather than serum to block with because the serum may not contain sufficient IgG of the correct isotype and subclass to effectively block FcR. We have found the effective blocking concentration is 30 μg IgG per 10^6 cells in a volume of 100 ml. This concentration provides enough IgG of each subclass to effectively block all FcR binding as well as "nonspecific" binding. In Table I, the approximate amounts of Ig isotypes and IgG subclasses in the serum of most mammals are shown. If 10% serum was used and a concentration of 30 μg (3 mg/ml/10^6 cells) is the effective blocking concentration, 10% serum would not be an effective blocking reagent for IgG2, IgG3, IgG4, IgM, or IgA. The problem would be even worse if lower serum concentrations were to be used.

The effect of blocking is shown in Fig. 4. A population of both small cells and large cells are positive in the histograms on the top of each panel. The antibody staining for the cells is shown in the histogram along side of each panel. By projecting each histogram event into the bivariate display, the correlation of

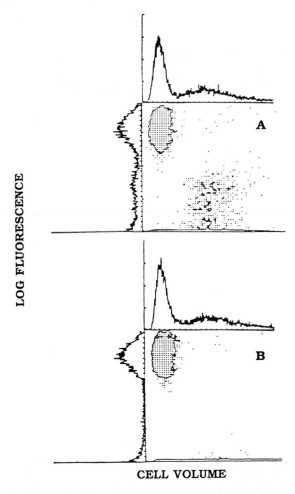

Fig. 4 Effect of blocking on mAb binding to mononuclear cells (MNC). Human cells were isolated from blood using neutrophil isolation medium (NIM, Cardinal Associates). Cells were adjusted to 20 × 10^6/ml in PAB and 50 μl was used for labeling. (A) Cells were incubated with fluoresceinated CD3 for 15 min. (B) Cells were first incubated for 10 min with 30 μg mouse IgG and then fluoresceinated CD3 was added for 15 min. After being washed, the samples were analyzed.

cell size with antibody staining can be visualized. In this example, directly fluoresceinated anti-CD3 (FL-CD3) was used to label mononuclear cells (MNC). When cells are incubated with FL-CD3 many large cells are dimly stained (Fig. 4A). When the cells are incubated for 10 min with mouse Ig (mIg) prior to the addition of the FL-CD3, no large cells are dimly stained (Fig. 4B). Thus, blocking the cells with mIg prior to staining them had a profound effect on the type of cells that were stained by the antibody.

When a second fluorochrome-conjugated antibody is used, it is important to use the $F(ab')_2$ fragment. In Fig. 5 MNC labeled either with the $F(ab')_2$ of fluoresceinated goat anti-mouse IgG (top panel) or intact fluoresceinated goat anti-mouse IgG (bottom panel) were analyzed. Both analyses were performed in the absence of any blocking Ig. Almost all the large cells and a small proportion of small cells are stained by the intact anti-IgG but not by the $F(ab')_2$. This simple test should always be performed when using a new second antibody and only those that behave like the $F(ab')_2$-marked cells illustrated in Fig. 6 should be used.

There is yet another problem when a myeloma protein is used as the isotype control. The myeloma protein, like the mAb, is derived from neoplastic cells. These cells produce structurally abnormal immunoglobulins. Because they are abnormal, myeloma proteins of the same isotype and subclass may behave differently in the way they bind to Fc receptors. Second, some suppliers may offer isotype control reagents that are actually monoclonal antibodies with specificities to cellular products believed to be absent in leukocytes. Illustrated in Fig. 6 are six isotype controls of the IgG1 subclass. IgG1 was selected for this illustration because it has a weak affinity for the murine macrophage IgG2b FcR. Two myeloma proteins did not bind to macrophages, others bound to only a small percentage of them, and others stained a high percentage of the macrophages. Which one should be chosen for the isotype control? Which one will behave like the primary mAb antibody? The best way to answer these two questions is to block their binding. This is most important to do when multiple antibodies for immunophenotyping are used together because the problem becomes increasingly more complex as the number of antibodies used together increases.

D. Directly Labeled Antibodies

Primary antibodies are currently available labeled with fluorescein (F), phycoerythrin (PE), the tandem complexes of phycoerythrin–Texas red (PETR), or phycoerythrin–Cy5 (PECY) and biotin (B). Using a directly labeled primary antibody the cells can be preincubated with murine IgG as the block prior to staining with a labeled murine mAb. Rat IgG would be used if the labeled primary mAb were of rat origin. When using a biotinylated antibody, labeled avidin is used as the second reagent. Avidins can be purchased conjugated with F, PE, PETR, PECY.

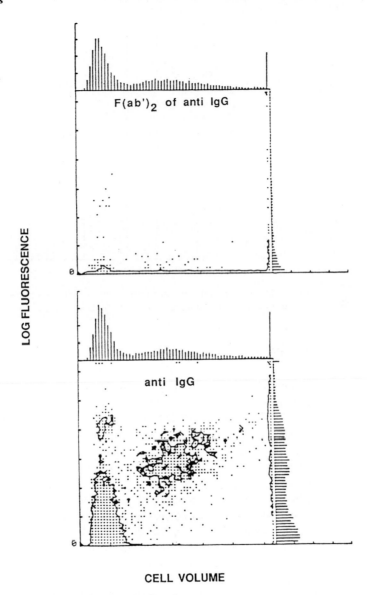

Fig. 5 Second-reagent quality. Human MNC were isolated from blood using NIM. Cells were adjusted to 20×10^6/ml in PAB and 50 μl was used for labeling. The MNC were incubated for 15 min at 4°C with either fluoresceinated goat F(ab')$_2$ anti-mouse IgG or fluoresceinated goat anti-mouse IgG. The cells were washed, resuspended, and analyzed. The cell volume histograms are shown on top and the antibody-staining histograms are shown alongside each bivariate distribution.

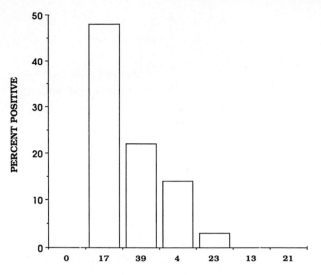

Fig. 6 Myeloma protein binding to macrophages. Murine peritoneal macrophages were incubated with 10 μg of each myeloma protein for 15 min, washed, and stained with FGAM.

III. Titering Antibodies

The proper antibody titer is most important if good immunophenotyping data are to be obtained. Since mAbs are myeloma proteins whose antibody specificity is known, they behave both like myeloma proteins (binding by Fc and nonspecifically) and like specific antibodies (binding to epitope). For any given mAb, there may be a different degree of Fc and nonspecific binding. Proper titering can both reduce or eliminate the problem of Fc and nonspecific binding as well as provide the desired specific epitope binding. To optimize specific binding both the fraction of positive cells and the mean channel fluorescence must be measured. As illustrated in Fig. 7, when unblocked cells are stained with too high an antibody concentration (Fig. 7A), there is a uniform shift to the right of the negative cells so some of them are above the marker established for positive cells using an isotype control. When the mean cellular fluorescence (MCF) is computed for cells below the selected marker separating positives from negatives, it will be higher than the MCF for the isotype control. As the antibody is diluted the negative cells shift to the left producing a decrease in the fraction of positive cells, a decrease in the MCF of the cells below the marker, and decrease in the MCF of cells above the marker (Fig. 7B). If, however, the ratio of MCF values above (signal) and below (noise) the marker is computed, as shown in Fig. 7C, this ratio will increase to a maximum and then decrease. The optimal antibody titer is the dilution that produces the maximum signal to noise ratio and it is not necessarily the amount that produces the highest percentage of positive cells.

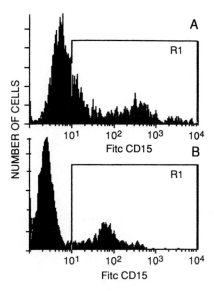

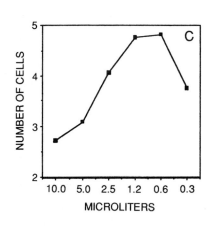

Fig. 7 Titering antibodies. Cells are stained with increasing 1/2 or 1/3 dilutions of antibody. The marker for positive and negative cells is set using cells stained with an isotype control antibody so that no more than 1% are above it. In A, not only is a clearly positive population of stained cells found but the entire cellular distribution is shifted to the right. In B, the negative cells are only slightly shifted to the right and well below the marker. By computing the MCF and plotting the values for each dilution (C), the dilution producing the maximum signal to noise ratio (4.8) can be determined.

IV. Third Color Reagents

Three fluorochromes are available to provide for a third and fourth color excitable by a single laser. The properties of these reagents have been previously described (Stewart, 1993) and are summarized in Table II.

The PECY5 fluorochrome is the brightest currently available because the CY5 absorption bandwidth is optimal for PE emission. This is reflected in the low amount of compensation required (usually <3% and for some constructs, <0.5%). PerCP requires no compensation, but its emission energy is less than that found for fluorescein. It should be used for antibodies to highly expressed cellular epitopes. PerCP also rapidly degrades at high laser power density.

The PETR tandem is intermediate in its fluorescence emission intensity because the TR absorption bandwidth is not as ideally matched to PE emission as CY5. This energy mismatch results in a high degree of PE emission (20–30%) requiring considerable compensation similar to the overlap found for fluorescein in the PE channel. As shown in Fig. 8, both the PETR and PECY5 tandems can exhibit unacceptable variation in their compensation requirements among different directly conjugated antibodies or on different batches of the same antibody. This problem has been solved by some suppliers of these reagents.

Table II
Comparison of Third Color Reagents

	PerCP	PETR	PECY5
Relative intensity	Dim	Medium	Bright
Compensation: amount	None	High	Low
Batch variability	None	Some	Significant
Light sensitivity	Stable	Stable	Unstable
Nonspecific binding			
Lymphocytes	None	None	None
Monocytes	None	None	High
Granulocytes	None	None	Low
Availability of direct conjugates	Fair	Poor	Good
Emission (nM)	673	613	670

While otherwise an ideal fluorochrome because it is so bright, the PECY5 tandem exhibits two other problems not found for PETR or PCP. This tandem is exquisitely sensitive to light and the CY5 molecule will become irreversibly degraded upon short-term exposure (<1 hr) to ambient light. This problem can be easily recognized as an increase in PE signal like that shown in Fig. 8D, requiring increasingly more compensation with exposure time and decrease in CY5 signal with light exposure. To prevent this problem, all staining should be performed in subdued light and samples stored in the dark.

The second problem is caused by this fluorochrome's ability to specifically bind to myelomonocytic cells. While there is currently no solution to this problem, it is exacerbated by the tendency for some suppliers to overconjugate antibodies with this tandem. This causes an increase in both nonspecific binding to all cells and specific binding to myelomonocytic cells. A PECY5 isotype control is essential as blocking is ineffective in eliminating this fluorescence.

V. Cell Preparation

We recommend a whole-blood (or bone marrow) lysing method to remove erythrocytes and there are several reagents commercially available. While all of them work with blood, some are better than others for lysing bone marrow erythroid cells. Lysing erythrocytes is not recommended before staining because the platelets and erythrocyte debris are concentrated along with the leukocytes during the subsequent washing steps causing difficulty in the analysis of the data. Therefore, postlabeling lysis is favored.

In the following procedures a "wash" means to add 4-ml phosphate-buffered saline to the tube, centrifuge the cells at 1200g for 3 min at 4°C, and decant the supernatant by inverting the tubes and blotting the lip by touching its rim

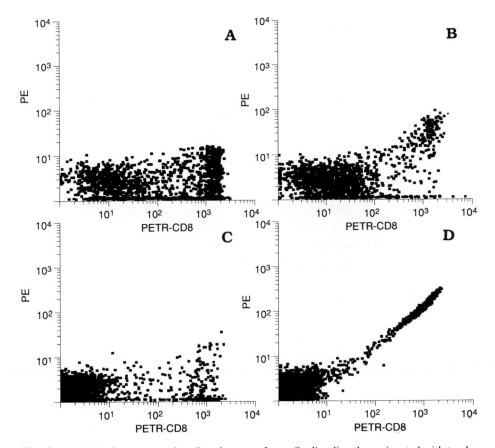

Fig. 8 Variation in compensation. Requirements for antibodies directly conjugated with tandem fluorochromes. Blood cells were stained with two separate batches of PETR-CD8 or PECY5-CD8. In A, the PETR-CD8-stained cells were properly compensated. When stained with a second batch (B), a different compensation would be required. The same effect is seen for the PECY5-CD8, where proper compensation for one batch (C) produces unacceptable compensation for the second batch (D).

to an absorbent towel. Thoroughly resuspend the cells in the residual buffer that has drained back to the bottom of the tube (not to exceed 100 μl). All labeling procedures are carried out on ice. Room temperature may be fine for lymphocytes but can be problematic for other hematopoietic cells due to internalization of the antibody.

Each antibody should be titered prior to use so that the correct amount to optimally stain the cells is used. Our general practice is to have the optimal amount in 10 μl. Commercially available antibodies that have already been titered may be diluted differently by the supplier so that more or less than 10 μl is required. In these cases, use the suppliers recommended amount (unless it is found to be incorrect by titering according to Section III).

VI. Labeling Procedures

For all the following procedures, put 50 μl of a cell suspension (do not exceed a concentration of 20 $\times$ 10^6 cells/ml blood, or bone marrow) into 13 $\times$ 75-mm tubes containing 30 μg of blocking immunoglobulin. Do not block when measuring Fc receptors.

A. Labeling Cells with a Single Antibody

Indirect labeling with one antibody and a second antibody:

1. Add 10 μl goat IgG (3 mg/ml) for 10 min
2. Add primary Ab for 15 min
3. Wash
4. Add fluoresceinated goat anti-mouse IgG F(ab')$_2$ (FGAM) second Ab for 15 min
5. Wash.

For the control tube, add the isotype control antibody instead of the first antibody in step 2. Also prepare a tube containing unstained cells; this will be used to measure cellular autofluorescence.

Indirect labeling with a biotinylated antibody and avidin:

1. Add 10 μl xIgG (3 mg/ml) for 10 min
2. Add B-Ab for 15 min
3. Wash
4. Add F-avidin for 15 min
5. Wash.

x is the species of IgG that is the same as the antibody. For murine antibodies block with normal mouse IgG; for rat antibodies block with normal rat IgG.

Direct labeling with one antibody:

1. Add 10 μl xIgG (3 mg/ml) for 10 min
2. Add F-Ab for 15 min
3. Wash.

FOR INDIRECT LABELING WITH BIOTINYLATED ANTIBODIES OR FOR DIRECT LABELING, BLOCK WITH THE IMMUNOGLOBULIN FRACTION FROM THE SAME SPECIES AS THE ANTIBODY. NEVER BLOCK IF THE EPITOPE BEING MEASURED IS THE Fc RECEPTOR.

For example, CD16 is the FcRIII receptor on granulocytes and NK cells. Some CD16 antibodies will not bind to this receptor if it is blocked.

B. Labeling Cells with Two Antibodies

One directly conjugated antibody with one unconjugated antibody:

1. Add 10 μl goat IgG (3 mg/ml) for 10 min
2. Add Ab for 15 min
3. Wash
4. Add PEGAM second Ab for 15 min
5. Wash
6. Add 10 μl xIgG (3 mg/ml) for 10 min
7. Add F-Ab for 15 min
8. Wash.

It is absolutely essential in step 6 that the blocking IgG be from the same species as the unlabeled antibody. This is because the second antibody will have free binding sites that must be blocked so it will not bind the directly conjugated second antibody when it is added subsequently in step 7. For example, suppose that both the unconjugated and conjugated antibodies are murine mAbs. After step 5, some cells have bound mAb1 and PEGAM to them. Because free binding sites remain on the PEGAM the F-mAb may bind to them when it is added. To prevent this binding, these free sites must be blocked by the addition of mouse IgG shown in step 6.

If a rat mAb had been used then rat IgG would be used for the block. If one mAb is rat and the other mouse then the appropriate second antibody and block would be used: Unlabeled rat, use PEGAR (goat anti-rat) and block with rat IgG; unlabeled mouse, use PEGAM and block with mouse IgG. FGAM or FGAR could also be used in combination with PE-labeled antibodies.

For two antibodies:

1. Always perform the indirect labeling step first.
2. Block with normal IgG of the first antibody species before adding the directly conjugated antibody.

Two directly conjugated antibodies:

1. Add 10 μl xIgG (3 mg/ml) for 10 min
2. Add F-Ab and PE-Ab for 15 min
3. Wash.

One antibody conjugated with fluorescein, one with biotin:

1. Add 10 μl xIgG for (3 mg/ml) for 10 min
2. Add F-Ab and B-Ab for 15 min
3. Wash
4. Add PE-avidin for 15 min

5. Wash.

xIgG is derived from the species in which the conjugated antibodies were produced.

C. Labeling Cells with Three Antibodies

There are several strategies that can be used for labeling cells with three antibodies. For the reagents, TC (third color) refers to either PerCP, PETR, or PECY5.

Two fluorochrome-conjugated antibodies and one unconjugated antibody:

1. Add 10 μl goat IgG (3 mg/ml) for 10 min
2. Add unlabeled Ab for 15 min
3. Wash
4. Add BGAM for 15 min
5. Wash
6. Add 10 μl xIgG (3 mg/ml) for 10 min
7. Add F-Ab, PE-Ab, and TC-avidin for 15 min
8. Wash.

xIgG is derived from the same species as the unlabeled Ab. In this procedure a biotinylated second antibody was used followed by TC-avidin to illustrate one of the many strategies that can be used to provide for the desired fluorochrome combination. A fluorochrome-conjugated second antibody could also have been used instead of the biotinylated one.

Three antibodies—two directly conjugated and one biotinylated antibody:

1. Block with 10 μl xIgG for (3 mg/ml) 10 min
2. Add Fl-Ab, PE-Ab, and B-Ab
3. Wash
4. Add TC-avidin
5. Wash.

xIgG is derived from the same species as the three directly labeled antibodies. Three directly conjugated antibodies:

1. Block with 10 μl xIgG (3 mg/ml) for 10 min
2. Add F-Ab, PE-Ab, and TC-Ab for 15 min
3. Wash.

D. Staining Cells with Four Antibodies

It is now possible to excite the four fluorochromes F, PE, PETR, and PECY5 or PerCP with a single laser. This provides for the opportunity to use four antibodies simultaneously.

Two directly conjugated, one biotinylated, and one unconjugated antibody:

1. Add 10 μl goat IgG (3 mg/ml) for 10 min
2. Add Ab for 15 min
3. Wash
4. Add PETRGAM
5. Wash
6. Block with xIgG (3 mg/ml) for 10 min
7. Add F-Ab, PE-Ab, and B-Ab for 15 min
8. Wash
9. Add PECY5-avidin or PerCP-avidin for 15 min
10. Wash.

xIgG is derived from the same species as the unlabeled antibody.
Three directly conjugated and one biotinylated antibody:

1. Add 10 μl xIg (3 mg/ml) for 10 min
2. Add F-Ab, PE-Ab, PECY5 (or PerCP)-Ab, and B-Ab for 15 min
3. Wash
4. Add PETR-avidin for 15 min
5. Wash.

Other combinations for the third color are also possible. For example, an antibody directly conjugated with PETR and a PECY5 or PerCP-avidin could have been used.
Four directly conjugated antibodies:

1. Add 10 μl xIgG (3 mg/ml) for 10 min
2. Add F-Ab, PE-Ab, PETR-Ab, and PECY5 (or PerCP) -Ab for 15 min
3. Wash.

The four-color panel for immunophenotyping human cells offers the advantage of combining three antibodies for subset identification with CD45 as the fourth antibody used for resolving leukocytes from debris and other cells in general and for resolving lymphocytes in particular (Stelzer *et al.*, 1993). The data analysis strategy for this application is discussed in Chapter 4.

VII. Lysing Erythrocytes

1. Thoroughly resuspend cells in residual PBS
2. Add 3 ml of lysing buffer and agitate cells for 3 min (rock, roll or tumble)
3. Centrifuge and wash.

VIII. Detecting Dead Cells

A. Unfixed Cell Suspensions

1. Resuspend cell pellet in 1 ml PBS containing 2 μg propidium iodide per ml
2. Analyze samples after 5 min incubation. Alternative: Add 10 μl of propidium iodide (200 μg/ml) to 1 ml of cell suspension.

B. Fixed Cell Suspensions (Riedy *et al.*, 1991)

1. Resuspend cell pellet (prior to fixation) in residual PBS and add enough ethidium monoazide (EMA) for a final concentration of 5 μg/ml[2]
2. Put samples 18 cm from 40-W fluorescent light for 10 min
3. Wash and fix cells.

IX. Fixation

1. Thoroughly resuspend cells in residual PBS
2. Add 1 ml 1% ultrapure formaldehyde
3. Analyze samples within 5 days.

If desired, cells may be centrifuged and resuspended in less formaldehyde to have them more concentrated for data acquisition. The formaldehyde concentration should not be less than 1% for reproducible FSC vs SSC characteristics. Use only ultrapure formaldehyde because other formulations will produce increased autofluorescence.

X. Solutions and Reagents

A. PBS (with Sodium Azide)

Phosphate-buffered saline containing no calcium or magnesium and supplemented with 0.1% sodium azide is used for all dilutions and washes. The pH should be adjusted to 7.2. While not necessary, PBS may be supplemented (PAB) with 0.5% bovine serum albumin fraction V (Sigma Chemicals, St. Louis, MO). Be sure to check osmolality of the final solution (290 $\pm$ 5 mOsm). Note: Handle sodium azide with extreme caution.

B. 1% Ultrapure Formaldehyde

A 10% solution of ultrapure formaldehyde can be obtained from Polysciences. This solution is diluted 1/5 with PBS to create the working solution. Do not

[2] Some batches require less EMA to reduce nonspecific fluorescence.

use impure formaldehyde or formylin as these solutions will markedly increase autofluorescence.

C. Lysing Reagent

> 1.6520 g ammonium chloride
>
> 0.2000 g potassium bicarbonate
>
> 0.0074 g EDTA (tetra)
>
> q.s. to 200 ml with distilled water, use at room temperature.

This reagent *must* be prepared daily because the dissolved CO_2 in water combines with NH_4Cl to form the carbonate $(NH_4)2CO_3$ that is ineffective in lysing erythrocytes. We recommend weighing the reagents and storing them as packets. The dry reagents are dissolved in water as required.

D. Propidium Iodide Stock Solution

Note: Handle propidium iodide with extreme caution. A total of $7.4 \times 10^{-5} M$ (20 mg/100 ml) propidium iodide (Sigma Chemicals) is prepared in PBS. Store in a dark foil wrapped container to protect from light. Propidium iodide MW = 668.

E. Ethidium Monoazide Stock Solution

EMA (Molecular Probes, Inc., Eugene, OR) is very light sensitive and must be stored at $-20°C$ in a dark vial or foil-wrapped container. The stock solution is prepared in PBS at 5 mg/ml. This can then be diluted further into small aliquots of 50–100 μg/ml and stored at $-20°C$. These small aliquots can then be thawed one at a time as needed to do the assays. Discard remaining EMA after thawing.

References

Bender, J. G., Unverzagt, K. L., Walker, D. E., Lee, W., Van Epps, D. L., Smith, D. H., Stewart, C. C., and Kik, T. (1991). *Blood* **77**, 2591–2596.

Landay, A. L. (1992). *NCCLS* **12**(6), 1–76.

"Pierce-Immunotechnology Catalogue and Handbook," Section C, 1989.

Riedy, M. C., Muirhead, K. A., Jensen, C. P., and Stewart, C. C. (1991). *Cytometry* **12**, 133–139.

Roitt, I., Brostoff, J., and Male, D. (1989). "Immunology," Chapter 5, pp. 5.1–5.11. St Louis: Mosby.

Sneed, R. A., Stevenson, A. P., and Stewart, C. C. (1989). *J. Leuk. Biol.* **46**, 547–555.

Stelzer, G. T., Shults, K. E., and Loken, M. R. (1993). *In* "Clinical Flow Cytometry," (A. Landay, K. Ault, K. Bauer, and P. Rabinovitch, eds.), pp. 265–280. New York: New York Academy of Science.

Stewart, C. C. (1990a). *In* "Methods of Cell Biology," (Z. Darzynkiewicz and H. Crissman, eds.), Vol. 33, pp. 411–426. San Diego: Academic Press.

Stewart, C. C. (1990b). *In* "Methods of Cell Biology," (Z. Darzynkiewicz and H. Crissman, eds.), Vol. 33, pp. 427–450. San Diego, Academic Press.

Stewart, C. C. (1992). *Cancer* **69,** 1543–1552.

Stewart, C. C., and Stewart, S. J. (1993). *Ann. N. Y. Acad. Sci.* **677,** 94–112.

Stewart, C. C., Stevenson, A. P., and Habbersett, R. C. (1988). *Int. J. Radiat. Biol.* **53,** 77–87.

Stewart, C. C., Stewart, S. J., and Habbersett, R. C. (1989). *Cytometry* **10,** 426–432.

CHAPTER 4

Multiparameter Analysis of Leukocytes by Flow Cytometry

Carleton C. Stewart and Sigrid J. Stewart

Laboratory of Flow Cytometry
Roswell Park Cancer Institute
Buffalo, New York 14263

I. Introduction

Multiparameter flow cytometry can be used to define the unique repertoire of epitopes on cells using forward scatter (FSC),[1] side scatter (SSC), and several antibodies labeled with a different-colored fluorochrome (Shapiro, 1988; Willman and Stewart, 1989; Stewart, 1992, Terstappen *et al.*, in press; Hoffman,

[1] Abbreviations used: AB, antibody; B, biotin; BP, band pass; CD3, antibody to T cells; CD8, antibody to suppressor T cells; CD14, antibody to monocytes; CD45, antibody to leukocytes; DM, dichroic mirror; EMA, ethidium monoazide; F(ab')$_2$, part of the antibody that has the epitope binding site; FITC, fluorescein isothiocyanate; FL, fluorescein fluorescence; FL-BrdU, fluoresceinated antibody to bromodeoxyuridine; FSC, forward scatter; Ig, immunoglobulin; IgG, immunoglobulin isotype gamma; L, lymphoid cells; LLy, large lymphocytes; LP, long pass; Ly, small lymphocytes;

1988; Stewart *et al.*, 1989). No other analysis system can obtain this information as rapidly on large numbers of cells while sorting the unique population. By using more than one antibody simultaneously it is possible to perform single-antibody histogram analysis and multiparameter analysis. The simplest is single-antibody histogram analysis in which each of the individual fluorescence histograms is treated as single-parameter data. Data analysis is rapid because only the percentage of positive cells for each marker is considered.

Multiparameter analysis is much more complex and time consuming because populations that are labeled with more than one antibody must be individually identified. Each file must be analyzed separately because current software is not capable of analyzing data automatically in a reliable fashion. This is because the relative positions of populations resolved by multiple antibodies can move relative to one another when the same panel is used with different subjects. Regions established to resolve a particular population in one sample may have to be moved slightly to resolve the same population in another sample.

Multiple antibodies, however, provide the opportunity to resolve unique cell populations that cannot otherwise be identified. These populations may be most important in understanding normal and disease processes. Strategies for analyzing this type of data and the pitfalls encountered in multiparameter data acquisition and analysis are discussed. Whereas the various aspects of analysis are illustrated using human peripheral blood, the principles discussed can be applied to any type of cell population.

II. Correlated List–Mode Data

For purposes of discussion, we use "acquisition" to mean obtaining the data of labeled cells using a flow cytometer. We use "analysis" to mean the analysis of that data. List-mode data should always be acquired when performing multiparameter sample analysis to provide retrospective data analysis. A simple correlated list-mode file is illustrated in Table I.

Suppose we wish to acquire FSC which reflects cell size and one color of fluorescence. When cells are analyzed, the measured values are recorded sequentially in the computer. For the first cell, the first number in the list is FSC, the second number is fluorescence. The second cell produces the third and fourth numbers in the list, the third cell produces the fifth and sixth numbers and so forth until all the cells have been measured. In this example, if 10,000 cells had been measured there would be 20,000 numbers in the list in exactly

M, monocytes; mAb, monoclonal antibody; mIg, mouse immunoglobulin; MNC, mononuclear cells; NG, nongranulated granulocytes; NIM, neutrophil isolation medium; NK, natural killer cells; OE, other events; PAB, phosphate-buffered saline supplemented with 0.1% sodium azide and 0.5% bovine albumin; PE, phycoerythrin; PECY5, PE/CY5 tandem; PerCP, perdinin chlorophyll; PETR, PE–Texas red tandem; PI, propidium iodide; PMT, photomultiplier tube; SP, short pass; SSC, side scatter; TT, trigger threshold.

Table I
Simple List-Mode Data File

Data list	Explanation
200	Forward scatter cell 1
400	Fluorescence cell 1
100	Forward scatter cell 2
700	Fluorescence cell 2
200	Forward scatter cell 3
500	Fluorescence cell 3
FSCn	Forward scatter cell n
Fn	Fluorescence cell n

the same order as they were acquired. The first and every third number thereafter would be FSC, the second and every fourth number thereafter would be fluorescence. Thus, measured parameters for each cell can be reanalyzed by the computer over and over again in the same order that the cells were measured the first time by the flow cytometer.

By selecting regions utilizing one set of parameters that contain the values for a particular cell population, the location of these cells in the other parameters can be found and displayed. The display could be a simple histogram or bivariate plot of one parameter versus the other.

There are three common methods for bivariate plots, contour plots, density plots, and dot plots. Contour plots provide a series of isometric lines that are intended to show the number of cells at selectable levels. Density plots are like contour plots except different colors are used to display different cell numbers. Contour plots are useful when there is a high frequency of cells in a population in two-parameter space. For dot plots, each cell is plotted as a dot. The greater the size of any given population the more dots plotted so that they may be plotted on top of one another. This causes a loss in the ability to see differences among populations with high cell frequencies. This method is best for displaying populations that contain few cells. For multiparameter data where more than one bivariate display is necessary, dot plots are the preferred method because the location of different cell populations in each bivariate display can be identified by assigning a specific color to them. All bivariate plots used in this chapter are dot plots.

III. Single-Label Analysis

For immunophenotyping cells, single-label analysis using the indirect method or directly labeled primary antibodies is most often used. Figure 1 shows human peripheral blood mononuclear cells obtained from a patient who had been infused with interleukin 2 and stained with fluoresceinated anti-CD4 (FL-CD4).

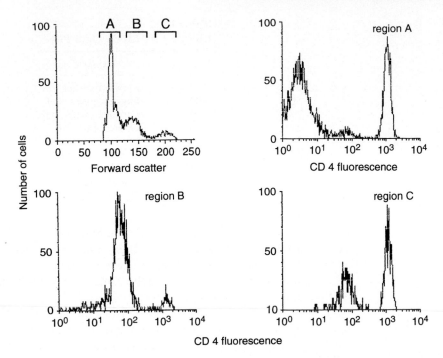

Fig. 1 Single-color analysis. Human cells were isolated from blood using neutrophil isolation medium (NIM, Cardinal Associates, Santa Fe, NM). The cells were adjusted to 20×10^6/ml in PAB and 50 μl were used for labeling. The mononuclear cells were incubated 10 min with 30 μg mouse IgG, then for 15 min with fluoresceinated CD4, washed, and analyzed. Using forward light scatter, region A was selected to include small cells, region B intermediate cells, and region C large cells. The CD4 staining characteristics of the cells in each of these regions is shown in the appropriately labeled histograms.

The purpose of this example is to show the importance of including all lymphocytes in the analysis region and the importance of list-mode data for performing this task. In general, any single property of a uniform group of cells is distributed in a Gaussian (bell-shaped) fashion. This is shown in Fig. 1 in the top left panel where three populations of cells with different FSC properties can be seen. The FSC measurement reflects cell size and events to the left represent the smallest and those to the right represent the largest cells. Those cells in region A are smallest and the fluorescence histogram is determined by processing the list-mode file for fluorescences associated with cells meeting the size criteria. The cells in the left peak of the region A fluorescence histogram (upper right) are the unstained cells that have a low innate fluorescence called autofluorescence. A population of brightly stained cells are in the right peak and a small population of cells between the two major peaks are dimly stained with CD4. Note the Gaussian nature of all of the populations.

By processing the list-mode file for cells meeting the size criteria for the cells in region B, the fluorescence histogram (lower left) shows these cells are dimly stained with CD4, but a few brightly stained cells are also found in this region. Finally, there is a small population of very large cells in region C (upper left). When their fluorescence histogram is displayed (lower right) most of these cells are brightly stained with CD4 characteristic of the positive cells found in region A. While this population of large lymphocytes is not a prominent one in normal individuals, it may be frequently found in sick patients. Had the common practice of acquiring only the small cells found in region A been employed, the population of large cells would not have been found. When the cells in each of these populations are sorted and stained with Wrights blood stain, the cells in region A are lymphoid; those in region B, with intermediate fluorescence intensity, are monocytes; and those in region C, with bright fluorescence intensity, are large lymphocytes. It is better to acquire more data for retrospective analysis than to never acquire it at all.

The presence in each of the fluorescence histograms of cells with the fluorescence characteristic of a dominant population found in another size region is due to the overlap of cell sizes that were used to establish the analysis regions. As shown later, by using both size and fluorescence together in a bivariate display, contamination of one population by the other can be significantly reduced.

IV. Optical Filtration

Before considering the use of more than one antibody, the problem of overlapping fluorescence spectra of fluorochromes used to label monoclonal antibodies needs to be addressed. As shown in Fig. 2, the emission from fluorescein (FL) overlaps that of phycoerythrin (PE), whereas phycoerythrin emission does not significantly overlap fluorescein. The PETR also overlaps the PECY5 (or PerCP from Becton–Dickinson) emission spectrum. The emission from each fluorochrome can be optimized to reduce but not eliminate unwanted fluorescence using optical filtration. Table II lists the optical filters commonly used for fluorescence detection when the fluorochromes are excited at 488 nm.

Optical filters are identified by the wavelengths they process. Long-pass (LP) filters will transmit wavelengths that are longer than the number given whereas short-pass (SP) filters transmit shorter wavelengths. For a band pass (BP) filter the first number refers to the wavelength that is transmitted at or near 100% and the second number refers to the bandwidth (50% transmission). Thus, 530-30 means that the filter transmits 50% of light from 515 to 545 nm (30 nm bandwidth) and the maximum transmission is at 530 nm. Dichroic mirrors (DM) are used to both transmit and reflect selective wave lengths of light along the optical path. An LP dichroic mirror transmits the longer and reflects the shorter designated wavelengths while a SP dichroic mirror does the opposite. An LP560

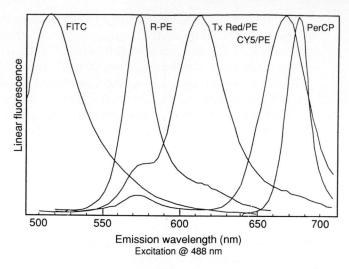

Fig. 2 Emission spectra of various fluorophores. The emissions of FITC, R-PE (phycoerythrin), PETR, PECY5, or PerCP (perdinin chlorophyll) are sufficiently well separated that they can be used together and excited by a single laser operating at 488 nm. (Courtesy of Dr. D. Rechtenwald, BDIS.)

dichroic mirror transmits wavelengths longer than 560 nm (e.g., PE fluorescence) and reflects shorter wavelengths (e.g., FITC fluorescence).

V. Compensation

Because the optical filters cannot remove all the overlap from the unwanted fluorochrome, electronic subtraction, called spectral compensation, is used to eliminate the remaining fluorescence. Figure 3 shows a schematic diagram of how this is accomplished.

Table II
Filters Commonly Used for
Immunofluorescence Measurements

Fluorochrome	Dichroic mirror (nm)	Band pass filter (nm)
Fluorescein	SP 560	530-30
Phycoerythrin	SP 560	575-26
PETR	SP 610	630-22
PECY5	SP 640	680-18
PerCP	SP 640	680-18

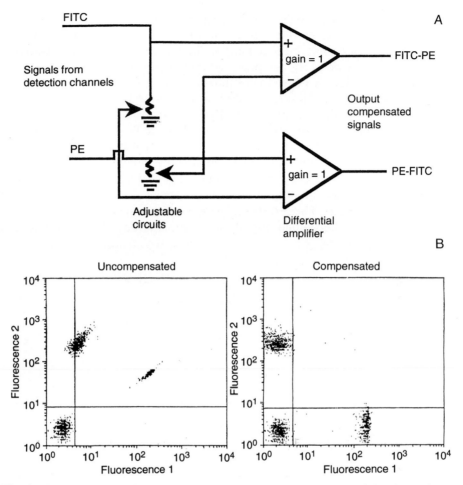

Fig. 3 Electronic compensation. (A) Simplified representation of the circuit for electronic compensation. (B) Bivariate distribution of uncompensated (left panel) and properly compensated (right panel) microspheres. The PMT high voltage was adjusted so that the unlabeled microspheres were in the lower left corner. The fluorescence 1 boundary limit is shown by the line parallel to the *x*-axis and the fluorescence 2 boundary limit by the line parallel to the *y*-axis. Events beyond these boundaries represent spectral overlap.

The signal from the fluorescence 1 (FITC) photomultiplier tube is routed directly into the positive side of the fluorescence 1 differential amplifier and via an adjustable circuit into the negative side of the fluorescence 2 (PE) differential amplifier. Similarly, the fluorescence 2 photomultiplier signal is routed into the positive side of its differential amplifier and, via an adjustable circuit, into the negative side of the fluorescence 1 differential amplifier. By adjusting the amount of signal fed into the negative side of the appropriate differential amplifier from particles labeled with a single fluorochrome, the unwanted fluorescences can be removed. For more than two fluorochromes several compensation networks can be ganged together to achieve the correct compensation.

A properly compensated instrument will provide good resolution of single- or double-labeled cells independent of the fluorochrome's intensity. Improper compensation can lead to false-positive or false-negative cell populations.

In Fig. 4, the effect of overcompensation is illustrated using lymphocytes labeled with either FL-CD4 (FL-1) or PE-CD4 (FL-2), separately or both together. In A, the negative cells are separated from the x- and y-axis as are the labeled cells and, illustrating good compensation, their leading edge is in line with the negative control boundary limits (dashed line). (Some instruments do not have an offset so that all events lie along the x- or y-axis and events cannot be seen. These instruments are difficult to properly compensate.) As shown in B, cells stained brightly with one antibody (PE-CD4) and dimly with another

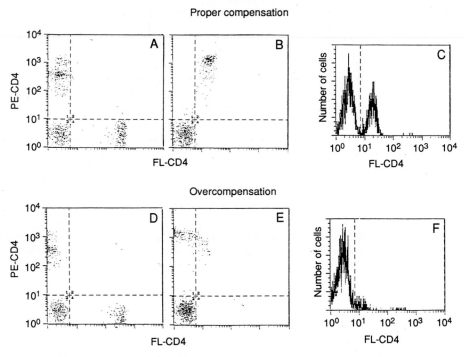

Fig. 4 Effect of overcompensation. Mononuclear cells (MNC) were labeled separately with FL-CD4 or PE-CD4 and mixed together. To obtain double-labeled cells with bright PE and dim FL fluorescence, a mixture of nine parts of PE-CD4 and four parts of FL-CD4 was prepared and the combination was used to label the cells. Cells were analyzed using a "lymphogate." The analysis of a mixture of cells that were separately labeled and mixed together is shown using (A) a proper compensation and (D) overcompensation. (B) Cells stained brightly with PE-CD4 and dimly stained with FL-CD4 analyzed with proper compensation; (C) histogram of fluorescence 1. (F) Histogram of fluorescence 1 of the overcompensated cells.

are well resolved and, in C, the histogram shows the dim cells clearly resolved. Overcompensation can result in artifacts, and, as shown in D, the negative cells are correctly positioned, but too much compensation has been used on the stained cells so many of the cells are against the axis itself. As shown in E, when the same cells shown in D that were stained brightly with one mAb and dimly with a second mAb are reanalyzed they actually appear as negatively stained cells for the second mAb. The phenomenon results in a significant loss of the dimly stained cells, illustrated in the histogram shown in F.

Often propidium iodide (PI) has been used in combination with a fluoresceinated antibody in fixed preparations. One example of this combination is the measurement of cellular incorporation of bromodeoxyuridine into S phase cells using a fluoresceinated antibody (FL-BrdU) and PI to measure DNA content (Stevenson *et al.*, 1985). To stain cells for DNA content, sufficient PI (usually >5 μg/ml) must be used to obtain good coefficients of variation in the DNA histogram; this amount stains cells very brightly and there is a significant amount of PI leakage into the fluorescein channel. This leakage can be as bright as the fluorescein from the FL-BrdU itself.

It has been the practice to compensate for this PI leakage into the fluorescein channel, but this is not an appropriate use of compensation. As mentioned above the purpose of compensation is to remove the one fluorochrome's fluorescence in the other's channel so that when cells are labeled with only one Ab, they can be resolved from those labeled with both antibodies. Since all cells are labeled with PI, compensation is being used to subtract PI from FITC-labeled cells when PI is actually present in the cells. When this is done, the sensitivity for detecting dimly FITC stained cells is markedly reduced.

We will perform compensation for up to four separate colors. Since there are no compensation standards available they must be prepared for each color. It is convenient to prepare these standards using biotinylated CD45 (or other antibody meeting the above criteria). Follow the method for staining cells outlined in Chapter 3 of this volume (using normal blood or other cells commonly used in the laboratory). The cells are divided into four separate tubes and then stained with FITC avidin, PE-avidin, PT-avidin or PC-avidin. One tube is prepared containing unlabeled cells.

There are two simple rules that should be followed to prevent problems from occurring and to establish the correct spectral compensation:

1. Use particles that have lower autofluorescence compared to that of the cells to be acquired and adjust the photomultiplier tube voltages so these events are in the lower left 10 $\times$ 10 relative fluorescence unit region on the log bivariate display.

2. Adjust compensation using particles that are as bright or brighter than any stained cells that are to be acquired.

Any stained cells that are brighter than the particles used for compensation will not be properly compensated. These cells can appear as doubly stained cells but are, in fact, merely undercompensated.

The following procedure is recommended for adjusting instrument compensation: Refer to Color Plate 2 (see p. 44), rows 1 and 2, for initial instrument adjustment and creation of a compensation reference file for the fluorochromes that will be used. The sample contains a mixture of one part of each separately labeled cell and unstained cell.

1. Display FSC vs SSC and adjust settings for the desired location of the cell pattern. Circumscribe the dense portion of the lymphocyte cluster as shown in Color Plate 2.

2. Display FL1 vs FL2, FL3 vs FL2, FL4 vs FL2, FL1 vs FL3, FL4 vs FL3, and FL4 vs FL1 gated on lymphocytes.

3. With all compensations set to 0%, run the unstained cells and adjust the high voltage for each PMT so that autofluorescence is in the lower left quadrant of each bivariate display. Display quadrant markers so the cells touch the top and right quadrant boundaries for each bivariate display.

4. Run cells stained with each fluorochrome, one at a time, and adjust compensation so the edge of the stained cells touch the appropriate quadrant boundary. There may be a slight shift of the autofluorescent cell distribution toward the axis. This is a normal response, do not readjust the quadrant markers.

5. Acquire data from a cellular compensation mixture containing one part of each color. Save for use as a reference file to be used for future instrument adjustments.

A cellular compensation mixture of stained cells can be prepared daily for verifying instrument compensation as follows:

1. Enter settings from the compensation reference file (previously generated).

2. Using microspheres, evaluate the instrument performance and adjust as necessary.

3. Reacquire the previous day's cellular compensation mixture and compare it to the previous day's reference files.

4. Acquire the newly produced cellular compensation mixture and compare it to the previous day's mixture.

There are several reasons for preparing the cellular compensation mixture daily.

1. If, after acquisition, the previous day's mixture do not compare favorably with its reference file then there is most likely an instrument problem.

2. If the mixture from the previous day looks like it did before, but the new mixture looks poor, the new compensation mixture is faulty and should be prepared again.

Daily compensation reference files provide a continuous log of instrument performance. It is essential that the instrument itself is in proper alignment when these files are created. By comparing the minor changes in settings over time, using both microspheres (e.g., QC beads from Flow Cytometry Standards Co., San Juan, PR) and the cellular compensation mixture, instrument performance can be monitored daily, trends established, and deterioration in instrument or sample preparation performance detected. A lymphocyte gate is used for acquisition to ensure the best homogeneity of each measurement. The need for excessive adjustment on any given day indicates a possible instrument problem.

The use of a single large cellular compensation mixture that can be used over a period of time as a compensation standard is not recommended because fixation results in a slowly increasing autofluorescence with time (described below). By preparing the mixture daily, this problem does not occur. The increase in cellular autofluorescence can be documented by comparing an old (14 days) cellular compensation mixture with a freshly prepared mixture or with the more stable microsphere standards (e.g., Flow Cytometry Standards Co.) used to evaluate instrument performance.

VI. Two-Color Fluorescence

To illustrate the amount of information that can be generated from a single sample using two antibodies simultaneously, a leukocyte differential can be obtained by staining whole blood with FL-CD45 and PE-CD14. While erythrocytes are usually lysed prior to analysis, this is not necessary in the presence of FL-CD45 because the sample can be analyzed by triggering the flow cytometer on CD45 fluorescence. Lysing erythrocytes is still the recommended procedure. A whole-blood sample in which FSC, SSC, FL-CD45, and PE-CD14 were all measured after triggering on CD45 fluorescence is shown in Fig. 6. In Fig. 5A, the FSC vs SSC bivariate distribution for CD45-positive events above the trigger threshold TT, shown in Fig. 5B, is displayed; region R1 contains predominantly small cells, region R2, large cells, and region R3, granulated cells. Note that there are not many other events (OE) in the lower left corner of this display. The bivariate distribution shown in Fig. 5B is also sectioned into regions R4 containing CD45-dull cells, R5 containing CD45-dim cells, R6 containing CD45-bright cells, and R7 containing CD14-bright cells. The forward vs side scatter distribution of the CD45-dull cells (Fig. 5D) after processing the list mode file for events in R4 is shown in Fig. 5C. These events are erythrocytes, large and aggregated platelets, dead cells, and debris. Some leukemias may express dull CD45 fluorescence and may appear in this region.

By analyzing the events that are coincident in R3 and R5, neutrophils are resolved as highly granulated cells shown in a view with dim CD45 fluorescence (Fig. 5F). Similarly, eosinophils can be resolved by requiring coincidence with R3 and R6. Eosinophils have a granularity similar to that of neutrophils (Fig. 5G) but they exhibit a higher CD45 fluorescence (Fig. 5H) than do neutrophils.

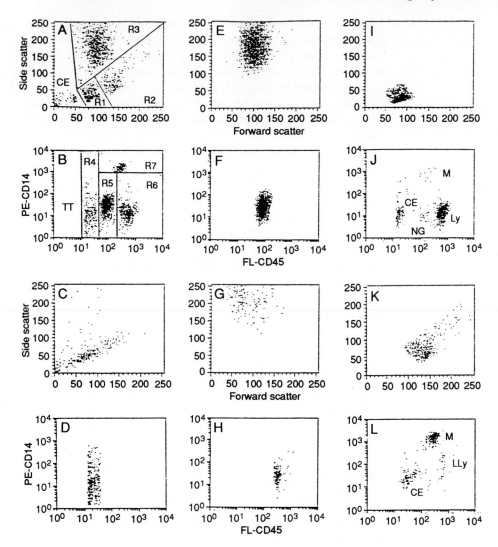

Fig. 5 Two-color fluorescence. Blood (50 μl) was incubated 10 min with 30 μg mouse IgG, and then FL-CD45 and PE-CD14 were added for 15 min. The cells were washed, fixed in 1% paraformaldehyde, and, without lysing the erhthrocytes, were analyzed using the trigger threshold (TT) shown in (B). (A) Small cells (R1), large cells (R2), granular cells (R3), and other events (OE) outside R1 or R2 or R3, consisting of erythrocytes, aggregated platelets, dead cells, and debris are shown. (B) The FL-CD45 bivariate distribution shows CD45-dull cells (R4), CD45-dim cells (R5), CD45-bright cells (R6), and CD14$^+$ cells (R7). (C) The forward versus side scatter distribution of CD45-dull events in R4, shown in (D). (E, F) Events coincident in both R3 and R5; these cells are neutrophils. (G, H) Events coincident in both R3 and R6; these cells are eosinophils. (J) The fluorescence of small events in R1, (I), including the location of small lymphoid cells (Ly), monocytes (M), nongranulated granulocytes (NG), and other events (OE). (L) The fluorescence of large events in R2 (K). Note CD45-bright large lymphoid cells (LLy).

It is common to use a lymphogate to exclude large and granulated cells from the analysis. Events found in R1, a typical lymphogate, are displayed in Figs. 5I and 5J. The events in this region are not all lymphocytes (LY) expressing bright CD45 fluorescence. There are also nongranulocytes (NG), predominantly basophils (Terstappen *et al.*, in press; Loken *et al.*, in press), small monocytes (M), and other events (OE). Similarly, as shown in Figs. 5K and 5L the large cells from (R2) are not all monocytes (M) as NK cells and large lymphoid cells (LLY) and other events (OE) are also present.

VII. Three-Color Immunophenotyping

When two color data are acquired, the usual practice for analysis is to use quadrant markers to separate positive from negative cells. The position of these markers is determined using an isotype control. This method, however, may not be applicable for more than two colors because the cell populations will appear in other dimensions as well. Using more than two colors also provides for new strategies for immunophenotyping that offer a better visualization of the acquired information than using quadrant markers alone. A few of these strategies are discussed.

Several software programs (listed in Table III) are now available for the analysis of multiparameter data. Except for AutoGate that utilizes a clustering algorithm and provides for Classification and Regression Tree (CART) analysis, the other programs require the user to provide an analysis template before batch processing can be performed.

A. Lymphogating

Historically, immunophenotyping has primarily been applied to lymphoid cells and there is general agreement on the data analysis method by selection

Table III
Multiparameter Analysis Software

Name	Supplier
Elite	Coulter Co., Miami, FL
Lysys	Becton–Dickinson Immunocytometry Systems (BDIS) San Jose, CA
Paint-A-Gate	BDIS
Attractors	BDIS
List View	Phoenix Flow Systems, San Diego, CA
Winlist	Verity Software House, Topham, ME
AutoGate	LANL Los Alamos, NM. (Salzman *et al.*, in press; Beckman *et al.*, in press)
Super CY + Analyst	Sierra Cytometry (Redelman *et al.*, in press)

of an objective lymphogate using CD45 vs CD14. This approach is shown in Color Plate 3 (see p. 44) and provides an objective means for both drawing the optimal lymphocyte region in FSC vs SSC to provide an evaluation of cells excluded from analysis and determining the purity of the gate.

The percentage of lymphocytes that will be excluded from analysis are all CD45-bright cells that fall outside R1 in Color Plate 3B. There are also some eosinophils (CD45 bright) present, shown by events with high side scatter (Color Plate 3B) and, because B cells are between eosinophils and other lymphocytes in the CD45 expression, including all of them can affect the frequency of eosinophils that are also included. While this approach is still a good way to determine the "proportion of lymphocytes" in the gate, it is erroneous to assume that all cells in R2 are lymphocytes as shown in (Color Plate 3B).

The purity of the lymphogate (R1) is found by determining the percentage of events that fall outside region R2 in Color Plate 3C. When cells are gated on R1, the lymphogate, once defined, is then used to analyze the samples containing the immunophenotypic antibodies. Several agencies are now requiring performance standards for these two values (NCCLS, 1992).

While these standards are designed to improve the quality and reproducibility among laboratories of lymphocyte immunophenotyping data, they have been based on "normal" individuals and may not be valid for sick patients. For the example shown in Color Plate 3, there are small myelomonocytes present in the lymphogate. High frequencies of these cells will produce a drop in the purity that cannot be improved by changing the lymphogate. Since these contaminating cells as well as other events will be in the denominator of the equation used to calculate the percentage of positive lymphocytes, it has been customary to perform a correction:

% Positive lymphocytes

$$= \frac{100 \times \text{number of marker positive cells}}{\text{Total events in lymphogate} \times \text{purity of lymphogate}}.$$

While correction of the lymphogate is the current recommended practice (NCCLS, 1992) as shown in this example, the non-lymphoid events (small myelomonocytes) may have clinical significance and, if they are not reported, immunophenotyping data will be biased. In lymphoid malignancy, it is common for the cells to express CD45 at lower levels than normal and none will be in the lymphogate. In these situations, the use of a lymphogate as defined here should not be used.

Stelzer *et al.* (1993) have suggested an improvement to the above strategy that markedly enhances the objectivity of the lymphogate by using CD45 as a third antibody in all the panels containing two other antibodies. This strategy can also be applied to any cell population as shown in Color Plate 4 (see p. 44). Log CD45 vs log SSC (rather than FSC vs SSC) is viewed to resolve all the cell clusters. This approach has the advantage that large lymphocytes are included and they are well resolved from monocytes and granulocytes. Even eosinophils are not a problem because they will be separately resolved based on their side scatter properties. Immunophenotyping antibodies can be combined with CD45 for resolving discrete clusters of lymphocytes.

It is also possible to use this strategy for immunophenotyping leukemias and lymphomas because their CD45 expression is almost always like that of immature progenitor cells. As shown in the middle row for B lineage acute lymphocytic leukemia and the last row for an acute myeloid leukemia, the leukemic cells are well resolved from the normal cells. This approach can provide for a greater appreciation for the frequency of normal cells which are present in leukemias as well as a means for significantly reducing them from the analysis so that the true leukemic phenotype is determined. This approach is also the first step toward the use of flow cytometry in monitoring residual leukemic cells.

B. Cell Gating

We originally introduced the concept of cell gating in an earlier edition (Stewart, 1990) and Mandy (Mandy *et al.*, 1992) has pioneered its use using CD3 to identify T cells (T gating) and then determining their subsets using other antibodies. A more generic approach is to specifically gate the cell of interest first and then display the other parameters. Common gating antibodies might be CD3 for T cells, CD19 for B cells, CD56 for NK cells, CD14 for monocytes, and CD34 for progenitor cells. In this paradigm, one antibody is used to identify the cell of interest and the others to resolve its subsets. When combined with data acquisition, the procedure is especially useful for phenotyping cell populations in low frequency. As shown in Color Plate 5 for CD34 progenitor cells, usually SSC vs the cell-specific antibody is displayed and a gate is created around the cells of interest. This strategy is similar to that just described for CD45. Then the other two antibodies are displayed. Using this approach, all the cells of interest are resolved with the highest specificity and purity that can be obtained because none are excluded from analysis.

Many instruments do not have the capability of four-color compensation and, even though the increased dimensionality of four-color data may not require it (Color Plate 2) compensation may still be desired as it improves the space available in any given bivariate display for viewing a cell population coexpressing different epitopes that have been stained. Using the software features of Winlist (Verity Software House, Topsham, ME), it is possible to produce the compensated data illustrated in the last row of Color Plate 2. The cells are stained individually with each of the four colors and along with an unstained sample the data are acquired without any compensation. These cells are used by the software to establish the compensation matrix. Subsequent samples are automatically compensated as they are analyzed.

Whether uncompensated, software-compensated, or electronic-compensated data are acquired and analyzed, four-color single-laser-based immunophenotyping offers a new approach for the resolution of important subsets of leukocytes. While illustration of the many applications of four-color immunophenotyping goes well beyond the scope of this chapter, consider the improved resolution of leukocytes when CD45 is combined with three other antibodies such as CD3, CD4, and CD8. Indeed, any combination of three antibodies with CD45 would also improve the quality of three-antibody immunophenotyping.

VIII. Fixing Cells

It is important to fix human material prior to flow cytometric analysis because of the health risk to the technical staff (NCCLS, 1992). A 1% solution of buffered ultrapure formaldehyde is the fixative of choice. When cells are fixed, they should be analyzed as soon as possible but no more than after 1 week of storage. As shown in Fig. 6, prolonged storage of cells in formaldehyde causes an increase in autofluorescence. This problem can be especially noticeable if lysed whole blood is used because the fixed leukocytes bind fluorescent products released by the lysed erythrocytes. This problem is not as noticeable for cells that have been separated from erythrocytes.

IX. Detecting Dead Cells

Dead cells can bind antibodies nonspecifically leading to an erroneous percentage of positive cells for some antibodies. Shipping cells, overnight storage, and cryopreservation are a few procedures that can result in increased cell death. It is important to account for these dead cells.

As shown in Fig. 7A, dead cells stain with propidium iodide but live cells do not. The addition of PI to samples 5 min prior to analysis provides excellent live/dead cell discrimination. PI can be analyzed using the fluorescence 2 detector for instruments with two photomultiplier detectors (in combination with a single antibody) or in combination with two antibodies in the fluorescence 3 position for instruments with three detectors. By gating against the dead cells during sample analysis or by analyzing the file on live cells, any PI overlap into the channels detecting antibody will be eliminated.

Due to the health risk in analyzing viable human cells, fixation is recommended and fixed cells will all be labeled with PI making it useless for evaluating cell viability. Ethidium Monoazide (EMA) can be used to label dead cells prior to fixation (Riedy *et al.*, 1991). The EMA is first added to the viable cell suspension. The cell suspension is exposed to fluorescent light so that the photoactive EMA can irreversibly bind to DNA of dead cells. Unbound EMA is washed out of the cells and they are fixed. As shown in Fig. 6B and 6C for samples containing different numbers of dead cells, they are resolved from live cells by EMA fluorescence.

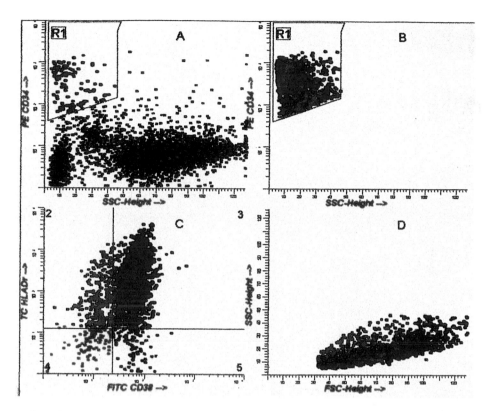

Color Plate 5 (Chapter 4) Cell gating. Human bone marrow was stained with FCD38, PECD34, and TC-HLADr. After an ungated file of only 5000 events is acquired (A), the location of the low frequency CD34$^+$ subset (blue) is found. A cell-specific live gate (R1) is then created and 5000 additional events (or as many as possible) in a second file are acquired. The gated events are shown in B. The phenotype of CD34$^+$ cells can then be determined by displaying the bivariate plot of the other two antibodies (C) as well as their FSC vs SSC characteristics (D). Utilizing this strategy, the number of acquired events can be significantly reduced and the frequency of CD34$^+$ subsets determined by multiplying the fraction of cells in R1 (ungated) by the fraction found in R2, R3, or R4 in C.

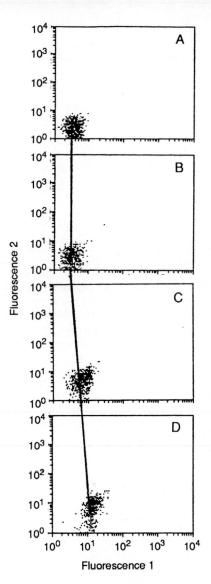

Fig. 6 Storage of lysed whole blood in 1% paraformaldehyde. Human blood was lysed and fixed in 1% paraformaldehyde. The same sample was analyzed (A) 1 hr, (B) 7 days, (C) 14 days, and (D) 21 days after fixing, using the same instrument settings. There is a fivefold increase in the mean autofluorescence from Day 7 to Day 21.

It is important to recognize that dying cells (late-stage apoptosis) dimly stain with PI or EMA. Therefore, the same fluorescence channel should not be used to measure both an antibody and EMA (or PI) fluorescence by utilizing fluorescence intensity of PI or EMA staining to discriminate live and dead cells if the latter are in high frequency. Thus, cells exhibiting dim fluorescence for

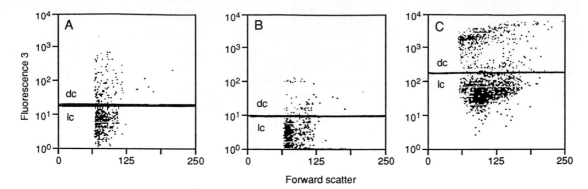

Fig. 7 Resolving dead cells. (A) Human blood cells were stained with 1 μg/ml propidium iodide (PI) and analyzed 5 min later. (B) Cells were stained with EMA and then fixed in 1% paraformaldehyde for 1 hr prior to analysis. Dead cells (dc) are found above the line and live cells (lc) below it. EMA is not as bright as PI because the emission maximum of EMA is at 600 nm and the measured fluorescence was above 650 nm. (C) Leukemic bone marrow was stained with four times the optimal concentration of EMA. Whereas dead cells are resolved better, there is a considerable amount of nonspecific staining of the live cells by EMA.

PI or EMA will appear in the same region as antibody positive cells unless these dim cells are explicitly resolved in a separate dedicated channel. In addition, as illustrated in Fig. 6C, EMA may nonspecifically bind to some cell types, causing increased fluorescence of live cells even though the dead ones can still be well resolved.

X. Summary

The flow cytometry described can be performed using a single laser. Each laboratory has to establish its own experience base and standard operating procedures. The intent of this discussion has been to illustrate the procedures that will lead to good flow cytometry data acquisition and analysis and to illustrate problematic areas. The most important rule of all is to recognize when there is a problem. It is hoped the information provided herein will be of help in the recognition process.

References

Beckman, R. J., Salzman, G. C., and Stewart, C. C. *Cytometry* (in press).
Center for Disease Control (CDC). (1992). Guidelines for the performance of CD4 + T-cell determinations in persons with human immunodeficiency virus infection. *MMWR* **41**(RR-6), 1–17.
Hoffman, R. A. (1988). *Cytomet. Suppl.* **3,** 18–22.
Lipson, J. P., Sasaki, D. J., and Singleman, E. G. (1985). *J. Immunol. Methods* **186,** 143–149.
Loken, M. R., Brosnan, J. M., Back, B. A., and Ault, K. A. (1990). *Cytometry* **11,** 453–459.

Lund-Johansen, F., and Terstappen, L. W. M. M. (1993). *J. Leuk. Biol.* **54,** 47–55.

Mandy, F. F., Bergeron, M., Recktenwald, D., and Izaguirre, C. A. (1992). *J. Immunol. Methods* **156,** 151–162.

National Committee for Clinical Laboratory Standards (NCCLS) (1992). ''Clinical Applications of Flow Cytometry: Quality Assurance and Immunophenotyping of Peripheral Blood Lymphocytes.'' Vol. 12, p. 6.

Redelman, D., and Coder, D. M., in *Cytometry* (in press).

Riedy, M. C., Muirhead, K. A., Jensen, C. P., and Stewart, C. C. (1991). *Cytometry* **12,** 133–139.

Salzman, G. C., Beckman, R. J., Parson, J. D., Nauman, A. M., Stewart, S. J., and Stewart, C. C., *Cytometry* (in press).

Shapiro, H. M. (1988). ''Practical Flow Cytometry.'' New York: A. R. Liss.

Stelzer, G. T., Shults, K. E., and Loken, M. R., (1993). *In* ''Clinical Flow Cytometry'' (A. Landay, K. Ault, K. Bauer, and P. Rabinovitch, eds.), pp. 265–280. New York: New York Academy of Sciences.

Stevenson, A. P., Crissman, H. A., and Stewart, C. C. (1985). *Cytometry* **6,** 578–583.

Stewart, C. C. (1990). *In* ''Methods of Cell Biology'' (Z. Darzynkiewicz and H. Crissman, eds.), Vol. 33, pp. 411–426. San Diego: Academic Press.

Stewart, C. C. (1992). *Clin. Immunol. Newslett.* **1,** 2.

Stewart, C. C., and Stewart, S. J. (1993). *Ann. N. Y. Acad. Sci.* **677,** 94–112.

Stewart, C. C., Stewart, S. J., and Habbersett, R. C. (1989). *Cytometry* **10,** 426–432.

Terstappen, L. W. M. M., Shaw, V. O., Conrad, M. P., Recktenwald, D., and Loken, M. R. (1988). *Cytometry* **9,** 477–484.

Willman, C. L., and Stewart, C. C. (1989). *Semin. Diagn. Pathol.* **6,** 3–12.

CHAPTER 5

Immunophenotyping Using Fixed Cells

George F. Babcock and Susan M. Dawes

Department of Surgery
University of Cincinnati Medical Center and
Shriners Hospital for Crippled Children-Burns Institute
Cincinnati, Ohio 45267

I. Introduction

Although most laboratories process blood samples for immunophenotyping soon after receiving them, some laboratories have the need to store their samples for various periods of time before staining with fluorochrome-labeled antibodies followed by analysis by flow cytometry. In addition, the use of cells as controls for staining is preferred to using beads as controls by many laboratories. The additional time required to prepare a fresh sample of cells and the day to day differences which occur in the cells from one individual to another has discouraged many laboratories from using this methodology. In this Chapter, methods for the use of fixed leukocytes are described for immunophenotyping with a variety of commonly used monoclonal antibodies.

Traditionally, the prefixation of samples prior to cells-surface staining has not been widely used. This concept comes in part from the belief that fixation with formaldehyde or other agents causes loss of antigenicity. Although the aldehyde crosslinkages formed by formaldehyde fixation can mask or alter the structure of certain antigens and prevent access to the antibody, these phenomena do not occur in all cases. Much of this information has been obtained from the use of fixed and embedded tissues. Because there are many steps in the embedding process, the effects seen on antigen expression are cumulative and do not necessarily represent the effects of fixation alone. Other problems which can be associated with the fixation process are an increase in autofluorescence and/or nonspecific fluorescence.

The fixation of cells used for immunophenotyping by flow cytometric analysis has traditionally been employed after the cells are reacted with the antibody (postfixation) especially when monoclonal antibodies are used. The most commonly used fixation method for leukocytes which are to be analyzed by flow cytometry is that reported by Lanier and Warner (1981) for murine lymphocytes. This procedure ultilizes 1% paraformaldehyde in saline for postfixation. The authors indicated that this procedure was not useful when employed as a prefixation method. It is possible that the lack of buffering in the solution could, at least in part, account for the loss of antigenicity. It has been reported that the stability of antigens to fixation, including several major T cell antigens, are dependent upon the fixation procedure occurring at neutral pH (Pollard and Holgate, 1987). Jones et al. (1986) obtained similar results for the preservation of human T cell antigens using prefixation with paraformaldehyde. Paraformaldehyde containing fixatives have also been shown to be superior for the preservation of T cell, B cell, and accessory cell antigens in paraffin-embedded human tissues (Collings et al., 1984; Holgate et al., 1986).

Several investigators have use prefixed cells for analysis by flow cytometry. In a study comparing a battery of fixatives including acetone, acid:alcohol, glutaraldehyde, and paraformaldehyde, it was found that 2% paraformaldehyde demonstrated the best preservation of the Thy-1 surface antigen on mouse thymocytes (Kerr et al., 1988). Slaper-Cortenbach et al. (1988) used a buffered solution of acetone and formaldehyde to prefix leukemic cells for the simultaneous detection of cell-surface and intracellular antigens by flow cytometry. This procedure was found useful for the preservation of many antigens including CD3, CD4, CD5, CD7, CD8, CD9, CD10, CD24, CD38, HLA-DR, and surface immunoglobulin. Similar results were obtained by Hallden et al. (1989) using paraformaldehyde fixation prior to permeabilization. These authors found no changes in the detection of the antigens used including CD3, CD20, CD16, or CD38. Others have used paraformaldehyde fixation following permeabilization to detect intracellular antigens (Labalette-Houache et al., 1991). Methanol prefixation has also been used successfully for the detection of intracellular antigens by flow cytometry (Levitt and King, 1987).

When employing prefixation methodology, a number of variables must be considered. The most obvious is the type of fixative. The most widely used compound is paraformaldehyde, generally in a concentration of 1 or 2%. Other variables include the length of the fixation period, the temperature at which fixation occurs, and the pH of the fixation solution. The time of fixation varies from several minutes to overnight. Both over- and underfixation usually results in a loss of antigenicity. Temperature influences speed of fixation with warmer temperatures reducing the length of time required. Most authors, however, report obtaining better results at 4°C. Almost all investigators report poor antibody staining when fixation was performed at other than a neutral pH.

II. Application

There are several instances where fixation of cells prior to immunophenotyping can be very advantageous for many laboratories. One of the most obvious uses is the inactivation of potentially dangerous infectious agents. Most infectious agents are inactivated following fixation especially when formaldehyde or paraformaldehyde is used. Studies by Lifson *et al.* (1986) reported that 0.37% formaldehyde or 0.5% paraformaldehyde completely inactivated HIV in 1 hr. It should be noted that shorter fixation times were not tested.

Another potential use for prefixation of cells is to relieve the busy laboratory from having to stain the cells within a short time after obtaining the blood sample. The small laboratory would also benefit from being able to collect several samples and immunophenotype them simultaneously at a later date. For laboratories which must ship their samples to a distant site for analysis, it would be advantageous to know that the samples would not lose antigenicity and would be protected from the potential spread of infectious agents if the container were broken in transit. Under the appropriate conditions prefixed cells can also be used as a pool of ''ready to stain'' control cells.

III. Materials

1. *Dulbecco's phosphate-buffered saline (DPBS):* This reagent can be prepared by combining the individual components or it can be purchased in powdered form at a reasonable price from several suppliers such as Gibco/BRL (Grand Island, NY). It is prepared according to the manufacturer's instructions.

2. *DPBS-gel-azide (DPBS$^+$):* This is a DPBS solution containing 0.1% gelatin and 0.1% sodium azide. The DPBS solution should be heated to a temperature less than boiling (approximately 85–90°C) with stirring. Sodium azide (0.1%) can be added at any time during the heating process. Sodium azide is classified as being extremely toxic and should be handled with great care. Once the solution has reached the desired temperature, 0.1% gelatin (type B from bovine skin, Sigma Chemical Corp., St. Louis, MO) is slowly added. Continue stirring at the elevated temperature until all the gelatin goes into solution. This may take 30 or more minutes. After cooling to room temperature, the pH of this solution should be checked and adjusted to 7.2 to 7.4. The solution should then be stored at 4°C.

3. *Blocking Solution:* This solution is used to prevent excessive nonspecific binding of immunoglobulin with the Fc receptor. Add human γ-globulin (Cohn fraction II, III, Sigma Chemical Corp.) to DPBS$^+$ at a final concentration of 10 μg/ml. Purified immunoglobulins can be used but γ-globulin works well and is considerably less expensive. The γ-globulin added should be nonreactive with any of the antigens used for phenotyping. This solution should be stored at 4°C.

4. *Fixative:* Fixative consists of a buffered 1% paraformaldehyde-saline solution. Paraformaldehyde is a strong corrosive and a suspected carcinogen, so it should be used with great caution. This reagent should be prepared in an approved fume hood. Heat 400 ml of deionized water to near boiling using a heating plate or a double-boiler. Do not boil. Add the following while stirring: 5.35 g sodium cacodylate, 3.8 g sodium chloride, and 5.9 g of paraformaldehyde. Stir gently until all the compounds are dissolved, cool, bring the volume up to 500 ml, and adjust pH to 7.2–7.4. This solution should be stored at 4°C and discarded after approximately 1 month.

Table I
Specificity and Source of Antibodies Used

Antigen detected	Source	Distribution
CD2	BD[a], DA	T[e]
CD3	BD, CO[b]	T
CD4	BD, CO	T sub[f]
CD5	BD	T, B[g] sub
CD8	BD, CO	T sub,NK[h] sub
CD11b	BD, DA[c]	Mo[i],P[j],T sub
CD14	BD, GT[d]	Mo, Psub
CD16	BD	NK, P, Mo
CD19	CO, GT	B
CD20	BD, CO	B
CD23	BD, CO, GT	B sub
CD29	CO	Tsub, B sub
CD45	BD	All
CD45RA	CO, GT	T sub,B,Mo,P
HLA-DR	BD, GT	B,Mo,Tsub,NKsub
Leu-8	BD	sub All

[a] Becton–Dickinson.
[b] Coulter Immunocytometry.
[c] Dako Corp.
[d] Gen Trak.
[e] T cells.
[f] Subset
[g] B cells.
[h] Natural killer cells.
[i] Monocytes.
[j] Polymorphonuclear leukocytes.

5. *Blood lysing solution:* A commercial whole-blood lysing solution can be purchased from Gen Trak, Inc. (Plymouth Meeting, PA) and is prepared according to manufacturer's instructions. The lysing solution should be diluted 1:25 in PBS to make the final working solution. DPBS$^+$ will work fine. This solution should be made up fresh each day. Although this solution was originally designed to lyse red blood cells (RBC) following antibody staining, it can be used to lyse rbc prior to both fixation and antibody staining.

6. *Blood separation media:* A number of media such as Ficoll-Hypaque or neutrophil isolation media are available commercially or can easily be prepared in the laboratory. These compounds, which allow the separation of leukocytes from red blood cells, can be used if desired in place of a red cell lysis procedure.

7. *Antibodies:* Antibodies which have been used for immunophenotyping following fixation of the cells were purchased from Becton–Dickinson Immunocytometry Systems (San Jose, CA), Coulter Cytometry (Hialeah, FL), Dako Corp. (Carpinteria, CA), and Gen Trak, Inc. The antibodies used and their specificities appear in Table I.

IV. Cell Preparation and Staining

1. Blood used for immunophenotyping should be collected with sodium heparin or ethylenediaminetetraacetic acid (EDTA) as the anticoagulant. Many of the isolation medium and blood lysing procedures suggest using only EDTA, thus this compound represents a safe choice. Following either the whole-blood lysis procedure or a gradient separation method the cells should be washed two times in PBS. The cells can then be used for immunophenotyping, or fixed if the prefixation procedure is to be used.

2. To fix cells prior to staining with antibody, washed cells should be resuspended in the fixative at a concentration of 2×10^7/ml or less. The time of fixation required varies with the cell type and the antigen(s) of interest but should not be longer than 24 hr (see Section VI, Results and Discussion). For most antigens on leukocytes an overnight fixation is well tolerated.

3. Following fixation the cells should be washed two times in DPBS$^+$ to remove the surplus fixative and should be adjusted to a concentration of 1×10^6 cells/100 μl in DPBS$^+$. Fifty microliters of the blocking solution is added to each tube and the tubes are incubated for 5 min at 4°C. *Note:* Do not add the blocking solution if the antigens of interest include the Fc receptors, CD16, CD32, or CD64, as this solution may interfere with antibody binding.

4. The appropriate antibody of interest is added to each tube of cells without removing the blocking solution. Antibodies should be titered prior to sample testing to determine the proper concentration necessary for optimal staining of the cell type being used. In most cases an antibody concentration of 1 μg/10^6 cells has been found to be optimal. The optimal concentration of antibody is

added in 20 μl if possible. This procedure can be used for single- or multiple-color fluorochromes using either direct or indirect staining methods. Single-color staining is described and the results are presented in that manner but the method works well for two-, three-, or four-color immunofluorescence. Following the addition of the primary antibody the cells are incubated for 20 min at 4°C and then washed in DPBS$^+$. If the antibody is fluorochrome labeled (direct method) the cells are washed a second time. If indirect staining is to be performed (the primary antibody is not fluorochrome labeled), then the cells are washed once followed by the addition of 25 μl of heat-inactivated (56°C for 30 min) goat serum (or serum of the appropriate species if the secondary antibody was not made in goats). This serves as an additional Fc receptor blocking reagent for the second antibody. The secondary antibody is then added at the appropriate concentration (this antibody should also be titered) without further washing. Following a second 20-min incubation the sample is rinsed twice in DPBS$^+$ and stored in 0.5 ml of fixative. The stained cell preparations can then be treated as if they had not been fixed previously and can be stored for a month or more without significantly changing.

5. Samples treated in the above manner can then be analyzed in the normal manner for that laboratory (Babcock *et al.*, 1987,1990).

V. Critical Aspects of the Procedure

There are several factors which need to be considered before using prefixed cells for immunophenotyping. Many of the variables which normally apply to the fixation of cells, however, have been optimized for leukocytes and thus the number of critical aspects which must be considered are relatively few. A variety of compounds have been used to fix leukocytes prior to staining with fluorochrome-labeled antibodies but paraformaldehyde at 1 to 2% is widely used and is preferred. The pH at which fixation occurs is also critical and the fixative should be buffered to remain at 7.2 to 7.4. Maintaining pH is important not only for optimal cell fixation but because many fluorochromes, especially fluorescein isothiocynate, are quenched at acidic pH. The temperature at which fixation occurs affects both the speed of the fixation process and the preservation of the antigenic structures. For most leukocyte antigens 4°C is optimal and this temperature is suggested as a starting point when the optimal temperature has not previously been determined.

The parameter which varies the most from antigen to antigen is the time in which the cells are allowed to remain in the fixative. As is evident from the results presented in the following section, many leukocyte antigens are destroyed or masked by overfixation with the majority of leukocyte antigens being barely detectable following 7 days of fixation. Most of these antigens, however, withstand 24 hr of fixation and this should be considered the practical maximum fixation time for most cells. Unfortunately, it is not possible to predict the optimal time or even the stability of any particular antigen in the fixation process. Therefore, each laboratory must develop their own protocol for the preservation of the antigens of interest. It has also been reported that for certain antigens paraformaldehyde fixation actually enhances antibody binding (Milstein *et al.,* 1983; Magee, 1985; von dem Borne and van der Lelie, 1986). It is not known how often this occurs but one suspects that it is fairly rare.

VI. Results and Discussion

Leukocytes from 12 normal individuals which were fixed in 1% paraformaldehyde for 1, 7, 14, or 28 days have been examined using immunofluorescence and flow cytometry for a variety of common antigens found on T cells, B cells, NK cells, monocytes, and neutrophils (Babcock *et al.,* 1991). Whenever possible several different clones of each antibody were tested to determine if different epitopes on the same antigen were altered. The antibodies tested and their suppliers appear in Table I. The results indicated that no universal statements can be applied to the fixation of leukocytes with paraformaldehyde. The light scatter patterns of leukocytes remained relatively stable following all fixation protocols tested. The forward versus 90° light scatter pattern for leukocytes fixed for 4 weeks appears in Fig. 1. The typical leukocyte pattern is generally well maintained following this procedure. The dead cells, debris,

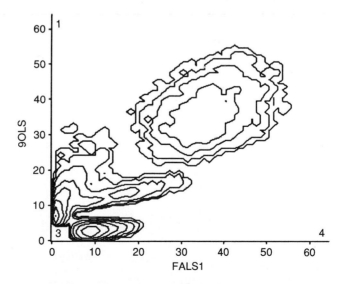

Fig. 1 The forward angle (FALS) versus 90° light scatter (90LS) pattern of leukocytes fixed in buffered 1% paraformaldehyde for 4 weeks. A normal light scatter pattern is apparent for the lymphocyte, monocyte, granulocyte populations as well as dead cells and debris.

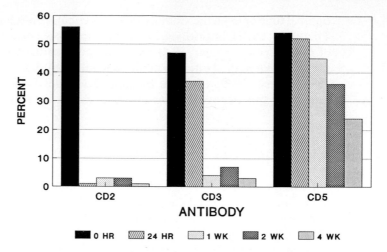

Fig. 2 A graphic representation of the percentage of cells positive for CD2, CD3, and CD5 following fixation in buffered 1% paraformaldehyde for 24 hr and 1, 2, and 4 weeks.

and degranulated neutrophils tended to infringe on the monocytic population a bit more than is normally expected.

Following 24 hr of paraformaldehyde fixation, the leukocyte cell-surface antigens examined fall into three groups: those which are undetectable; those which are detectable but at a much lower percentage; and those which are essentially unchanged. Of the pan T cell antigens examined, the percentage of

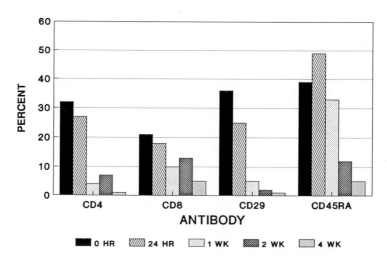

Fig. 3 A graphic representation of the percentage of leukocytes positive for CD4, CD8, CD29, and CD45Ra following fixation in buffered 1% paraformaldehyde for 24 hr and 1, 2, and 4 weeks.

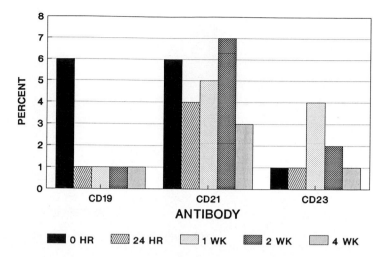

Fig. 4 A graphic representation of the percentage of leukocytes positive for CD19, CD21, and CD23 following fixation in buffered 1% paraformaldehyde for 24 hr and 1, 2, and 4 weeks.

lymphocytes expressing CD3 was slightly lower while CD5 was unchanged (Fig. 2). However, CD2 was almost undetectable following 24 hr of fixation. The T cell subset markers CD4, CD8, and CD45Ra were essentially unchanged while the percentage of CD 29$^+$ lymphocytes was slightly lower (Fig. 3).

The B cell antigens appeared to be less predictible than the T cell antigens. The percentage of CD19$^+$ lymphocytes was significantly reduced following

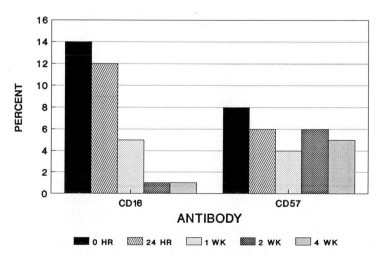

Fig. 5 A graphic representation of the percentage of leukocytes positive for CD16 and CD57 following fixation in buffered 1% paraformaldehyde for 24 hr and 1, 2, and 4 weeks.

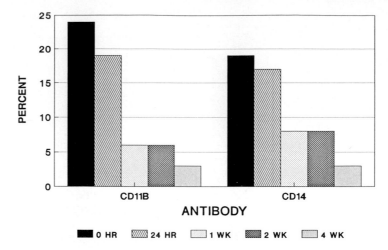

Fig. 6 A graphic representation of the percentage of leukocytes positive for CD11b and CD14 following fixation in buffered 1% paraformaldehyde for 24 hr and 1, 2, and 4 weeks.

fixation, while CD21 expression was slightly reduced after 24 hr of fixation but returned to normal after 7 days of fixation (Fig. 4). It is not clear if the 24-hr observation represented an abnormal incident with the cell preparations used in this study, or if CD21 is always reduced. CD23 expression was normal following 24-hr fixation but the percentage of positive cells actually increased after 7 days.

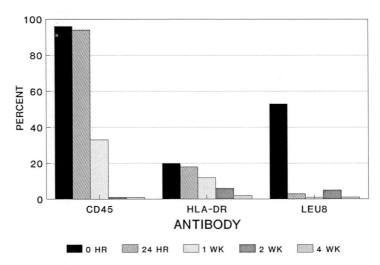

Fig. 7 A graphic representation of the percentage of leukocytes positive for CD45, HLA-DR, and Leu 8 following fixation in buffered 1% paraformaldehyde for 24 hr and 1, 2, and 4 weeks.

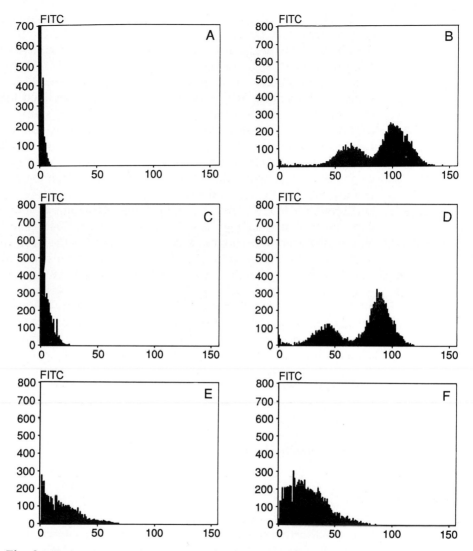

Fig. 8 Single-parameter histograms of unfixed and fixed cells stained with CD45 or nonreacting isotype-matched immunoglobulin (isotypic controls). The light scatter gates (FALS vs 90°LS) were set to encompass the mononuclear cell fraction. The fluorescence intensity measured on a logarithmic scale appears on the ordinate, while the relative number of cells appears on the abscissa. (A) The isotypic control for the unfixed leukocyte population. (B) Unfixed cells stained with CD45. (C) The isotypic control for leukocytes fixed for 24 hr. (D) Leukocytes stained for CD45 following 24 hr of fixation. (E) The isotypic control for leukocytes fixed for 1 week. (F) Leukocytes stained for CD45 following 1 week of fixation.

The NK cell antigens CD16 and CD57 were stable following 24-hr fixation (Fig. 5), with the CD57 antigen remaining stable throughout the 4-week fixation period. CD16 expression on neutrophils was also found to be stable after 24 hr of fixation (data not shown).

The CD14 antigen on monocytes was stable for the 24-hr fixation period while the percentage of CD11b mononuclear cells declined following fixation (Fig. 6). The common leukocyte antigen, CD45, was stable during the 24-hr fixation as was HLA-DR (Fig. 7). The L-selectin molecule detected by anti-Leu 8 was lost, however, following 24 hr of fixation. An example of the typical results observed when using prefixed leukocytes appears in Fig. 8. Histograms of nonfixed leukocytes stained with isotypic antibodies or with CD45 appear in Figs. 8A and 8B, respectively. Essentially all the cells are positive and the background is low. Figures 8C and 8D are histograms of leukocytes fixed for 24 hr prior to staining with isotypic antibodies or CD45. The samples appear almost identical to the nonfixed samples except for the very slight increase in fluorescence in the isotype control. Histograms of cells fixed for 2 weeks and stained with isotypic antibody or CD45 appear in Figs. 8E and 8F, respectively. A decrease in the fluorescence of cells stained with CD45 can be clearly observed while the fluorescence in the isotypic controls is greatly increased. The increase in the controls was shown to be almost entirely attributable to a large increase in autofluorescence. All the antibody clones tested appeared to give results similar to that of other corresponding clones.

The results indicate that for many antigens prefixation can be used for the successful preservation of these cell-surface structures. It is important to note that the fixation should be kept to a period of less than 24 hr. Also, some antigens are lost after even short periods of fixation. Each laboratory must determine the stability of the antigens in which they are interested before routinely employing this technique.

Acknowledgments

This work was supported in part by a grant from the Shriners of North America. The authors thank R. Michael Sramkoski and Sharon Hartmann for their excellent technical assistance. The authors also thank Dr. Anne L. Jackson for her insightful discussions concerning this project.

References

Babcock, G. F., Taylor, A., Hynd, B., Sramkoski, R. M., and Alexander, J. W. (1987). *Diagn. Clin. Immunol.* **5,** 175–179.

Babcock, G. F., Alexander, J. W., and Warden, G. (1990). *J. Clin. Immunol. Immunopathol.* **54,** 117–125.

Babcock, G. F., Hartmann, S., Sramkoski, R. M., and Jackson, A. L. (1991). *Cytometry, Suppl.* **5,** 102.

Collings, L., Poulter, L., and Janossy, G. (1984). *J. Immunol. Methods* **75,** 227–239.

Hallden, G., Andersson, U., Hed, J., and Johansson, S. (1989). *J. Immunol. Methods* **124,** 103–109.

Holgate, C., Jackson, P., Polland, K., Lunny, D., and Bird, C. (1986). *J. Pathol.* **149,** 293–300.

Jones, H., Hughes, P., Kirk, P., and Hoy, T. (1986). *J. Immunol. Methods* **92,** 195–200.

Kerr, R., Kaye, A., and Bartlett, P. (1988). *Neurosci. Lett.* **87,** 259–265.

Labalette-Houache, M., Torpier, G., Capron, A., and Dessaint, J. (1991). *J. Immunol. Methods* **138,** 143–153.

Lanier, L., and Warner, N. (1981). *J. Immunol. Methods* **47,** 25–30.

Levitt, D., and King, M. (1987). *J. Immunol. Methods* **96,** 233–237.

Lifson, J., Sasaki, D., and Engleman, E. (1986). *J. Immunol. Methods* **86,** 143–149.

Magee, J. (1985). *Br. J. Haematol.* **61,** 513–516.

Milstein, C., Wright, B., and Cuello, A. (1983). *Mol. Immunol.* **20,** 113–123.

Pollard, K., and Holgate, C. (1987). *J. Immunol. Methods* **96,** 145.

Slaper-Cortenbach, I., Admiraal, L., Kerr, J., Van Leeuwen, E., von dem Borne, A., and Tetteroo, P. (1988). *Blood* **72,** 1639–1644.

von dem Borne, A., and van der Lelie, J. (1986). *Br. J. Haematol.* **64,** 205–210.

CHAPTER 6

Simultaneous DNA Content and Cell-Surface Immunofluorescence Analysis

**Paul P. T. Brons,★ Piet E. J. Van Erp,† and
Arie H. M. Pennings★**

Departments of ★Hematology and †Dermatology
University Hospital Nijmegen
6500 HB Nijmegen
The Netherlands

I. Introduction

Flow cytometric measurement of relative DNA content per cell using intercalating dyes (Crissman and Steinkamp, 1973) has become a well-established method to study proliferation of tissues and cell populations (Crissman and Hirons, this volume, Chapter 13). In particular S phase DNA content can be used as a measure of proliferative activity. However, in the case of heterogeneous cell populations simultaneous detection of subpopulation-specific antigens is required. The protocol described here is a single-laser method for bivariate analysis of DNA content and cell-surface immunofluorescence. It uses a standard immunofluorescence technique which is followed by propidium iodide (PI) staining of double-stranded DNA under hypotonic incubation conditions.

S phase DNA content has been extensively used to study cell kinetics in hematological malignancies, since it is simple, rapid, and reproducible. Because of these characteristics, this technique is often preferred over [³]thymidine incorporation. Simultaneous detection of DNA content and DNA synthesis using halogenated thymidine analogues (BrdUrd, IdUrd) can be used to obtain additional cell kinetic information (Brons *et al.*, 1992). However, this method is relatively laborious, necessitates the *in vivo* administration of IdUrd, and requires an optimal sampling time.

A limitation of single measurements of S phase DNA content (and also of the BrdUrd/DNA technique) is that it cannot deal with the heterogeneous composition of bone marrow, peripheral blood, and lymphoma tissue. Relevant information about distinct cell subpopulations is required. The availability of monoclonal antibodies directed against specific cell-surface antigens of hemato-poietic cells has enabled the detection of these subpopulations (Civin and Loken, 1987).

II. Materials and Methods

A. Reagents

1. Sterile Buffered Acid–Citrate Dextrose (ACD) Solution

Dissolve 2.82 g trisodium citrate, 1.07 g citric acid, 3.65 g glucose, 4.96 g sodium chloride, 3.10 g sodium monohydrogen phosphate, and 2.62 g sodium dihydrogen phosphate in 1 liter distilled water. Adjust pH to 7.4 with 4 *N* NaOH. The solution is sterilized 20 min at 120°C.

2. Hypotonic PI Solution

Stock solutions (stored at 4°C.): 1 g trisodium citrate in 1 liter distilled water; 100 mg ribonuclease A (RNase, Sigma R-5503) in 100 ml PBS; 10 ml Triton X-100 added to 90 ml distilled water; 100 mg PI (Sigma P-5264) in 100 ml PBS.

Working solution: Add 5 ml RNase solution, 500 μl diluted Triton X-100, and 1 ml PI solution to 43.5 ml trisodium citrate solution. This hypotonic PI working solution must be freshly prepared each week and should be stored at 4°C.

B. Cell Separation

Bone marrow aspirates are mixed with five parts of ACD solution, filtered through a 70-μm filter, and transferred to a 50 ml tube. Peripheral blood samples are taken in EDTA containing glass tubes and transferred to a 50-ml tube. PBS is added to a final volume of 35 ml and the cells are counted. Cell concentration should not exceed 2 $\times$ 10^8 cells per tube. Low density blood and bone marrow cells are collected by density centrifugation (20 min, 18°C, 700g) on 15 ml of Ficoll-Isopaque (1.077 g/ml; Pharmacia, Uppsala, Sweden, 17-0840-03). Fresh unfixed tissue samples of lymphoma biopsies are suspended by mincing and washing in glucose containing phosphate buffer. Bone marrow, peripheral blood, and lymphoma cells should be washed at least once with PBS. Finally, the cells are resuspended in PBS containing 20% human pooled serum.

C. Monoclonal Antibodies

The monoclonal antibodies used in this study are OKT11 (CD2, Ortho Diagnostics) and MCA 233 (anti-glycophorin A, Instruchemie, Hilversum, The Netherlands). My7 (CD13), My4 (CD14), B4 (CD19), and My9 (CD33) are from Coulter Corp.

D. Staining Procedure

A saturating concentration of monoclonal antibody according to the manufacturer's instructions is added to 10^6 cells (100–200 μl of cell suspension) in a 4-ml tube. The cells are incubated for 30 min at 4°C. After being labeled with the primary antibody, the cells are washed by adding 3 ml PBS and 5 min centrifugation at 700g and 4°C. The pellet is resuspended in a small volume of PBS and incubated for 30 min at 4°C with 200 μl FITC-conjugated F(ab')$_2$ fragments of goat anti-mouse IgG (H&L) (American Qualex International, La Mirada, CA, (714) 521–3753; diluted 1 : 20 in PBS containing 20% human pooled serum). After the second labeling step 3 ml PBS is added. After being mixed, the cells are centrifuged (5 min, 700g, 4°C). The pellet is resuspended and put on ice. After 10 min, 500 μl ice-cold hypotonic PI solution is added. After being mixed, the suspension is kept overnight on ice. The samples are analyzed for S phase DNA content and s-IF the next day.

III. Instruments

Any flow cytometer equipped with an argon ion laser tuned at 488 nm and with the capability of detecting two fluorescent signals together with forward and right angle scatter is suitable. Early experiments in our laboratory were carried out on an Ortho 50H system using the 488-nm line (200–400 mW) of a water-cooled argon ion laser (Model 2020, Spectra Physics). More recent measurements were performed on a Coulter Elite flow cytometer equipped with a 40-mW air-cooled argon ion laser providing excitation at fixed wavelength (488 nm). A high-pass filter for red fluorescence (DNA content, >630 nm) and a band pass filter for green fluorescence (s-IF, 515/30 nm) were used. For bivariate measurements of DNA content and s-IF the fluorescent signals were separated using a 580-nm dichroic mirror.

IV. Critical Aspects of the Method

Standardized fixation conditions and precise handling of the cells are essential in order to retain scatter parameters to a certain extent. Although scatters are not optimal under these conditions, they are still sufficient to gate on these parameters.

In principle, fresh bone marrow, peripheral blood cells, or lymphoma cells should be used for s-IF labeling. However, we found that optimal cryopreserved material after careful thawing could also be used.

We noted that after being labeled with FITC-conjugated monoclonal antibodies, the fluorescent signal decreased dramatically after hypotonic PI staining and, with some antibodies, disappeared totally.

One should be particularly cautious with minor subsets (e.g., <5%, for instance, CD34 in bone marrow can be very low). These small percentages will introduce much noise, and, therefore, false positives and false negatives. An enrichment step prior to the labeling procedure or a multiparameter approach (Stewart and Stewart, Chapter 4, this volume) is recommended in such cases.

Optimal stoichiometric DNA staining is required for reproducible and accurate measurements of percentage S phase. This is achieved by overnight incubation on ice in hypotonic PI using a relatively high dye concentration of 20 μg/ml. A correction was therefore made for crossover of the PI emission in the FITC channel. Suboptimal conditions may result in differences in PI staining of DNA in some diploid cells, which may produce small spurious aneuploid peaks in normal cells.

Also cell membrane permeabilization with saponin and the use of lysolecithin have been proposed as techniques to establish a combined s-IF/DNA measurement. In our hands these techniques were less reproducible and resulted in higher DNA CVs compared to the indirect s-IF/hypotonic PI labeling method.

For accurate S phase DNA content measurements, not only the area of the PI signal, but also the peak height of the red fluorescence was recorded (with the Coulter Elite the ratio of area and peak signal) in order to discriminate between single cells and cell clumps.

V. Results

Evaluation of the technique was performed by assessing the percentage of immunofluorescent cells before and after hypotonic DNA staining. For this purpose 10 separate bone marrow samples of normal individuals were used. The monoclonal antibodies applied were specific for the erythroid (anti-glycophorin A), lymphoid (CD19 and CD2), monocytic (CD14), and myelomonocytic lineages (CD13 and CD33). Labeling of most monoclonal antibodies revealed a clear discrimination between positive and negative populations. The

Table I

Percentages of Positive Cells before and after Hypotonic Treatment of Bone Marrow Aspirates Labeled with Different Antibodies

	MoAb					
	Glycophorin A	CD2	CD19	CD14	CD13	CD33
Before treatment mean (%)	5.7	17.7	8.8	10.2	18.5	58.7
After treatment mean (%)	6.2	17.5	8.0	11.8	18.7	56.4
Difference mean ± SD (%)	+0.5 ± 1.2	−0.2 ± 1.8	−0.8 ± 1.5	+1.6 ± 1.3	+0.2 ± 4.5	−2.3 ± 5.4

dimly stained CD13$^+$ cells were discriminated from the negative subpopulation according to the level set in the control sample. Control samples were cell suspensions incubated with the second step antibody only. The mean values of false-positive cells in these controls before and after DNA staining were 1.5 and 1.1%, respectively. The results obtained with the different antibodies and for the controls are listed in Table I.

DNA staining was performed after the indirect immunofluorescent labeling step, by overnight incubation with $20\mu g/ml$ PI on ice. This resulted in excellent DNA histograms with low CVs (mean, 1.8%). The procedure allows the assessment of very low percentages of S phase cells, even in a heterogeneous population such as low-density blood cells. S phase DNA content measured in normal low-density blood cells of eight healthy individuals varied from 0.06 to 0.09% (mean 0.07%). Only few doublets, interfering with the calculation of G_2M phase DNA, were seen; they were usually less than 3% of all counted cells.

The method described here proved relatively simple and was used with a single-laser flow cytometer. Other monoclonal antibodies, such as WT1 (CD7), OKT1 (CD5), OKT3 (CD3), OKT4 (CD4), OKT8 (CD8), B1 (CD20), J5 (CD10), EA3 (CD10), BA1 (CD24), OKDR (HLA-DR), OKT10 (CD34), OKT9 (transferrin), BOM, My10 (CD34), glycophorin IIB/IIIA, WT14 (CD14), and OKM5 also work with this protocol, and therefore it is now often implemented in routine cell-surface immunophenotyping in leukemia at our institute (see Section VI, Applications and Future Directions).

VI. Applications and Future Directions

The technique described above was used to analyze S phase DNA content and aneuploidy of immunophenotypic defined subpopulations in acute myeloid leukemia determined by multiparameter flow cytometry (Brons *et al.*, 1991).

One of the major limitations of DNA FCM in hematological malignancies is the lack of information about the proliferative activity of subpopulations of the heterogeneous bone marrow compartment. We studied the S phase DNA content of immunophenotypically defined bone marrow subpopulations (CD2[+]; CD19[+]; glycophorin A[+]; CD14[+]; CD13[+], and CD33[+]) in 18 patients with acute myeloid leukemia. Results were compared with normal bone marrow. In patients with acute myeloid leukemia the proportion of bone marrow cells expressing myelomonocytic and monocytic markers (M1–M5 acute myeloid leukemia) or erythroid marker (M6 acute myeloid leukemia) was expanded. However, in many patients other subpopulations were 4% or higher permitting the calculation of their S phase DNA content. Comparing normal and leukemic bone marrow no significant differences in S phase DNA content were observed, although the ranges in patients with acute myeloid leukemia were much wider. This suggests that acute myeloid leukemia is not characterized by an increased nor a decreased proliferative activity, but rather by a situation of uncontrolled cell growth.

Additional information was obtained upon DNA aneuploidy using CD2[+] cells as an intrinsic DNA standard which allowed us to define differences in the DNA index as small as 2% as aneuploid. This approach appeared suitable for detecting small-degree numerical chromosomal aberrancies, as confirmed by cytogenetics, in 4/6 cases.

We applied this method to cell suspensions of malignant B cell lymphomas. In 10/28 patients a DNA aneuploidy was found but usually with a considerable admixture of normal diploid cells (median 56%). The admixture of diploid cells interfered with the S phase DNA content analysis of the malignant population. We therefore applied a bivariate staining of B cell and T cell markers versus DNA content. This allowed a selective measurement to be made of the S phase DNA content in T cell fractions enriched for diploid cells (median 97% diploid cells) and B cell fractions enriched for aneuploid cells (median 90% aneuploid cells). But also in diploid B cell lymphomas with a high admixture of T cells, a selective measurement of S phase DNA content of the B cell population is more accurate. Figure 1 shows a malignant lymphoma with 6.0% S phase cells. Bivariate measurements of s-IF and S phase DNA content reveal that this lymphoma is composed of a small B cell population (27.1%) with a high proliferative activity (S phase DNA content is 16.7%) and a large T cell population (63.5%) with a low proliferative activity (S phase DNA content is 3.3%). These analyses were performed on a cryopreserved cell suspension. This caused slightly larger but still acceptable CVs.

Until now only PI has been used for measuring S phase DNA content. When combined with a DNA stain with an emission peak wavelength higher than PI, it is possible to perform triple labeling. An example of such a dye is 7-amino actinomycin D, but members of a recently introduced group of new membrane-impermeant nuclear stains by Molecular Probes (YO-PRO-3 and TO-PRO-3) are also acceptable.

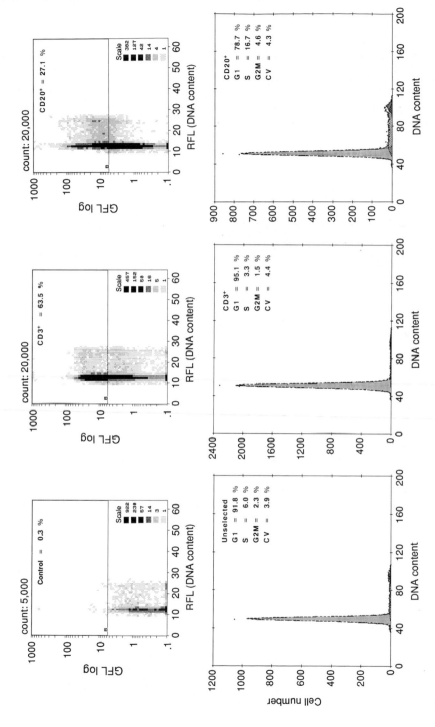

Fig. 1 Bivariate measurements of s-IF (GFL-log) and S phase DNA content of cryopreserved malignant lymphoma cells. The cell suspension was carefully thawed and aliquots were stained for CD3 and CD20 using an indirect immunofluorescent technique. The control sample was stained with the FITC-conjugated second step antibody only. DNA was subsequently stained with PI under hypotonic incubation conditions. DNA histograms given in the lower panels are derived from the bivariate plots above.

References

Brons, P., Pennings, A., Haanen, C., Wessels, H., and Boezeman, J. (1991). *Leuk. Res.* **15,** 827–835.

Brons, P. P. T., Raemaekers, J. M. M., Bogman, M. J. J. T., van Erp, P. E. J., Boezeman, J. B. M., Pennings, A. H. M., Wessels, H. M. C., and Haanen, C. (1992). *Blood* **80,** 2336–2343.

Civin, C. I., and Loken, M. R. (1987). *Int. J. Cell Cloning* **5,** 267–288.

Crissman, H. A., and Steinkamp, J. A. (1973). *J. Cell Biol.* **59,** 766–771.

CHAPTER 7

Flow Cytometry Crossmatching for Solid Organ Transplantation

Robert A. Bray

Department of Pathology
Emory University Hospital
Atlanta, Georgia 30322

I. Introduction

Independent of the level of human leukocyte antigen (HLA) matching, the single most important test employed in solid organ allocation is the lymphocyte crossmatch. This test assesses the level of circulating antibodies in the potential recipient that are directed against the HLA Class I and/or Class II molecules of the donor. Traditionally, the complement-dependent lymphocytotoxicity (CDC) crossmatch has been the method employed for assessing circulating alloantibody. However, because of its increased sensitivity, flow cytometry is now being utilized as an alternative methodology. The flow cytometric crossmatch (FCXM) uses indirect immunofluorescence to detect human alloantibody that has bound to target lymphocytes. Although IgG class antibodies appear to be most associated with early rejection and accelerated graft loss, the detection of other immunoglobulin isotypes may also be informative. In contrast to the CDC, the FCXM can easily discriminate between the isotypes and subclasses of immunoglobulins. Additionally, due to the differential expression and distribution of HLA molecules, it is beneficial to discriminate between T cell and B cell reactivities. T cells bear predominantly HLA Class I molecules while B cells express Class II molecules and possess a higher density of Class I molecule expression.

In the FCXM, individual cell populations are discriminated by multicolor staining using directly conjugated monoclonal antibodies directed against T cells (CD3) and/or B cells (CD20 or CD19). Since most newer cytometers have the capability of simultaneously evaluating three distinct fluorochromes, a three-color crossmatch can routinely be performed. One fluorescence parameter is used to assess alloantibody binding via indirect immunofluorescence while the other two fluorescent parameters are used to delineate desired cell populations (e.g., identify T cells from B cells). Appropriate postacquisition analyses then allow for the independent evaluation of the reactivity of a given cell population. This capability is in direct contrast to the standard CDC crossmatch which requires the physical separation and purification of a given cell population (e.g., T and B lymphocytes) or requires the use of fluorescence microscopy and selective cell tagging to discriminate between cell populations. Interpretation of the FCXM is done by directly comparing the fluorescence intensity of the test cell population (e.g., potential donor cells) treated with the patient's serum to the fluorescence intensity of the same cell population treated with a known negative control serum (i.e., a serum which lacks HLA antibodies). This interpretation is in direct contrast to the subjective visual assessment used in the CDC technique. Hence, the FCXM is more objective and has the additional potential to quantitatively assess bound alloantibody.

II. Applications

Currently, the complement-dependent lymphocytotoxicity crossmatch (CDC) test is the standard methodology employed in a majority of transplant laboratories. In this test, donor lymphocytes are mixed with serum from potential recipient(s). This primary incubation is followed by the addition of complement, a second incubation period, and the addition of a vital dye to assess cell death. Significant cell death above background is interpreted as a positive crossmatch and, in most instances, is a contraindication to transplantation. Although the CDC assay is the standard method used, numerous adaptations of the test exist. These variations employ combinations of multiple cell washes, alterations in the incubation times, the use of anti-human globulin (AHG), and reducing agents such as DDT/DTE and/or 2-mercaptoethanol. The ultimate goals of these variations are to impart increased sensitivity and to discriminate IgG from IgM antibodies. A negative crossmatch is an indication that the recipient does not possess antibodies that will result in an immediate or "hyper-acute" allograft rejection. Despite its many permutations, pitfalls, and subjective nature, the CDC crossmatch remains a mainstay of transplant laboratories and has virtually eliminated hyperacute rejection.

Although hyperacute rejections are rare, studies have demonstrated that some patients with a negative CDC test experienced early graft loss consistent with humoral rejection. These findings indicated that low levels of alloantibody, undetectable by the CDC assay, were present in the recipient at the time of transplantation. Within the past few years, several reports have shown that these low levels of alloantibody could be detected by flow cytometry (Garovoy et al., 1983; Chapman et al., 1985; Bray et al., 1989). More importantly, renal transplant recipients exhibiting a negative CDC assay but a positive FCXM were more likely to experience early accelerated rejection episodes and graft loss (Cook et al., 1987; Lazda et al., 1988; Talbot et al., 1989; Mahoney et al., 1990; Karuppan et al., 1992). The patient groups most likely to experience this type of rejection were those individuals who were previously alloimmunized from a rejected transplant, multiple pregnancies, or multiple blood transfusions. Hence, the use of the FCXM for the semi-quantitative assessment of alloantibody in these potentially high-risk renal and heart patients is quickly replacing the CDC assay as the standard pretransplant crossmatch.

The FCXM offers additional advantages over the CDC as well. For example, the FCXM takes less time to perform than the CDC. This is an important difference since time is a most critical factor when allocating cadaveric organs. Also, since IgG class antibody is the predominant immunoglobulin responsible for the observed early graft loss, the FCXM can specifically identify IgG antibodies without the additional manipulations of patient's serum (e.g., DTT treatment) needed for the CDC. This factor also can reduce the time involved in resulting out the crossmatch. Since CDC crossmatches use cell death as the assay end point, they require that the initial cell viability be at least $>85\%$. Some cell preparations, such as those derived from cadaveric blood, spleen, or lymph nodes, may have low initial viability which requires additional manipulations. Additionally, even though initial viability may be acceptable, due to the fragile nature of these cell preparations, significant cell death may occur during the CDC assay making reading and interpretation quite difficult. In contrast, the FCXM can take advantage of light scatter and/or viability stains to select the appropriate cell population for analysis. FCXMs have been successfully performed with cell viabilities $<50\%$. Lastly, since the FCXM directly measures alloantibody bound to cells, interpretation of the FCXM tends to be more objective than the CDC. CDC assays rely on visual scoring of stained cells and the results can be influenced to a significant degree by reader bias.

III. Materials

A. Reagents

1. Secondary Antibody

The most critical reagent for the FCXM is the secondary antibody. This reagent is used to identify the human immunoglobulin that is specifically bound to the target cell. As such, this reagent should be selected to provide optimum specificity with minimum background staining. Antibodies can be selected that are specific for the various isotypes; however, each lot must be verified to confirm specificity. A preferred secondary antibody would possess the following properties: (a) well-defined specificity (e.g., Fc specific); (b) a $F(ab')_2$ or Fab fragment to reduce background staining; and (c) no cross-reactivity with mouse or horse immunoglobulins. Currently, the secondary reagents utilized in our laboratory are FITC anti-IgG, Fc specific (Jackson ImmunoResearch, West Grove, PA; Cat. No. 109-016-098) and anti-IgM, $Fc_{5\mu}$ (Jackson ImmunoResearch, Cat. No. 109-16-043). Most importantly, each lot of secondary antibody should be titered to determine the optimum working dilution. A sample titration is shown in Table I. In this titration, the FITC secondary reagent is titered from $1:4$ through $1:20$. Dilutions of the FITC reagent are tested against a lymphocyte cell population that has been preincubated with various dilutions of a known positive control serum. From the resulting chessboard titration one can identify the appropriate dilution of the secondary reagent. In this example, it appears that FITC dilutions of between $1:8$ and $1:10$ would be appropriate (shaded area). The actual working dilution for this lot was $1:8$. Although higher channel displacements are seen with lower (more concentrated) dilutions of the secondary reagent, higher channel displacements are observed at greater dilutions of the pooled positive serum (PPS) when higher dilutions of the FITC are employed. Primarily, this is a result of reducing the background noise as can be observed in the top line of the table which lists the actual median channel value for each dilution of the secondary reagent tested with normal human serum (NHS). After determining the appropriate working dilution of the reagent, testing against mouse and horse immunoglobulins may be performed.

2. Monoclonal Antibodies

Depending upon the cell population to be evaluated, any number of monoclonal antibodies that react with the desired cluster antigen are acceptable. However, for most solid organ-related crossmatches, CD3, for identifying T cells, and CD19 or CD20, for identifying B cells, are the principal choices. Antibodies such as CD4 or CD8 should be avoided since these molecules may associate with HLA molecules on the cell surface and one runs the risk of stearic hinderance in antibody binding. The choice of fluorochrome for the

Table I
Example of an IgG Titration

Sera[a]	IgG dilutions[b]				
	1 : 4	1 : 6	1 : 8	1 : 10	1 : 20
NHS	247[c]	241	186	175	150
PPS					
1 : 5	283[d]	273	260	251	175
1 : 10	152	155	188	190	140
1 : 20	113	108	147	133	105
1 : 40	80	87			90
1 : 80	80	78			48

Note. The highlighted box indicates the region where dilutions in the FITC and PPS were informative for determining the appropriate dilution of the FITC reagent. The actual working dilution for this lot of FITC was 1 : 8.

[a] NHS, normal human serum control; PPS, pooled positive serum control.

[b] FITC-conjugated, anti-human IgG, Fc specific (Jackson ImmunoResearch, West Grove, PA). Reagent was reconstituted from lyophilized form with distilled water and dilutions were made with ''wash buffer''.

[c] Values in this line are the actual median channel values for the IgG reagent when incubated with NHS.

[d] Values in this section are channel displacements, calculated by subtracting the actual median channel value of the PPS dilution from the median channel value of the NHS for the given dilution of IgG.

secondary antibody will determine the fluorochromes for the monoclonal antibodies. At present, most laboratories utilize a fluorescein-conjugated secondary antibody thus relegating phycoerythrin or other longer wavelength emitters such as PerCP for identifying cell populations.

3. Control Sera

The FCXM requires both a positive and a negative control serum. The positive control serum usually consists of pooled sera from patients with very high levels of anti-HLA antibodies. The PPS should be diluted such that a reasonable shift in fluorescence is seen. Using PPS undiluted will produce very strong positive results; however, most actual crossmatches do not exhibit strong reactions. Hence, using a positive control serum that exhibits a moderate positive reaction can help in assessing the performance of the test. For example, if the laboratory's positive control regularly exhibits a shift in fluorescence of 100 channels over the negative control serum, then a test sample where the positive control only exhibits a 40- or 50-channel shift may be an indication of an invalid test.

The negative control serum is, by far, the most important control element in the FCXM. This reagent can be either a pooled or a single-donor serum. However, this serum must be tested thoroughly with several (>10) normal cell donors to ensure that the reagent has no reactivity with human lymphocytes. This exercise will not only ensure that the negative control serum is appropriate but will also provide the information needed for proper interpretation of the FCXM.

4. Wash Buffer

The standard wash buffer used throughout the crossmatch consists of phosphate-buffered saline (PBS) containing 2% fetal bovine serum (FBS) and 0.1% sodium azide (NaN_3). Wash buffer should be stored at 4°C and appropriate precautions should be taken for disposal of NaN_3 containing solutions.

5. Paraformaldehyde

A standard 2% paraformaldehyde solution in PBS (final pH 7.2 ± 0.2) is used as the final fixative. Paraformaldehyde should be prepared in a fume hood with proper ventilation. Paraformaldehyde solution should be stored at 4°C and protected from exposure to light. Shelf life is approximately 2 weeks.

B. Miscellaneous Supplies

The following is a list (by no means inclusive) of additional supplies that will be needed to perform the FCXM.

• Eppendorf pipettor (or similar pipetting device) capable of delivering volumes from 10 to 100 μl.

• Small volume test tubes. Our laboratory utilizes plastic and glass, 6 × 50-mm tubes for the FCXM (glass, Baxter Scientific Products, Stone Mountain, GA., product T12901-1; polypropylene, Evergreen Scientific, Los Angeles, CA., product 2142301-030). These types of tubes hold a total volume of 450 μl and permit the use of small numbers of cells while minimizing cell loss.

• Table top centrifuge.

• Airfuge (Beckman Instruments, Inc.).

• Vacuum aspiration system.

IV. Cell Preparation and Staining

A. Cell Preparation

Several cell types can be used in the FCXM. Essentially, any single-cell suspension can serve as a target cell in the FCXM. However, the most common sources of targets are mononuclear cells derived from peripheral blood, lymph nodes, or spleen. Detailed procedures for preparing cell populations can be found in a number of other references (Zachari and Teresi, 1990).

Once an adequate cell preparation is obtained, only small numbers of cells are needed for the crossmatch. The cells should be resuspended in the "wash buffer" and held at 4°C until ready to stain. The actual cell concentration of the preparation is not critical; however, the actual number of cells used per tube is extremely critical. In general, the cell number need not exceed 2.5×10^5 cells per tube. By far the most important variable is the ratio of cells to volume of serum used in the FCXM. By utilizing the small-volume test tubes, cell numbers as low as 10^5/tube can be easily used. Since patient serum is usually limiting, volumes between 20 and 50 μl/tube are recommended. Increasing the volume of patient serum used per tube or decreasing the number of cells per tube will increase the sensitivity of the FCXM. Conversely, using less serum or, more importantly, increasing the number of cells used per test can significantly decrease the sensitivity. Hence, it is important that the laboratory establish its staining protocol and insist that accuracy and consistency of cell concentrations and serum volumes used be strictly adhered to.

B. Staining Procedure

(See Fig. 1 for procedure flow chart and Table II for tube layout.)

1. Place appropriate quantity of cells in 6 × 50-mm tubes (or other small volume tube) and centrifuge to pellet (700g for 5 min).
2. Aspirate supernatant, being cautious not to aspirate the cell pellet.
3. Add the appropriate patient or control serum (25–50 μl) to the corresponding labeled tubes (see Table II). Vortex to ensure proper mixing of serum and cells.
4. Incubate for 30 min at 4°C.
5. Wash cells with 400 μl of COLD wash buffer. (Note: add 200 μl, vortex, and then add the additional 200 μl.) Centrifuge at 700g for 5 min.
6. Aspirate the supernatant and repeat step 5.

NOTE: BE CAREFUL WHEN ASPIRATING SAMPLES TO PREVENT CROSS-CONTAMINATION.

7. Following the second wash step and aspiration, add 20 μl of pretitered secondary antibody and the appropriate quantity of directly conjugated CD3, CD20, or other commercial monoclonal antibody directly to the dry cell pellet.

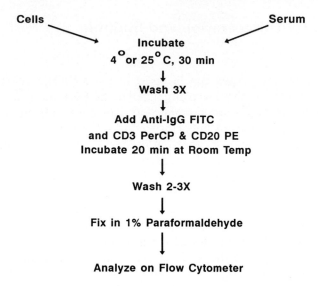

Fig. 1 Flow chart illustrating a three-color flow cytometric crossmatch.

Table II
Suggested Setup for Performing Three-Color Flow Cytometric Crossmatch

Tube no.	Primary antibody/serum	Secondary antibody(s)
1	NHS or autologous[a]	FITC α-IgG
		CD3 PerCP and CD20 PE[b]
2	Pooled positive serum	FITC α-IgG
		CD3 PerCP and CD20 PE
3	Patient sample 1	FITC α-IgG
		CD3 PerCP and CD20 PE
4	Patient sample 2	FITC α-IgG
		CD3 PerCP and CD20 PE
5-n	Repeat setup as in tubes 3 or 4 for each additional serum	

[a] NHS, normal human serum; autologous, patient's own serum may be used to identify self-reactive antibodies.

[b] For performing a two-color crossmatch, individual tubes must be set up for CD3 and CD20. In this instance, phycoerythrin would be the conjugate for both CD3 and CD20. In addition, the total volume of each serum used and number of cells required will have to be doubled.

NOTE: USE THE MANUFACTURER'S RECOMMENDED CONCENTRATION FOR ALL MONOCLONAL ANTIBODIES UNLESS OTHERWISE TESTED.

8. Incubate antibodies for 20 min at 4°C and protect from light.

9. Following incubation, wash cells twice with 400-μl volumes of COLD wash buffer. Follow wash procedure as indicated in steps 5 and 6.

10. After final wash and aspiration, resuspend cells in 100 μl of COLD wash buffer.

11. While vortexing, add 100 μl of COLD 2% paraformaldehyde solution to each tube. Fixed cells can be held at 4°C, in the dark, for up to 7 days.

V. Critical Aspects of the Procedure

Although the FCXM is a relatively straightforward technique, there are several critical aspects that can significantly effect the results and performance.

The first and most critical aspect of the crossmatch is that a multicolor method be used. Figure 2 clearly demonstrates the drawbacks of evaluating an ungated FCXM. The presence of natural killer (NK) cells in peripheral blood contribute to a high background fluorescence. NK cells bear the CD16 molecule which is the Fc receptor for IgG. Unfortunately, no secondary anti-IgG reagent can discriminate specifically bound IgG from cytophilic IgG. As illustrated in Fig. 2, cytophilic IgG bound to NK cells can generate a significant signal. Hence, the need for a multicolor technique to identify only the cell populations of interest. Additionally, and depending upon the cell preparation, other cell populations can contaminate the lymphocyte gate.

A second important aspect of the FCXM is the ratio of serum volume to number of cells. The lymphocytes are, in fact, small antigen bearing particles that are used to assess the levels of alloantibody. The ability to detect alloantibody in a patient's serum sample is dependent upon the level of antibody and the number of cells tested. Thus, it would seem prudent to use as much serum as possible with as few cells as possible to obtain maximum sensitivity. This, however, is not feasible and most laboratories use between 20 and 100 μl of serum and 0.25 to 1 $\times$ 10^6 cells per tube. The important point here is that fluctuations in the serum–cell ratio can significantly alter the crossmatch results essentially producing false-negative results if too many cells are used. (Fig. 3). For flow cytometric crossmatching, no more than 0.5 $\times$ 10^6 are needed per tube. Utilizing the small 6 $\times$ 50 polypropylene tubes, as few as 10^5 cells can be easily stained with only minimal cell loss. Thus, one cannot overemphasize the importance of strict adherence to the laboratory's established procedure specifying the numbers of cells and volume of serum used.

A third critical aspect relates to the number of wash steps that are used following the primary incubation. Within a patient's serum, only a small portion of the total immunoglobulin will constitute the anti-HLA antibody. This is true

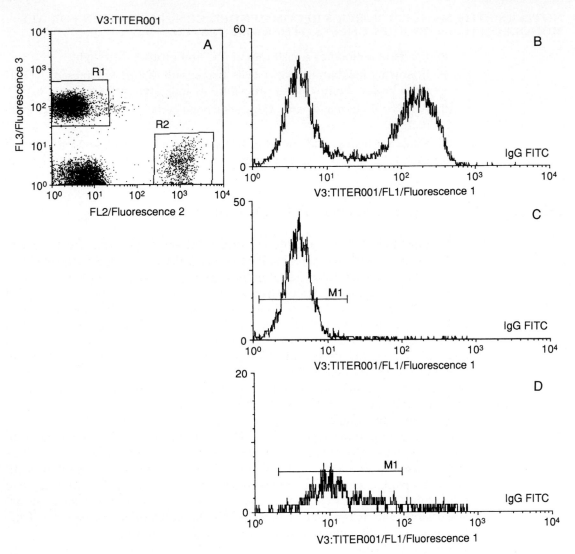

Fig. 2 Cytophilic immunoglobulin binding to NK cells. Peripheral blood lymphcytes were preincubated with normal human serum that was devoid of HLA antibodies. Cells were subsequently stained with FITC anti-human IgG, phycoerythrin (PE)-conjugated CD20 and PerCP CD3. (A) Dual-parameter dot plot of PE CD20 (x-axis; R2) versus PerCP CD3 (y-axis; R1). (B) FITC anti-immunoglobulin staining pattern of the total mononuclear cell population (ungated). (C and D) FITC antiimmunoglobulin staining of selected T cells (R1) and B cells (R2), respectively.

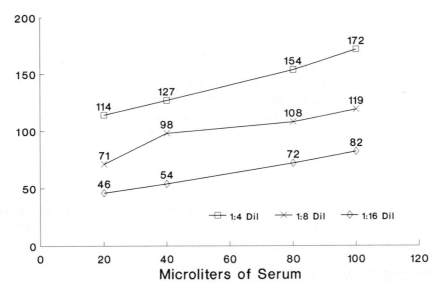

Fig. 3 Effects of alterations in serum : cell ratio on the sensitivity of the FCXM. Three dilutions of a known positive control serum were tested against a fixed number of cells (5×10^5/tube). Each dilution of serum was then evaluated by increasing the volume added per tube while holding the cell number constant. Thus, each line represents an experiment wherein four individual tubes were incubated with increasing volumes of the indicated serum dilution. As the results show, as one increases the volume of serum (antibody), there is a corresponding increase in the channel displacement. MDCF indicates channel displacement as calculated from the median channel fluorescence value of each tube. The numbers listed are the actual channel displacement values. Results reported are for CD3-gated T lymphocytes.

even in patients with high-titer anti-HLA antibodies. Thus, it is important to use a sufficient wash protocol following the primary incubation in order to ensure that residual immunoglobulin is removed prior to the addition of the secondary antibody. Insufficient washes may result in a false-negative FCXM.

A fourth critical aspect of the procedure deals with the nature of the secondary antibody used. Since the goal of the FCXM is the detection of specifically bound antibody of a defined subclass, it is important to use a secondary antibody with a very narrow specificity. For example, a conjugated anti-human IgG that is Fc specific and preabsorbed against mouse and horse serum proteins is desirable for the antibody. Additionally, if the immunoglobulin is an F(ab')$_2$ or Fab fragment, nonspecific background staining can be reduced. If the secondary antibody is a mouse monoclonal, then demonstrating lack of cross-reactivity with horse immunoglobulins would be important. These cross-reactivities are notable since the directly conjugated monoclonal antibodies are of mouse origin and cross-reactivity with these immunoglobulins will render the crossmatch uninterpretable. Additionally, many HLA laboratories are now engaging in post-transplant monitoring in an attempt to quantify anti-donor antibodies. In many instances these patients are receiving some form of anti-rejection therapy which may include OKT3 (mouse) or ALG (anti-lymphocyte globulin; horse). In these instances, cytotoxicity assays are of no value since both OKT3 and ALG are lymphocytotoxic in the CDC assay. However, if one has chosen the secondary antibody appropriately, screening for the presence of anti-HLA antibodies can easily be performed on these individuals. Again, cross-reactivity with mouse or horse immunoglobulins would render the FCXM uninterpretable.

A final critical aspect of this procedure deals with the preparation of patient and control sera. In our laboratory we have found it most useful to airfuge (100,000g for 15 min) all sera prior to testing. Most of the sera used in cross-matching have undergone at least one freeze–thaw cycle. Ultracentrifugation helps remove immunoglobulin aggregates that may produce nonspecific background staining, particularly on B cells. This background is most likely a result of immune aggregates binding to Fc receptors (e.g., CD32).

VI. Standards

Unfortunately, there are no commercial standards available for the flow cytometric crossmatch. Hence, each laboratory must establish its own set of controls and standards. By far the most important control/standard is the normal human serum that is used as the negative control. For the FCXM either pooled male serum or single-donor, preferably male, serum is recommended. Serum from females may be used provided the individual has not been pregnant. During pregnancy, the female is exposed to paternal HLA antigens and has the potential to produce anti-HLA antibodies. Although the frequency of this sensitization is only about 1/20, care must be taken to ensure that the chosen normal human serum does not contain anti-HLA antibodies. Screening of a potential negative control serum should be accomplished by performing a FCXM using cells from a panel ($N \geq 10$) of individuals whose HLA types are known and which cover the majority of the HLA specificities. This testing should be done on the same day. A record of the actual median channel value for each individual's T cell and B cell populations should be made without changing any machine settings. Calculate the average channel value and the standard deviation for both T and B cell populations. Using a two standard deviation (2 SD) cutoff, one can then define a channel value which can be used as a discriminator for positivity. That is, within a certain degree of statistical confidence, a channel displacement in a patient's sample that is greater than the 2 SD cut point may be interpreted as positive. This, however, is not absolute and should be determined by each laboratory and reconfirmed with each now lot of normal human serum.

As an alternative to median channel values, one can also utilize a semi-quantitative method for determining relative immunofluorescence. This involves the use of fluorescent microbeads that have a defined number of fluoro-chrome molecules attached (Schwartz and Fernandez-Repollet, 1993). By running a mixture containing beads, each with a different level of fluorescence, one can perform a relative calibration of the flow cytometer. Subsequent channel values can be converted to the number of molecules of equivalent soluble fluorochrome (MESFs). Results are then reported in MESF units and the determination of a positive result is made by indicating a minimum MESF value above which a crossmatch is considered positive.

In contrast to the negative control serum, the positive control standard is simpler to obtain. One possible approach for obtaining a good positive control is to pool sera from patients that have high levels of cytotoxic anti-HLA antibodies. Such a pool can then be diluted several fold and used as the positive control for the FCXM. In determining an appropriate working dilution, it is beneficial to utilize this control at a dilution which yields a reasonable shift in fluorescence but not necessarily a maximum shift (saturating). It is important to remember that the FCXM should be optimized for the detection of low levels of circulating alloantibody. Thus, using a positive control that produces a 1000-channel displacement, when the 2 SD cutoff for determining a positive result is 40 channels, does not add any useful information to the assay. Significant errors could be made that would not necessarily result in a significant alteration in the channel displacement of the positive control. Additionally, using such a strong positive control increases the possibility that a minute amount of carryover contamination may produce a false-positive result.

Therefore, an appropriate working dilution of the positive control would be more sensitive to deviations in the system and less likely to contribute to a false-positive result. Diluting the positive control to yield a reproducible displacement of between 100 and 200 channels (i.e., well below the level of saturation) can be helpful in detecting errors in the numbers of cells added (Fig. 4). For example, in an actual crossmatch if the positive control serum only demonstrated a displacement of 50 channels, the crossmatch may be judged as uninterpretable or may require repeating.

Lastly, accreditation for flow cytometry crossmatching is available through the American Society for Histocompatibility and Immunogenetics (ASHI). ASHI has published standards for the performance of flow cytometric crossmatching for transplantation. These standards are available from the ASHI office in Lenexa, Kansas, and a copy may be obtained by calling (913) 541–0009.

VII. Instruments

At present, all manufacturers' instruments are suitable for performing the FCXM. The only restriction might be that some of the older instruments may not be fitted with a third photomultiplier tube for evaluating the long wavelength emitting fluorochromes and thus may be unable to perform a three-color crossmatch. In this instance, performing two separate dual-color crossmatches (e.g., CD3 and CD20) would be appropriate and would yield the same information

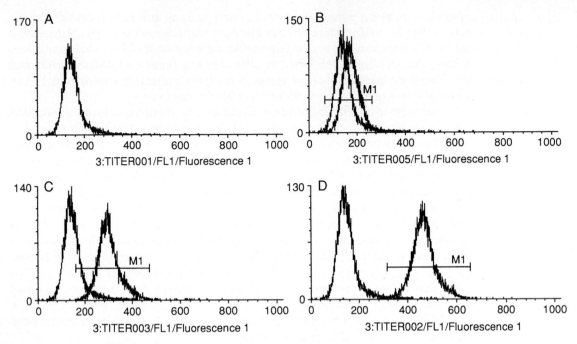

Fig. 4 Channel displacements observed by titration of a positive crossmatch serum. (A) Negative control, normal human serum (NHS). (B, C, D) Positive serum titration demonstrating a channel displacements (CD) of 40, 160, and 337 channels, respectively. Histogram C represents a recommended dilution for the positive control. (FACScan cytometer; 4-decade log scale; 1024-channel resolution; LYSYS II analysis.)

although the requirement for total numbers of cells and volume of serum would be doubled. Nonetheless, virtually all manufactures' instruments and their associated software are capable of producing acceptable and reproducible FCXM results. By far the greatest variable is the setup of the crossmatch.

VIII. Results and Discussion

Analysis of the FCXM is also quite straightforward. During data acquisition, samples may be collected either ungated or with gates set around the cell population of interest (e.g., monocytes or lymphocytes). The majority of FCXMs will benefit from acquiring lymphocyte-gated data. Postacquisition analysis is then performed by gating on a selected cell population based on fluorescence and then displaying the fluorescence histogram of the anti-immunoglobulin reagent. For example, in a standard three-color crossmatch, FITC is used as the conjugate for the anti-IgG reagent in conjunction with phycoerythrin-conjugated CD20 and PerCP-conjugated CD3. Following acquisition of lymphocyte-gated data, individual regions are set around CD3- and CD20-positive cells. This is performed by displaying a dot (or contour) plot of CD3 vs CD20 (Fig. 5). Subsequently, analysis of the FITC staining can be made by displaying a single-parameter histogram from the CD3 region (Region 1, Fig. 5) and/or from the CD20 region (Region 2, Fig. 5). This display format is employed for each tube of the FCXM. Following completion, the median channel value for each cell population is recorded. Median channel provides the best assessment of the distribution of the total population of cells. If a particular instrument cannot generate a medium channel value, then placing analysis cursors tightly around the peak distribution will produce a mean channel value that will approximate the median. Since HLA molecules are equally distributed on the appropriate cell populations, if an HLA antibody is present, all cells should be displaced on the histogram. A displacement in 10% of the

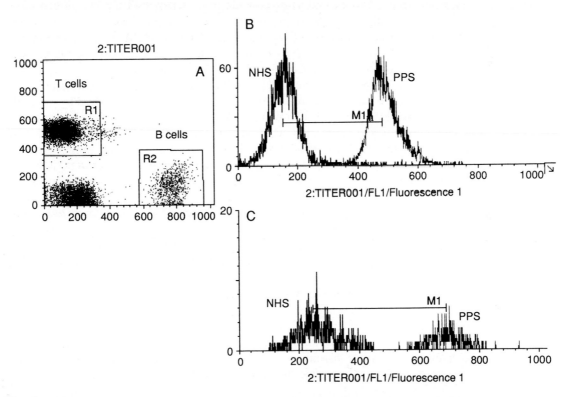

Fig. 5   Three-color flow cytometric crossmatch. (A) Dot plot displaying CD20 PE staining of B cells (x-axis) versus CD3 PerCP staining of T cells (y-axis). (B) Single-parameter histogram of T cells (R1). NHS, staining fluorescence with the normal human serum control; PPS, staining fluorescence observed with a positive serum sample. CD, channel displacement, is the distance between the median channel values of the two peaks (line). (C) Single-parameter histogram of B cells (R2). (FACScan cytometer; 4-decade log scale; 1024-channel resolution; LYSIS II analysis.)

total cells, for example, would not be consistent with an HLA antibody and hence would not be considered significant. The actual calculation of results is performed by subtracting the median channel value of the negative control serum from the value of the positive control and all the evaluated patient's serum samples. This difference in channel displacement (CD) is used as the criteria for determining whether a crossmatch is positive. As detailed above, the laboratory must establish its own cutoff for determining how great a channel displacement is considered a positive result.

While technically simple to evaluate, the actual clinical interpretation of the FCXM is quite a bit more complex. A complete discussion of the clinical implications of a positive flow crossmatch is beyond the scope of this chapter and the reader is referred to the literature for more information (Cook *et al.,* 1987; Lazda *et al.,* 1988; Talbot *et al.,* 1989, 1992; Mahoney *et al.,* 1990; Karuppan *et al., 1992;* Bray, 1993; Ogura, 1993). However, the clinical history of each patient needs to be taken into consideration when the FCXM is interpreted. More importantly, since there are no methods for confirming that the detected antibody is directed against a Class I and/or a Class II HLA antigen, knowledge of the patient's clinical history is most important. Thus, a positive FCXM will not always constitute a contraindication to transplantation.

Lastly, the utility of this method is not limited to the final crossmatch prior to transplantation. Several laboratories have begun to use this method to routinely screen serum from patients on transplant waiting lists. As part of the laboratory's ongoing monitoring of patients awaiting organ transplantation, periodic reevaluation of a patient's level of alloantibody is performed. A PRA (panel reactive antibody) value is obtained by screening patients against a panel of cells whose HLA type is known. From such a panel, the laboratory not only can assess the quantity of alloantibody (i.e., how many of the panel cells are reactive with the serum), but also may be able to identify the actual specificity (i.e., determine which HLA antigen the antibody is directed against). Our laboratory and others have reported significant levels of anti-HLA antibody in patients whose cyto-toxic PRA was 0% (i.e., no reactivity with panel donors in the standard CDC assay) (Shroyer *et al.,* 1991; Bray, 1993). Obviously, if antibodies can be detected and a specificity identified prior to crossmatching, a significant amount of time and resources may saved by initially selecting individuals who show no evidence of anti-HLA antibodies that are directed against the HLA antigens of the organ donor. Thus, performing flow cytometric PRAs of selected patients can aid in the proper allocation of donor organs which, in turn, saves time and resources and helps increase the chances for long-term graft survival.

References

Bray, R. A. (1993). *Ann. N.Y. Acad. Sci.* **677,** 138–151.
Bray, R. A., Lebeck, L. L., and Gebel, H. M. (1989). *Transplantation* **48,** 834–840.
Chapman, J. R., Deierhoi, M. H., Carter, N. P., Ting, A., and Morris, P. J. (1985). Transplant. Proc. **17,** 2480–2482.

Cook, D. J., Terasaki, P. I., Iwaki, Y., Terashita, G. Y., and Lau, M. (1987). *Clin. Transplant.* **1,** 253.

Garovoy, M. R., Rheinschmilt, M. A., Bigos, M., Perkins, H., Colombe, B., Feduska, N., and Salvatierra, O. (1983). *Transplant. Proc.* **15,** 1939–1942.

Karuppan, S. S., Lindholm, A., and Moller, E. (1992). *Transplantation* **53**(3), 666–673.

Lazda, V. A., Pollak, R., Mozes, M. F., and Jonasson, O. (1988). *Transplantation* **45,** 562.

Mahoney, R. J., Ault, K. A., Given, S. R., Adams, R. J., Breggia, A. C., Paris, P. A., Palomaki, G. E., Hitchcox, S. A., White, B. W., Himmelfarb, J., and Leeber, D. A. (1990). *Transplantation* **49**(3), 527–535.

Ogura, K., Terasaki, P. I., Johnson, C., Mendez, R., Rosenthal, J. T., Ettenger, R., Martin, D. C., Dainko, E., Cohen, L., Mackett, T., Berne, T., Barba, L., and Lieberman (1993). *Transplantation* **56,** 294–298.

Schwartz, A., and Fernandez-Repollet, E. (1993). *Ann. N. Y. Acad. Sci.* **677,** 28–39.

Shroyer, T. W., Diethelm, A. G., Burgamy, M. A., Harvey, L. R., Hudson, S. L., and Barger, B. (1991). *Hum. Immunol.* **32,** Suppl. 1, 100.

Talbot, D., Givan, A. L., Shenton, B. K., Stratton, A., Proud, G., and Taylor, R. M. (1989). *Transplantation* **47**(3), 552–555.

Talbot, D., Cavanaugh, G., Coates, E., Given, A. L., Shenton, B. K., Lennard, T. W., Proud, G., and Taylor, R. M. (1992). *Transplantation* **53**(4), 925–928.

Zachari, A., and Teresi, G., eds. (1990). "ASHI Laboratory Manual," 2nd ed. American Society for Histocompatibility, Kansas City.

CHAPTER 8

Cell Membrane Potential Analysis

Howard M. Shapiro

Howard M. Shapiro, M.D., P.C.
283 Highland Avenue
West Newton, Massachusetts 02165

I. Introduction

Cytoplasmic and mitochondrial membrane potential (MP) changes may occur during the early stages of surface receptor-mediated activation processes related to the development, differentiated function, and pathology of a large number of cell types and are thought to play a role in signal transduction between the cell surface and the interior. Investigations in this area have been facilitated by the development of methods for flow cytometric (FCM) estimation of MP in single cells (Shapiro *et al.,* 1979; Shapiro, 1981, 1982, 1988). This chapter discusses the principles, practical aspects, and limitations of such methods.

METHODS IN CELL BIOLOGY, VOL. 41
Copyright © 1994 by Academic Press, Inc. All rights of reproduction in any form reserved.

A. The Basis of Membrane Potentials

Electrical potential differences normally exist across eukaryotic cell membranes, due in part to concentration gradients of Na^+, K^+, and Cl^- ions across the cell membrane, and in part to the operation of electrogenic pumps. The potential differences across the cytoplasmic membranes of resting mammalian cells range from ~10 to 90 mV, with the cell interior negative with respect to the exterior. There is also a potential difference of $\geq$100 mV across the membranes of energized mitochondria, with the mitochondrial interior negative with respect to the cytosol; this potential is dependent upon energy metabolism. In prokaryotes, the enzymes responsible for energy metabolism are located on the inner surface of the cytoplasmic membrane, and most of the 100- to 200-mV, interior negative, potential difference across this membrane is generated by energy metabolism.

B. Potential Measurement by Indicator Distribution

Membrane potential can be estimated from the distribution of lipophilic cationic indicators or dyes between cells and the suspending medium. Lipophilicity, or hydrophobicity—that is, a high lipid:water partition coefficient—enables dye molecules to pass freely through the lipid portion of the membrane; the concentration gradient of a lipophilic cationic dye C^+ across the membrane is determined by the transmembrane potential difference according to the Nernst equation

$$[C^+]_i/[C^+]_o = e^{-F\Delta\Psi/RT},$$

where $[C^+]_i$ is the concentration of C^+ ions inside the cell, $[C^+]_o$ is the concentration of C^+ ions outside the cell, $\Delta\Psi$ is the membrane potential, R is the gas constant, T is the temperature in degrees Kelvin, and F is the Faraday. Indicators or dyes used in this fashion are referred to as distributional probes of membrane potential.

Once cells have been equilibrated with a cationic dye, a subsequent electrical depolarization of the cells (i.e., a decrease in membrane potential) will cause release of dye from cells into the medium, while a hyperpolarization (i.e., an increase in membrane potential) will make cells take up additional dye from the medium. The dye distribution will not adequately represent the new value of potential until equilibrium has again been reached; this process requires periods ranging from a few seconds to several minutes. Thus, while distributional probes may be suitable for detection of slow potential changes, they cannot be used to monitor the rapidly changing action potentials in excitable tissues such as nerve and muscle.

The use of cyanine dyes as MP probes began with the study of Hoffman and Laris (1974), who used 3,3'-dihexyloxacarbocyanine [DiOC$_6$(3) in the common notation introduced by Sims et al. 1974] to estimate membrane potential in red blood cell (RBC) suspensions based on the partitioning of the dye into the cells. They noted that addition of RBC to a micromolar dye solution in a spectrofluorometer cuvette produced a suspension with lower fluorescence than that of the original solution, indicating that—at these concentrations—the fluorescence of cyanine dyes taken into cells is quenched. When extracellular ion concentrations were manipulated so as to hyperpolarize the cells, increasing cellular uptake of dye, the fluorescence of the suspension decreased further; when the cells were depolarized, releasing dye into the medium, the fluorescence of the suspension increased. Membrane potential measurements of giant *Amphiuma* RBC using cyanine dyes were consistent with results obtained from microelectrode measurements.

In normal circumstances, the intracellular concentration of K^+ is considerably higher than the extracellular concentration, while the intracellular concentration of Na^+ is considerably lower than the extracellular concentration. Valinomycin (VMC), a lipophilic potassium-selective ionophore, forms complexes with K^+ ions and thus readily transports them across cell membranes, effectively increasing cells' potassium permeability to the point at which membrane potential is determined almost entirely by the transmembrane K^+ gradient. Addition of VMC hyperpolarizes cells if the external K^+ concentration is lower, and depolarizes them if it is higher, than the internal concentration. The ionophore gramicidin A, which forms transmembrane channels passing Na^+, K^+, and other ions, depolarizes cells in solutions with approximately physiologic ionic concentrations.

C. Single-Cell Measurements with Distributional Probes: Principles and Problems

Potential estimation by fluorometry of cell suspensions in cuvets requires dye concentrations sufficiently high that the fluorescence of intracellular dye is largely quenched, so that most of the measured fluorescence comes from free dye in solution. When cells are hyperpolarized, they take up more dye from the solution, decreasing total fluorescence. When cells are depolarized, dye molecules that were quenched when inside the cell are released into solution, increasing total fluorescence.

In cytometric estimation of MP, it is the fluorescence of intracellular dye which is measured. The dye concentrations used are lower than those used for bulk measurements, to minimize quenching of intracellular dye. In principle, MP could be calculated accurately from measurements of intracellular and extracellular dye concentrations using the Nernst equation (Ehrenberg *et al.*, 1988). However, the accuracy of the FCM procedure is limited, for several reasons.

The flow cytometer measures the amount, not the concentration, of dye in each cell. To find the intracellular concentration, it would be necessary to divide the fluorescence value for each cell by the cell's volume, obtained by an electronic (Coulter) volume sensor or estimated from forward scatter or extinction measurements. Measuring the extracellular concentration of dye is more problematic; this would require signal processing electronics of a type not normally used in flow cytometers. However, the large variances of fluorescence distributions obtained from conventional FCM measurements of cells at a known MP, even when cell size corrections are introduced (Shapiro, 1981; Seamer and Mandler, 1992), suggest that accuracy would not be significantly increased by the instrumental refinements just discussed.

One possible source of fluorescence variance is mitochondrial uptake of dye. Since the mitochondrial MP is typically ≥ 100 mV negative with respect to the cytosol, dye should be present in energized mitochondria at almost 100 times the concentration found in the cytoplasm. The mitochondrial MP, in fact, provides the basis for the accumulation in mitochondria of cationic dyes such as rhodamine 123 (Johnson et al., 1980, 1981; Darzynkiewicz et al., 1981), pinacyanol, and Janus green. However, the high concentration of dye in mitochondria should result in substantial quenching. Indeed, in at least some cell types, treatment with metabolic inhibitors and uncouplers which abolish the mitochondrial MP gradient, which should eliminate concentration of dye in mitochondria, reduces neither the intensity nor the variance of cyanine dye fluorescence.

In cells depolarized with gramicidin A, the MP should be zero; the intracellular and extracellular dye concentrations prediced by the Nernst equation should, therefore, be equal, and the very fact that the flow cytometer can detect the cells' fluorescence above background requires some explanation. One reason for the cells' increased fluorescence is that the fluorescence of cyanine dyes is enhanced [~6-fold in the case of $DiOC_6(3)$, less for other dyes] when the dye is in a hydrophobic environment (Sims et al., 1974) such as the membranous structures in which it can be observed in cells.

Perhaps even more important, the lipophilic, hydrophobic character of cyanine dyes causes them to be concentrated in cells even in the absence of a potential gradient. Similar problems with such "non-Nernstian" probe binding are also encountered when membrane potential is estimated with radiolabeled lipophilic cations such as [^{3}H]triphenylmethylphosphonium ($TPMP^+$) or with less hydrophobic cationic dyes (Ehrenberg et al., 1988). In order to get accurate values of cytoplasmic MP using these indicators, it is necessary to inhibit the mitochondria and to correct for probe uptake by cells in the absence of a potential gradient. The affinity of hydrophobic cyanine dyes for cells is sufficiently high that cells may take up most of the dye molecules even in a suspension in which the cells occupy only a fraction of a percent of the total volume. This makes it necessary to keep cell concentrations relatively constant from sample to sample in order to obtain reproducible results.

Flow cytometry of cells exposed to increasing concentrations of cyanine dyes shows that saturation occurs, i.e., that, eventually, further increasing the dye concentration does not increase fluorescence in the cells. For $DiOC_6(3)$, this happens when cells at a concentration of 10^6/ml are incubated with $2 \ \mu M$ dye.

The variance of the fluorescence distribution remains large. It is likely that most of the fluorescence measured in cells comes from dye in hydrophobic regions, which represent both the highest affinity binding sites and the sites in which dye fluorescence yield would be highest; the variance of fluorescence would then be explained by cell-to-cell variations in the number of binding sites. It has been observed (Shapiro, 1981) that, when dye binding sites are saturated, cellular fluorescence does not change when cells are depolarized or hyperpolarized by ionophores; however, such potential changes would be detectable by bulk fluorometry in a cell suspension exposed to the $2 \mu M$ saturating concentration of $DiOC_6(3)$. Thus, at this concentration, the cells contain dye which is essentially nonfluorescent due to quenching, as well as dye bound to the hydrophobic sites in which fluorescence is enhanced.

The dye concentration at which saturation of binding sites occurs is determined primarily by the hydrophobicity of the dye; the fluorescence of cells equilibrated (at 10^6/ml) with $2 \mu M$ diethyloxacarbocyanine [$DiOC_2(3)$], which is less hydrophobic than $DiOC_6(3)$, is less than the fluorescence of cells equilibrated with the C_6 dye and does change when MP is changed by ionophore addition or by manipulation of ion concentrations in the medium, indicating that the saturating concentration for the C_2 dye is higher than for the C_6 dye. When cells in $2 \mu M$ $DiOC_2(3)$ in NaCl (normal MP, higher fluorescence) are mixed with an equal volume of cells in $2 \mu M$ $DiOC_2(3)$ in KCl (depolarized, lower fluorescence), the cells and dye reequilibrate within a few minutes, yielding a fluorescence distribution that reflects the intermediate value of MP resulting from the ionic composition of the mixed suspending medium. Thus, the fluorescence of cell-associated cyanine dye can provide a reasonably rapid indication of substantial changes in MP, even if fluorescence variance limits overall accuracy and precision. Fluorescence variance also limits the capability of FCM measurements to detect heterogeneous responses within cell populations; two subpopulations of equal size with a mean difference of 50% in fluorescence intensity will be clearly resolved, while a 5% subpopulation with a mean 10% above the population mean might be undetectable.

II. Materials and Methods

A. Dyes and Reagents

1. Cyanine Dyes

The choice of dye is determined primarily by the excitation wavelength(s) available and the emission wavelengths desired. Dihexyl- or dipentyloxacarbocyanine [$DiOC_6(3)$ or $DiOC_5(3)$] or hexamethylindocarbocyanine [$DiIC_1(3)$] are both suitable for blue–green (488 nm) excitation. $DiIC_1(3)$ can also be used with green (515–546 nm) excitation. $DiOC_6(3)$ and $DiOC_5(3)$ are green fluorescent and can be used with the same detector/filter combination used for fluorescein;

$DiIC_1(3)$ is orange fluorescent and works well with filters designed for phycoerythrin. Hexamethylindodicarbocyanine [$DiIC_1(5)$; also known as HIDC] can be excited with a red (633) HeNe laser and measured through 665-nm glass filters using an R928 or other red-sensitive photomultiplier tube. The dyes are available from Molecular Probes (Eugene, OR) and other sources.

Stock solutions with a 1 mM concentration of any of the dyes mentioned in dimethyl sulfoxide (DMSO) may be kept for several weeks in the dark at room temperature (RT). Working solutions are made by diluting the stock solution with absolute ethanol or DMSO to allow the desired final dye concentration to be obtained by adding 5 μl of working solution to each 1 ml of cell suspension. For $DiOC_5(3)$ and $DiOC_6(3)$, a 10 μM working solution and a 50 nM final concentration are appropriate; for $DiIC_1(3)$ and $DiIC_1(5)$, a 20 μM working solution yields a 100 nM final concentration.

2. Ionophores

Valinomycin is used to hyperpolarize cells in high-Na^+, low-K^+ media and to depolarize cells in high-K^+ media. A 1 mM stock solution in DMSO is stable for several weeks at RT; adding 5 μl of stock solution to a 1-ml cell sample produces a 5 μM concentration of valinomycin.

Gramicidin D is used to depolarize cells; this material is a mixture containing a variable percentage of gramicidin A, which is the active ingredient. A stock solution of 1 mg/ml gramicidin D in DMSO is stable for several weeks at RT; addition of 5μl stock solution to 1 ml of cell suspension yields a final concentration of approximately 2 μM gramicidin A.

The proton ionophore carbonyl cyanide chlorophenylhydrazone (CCCP), an uncoupler and mitochondrial inhibitor, is an effective depolarizing agent for both aerobic and anaerobic bacteria. A final concentration of 5 μM is obtained by adding 5 μl of a 1 mM stock solution in DMSO to 1 ml of dilute bacterial suspension.

B. Staining Procedures

Cyanine dye equilibration with cells in protein-free media is usually complete after 15 min at RT. For cells in media containing protein, 30-min incubation at 37°C generally suffices. The cell concentration should be kept relatively constant from sample to sample, since the high affinity of the dye for cell constituents otherwise produces variations in fluorescence intensity per cell with cell concentration. Concentrations in the range of 10^6 cells/ml give satisfactory results. Cells are not washed prior to analysis.

C. Flow Cytometry and Data Analysis

The incubation temperature and the interval between dye addition and introduction of samples into the flow cytometer should be kept as nearly constant as possible; it is also advisable to run all samples in an experiment at the same flow rate to avoid artifacts due to dye diffusion between core and sheath and resultant changes in intracellular dye concentration.

While cyanine dye fluorescence signals are generally strong enough to be used as gating signals, it is preferable to use a forward scatter signal to avoid missing depolarized cells. Since MP is meaningful only in viable cells, although cyanine dyes will stain dead cells and debris, using the scatter signal for gating allows extraneous signals to be excluded during data collection. Pronounced changes in scatter signals following exposure of cells to some stimulus to be tested provide evidence that an apparent MP effect may be secondary to a more obvious change such as membrane lysis.

When using $DiOC_5(3)$ or $DiOC_6(3)$ as an MP probe, it is convenient to add propidium iodide (5–20 μg/ml final concentration) to the cell suspension, allowing dead (i.e., membrane-damaged) nucleated cells to be gated out of analyses on the basis of their strong red fluorescence.

The effects of many agents on MP of homogeneous cell populations may be appreciated by simple comparison of single-parameter fluorescence distributions such as those shown in Fig. 1. In other situations, e.g., observation of stimulated lymphocytes, multiparameter analysis that, for example, relates MP to nuclear DNA content in cells vitally stained with Hoechst 33342 (Shapiro, 1988) may be preferable. Rapid changes in MP are best appreciated when time is used as a measurement parameter.

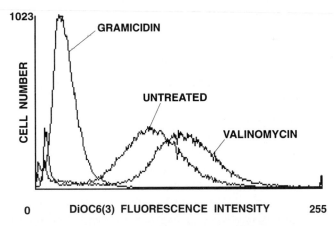

Fig. 1 Distributions of the fluorescence of $DiOC_6(3)$ in CCRF-CEM lymphoblasts in phosphate-buffered saline (pH 7.4), initially incubated with 50 nM dye for 15 min. Fluorescence histograms are shown for untreated cells, for cells hyperpolarized after an additional 10-min incubation with 5μM valinomycin, and for cells depolarized after a further 10-min exposure to 5 μg/ml gramicidin following the valinomycin treatment. Each histogram represents 25,000 cells; fluorescence measurements were made in a "Cytomutt" flow cytometer (Shapiro, 1988) with 21-mW excitation at 488 nm, using forward scatter signals for triggering.

D. Controls

When an experiment extends over a period of a few hours, control samples should be run during the course of the experiment as well as at the beginning and end. Controls should include an untreated cell sample, a sample of cells hyperpolarized by addition of 5–10 μM VMC, and another sample depolarized by addition of 5–10 μg/ml gramicidin.

A more dynamic control procedure, which verifies that cells are capable of response to hyper- and depolarizing stimuli, can be implemented by examining 1 aliquot of a sample of cells after they have equilibrated with dye, then adding valinomycin and analyzing a second aliquot after 5–10 min, and finally adding gramicidin and analyzing a third aliquot after an additional 5–10 min. The data shown in Fig. 1 are typical of the results of such a procedure. Valinomycin should increase the fluorescence of the cells; gramicidin should decrease the fluorescence well below the initial value (and, incidentally, render cells unresponsive to subsequent treatment with VMC or other hyperpolarizing agents).

While the large variance (CV values typically ~30%) of cyanine dye fluorescence distributions provides a strong argument against attempting to calibrate flow MP distributions to absolute voltages, the nature and magnitude of changes observed by different investigators in many cell systems have been fairly consistent. On several occasions, cyanine dye-stained lymphocytes from different donors, run on the same flow cytometer on successive days, have yielded distributions with peaks within 5% of one another. This suggests that both MP and the amount of dye binding sites per cell are relatively well regulated.

The control procedures just described make it possible to establish that the cells under study have a detectable MP and that the dye in use will respond to potential changes in either direction from the control value. All of this is prerequisite to any investigation of the effects of biologic, chemical, and/or physical agents on cell MP.

E. Pitfalls and Cautions

Addition of any substantial amount of protein to a cell suspension which has been equilibrated with dye in a protein-free medium will decrease the dye concentration in cells, because the dye will bind to the protein in solution. This artifactual apparent depolarization can be avoided by working in a medium with added protein (e.g., 1–10% albumin), when trying to determine effects of adding specific proteins to cells. Similar cautions apply to studies of the effects of solvents such as DMSO, which alter the partitioning of dye between cells and the medium. The total amount of added solvent should be maintained relatively constant from sample to sample and not exceed 2% of sample volume.

Other problems with cyanine dyes have been discussed at length elsewhere (Shapiro, 1988). At micromolar concentrations, cyanines have been observed to be toxic to bacteria and mammalian cells. The dyes themselves may perturb MP directly by altering membrane conductivity; their inhibition of energy metabolism might also result in potential changes. When used to monitor neutrophil responses to chemotactic peptides, $DiOC_6(3)$ and $TPMP^+$ were reported to give contradictory results, while the thiacyanine dye $DiSC_3(5)$ was found to be destroyed by oxidation following neutrophil activation.

Toxicity is a liability shared by the cyanines with other families of cationic dyes, such as acridines, safranins, oxazines, pyronins, rhodamines, and triaryl-methanes, and with lipophilic cations such as $TPMP^+$. When the dyes are used in flow cytometry, at concentrations of 5–100 nM, toxicity is less than when radiolabeled cations or dyes are used at micromolar concentrations for bulk measurements. Different cell types appear to have different degrees of suscepti-bility to cyanine dye toxicity. Crissman *et al.* (1988) found that simultaneous staining with $DiOC_5(3)$ improved Hoechst 33342 staining of live CHO cells, presumably by interfering sufficiently with energy metabolism to block efflux of the dye from cells via a pump mechanism. However, the cyanine dye did not affect cell viability following sorting, which remained ~90%. In this instance, at least, cyanine dye toxicity is evidently entirely reversible.

Oxonols, which are negatively charged, lipophilic dyes, bind to the cyto-plasmic membrane and some cellular constituents because of their lipophilicity, but are largely excluded from mitochondrial and cytoplasmic compartments because of their charge. This probably explains why oxonols are much less toxic than cyanines and other cationic dyes and why oxonol fluorescence is less affected by mitochondrial potential changes than are the fluorescence sig-nals from cationic dyes. These desirable characteristics of oxonols are offset somewhat by their weaker fluorescence, as compared to cyanines. As can be seen from Fig. 2, the variance of oxonol fluorescence distributions is no better than that of cyanine fluorescence distributions, and oxonol and cyanine dyes produce comparable results in most cases (e.g., Lazzari *et al.*, 1990).

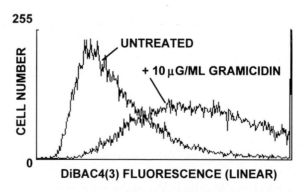

Fig. 2 Oxonol fluorescence distributions from CCRF-CEM cells. Cells in RPMI 1640 medium with 10% fetal calf serum were incubated for 15 min at RT with 100 nM bis-(1,3-dibutylbarbituric acid)trimethine oxonol [$DiBAC_4(3)$] with and without 10 μg/ml gramicidin; fluorescence was excited with 25 mW at 488 nm and measured at 525 nm using a 20-nm bandwidth interference filter, using forward scatter signals for triggering. Each histogram represents 20,000 cells.

F. Mitochondrial Staining with Rhodamine 123

The lipophilic cationic dye rhodamine 123 has been used for investigations of mitochondrial structure and function (Johnson *et al.*, 1980, 1981; Darzynkiewicz *et al.*, 1981); it accumulates in energized mitochondria as a result of their MP. Cells are equilibrated for 30 min with 10 μg/ml (~25 μM) dye, then washed, and examined; most of the retained dye is found in mitochondria. Safranin and the less-hydrophobic cyanine and styryl dyes show similar potential-dependent mitochondrial staining when cells are washed as they are for the rhodamine 123 procedure.

The glycoprotein efflux pump responsible for the multidrug resistance (MDR) phenotype in cells can transport a variety of neutral and positively charged compounds and dyes and, in some experimental situations (Kessel *et al.*, 1991), changes in pump function can be misinterpreted as changes in membrane potential and vice versa. Rhodamine 123 and cyanine dyes are now used for demonstration of MDR by FCM observation of loss of fluorescence from washed cells over time, and some observations in the literature relating low rhodamine 123 retention by cells to loss of mitochondrial activity may need to be reexamined in the light of these findings. Efflux pump activity has less effect on the fluorescence of cationic dyes when cells are kept in equilibrium with dye, as in the MP measurement techniques described here, than on fluorescence in washed cells. Since the anionic oxonol dyes are not substrates for the pump, oxonol fluorescence is not affected by its activity.

III. Improving Cytometry of Membrane Potentials

The limits of accuracy and precision with which cell MPs and changes can be measured by flow cytometry are due largely to the population variance of fluorescence intensity, presumably reflecting cell-to-cell differences in numbers of dye binding sites. This problem is largely eliminated when repeated measurements of the fluorescence of a cyanine dye or other distributional probe of MP are made of the same cell at intervals using a static low-resolution or image cytometer.

Even in the best circumstances, however, distributional probes could only be used to monitor relatively slow MP changes, occurring over periods of seconds to minutes. Better results would be expected using faster responding dyes, which sense MP by different mechanisms. A number of dyes developed for use by neurophysiologists respond to MP changes by changing their position and/or orientation in the membrane; most do not penetrate to the cell interior. Since all of the transmembrane potential difference is developed across the thickness of the membrane, the electric field strength in the membrane itself can be quite high. The fluorescence or absorption changes seen with fast response dyes are typically small, on the order of a few percent for a 60-mV change in potential, but such changes, which are difficult to detect reliably in a flow cytometer, are readily measured in a static apparatus.

At present, interest among investigators studying transmembrane signaling in eukaryotic cells has shifted from cytometry of MP to cytometry of intracellular pH and calcium. This is due in part to the relative inconstancy and variable magnitude of MP changes associated with surface ligand–receptor interactions, as compared to changes in calcium concentration and distribution and pH, and in part to better probe technology. Both pH and calcium probes suitable for ratiometric measurements of fluorescence using two excitation or emission wavelengths are now available; a ratiometric procedure eliminates the contribution of cell-to-cell differences in dye binding to measurement variance, yielding narrow distributions and making it easier to identify heterogeneity in populations.

In the particular area of lymphocyte activation, calcium flux measurements have proven more useful than MP measurements for analysis of changes occurring within the first minutes following stimulation by mitogens or antigens. Measurements of cytoplasmic or mitochondrial MP may be useful by 5–12 hr, when changes are sufficiently large to detect relatively small activated subpopulations, but measurements of the expression of early activation antigens can also be done at this time and do not require that the cells be maintained alive and in good condition until FCM analyses are done. Work by Waggoner and his colleagues (Hahn *et al.*, 1993) has, however, made it feasible to analyze mitochondrial MP in cells after fixation; these authors have developed a cyanine derivative which can be covalently bound to mitochondria by photo-induced crosslinking after live cells are incubated with dye and washed.

Bacterial MPs are primarily dependent upon metabolic activity, of which they are a sensitive indicator. Membrane potential changes substantially and rapidly in response to the availability or lack thereof of suitable energy sources and is rapidly dissipated when the organism is killed by drugs or other agents. Cyanine dye fluorescence in bacteria is illustrated in Fig. 3. It is possible to exploit potential-sensitive dyes in rapid FCM procedures for bacterial detection, identification, and antibiotic susceptibility testing (Shapiro, 1988). This application might be facilitated by the availability of ratiometric potential probes.

Loew and his co-workers have developed electrochromic probes of MP, which undergo spectral changes in responses to changes in the electric field in the membrane. One such dye, di-4-ANEPPS, has been used for dual-excitation beam ratiometric MP measurements in an image cytometer; it yields a 10% change in fluorescence ratio for a 90-mV potential change (Montana *et al.*, 1989). A flow cytometer with helium–cadmium (441 nm) and argon ion (515 nm) laser excitation could be used for di-4-ANEPPS measurements, but it is not clear that the precision needed to detect MP changes of a few tens of millivolts could be readily attained.

Chen and co-workers (Smiley *et al.*, 1991; Reers *et al.*, 1991) have described the fluorescence properties of the cyanine dye 5,5',6,6'-tetrachloro-1,1',3,3'-

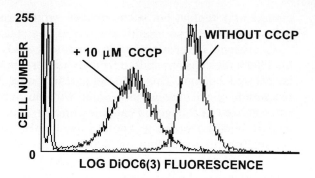

Fig. 3 Fluorescence distributions from *Staphylococcus aureus* in a Tris buffer containing glucose and EDTA. Cells were incubated for 2 min at RT with 50 nM DiOC6(3) with and without 10 μM CCCP, which depolarizes the bacteria. The instrument was set up as described in the legend to Fig. 2; fluorescence was measured on a logarithmic scale.

tetraethylbenzimidazolocarbocyanine iodide (JC-1), which emits green (527 nm) fluorescence in the monomeric form and orange (590 nm) when aggregates form. Although it has been suggested that JC-1 detects local differences in mitochondrial MP even within individual mitochondria, based on spectral differences in fluorescence emission, this dye is no less likely than other cyanines to be affected by local variations in the number and availability of binding sites.

Results of FCM measurements of JC-1 fluorescence in *Staphylococcus aureus* stained with 100 nM dye are shown in Fig. 4. When the organisms are depolar-

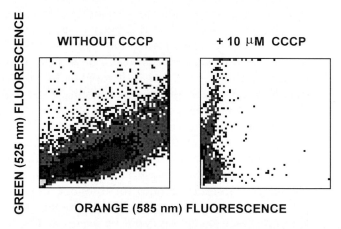

Fig. 4 Two-parameter distributions of orange (585 nm, 20 nm bandwidth) and green (525 nm, 20 nm bandwidth) from *Staphylococcus aureus* in a Tris buffer containing glucose and EDTA. Cells were incubated for 15 min at RT with 100 nM JC-1 (Molecular Probes), with and without 10 μM CCCP. Fluorescence was excited with 25 mW at 488 nm and measured on linear scales; the green fluorescence signal was used for triggering. Each distribution represents 20,000 cells.

ized with CCCP, orange fluorescence from the aggregates is greatly diminished, although green fluorescence is not changed significantly. The differences between untreated and CCCP-treated cells in Fig. 4 appear greater than those in Fig. 3, but this is due primarily to the use of a linear fluorescence scale in Fig. 4 rather than the logarithmic scale of Fig. 3. In mammalian cells, orange and green fluorescence from JC-1 are not as well correlated as in bacteria, presumably due to the influences of multiple (medium cytosol and cytosol mitochondrial interior) potential gradients as well as to variation in numbers of binding sites. Thus, the ultimate efficacy of JC-1 as a ratiometric probe remains to be determined.

Even if probes usable for ratiometric MP measurements become available, it is, obviously, more appropriate to assess effects of stimuli on cell MP by measuring the same cell at different times, as is done in a static system, than by measuring different cells at different times and assuming their behavior to be similar, a compromise forced on the experimenter by the nature of flow cytometry. It thus seems likely that further progress in understanding transmembrane signaling will come from the refinement of static cytometric techniques for measurement of MP and of other functional parameters such as intracellular pH, calcium flux, and redox state.

References

Crissman, H. A., Hofland, M. H., Stevenson, A. P., Wilder, M. E., and Tobey, R. A. (1988). *Exp. Cell Res.* **174,** 388–396.

Darzynkiewicz, Z. D., Staiano-Coico, L., and Melamed, M. R. (1981). *Proc. Natl. Acad. Sci. U. S. A.* **78,** 6696–6698.

Ehrenberg, B., Montana, V., Wei, M.-D., Wuskell, J. P., and Loew, L. M. (1988). *Biophys. J.* **53,** 785–794.

Hahn, K. M., Conrad, P. A., Chao, J. C., Taylor, D. L., and Waggoner, A. S. (1993). *J. Histochem. Cytochem.* **41,** 631–634.

Hoffman, J. F., and Laris, P. C. (1974). *J. Physiol.* (*London*) **239,** 519–552.

Johnson, L. V., Walsh, M. L., and Chen, L. B. (1980). *Proc. Natl. Acad. Sci. U. S. A.* **77,** 990–994.

Johnson, L. V., Walsh, M. L., Bockus, B. J., and Chen, L. B. (1981). *J. Cell Biol.* **88,** 526–535.

Kessel, D., Beck, W. T., Kukuraga, D., and Schulz, V. (1991). *Cancer Res.* **51,** 4665–4670.

Lazzari, K. G., Proto, P., and Simons, E. R. (1990). *J. Biol. Chem.* **265,** 10959–10967.

Montana, V., Farkas, D. O., and Loew, L. M. (1989). *Biochemistry* **28,** 4536–4539.

Reers, M., Smith, T. W., and Chen, L. B. (1991). *Biochemistry* **30,** 4480–4486.

Seamer, L. C., and Mandler, R. N. (1992). *Cytometry* **13,** 545–552.

Shapiro, H. M. (1981). *Cytometry* **1,** 301–312.

Shapiro, H. M. (1982). U. S. Pat. 4,343,782.

Shapiro, H. M. (1988). "Practical Flow Cytometry," 2 Ed. Wiley-Liss, New York.

Shapiro, H. M., Natale, P. J., and Kamentsky, L. A. (1979). *Proc. Natl. Acad. Sci. U. S. A.* **76,** 5728–5730.

Sims, P. J., Waggoner, A. S., Wang, C.-H., and Hoffman, J. F. (1974). *Biochemistry* **13,** 3315–3330.

Smiley, S. T., Reers, M., Mottola-Hartshorn, C., Lin, M., Chen, A., Smith, T. W., Steele, G. D. Jr., and Chen, L. B. (1991). *Proc. Natl. Acad. Sci. U. S. A.* **88,** 3671–3675.

CHAPTER 9

Measurement of Intracellular pH

Michael J. Boyer* and David W. Hedley†

*Department of Medical Oncology
Royal Prince Alfred Hospital
Sydney, 2050 New South Wales
Australia
†Departments of Medicine and Pathology
Ontario Cancer Institute/Princess Margaret Hospital
Toronto, Ontario, Canada M4X 1K9

I. Introduction

A. Maintenance of Intracellular pH

The level of intracellular pH (pH_i) is of considerable importance to the viability and normal function of mammalian cells. The passive diffusion of ions based on their electrochemical gradients is predicted by the Donnan equilibrium; if these calculations are performed for H^+, the predicted level of pH_i when extracellular pH (pH_e) is 7.4 is approximately 6.4, considerably lower than the measured pH_i under these conditions. It follows, therefore, that mechanisms exist which regulate pH_i and maintain it at a level above that predicted by the electrochemical gradient.

Two major mechanisms allow cells to regulate their pH_i. These are the buffering capacity of the cytosolic and organellar contents and membrane-based transport systems, including the Na^+/H^+ exchanger and anion exchangers. The buffering capacity of a cell is its ability to buffer a change in pH_i following the addition (or removal) of H^+ and is comprised of both bicarbonate-dependent and nonbicarbonate (mainly protein) components (Roos and Boron, 1981; Boron, 1989). Intracellular buffering provides substantial protection for the cell against the effects of an acid load, with most cells capable of buffering millimolar concentrations of H^+ (compared to the submicromolar concentrations that are normally present) (Roos and Boron, 1981).

The Na^+/H^+ exchanger is a membrane-based transport mechanism that is ubiquitous in mammalian cells. The exchanger is a 110-kDa protein whose gene has been cloned from several tissues in different species (Sardet *et al.*, 1989; Reilly *et al.*, 1991; Hildebrandt *et al.*, 1991; Tse *et al.*, 1991; Fliegel *et al.*, 1991). It uses the inwardly directed Na^+ gradient to pump H^+ out of cells, with a 1 : 1 stoichiometry. Its actions are not directly dependent on the utilization of ATP, although energy is required to maintain the 10-fold inwardly directed Na^+ gradient, which allows the pump to operate. The Na^+/H^+ antiport is reversibly inhibited by amiloride and its more potent analogues such as ethylisopropylamiloride and hexamethylene amiloride (HMA) (Grinstein and Rothstein, 1986, L'Allemain *et al.*, 1984).

At levels of pH_i close to the normal resting value, the Na^+/H^+ exchanger is relatively quiescent. It becomes activated as pH_i falls below ~7.0, and there is a rapid increase in its rate of activity at levels of pH between 7.0 and 6.6. Its activity is also modulated by several other stimuli including hormones, mitogens, and chronic exposure of cells to acidic environments (Horie *et al.*, 1990).

Anion exchange also contributes to regulation of pH_i and at levels of pH_i close to neutrality, it may be more active than Na^+/H^+ exchange in some cell lines. Two different anion exchangers exist, the cation-independent Cl^-/HCO_3^- exchanger and the Na^+-dependent Cl^-/HCO_3^- exchanger. Under physiological conditions, the cation-independent Cl^-/HCO_3^- exchanger allows Cl^- ions to enter the cell in exchange for HCO_3^- ions and therefore acts to decrease pH_i following cytoplasmic alkalination (Cassel *et al.*, 1988; Reinertsen *et al.*, 1988). By contrast, the Na^+-dependent Cl^-/HCO_3^- exchanger acts to protect cells from cytoplasmic acidification (Cassel *et al.*, 1988; Reinertsen *et al.*, 1988). Both of these exchangers are inhibited by the stilbene derivative DIDS.

B. Measurement of pH$_i$

1. Fluorescent pH$_i$ Probes

The measurement of pH$_i$ using fluorescent probes generated *in situ* was first carried out by Thomas *et al.*, using fluorescein and carboxyfluorescein diacetate (Thomas *et al.*, 1979). These early pH$_i$ probes suffered from considerable limitations including difficulties with cellular uptake, rapid leakage out of cells, and pK$_a$ that was too acidic to make accurate measurements of pH$_i$ at values close to neutrality. In addition to the flourescein derivatives, 2,3-dicyanohydroquinone (DCH) (Fig. 1) has been used to measure pH$_i$ using a ratiometric technique (Valet *et al.*, 1981; Musgrove *et al.*, 1986). DCH has a shift in emission wavelength and a decrease in emission intensity with increasing pH. Use of this probe, however, is associated with significant problems including the need for UV excitation, rapid leakage from cells, and the fact that it enters organelles, so that the values of pH$_i$ obtained include a component of organellar pH. Finally, its pK of 8.0 is not ideal for measurements of pH$_i$ in mammalian cells.

A derivative of carboxyfluorescein, biscarboxyethylcarboxyfluorescein (BCECF) (Fig. 1) overcomes many of these problems. It is well retained within cells for up to 2 hr (Musgrove *et al.*, 1986) and has a pK$_a$ of 6.98. After uptake into cells as the acetoxymethyl (AM) ester, BCECF is confined to the cytoplasmic compartment (Paradiso *et al.*, 1984), providing a measure of the pH of this compartment only and not some average value that includes organelles. Although originally introduced for use in fluorometers, BCECF has been used widely in the flow cytometric measurement of pH$_i$.

The acetoxymethyl ester of BCECF (BCECF-AM) is a nonfluorescent molecule that enters cells easily. Once within the cytoplasm the AM groups are cleaved by the action of nonspecific esterases, yielding the highly fluorescent molecule BCECF. Following excitation at 480–500 nm, the emission intensity of BCECF at 525–535 nm is pH dependent with greater intensity at higher pH. The excitation source can be either an argon laser or a mercury arc lamp. In order to make measurements of pH$_i$, a ratio is usually taken between a pH-dependent emission intensity (e.g., 525 nm) and a pH-independent emission intensity (e.g., 640 nm). The value obtained is independent of such factors as the amount of dye that was loaded into or leaked from cells.

New probes have become available for the measurement of pH$_i$. The carboxyseminaphthofluoresceins (SNAFLs) and the carboxyseminaphthorhodafluors (SNARFs) (Fig. 1) were designed to be long wavelength, dual-emission pH indicators with a pK$_a$ near neutrality. The most promising of these for flow cytometry is SNARF-1 which has a pK$_a$ of 7.4 (Haugland, 1992). They are loaded into cells as AM esters in a manner analogous to BCECF. These probes have dual emission wavelengths, one (587 nm for SNARF-1) which is maximal under acidic conditions and the other (636 nm for SNARF-1) which is maximal under alkaline conditions. This provides the theoretical advantage of increased sensitivity.

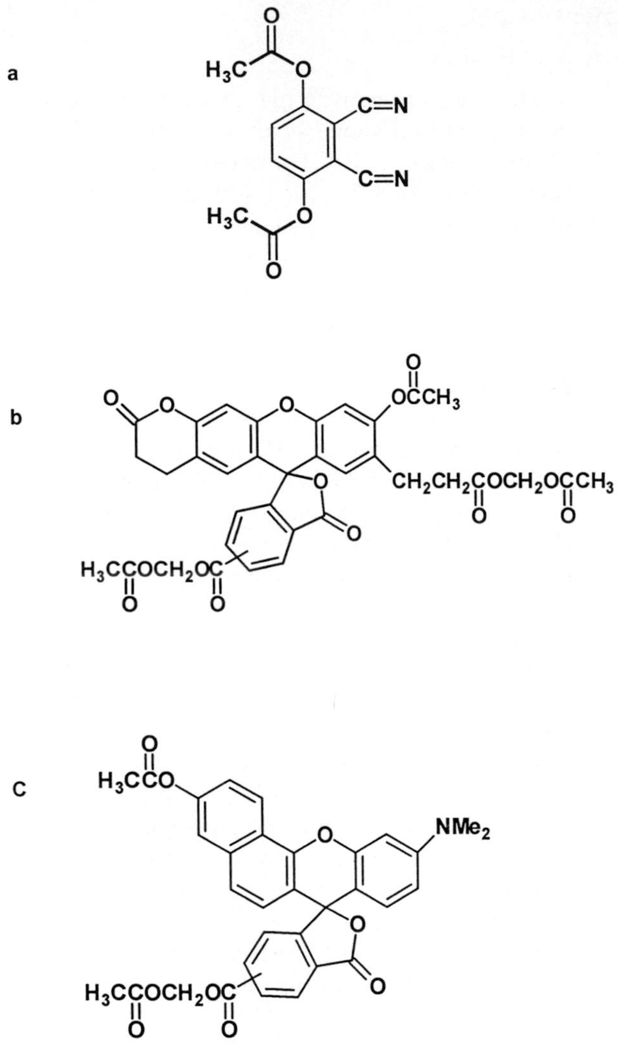

Fig. 1 Structures of (a) 1,4-diacetoxy-2,3-dicyanobenzene (the esterified form of 2,3-dicyano-hydroquinone), (b) BCECF-AM, and (c) carboxy-SNARF-1.

2. Calibration

Calibration of fluorescence ratio to pH is carried out by the use of the H^+/K^+ ionophore, nigericin. Nigericin causes the ratios of intracellular to extracellular hydrogen ion concentration ($[H^+]_i$ and $[H^+]_e$, respectively) to be equal to the ratios of intracellular to extracellular potassium ion concentration ($[K^+]_i$ and $[K^+]_e$)

$$[K^+]_i/[K^+]_e = [H^+]_i/[H^+]_e$$

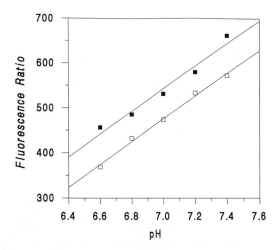

Fig. 2 Calibration curves obtained with BCECF and carboxy-SNARF-1, in EMT-6 cells, using a Coulter Epics Elite cytometer.

(Thomas *et al.*, 1979). If $[K^+]_i$ and $[K^+]_e$ are equal, then $[H^+]_i$ will be equal to $[H^+]_e$, and hence pH$_i$ can be estimated simply by measureing pH$_e$. The accuracy of this method of calibration depends on the equality of intra- and extracellular potassium concentrations which typically are not determined experimentally but are assumed to be of the order of 130–140 mM. This is a valid assumption for most cell types, and the method gives values of pH$_i$ which are close to those obtained using other techniques. However, if experiments are to be performed with cells whose $[K^+]_i$ is variable (e.g., excitable cells) or under conditions which might alter $[K^+]_i$, the value must be measured and the appropriate concentrations used for calibration.

3. Measurement Accuracy

The sensitivity of measurements of pH$_i$ is the ability to detect differences in the mean of a population distributed normally in pH$_i$. The theoretical limit of sensitivity is determined by the magnitude of difference in pH$_i$ represented by one channel number (i.e., the slope of the calibration curve). Calibration curves with BCECF obtained in this laboratory typically have slopes of ~250 channels per pH unit. Although SNARF-1 would be expected to have greater sensitivity than BCECF, because it has an obvious isosbestic point, in practice we have found it to have identical sensitivity (see calibration curves in Fig. 2). These values correspond to a maximum sensitivity of ~0.004 pH units.

The maximum sensitivity described above is not achieved in practice because of lack of resolution. The resolution of measurements of pH$_i$ is the ability to resolve subpopulations of different pH$_i$ within the measured sample and can be estimated by examining the overlap between the ratio histograms of populations differing in pH_i. The breadth of the ratio histogram is usually measured as a coefficient of variation (CV). The CV is a measure of the cell-to-cell variation in intracellular pH, together with variation introduced by the instrument and uncertainties inherent to the technique. The cell-to-cell variation can be virtually eliminated by the use of nigericin in high $[K^+]$ buffer, allowing the resolution of the technique to be assessed.

II. Application

These techniques are widely applicable and have been used to measure pH_i in cell lines, as well as in cells derived from multicellular spheriods and solid tumors. In addition to their use to measure steady-state values of pH_i, these techniques can be adapted to measure the effect of various stimuli on pH_i. They have also been employed to measure the operation of the Na^+/H^+ exchanger. The major limitation to the use of fluorescent pH probes is that some cell lines fail to retain the probes, although this is a relatively uncommon problem.

III. Materials

A. Chemicals

1. BCECF

BCECF-AM is available from Molecular Probes (Eugene, OR). Stock solutions at a concentration of 1 mg/ml in dimethyl sulfoxide are stored at $-20°C$.

2. SNARF

cSNARF-AM is available from Molecular Probes (Eugene, OR). Stock solutions at a concentration of 1 mM in dimethyl sulfoxide are stored at $-20°C$.

3. DCH

The esterified form of DCH, 1,4-diacetoxy-2,3-dicyanobenzene (ADB) is obtained from Boehringer–Mannheim (Indianapolis, IN). Stock solutions are made at 2 μl/ml in anhydrous dimethyl formamide and stored at $-20°C$.

4. Other Chemicals

Nigericin is available from Sigma Chemical Co. (St. Louis, MO). A stock solution of 1 mg/ml in absolute ethanol is stored at $-20°C$. Hexamethylene amiloride is available from Research Biochemicals, Inc. (Natick, MA) and a stock solution at a concentration of 1 mM in dimethyl sulfoxide is stored at 4°C.

B. Buffers

1. Physiological Saline Solution (PSS)

140 mM NaCl

5 mM KCl

5 mM glucose

1 mM CaCl$_2$

1 mM MgCl$_2$

20 mM Tris/MES

Adjust to pH 7.2–7.4 by mixing appropriate amounts of MES-buffered and Tris-buffered solution.

2. NMG Buffer

NMG buffer is identical to PSS, but with isoosmotic replacement of NaCl with N-methyl D-glucamine (NMG). In preparing this buffer, the NMG should be dissolved first and the pH adjusted to ~7.4 with 10 N HCl. Other components can be added, and the final pH adjusted as described above.

3. NH$_4$ Buffer

NH$_4$ buffer is prepared as NMG buffer, but with 130 mM NMG and 10 mM NH$_4$Cl.

4. High [K$^+$] Buffer

Solution 1:

130 mM KH$_2$PO$_4$

20 mM NaCl

Solution 2:

110 mM K$_2$HPO$_4$

20 mM NaCl

Mix solutions 1 and 2 to give buffers with a range of pH between 6.0 and 7.5.

IV. Cell Preparation and Staining: BCECF or SNARF

Following harvest, cells are rinsed once with PSS and then resuspended in the same solution at their final concentration (usually 10^6 cells/ml).

Add BCECF-AM to give a final concentration of 2 μl/ml (add 2 μl stock solution per 1 ml of cells + buffer). Phosphate-buffered saline or serum-free medium can be used in place of PSS; serum must not be present at this stage since the activity of esterases in serum will cleave BCECF-AM extracellularly, resulting in poor loading.

Incubate cells at 37°C for 20–30 min to allow cleavage of the AM ester. Although shorter periods can be used because the method is independent of the amount of dye loaded into the cells, we have obtained better results with incubations of this duration.

Remove aliquots of 10^6 cells and centrifuge. Discard the supernatant and resuspend the pellet (which will appear yellow) in 1 ml of PSS (or in another test solution as appropriate). For calibration samples, the pellet should be resuspended in high [K^+] buffer of differing pH. Nigericin 1 μl/ml (1 μl stock solution per 1 ml of cells + buffer) should be added 2–3 min prior to measurement of pH_i.

The procedure for SNARF is identical except that SNARF-AM is added rather than BCECF-AM. The final concentration of SNARF should be 5 μM (5 μl stock per 1 ml of cells + buffer). Cleavage of SNARF-AM seems to take a little longer than that of BCECF-AM so a minimum incubation of 30 min should be used. The cell pellet will appear pink.

V. Instruments

A flow cytometer capable of generating a ratio signal either in hardware or in software is needed to carry out these measurements. In addition, since measurements of the activity of the Na^+/H^+ exchanger is actually a measurement of rate, time becomes a component of these assays. This is most conveniently done on flow cytometers that are able to use time as a parameter, although it is possible to carry out these experiments by using multiple samples, if the software is not capable of collecting events over time.

A. Measurement of Steady-State pH_i

1. BCECF

Excitation of BCECF is provided by the 488-nm line of an argon laser. When used with an Epics Elite cytometer, power as low as 15–20 mW is adequate for excitation. The resulting fluorescence is separated into high- and low-wavelength components by a 550-nm dichroic filter. These components are further narrowed by passing through a 640-nm band pass filter (10 nm bandwidth) and 525-nm band pass filter (20 nm bandwidth) respectively. The ratio of 525/640 nm fluorescence is measured; this ratio increases with increasing pH_i (Fig. 2).

Measurements of pH$_i$ should be possible with a benchtop flow cytometer equipped with filters for both fluorescein and red wavelengths, provided that the ratio of these two signals can be obtained. However, we have not performed such measurements ourselves.

2. SNARF

As with BCECF, excitation of SNARF is provided by the 488-nm line of an argon laser, at a power of 20–25 mW. The emitted light is passed through a 625 dichroic filter, and the resultant beams are narrowed by passage through 640-nm band pass (10 nm bandwidth) and 580-nm band pass (10 nm bandwidth) filters. The ratio of 640/580 nm fluorescence is measured; this ratio increases with increasing pH (Fig. 2).

3. DCH

Although DCH has been largely superseded by more recent pH$_i$ indicators capable of excitation at 488 nm, the following procedure based on Cook and Fox (1988) can be used.

Cells are loaded with the esterified form of DCH, ADB. This is obtained from Boehringer-Mannheim and made up at 2 mg/ml in anhydrous dimethyl formamide. Add 5 μl to each milliliter of cells in buffer, giving a final concentration of 10 μg/ml (41 μM). Incubate at room temperature for 20 min, and then run on flow cytometer immediately. Instrument setup requires UV excitation, using either a mercury arc or the UV doublet of an argon ion laser, and band pass filters centered around 430 and 480 nm, separated by a suitable dichroic mirror. Calibration is as for BCECF or SNARF-1.

B. Measurement of Na$^+$/H$^+$ Exchange Activity

In order to make these measurements, a flow cytometer which allows the measurement of time as a parameter is desirable (though not essential). Cells are loaded with either BCECF or SNARF-1 using the procedure described above, but NMG buffer is substituted for PSS. Three minutes prior to the end of loading with fluorochrome, nigericin, at a final concentration of 2 μg/ml, is added (2 μl of stock solution per 1 ml of cells + buffer). This will produce intracellular acidification to a level of ~6.5 in most cell lines; the absence of extracellular Na$^+$ prevents activity of the Na$^+$/H$^+$ exchanger. Following the conclusion of incubation, the sample is centrifuged and resuspended in a small volume (100–200 μl) of NMG buffer. The action of the Na$^+$/H$^+$ exchanger is commenced by the addition of 1 ml of PSS, at which time measurement should begin, using the technique described above. Inhibition of the exchanger can be produced by adding HMA at a final concentration of 1–10 μM.

C. Measurement of Intracellular Buffering Capacity

Cells are loaded with fluorochrome as described above. Following centrifugation, the pellet is resuspended in 1 ml NMG buffer, and measurement of pH_i is commenced. After a sufficient number of cells have been measured, and (ideally) while the sample is still running, 1 ml of NH_4 buffer is added to the sample. A boost should be applied to the sample so that measurement of pH_i continues immediately after the addition of NH_4. Continue making measurements until sufficient events have been collected, but not for longer than 1 min.

Intrinsic (nonbicarbonate) buffering (β_I) capacity is given by the formula $\beta_I = \Delta[NH_4^+]_i/\Delta pH_i$, where $\Delta[NH_4^+]_i$ and ΔpH_i are the changes which occur in intracellular $[NH_4^+]$ and intracellular pH, respectively, following the addition of NH_4 buffer. The change in pH_i is measured. If it is assumed that $[NH_4^+]_i$ prior to the addition of NH_4 buffer is 0, then $\Delta[NH_4^+]_i$ is simply the value of $[NH_4^+]_i$ after the change in buffer and is given by the equation (Roos and Boron, 1981)

$$[NH_4^+]_i = ([NH_4Cl]_e \cdot 10^{pH_e - pK})/((1 + 10^{(pH_e - pK)}) \cdot 10^{(pH_i - pK)}).$$

The pK of NH_4^+ is 9.3.

D. Calibration

Cells are loaded with BCECF or SNARF in PSS, as described above. At the conclusion of incubation, the cell samples are centrifuged and the pellet is resuspended in high $[K^+]$ buffers at several different levels of pH. At least four high $[K^+]$ calibration solutions should be prepared with pH increasing in 0.2–0.3 pH unit increments; ideally these should have values of pH centered around the values expected to be measured in the experimental samples. Following addition of nigericin (to equilibrate pH_i and pH_e) fluorescence is measured.

The calibration curve is obtained by plotting the mean fluorescence ratio of samples measured in high $[K^+]$ buffer and nigericin against pH (Fig. 2). At levels of pH_i close to the pK_a of the probe, the calibration curve is linear, and its equation may be obtained by linear regression. Outside this range, the calibration curve is sigmoidal. Although computerized curve-fitting programs are available which will provide the equation of such a curve, the appropriate selection of pH probes based on knowledge of their pK can prevent this problem.

VI. Critical Aspects

The success of techniques to measure pH_i depends upon obtaining a fluorescence signal of adequate strength. The major causes of weak signal strength are poor intracellular loading or rapid leakage of the fluorescent probe. Poor

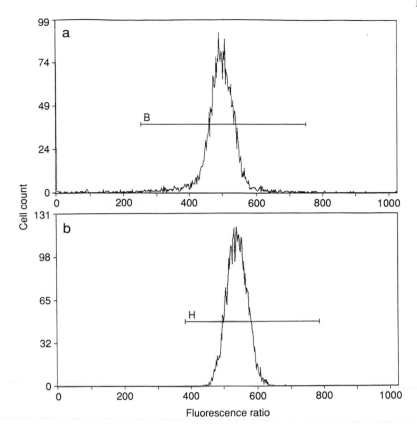

Fig. 3(a and b) Fluorescence ratio histograms for a population of EMT-6 cells, suspended in physiological saline buffer at extracellular pH 7.30, obtained with the use of (a) BCECF or (b) carboxy-SNARF-1.

loading of esterified probes may be due to extracellular deesterification which may occur if serum (which may contain esterases) is present in the buffer during the exposure of cells to the probes. The use of serum-free buffers should overcome this problem. Another cause of a weak signal is poor intracellular hydrolysis of esterified probes if experiments are carried out at temperatures below 37°C. If it is necessary to carry out experiments at low temperature, increasing the concentration of BCECF-AM or SNARF-AM may overcome the problems of slow or poor loading.

The accuracy of the values of pH_i obtained is dependent upon the calibration procedure. As pointed out above, calibration is based on the assumption that intra- and extracellular $[K^+]$ are equal. Under some conditions (e.g., in excitable cells, or in the presence of inhibitors of ion transporter) this assumption may not be valid, and use of the calibration procedure as described will result in a

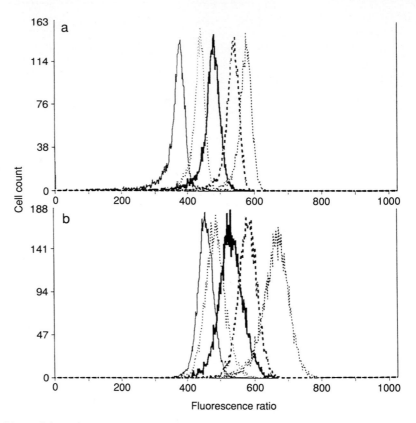

Fig. 4(a and b) Fluorescence ratio histograms from which the calibration curve in Fig. 2 is constructed. The peaks in each figure (from left to right) are at pH 6.6, 6.8, 7.0, 7.2, and 7.4, and are shown for (a) BCECF and (b) carboxy-SNARF-1.

systematic error in the values of pH_i. This difficulty can be overcome by measuring the intracellular concentration of $[K^+]$ and adjusting the calibration buffers appropriately.

The methods described in this chapter outline the measurement of pH_i under bicarbonate-free conditions. The presence of bicarbonate may result in changes in the steady-state level of pH_i as a result of the activation of additional pH_i regulating systems (Ganz *et al.*, 1989). Whether measurements of pH_i should be made in the presence of bicarbonate will depend on the purpose of the experiment; if an estimate of pH_i is desired under physiological conditions, experiments should be carried out with bicarbonate. In order to carry out experiments in the presence of bicarbonate, the composition of buffers must be altered accordingly. This will also have an impact on buffering capacity (Boron, 1989). Care should be taken when working with bicarbonate containing

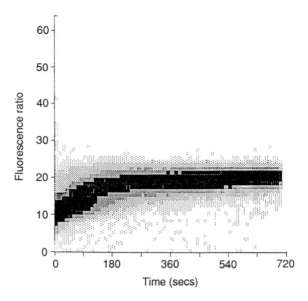

Fig. 5 Change in fluorescence ratio (which is proportional to intracellular pH) in acidified EMT-6 cells following the addition of Na^+. The slope of the curve is a measure of the activity of the Na^+/H^+ exchanger.

solutions to prevent its loss into the atmosphere, thus lowering the concentration of bicarbonate in the buffer.

VII. Results and Discussion

The fluorescence ratio histogram shown in Fig. 3a is from a population of EMT-6 cells stained with BCECF and suspended in PSS. The mean pH₁ (in the absence of bicarbonate) is 7.07 and the CV is 4.12%. These are typical values for mammalian cells, although under different environmental conditions (e.g., reduced extracellular pH) different values are obtained. The use of SNARF-1 to make the same measurements is shown in Fig. 3b; pH₁ is 6.99 and the CV is 4.39%.

Figure 4 shows histograms obtained during the construction of a calibration curve with BCECF (a), SNARF-1 (b); the calibration curve is shown in Fig. 2. Two points are of note. First, the slope of the calibration curves for BCECF and SNARF-1 are almost identical (253 and 254 channels/pH unit, respectively). Based on these data, the maximum sensitivity of the measurements are of the order of 0.003 pH units. This compares well with previously reported values. Second, the decrease in CV due to nigericin is apparent (compare with Figs. 3a and 3b). The resolution based on these data is dependent on the level of pH. The resolution with BCECF for values of pH_i below 7.0 is ~0.1 pH units, while for values above 7.0 it is ~0.2 pH units. For SNARF-1, the resolution above pH_i 7.0 is ~0.1 pH units, while below 7.0 it is ~0.2 pH units. Therefore, BCECF appears to be the probe of choice for when values of pH_i below 7.0 are being measured, while above this level, SNARF-1 is a better choice.

An application of the technique is shown in Fig. 5. This is a plot of fluorescence ratio (pH_i) against time for EMT-6 cells following experimental acidification in Na^+-free medium and then recovery as a result of the activity of the Na^+/H^+ exchanger following addition of Na^+. The slope of the curve is a measure of the activity of this exchanger. The slope can be obtained by measuring the mean fluorescence ratio during several time intervals.

References

Boron, W. F. (1989). *In* "The Regulation of Acid-Base Balance" (D. W. Seldin and G. Giebisch, eds.), pp. 33–56. Raven Press, New York.

Cassel, D., Scharf, O., Rotman, M., Cragoe, E. J., Jr., and Katz, M. (1988). *J. Biol. Chem.* **263**, 6122–6127.

Cook, J. A., and Fox, M. H. (1988). *Cytometry* **9**, 441–447.

Fliegel, L., Sardet, C., Pouyssegur, J., and Barr, A. (1991). *FEBS Lett.* **279**, 25–29.

Ganz, M. B., Boyarsky, G., Sterzel, R. B., and Boron, W. F. (1989). *Nature (London)* **337**, 648–651.

Grinstein, S., and Rothstein, A. (1986). *J. Membr. Biol.* **90**, 1–12.

Haugland, R. P. (1992). "Handbook of Fluorescent Probes and Research Chemicals," 2nd ed. Molecular Probes, Eugene, OR.

Hildebrandt, F., Pizzonia, J. H., Reilly, R. F., Reboucas, N. A., Sardet, C., Pouyssegur, J., Slayman, C. W., Aronson, P. S., and Igarashi, P. (1991). *Biochim. Biophys. Acta* **1129**, 105–108.

Horie, S., Moe, O., Tejedor, A., and Alpern, R. J. (1990). *Proc. Natl. Acad. Sci. U.S.A.* **87**, 4742–4745.

L'Allemain, G., Franchi, A., Cragoe, E. J., Jr., and Pouyssegur, J. (1984). *J. Biol. Chem.* **259**, 4313–4319.

Musgrove, E., Rugg, C., and Hedley, D. (1987). *Cytometry* **7**, 347–355.

Paradiso, A. M., Tsien, R. Y., and Machen, T. E. (1984). *Proc. Natl. Acad. Sci. U.S.A.* **81**, 7436–7440.

Reilly, R. F., Hildebrandt, F., Biemesderfer, D., Sardet, C., Pouyssegur, J., Aronson, P. S., Slayman, C. W., and Igarashi, P. (1991). *Am. J. Physiol.* **261**, F1088–F1094.

Reinertsen, K. V., Tonnessen, T. I., Jacobsen, J., Sandvig, K., and Olsnes, S. (1988). *J. Biol. Chem.* **263**, 11117–11125.

Roos, A., and Boron, W. F. (1981). *Physiol. Rev.* **61**, 296–434.

Sardet, C., Franchi, A., and Pouyssegur, J. (1989). *Cell (Cambridge, Mass.)* **56**, 271–280.

Thomas, J. A., Buchsbaum, R. N., Zimniak, A., and Racker, E. (1979). *Biochemistry* **18**, 2210–2218.

Tse, C. M., Ma, A. I., Yang, V. W., Watson, A. J., Levine, S., Montrose, M. H., Potter, J., Sardet, C., Pouyssegur, J., and Donowitz, M. (1991). *EMBO J.* **10**, 1957–1967.

Valet, G., Raffael, A., Moroder, L., Wunsch, E., and Ruhenstroth-Bauer, G. (1981). *Naturwissenschaften* **68**, 265–266.

CHAPTER 10

Intracellular Ionized Calcium

Carl H. June* and Peter S. Rabinovitch[†]

* Department of Immunobiology
Naval Medical Research Institute
Bethesda, Maryland 20889
† Department of Pathology
University of Washington
Seattle, Washington 98121

I. Introduction

Measurement of intracellular ionized calcium concentration ($[Ca^{2+}]_i$) in living cells is of considerable interest to investigators over a broad range of cell biology. Calcium has an important role in a number of cellular functions and, perhaps most interestingly, can transmit information from the cell membrane to regulate diverse cellular functions. An optimal indicator of $[Ca^{2+}]_i$ should span the range of calcium concentrations found in quiescent cells ($\sim$100 nM) to levels measured in stimulated cells (micromolar free Ca^{2+}), with greatest sensitivity to small changes at the lower end of that range. The indicator should freely diffuse throughout the cytoplasm and be easily loaded into small cells. The response of the indicator to transient changes in $[Ca^{2+}]_i$ should be rapid. Finally, the indicator itself should have little or no effect upon $[Ca^{2+}]_i$ itself or on other cellular functions.

Until 1982 it was not possible to measure $[Ca^{2+}]_i$ in small intact cells, and attempts to measure cytosolic free calcium were restricted mostly to large invertebrate cells where the use of microelectrodes was possible. Bioluminescent indicators such as aequorin, a calcium-sensitive photoprotein, are well suited for certain applications (Blinks et al., 1982). Their greatest limitation is the necessity for loading into cells by microinjection or other forms of plasma membrane disruption. $[Ca^{2+}]_i$ was first measured in populations of small nonadherent cells with the development of quin2 (Tsien et al., 1982). The indicator was easily loaded into intact cells using a chemical technique developed by Tsien (1981). Cells are incubated in the presence of the acetoxymethyl ester of quin2. This uncharged form is cell permeant and diffuses freely into the cytoplasm where it serves as a substrate for esterases. Hydrolysis releases the tetraanionic form of the dye which is trapped inside the cell. Unfortunately, quin2 has several disadvantages that limit its application to flow cytometry (Ransom et al., 1986). A relatively low extinction coefficient and quantum yield have made detection of the dye at low concentrations difficult; at higher concentrations, quin2 itself buffers the $[Ca^{2+}]_i$. Grynkiewicz et al. (1985) have described a family of highly fluorescent calcium chelators which overcome most of the above limitations. One of these dyes, indo-1, has spectral properties which make it especially useful for analysis with flow cytometry. In particular, indo-1 exhibits large changes in fluorescent emission wavelength upon calcium binding (Fig. 1). As described below, use of the ratio of intensities of fluorescence at two wavelengths allows calculation of $[Ca^{2+}]_i$ to be made independent of variability in cellular size or intracellular dye concentration. Although a single visible light-excited ratiometric dye has yet to be developed, several alternatives have been developed that have proven especially useful for analysis requiring visible excitation (Table I). Nonratiometric dyes that exhibit *increased* fluorescence upon binding of calcium include fluo-3 (Minta et al., 1989) and a series of dyes developed by Molecular Probes: calcium green, calcium orange,

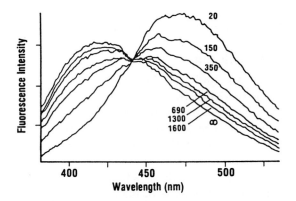

Fig. 1 Emission spectra for indo-1 as a function of ionized calcium. Indo-1 (3 μM) was added to a buffer consisting of 21.64 mM K$_2$H$_2$EGTA, 100 mM KCl, 20 mM Hepes, pH 7.20. Small aliquots of a buffer that contained equimolar calcium and EGTA as K$_2$CaEGTA and was otherwise identical to the first buffer were added. Fluorescence excited at 356 nm was measured with a spectrofluorimeter.

and calcium crimson. Fura-Red exhibits *decreased* fluorescence intensity upon calcium binding and, as described below, can be combined with simultaneous cellular loading of fluo-3 to provide a sensitive form of ratiometric analysis using 488-nm excitation. Improvement in available calcium probes that exhibit different fluorescence properties now permits single-cell measurements of [Ca^{2+}]$_i$ in large number of cells with flexibility that was not possible previously. The use of a flow cytometer allows the correlation of [Ca^{2+}]$_i$ with other cell parameters such as surface antigen expression and cell cycle and, furthermore, allows one to electronically sort cells based upon [Ca^{2+}]$_i$.

Table I
Calcium Indicator Dyes Useful for Flow Cytometric Applications

Indicator	Emission response to elevated calcium	Excitation wavelength (nm)	Emission wavelength(s) (nm)	Calcium affinity, K_d (nm) 22°C	37°C
Indo-1	Δ Ratio	325–360	390 / 520		~250
Fluo-3	Increase	488	530	~400	~860
Calcium green-1	Increase	488	530	~250	
Calcium orange	Increase	550	575	~330	
Calcium crimson	Increase	590	610	~200	
Fura-Red	Decrease	488	660	~400	
Fluo-3 / Fura-Red	Δ Ratio	488	530 / 660	~400	

II. Flow Cytometric Assay with Indo-1

A. Loading of Cells with Indo-1

Uptake and retention of indo-1 ([1-[2 amino-5-[6-carboxylindol-2-yl]-phenoxy]-2-[2'-amino-5'-methylphenoxy]ethane N,N,N',N'-tetraacetic acid] are facilitated by the use of the pentaacetoxymethyl ester of indo-1, using the scheme described above. Approximately 20% of the total dye is trapped in this manner during typical loadings. After loading, the extracellular indo-1 should be diluted 10- to 100-fold before flow cytometric analysis (Rabinovitch et al., 1986). One incidental benefit of this loading strategy is that this procedure, like the more familiar use of fluorescein diacetate or carboxyfluorescein diacetate, allows one to distinguish between live and dead cells. The latter will not retain the hydrophilic impermeant dye and can be excluded during subsequent analysis.

The lower limit of useful intracellular loading concentrations of indo-1 is determined by the sensitivity of fluorescence detection of the flow cytometer and the upper limit is determined by avoidance of buffering of $[Ca^{2+}]_i$ by the presence of the calcium chelating dye itself. In practice, one should use the least amount of indo-1 that is necessary to reliably quantitate the fluorescence signal. Fortunately, indo-1 has excellent fluorescence characteristics [30-fold greater quantum yield than quin2 (Grynkiewicz et al., 1985)] and useful ranges of indo-1 loading are much lower than the millimolar amounts required with quin2. For human peripheral blood T cells, we have found adequate detection at or above 1 μM indo-1 ester, under conditions that achieve ~5 μM intracellular indo-1. Buffering of $[Ca^{2+}]_i$ in human T cells was observed as a slight delay in the rise in $[Ca^{2+}]_i$ and a retarded rate of return of $[Ca^{2+}]_i$ to baseline values when loading concentrations above 3 μM (22 μM intracellular concentration) were used. A reduction in peak $[Ca^{2+}]_i$ occurred at even higher indo-1 concentrations (Rabinovitch et al., 1986). Chused et al. (1987) have observed slightly greater sensitivity of murine B cells to indo-1 buffering, recommending a loading concentration of no greater than 1 μM. In side-by-side comparisons, we have found that calcium transients in B cells are much more sensitive to the effects of buffering by indo-1 than T cells. For human platelets, a 2 μM loading concentration has been reported (Davies et al., 1988). Rates of loading of the indo-1 ester can be expected to vary between cell types, perhaps as a consequence of variations in intracellular esterase activity, as well as varying abilities of cells to extrude the cell-impermeant form. In peripheral human blood, more rapid rates of loading are seen in platelets and monocytes than in lymphocytes. Even within one cell type, donor or treatment-specific factors may affect loading; for example, lower rates of indo-1 loading were seen in splenocytes from aged than from young mice (Miller et al., 1987).

Indo-1 has been found to be remarkably nontoxic to cells subsequent to loading. Analysis of the proliferative capacity of human T lymphocytes loaded with indo-1 (Table II) has shown no adverse effects on the ability of cells to enter and complete three rounds of the cell cycle. Similar results have been obtained with murine B lymphocytes (Chused *et al.*, 1987). This is especially pertinent to the sorting of indo-1 loaded cells based on $[Ca^{2+}]_i$ as described subsequently.

B. Instrumental Technique

The absorption maximum of indo-1 is between 330 and 350 nm, depending upon the presence of calcium (Grynkiewicz *et al.*, 1985); this is well-suited to excitation at either 351–356 nm from an argon ion laser or 337–356 nm from a krypton ion laser. Laser power requirements depend upon the choice of emission filters and optical efficiency of the instrument; however, it is our experience that although 100 mW is often routinely employed, virtually identical results can be obtained with 10 mW. The use of helium-cadmium lasers has been compared to argon laser excitation, and little difference observed for analysis in the case of indo-1 (Goller and Kubbies, 1992). Stability of the intensity of the excitation source is less important in this application than in many others, because of the use of the ratio of fluorescence emissions. Analysis with indo-1 has also been performed using excitation by a mercury arc lamp (FACS Analyzer, Becton–Dickinson). It is instructive to consult previous work using quin2 for flow cytometry in order to appreciate how much the current probes have improved over previous generations (Ransom *et al.*, 1987).

The cellular $[Ca^{2+}]_i$ signaling response is an active process and is highly temperature dependent in most cell types. Thus, the sample chamber must usually be kept warm, and the time that cells spend in cooler tubing in transit to the flow cell must be kept minimal (less than 10 sec), or else the sample

Table II
Proliferation of Peripheral Blood Lymphocytes Loaded with Indo-1[a]

Day of culture	Indo-1/AM (μM)	Percentage of cells			
		Noncycling	1st cycle	2nd cycle	3rd cycle
2	0	69.8 ± 1.3	28.4 ± 0.8	1.9 ± 0.7	—
	3	70.7 ± 1.8	27.5 ± 1.2	1.5 ± 0.8	—
3	0	58.1 ± 2.5	14.3 ± 0.5	26.5 ± 2.7	1.2 ± 0.4
	3	59.5 ± 2.6	14.9 ± 1.1	24.7 ± 3.4	0.9 ± 0.3
4	0	43.8 ± 0.7	8.4 ± 0.5	29.5 ± 1.0	18.4 ± 1.7
	3	44.6 ± 0.8	9.5 ± 0.9	28.6 ± 1.4	17.3 ± 3.0

[a] Peripheral blood lymphocytes were either loaded or not loaded with indo-1 ester and cultured with PHA (10 μg/ml) and BrdU (1 × $10^{-4}M$). The cells were harvested on the indicated days, stained with Hoechst 33258, and analyzed by flow cytometry, and the percentage of cells (mean ± SEM, $n = 4$) in each cell cycle was quantitated as described (Kubbies *et al.*, 1985).

tubing must be kept warm also. The agonist is ordinarily introduced into the sample by quickly ceasing flow, removing the sample container, adding agonist, and restarting flow, "boosting" the new sample to the flow cell quickly. With practice, this procedure can be completed in less than 20 sec. If more rapid $[Ca^{2+}]_i$ transients after agonist addition must be analyzed, then various agonist injection methods may be utilized so that disruption of sample flow is not required. A commercial device is available for on-line addition of agonist using a syringe (Cytek Development, Fremont, CA).

An increase in $[Ca^{2+}]_i$ is detected with indo-1 as an increase in the ratio of fluorescence intensity at a lower to a higher emission wavelength. The optimal strategy is to select band pass filters so that one minimizes the collection of light near the isosbestic point and maximizes collection of fluorescence that exhibits the largest variation in calcium-sensitive emission. The choice of filters used to select these wavelengths is dictated by the spectral characteristics of the shift in indo-1 emission upon binding to calcium (Fig. 1). The original spectral curves published for indo-1 (Grynkiewicz *et al.*, 1985) did not depict the large amounts of indo-1 emission in the blue–green and green wavelengths; in practice, on flow cytometers, we find that there is more light available in the blue region than in the violet region, although the "information" content of the light from the violet region exceeds that of the blue region (Fig. 1). Commercially available band pass filters for analysis with indo-1 are usually centered on the "violet" emission of the calcium-bound indo-1 dye (405 nm) and calcium-free indo-1 dye "blue" emission (485 nm). However, we have found these wavelengths to be suboptimal and that a larger dynamic range in the ratio of wavelengths is obtained if "blue" emission below 485 nm is not collected and the center of the blue emission band pass filter is instead moved upward. Similarly, the violet band pass filter should be chosen to minimize the collection of wavelengths above 405 nm. Thus, in order to optimize the calcium signal, it is important to exclude light from analysis that is near or at the isosbestic point. This effect is summarized in Table III by values R_{max}/R, which

Table III
Effect of Wavelength Choice on Calcium-Sensitive Indo-1 Ratio Shifts[a]

Wavelength pair (nM)	R_{max}	R_{min}	R	R_{max}/R_{min}	R_{max}/R
475/395	2.33	0.040	0.352	58.2	6.62
475/405	2.38	0.100	0.410	23.8	5.80
495/395	3.51	0.048	0.429	73.1	8.18
495/405	3.59	0.119	0.501	30.2	7.17
515/395	5.75	0.070	0.644	82.1	8.93
515/405	5.88	0.176	0.752	33.4	7.82
530/395	**9.68**	**0.117**	**1.073**	**82.7**	**9.02**
530/405	9.89	0.292	1.252	33.9	7.90

[a] By spectrofluorimetry, 2-nm slit width; uncorrected fluorescence. The wavelength choice shown in bold demonstrated the best ratio shift.

indicates the range of change in indo-1 ratio observed from resting intracellular calcium to saturated calcium.

C. Calibration of Ratio to $[Ca^{2+}]_i$

Prior to the development of indo-1, $[Ca^{2+}]_i$ determination with quin2 fluorescence was sensitive to cell size and intracellular dye concentration as well as $[Ca^{2+}]_i$. This made necessary calibration at the end of each individual assay by determination of the fluorescence intensity of the dye at zero and saturating Ca^{2+}. In contrast, with indo-1 use of the Ca^{2+}-dependent shift in dye emission wavelength allows the ratio of fluorescence intensities of the dye at the two wavelengths to be used to calculate $[Ca^{2+}]_i$,

$$[Ca^{2+}]_i = K_d \cdot \frac{(R - R_{min}) S_{f2}}{(R_{max} - R) S_{b2}}, \tag{1}$$

where K_d is the effective dissociation constant (250 nM); R, R_{min}, and R_{max} are the fluorescence intensity ratios of violet/blue fluorescence at resting, zero, and saturating $[Ca^{2+}]_i$, respectively; and S_{f2}/S_{b2} is the ratio of the blue fluorescence intensity of the calcium-free and bound dye, respectively (Grynkiewicz et al., 1985). Because this calibration is independent of cell size and total intracellular dye concentration as well as instrumental variation in efficiency of excitation or emission detection, it is not necessary to measure the fluorescence of the dye in the calcium-free and saturated states for each individual assay. In principle, it is sufficient to calibrate the instrument once after setup, and tuning by measurement of the constants R_{max}, R_{min}, S_{f2}, and S_{b2}, after which only R is measured for each subsequent analysis on that occasion. It is important to note that the apparent K_d of indo-1 was measured at 37°C at an ionic strength of 0.1 at pH 7.08; this value will change significantly at different temperatures, pH, and ionic strength. For example, the K_d of Fura-2 changes from 225 to 760 nM when ionic strength is changed from 0.1 to 0.225. Similar values for indo-1 have not been published.

One strategy to obtain the R_{max} and R_{min} values for indo-1 is to lyse cells in order to release the dye to determine fluorescence at zero and saturating $[Ca^{2+}]_i$, as is carried out in fluorimeter-based assays with quin2 and Fura-2. However, this is not possible with flow cytometry, due to the loss of cellular fluorescence. Another strategy is the use of an ionophore to saturate or deplete $[Ca^{2+}]_i$ in order to allow approximation of the true end points without cell lysis. For this approach the ionophore ionomycin is best suited, due to its specificity for calcium and low fluorescence. When flow cytometric quantitation of fluorescence from intact cells treated with ionomycin or ionomycin plus EGTA was compared with spectrofluorimetric analysis of lysed cells in medium with or without EGTA, the indo-1 ratio of unstimulated cells (R) and the ratio at saturating amounts of Ca^{2+}, R_{max}, were similar by both techniques (Rabinovitch et al., 1986). The latter indicates that ionomycin-treated cells reach near-saturating levels of $[Ca^{2+}]_i$. The value of R_{min} which is obtained by treatment of intact cells with ionomycin in the presence of EGTA, however, is substantially higher than either that predicted from the spectral emission curves (Fig. 1) or that obtained by cell lysis and spectrofluorimetry. Spectrofluorimetric quantitation with either quin2 or indo-1 indicates that $[Ca^{2+}]_i$ remains at approximately 50 nM in intact cells treated with ionomycin and EGTA. Thus, due to the inability to obtain a valid flow cytometric determination with calcium-free dye, we have used for calibration the values of R_{min} and S_{f2} or S_{b2} derived from spectrofluorimetry, of either the indo-1 pentapotassium salt or, after lysis, cells loaded with indo-1 acetoxymethyl ester in the presence of EGTA. It is essential that the same optical filters be used for flow cytometry and spectrofluorimetry, since the standardization is very sensitive to the wavelengths chosen. Typical values of R_{max}, R_{min}, R, and S_{f2}/S_{b2} are shown for different emission wavelength combinations in Table III. Even if careful calibration of the fluorescence ratio to $[Ca^{2+}]_i$ is not being performed for a particular experiment, ordinary quality control can include a determination of the value of R_{max}/R. A limited range of R_{max}/R values should be obtained since unperturbed cells are routinely found to have a reproducible value of $[Ca^{2+}]_i$ and day-to-day optical variations in the flow cytometer are usually minimal (with the same filter set). The R_{max}/R values obtained on the Ortho Cytofluorograph and on Coulter instruments are typically in the range of six to eight; for unexplained reasons, several groups have reported that the value obtained on Becton–Dickinson instruments is only about four.

Chused *et al.* (1987) have suggested that metabolic inactivation of cells by the use of a cocktail of nigericin, high concentrations of potassium, 2-deoxyglucose, azide, and carbonyl cyanide *m*-chloro-phenylhydrazone and the calcium ionophore ionomycin can collapse the calcium gradient to zero and, therefore, that $[Ca^{2+}]_i = [Ca^{2+}]_o$. To avoid the apparent impossibility of assessing R_{min} in intact cells, the calibration is based upon a regression formula that relates R to ionomycin-treated cells suspended in a series of precisely prepared calcium buffers. Thus, this technique allows one to estimate $[Ca^{2+}]_i$ without the need to determine R_{min}, S_{f2}, S_{b2}, although it is subject to the limitation that calcium concentrations that are less than those found in resting cells cannot be quantitated and, further, by the precision with which one can prepare a series of calcium buffers that yield known and reproducible free calcium concentrations.

Accuracy of prediction of ionized Ca^{2+} concentration in buffer solutions is dependent upon a variety of interacting factors, so that care must be exercised in formulating Ca^{2+} standards. The ionized calcium concentration in an EGTA buffer system is dependent upon the magnesium concentration; other metals such as aluminum, iron, and lanthanum also bind avidly to EGTA. In addition, the dissociation constant of Ca^{2+}-EGTA is a function of pH, temperature, and ionic strength (Harafuji and Ogawa, 1980; Blinks *et al.*, 1982). For example, in an EGTA buffer (total EGTA 2 mM, total Ca^{2+} 1 mM, ionic strength 0.1 at 37°C), changing the pH from 7.4 to 7.0 can result in the ionized calcium increasing by more than 200 nM, a change that is approximately twice the magnitude of that found in resting cells, and is easily measured on a flow cytometer. Finally, it is important to prepare the buffers using the "pH metric technique," in part because of the varying purity of commercially available EGTA. Buffer solutions suitable for calibration are available (Molecular Probes, Eugene, OR).

As an alternative to the above approaches, there are several potential schemes for use of the flow cytometer to directly analyze the fluorescence of dye in solution; the fluorescence measurement must be converted to a brief pulse for processing by the flow cytometer, either by strobing the exciting laser beam or by chopping the PMT output signals (Kachel et al., 1990).

D. Display of Results

It is possible to display the data as a bivariate plot of the inversely correlated "violet" vs "blue" signals derived from each cell. Thus, the increase in ratio seen with increased $[Ca^{2+}]_i$ will be observed as a rotation about the axis through the origin. This method of ratio analysis is cumbersome, and fortunately, commercial flow cytometers all have some provision for a direct calculation of the value of the fluorescence ratio itself, either by analog circuitry or by digital computation.

Plotted as a histogram of the ratio values, quiescent cell populations show narrow distributions of ratio, even when cellular loading with indo-1 is very heterogeneous, and coefficients of variation of less than 10% are not uncommon (Fig. 2). The effects of perturbation of $[Ca^{2+}]_i$ by agonists can be noted by changes in the ratio histogram profiles by storing histograms sequentially with subsequent analysis of data. The above approach results in a linear display of indo-1 blue and violet fluorescence. If cellular indo-1 loading is extremely heterogenous, it may be desirable to work with a logarithmic conversion of "violet" and "blue" emission intensities in order to observe a broader range of cellular fluorescence. In this case, the hardware must permit the logarithm of the ratio to be calculated by *subtraction* of the log blue from the log violet signals (Rabinovitch et al., 1986).

A more informative and elegant display is obtained by a bivariate plot of ratio vs time. The bivariate histogram can be stored and the data displayed as "dot plots" on which the indo-1 ratio or each cell (proportional to $[Ca^{2+}]_i$) is plotted on the y-axis versus time on the x-axis (Fig. 3A). Alternatively, the data can be subjected to further analysis for presentation as "isometric plots" in which the x-axis represents time; the y-axis, $[Ca^{2+}]_i$; and the z-axis, number of cells (Fig. 3B). In these bivariate plots, kinetic changes in $[Ca^{2+}]_i$ are seen with much greater resolution, limited only by the number of channels on the time axis, the interval of time between each channel, and the rate of cell analysis. Changes in the fraction of cells responding, in the mean magnitude of the response, and in the heterogeneity of the responding population are best observed with these displays. For example, it can be seen in Fig. 3 that not all murine thymocytes respond to stimulation by anti-CD3, and of those that do, the values of $[Ca^{2+}]_i$ are quite heterogeneous.

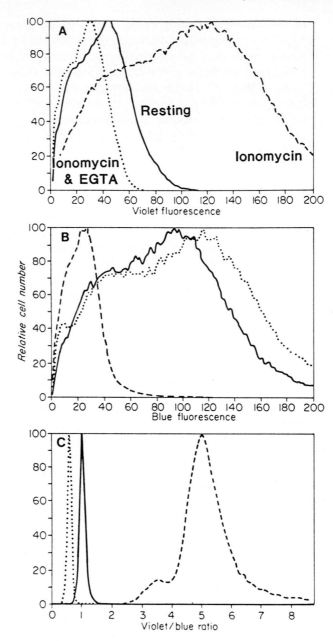

Fig. 2 Histograms of (A) violet (405 nm) fluorescence, (B) blue (485 nm) fluorescence, and (C) the ratio of violet/blue fluorescence of peripheral blood lymphocytes loaded with indo-1. Cells were analyzed in a basal state (solid lines), in the presence of 3 μM ionomycin (dashed lines), or in the presence of ionomycin and EGTA, used to reduce free calcium in the medium to $\sim$20 nM (dotted lines). Data are plotted on linear scales; the violet/blue ratio was normalized to unity for resting cells. Reproduced from Rabinovitch *et al.* (1986), by copyright permission of the American Association of Immunologists.

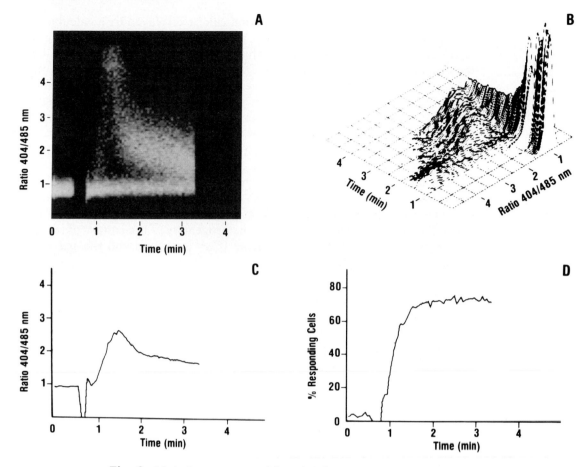

Fig. 3 Methods to express calcium signaling measured in single cells as a function of time. Thymocytes from a C57Bl/6 mouse were loaded with indo-1 and stimulated with anti-CD3 antibody 145-2C11. The cells were analyzed at 300 cells/sec. (A) Results were displayed on a "dot plot" in which $[Ca^{2+}]_i$ is plotted for each cell analyzed on a 100×100-pixel grid. The number of cells per pixel is displayed by intensity which ranges over 16 levels. (B) Isometric plots of the same experiment as in panel A are shown. Sequential histograms are plotted in which the x-axis represents time, the y-axis $[Ca^{2+}]_i$, and the z-axis number of cells. (C and D) Mean calcium vs time and percentage cells responding to two standard deviations above the mean of the cells before antibody stimulation are plotted for the same experiment depicted in panels A and B.

Calculation of the mean y-axis value for each x-axis time interval allows the data to be presented as mean ratio vs time (Rabinovitch *et al.*, 1986). Calibration of the ratio to $[Ca^{2+}]_i$ (see below) allows data to be presented in the same manner as traditionally displayed by spectrofluorimetric analysis, i.e., mean $[Ca^{2+}]_i$ vs time. While this presentation yields much of the information of interest in an easily displayed format, data relating to heterogeneity of the $[Ca^{2+}]_i$

response is lost. Some of this information can be displayed by a calculation of the "proportion of responding cells." If a threshold value of the resting ratio distribution is chosen, e.g., one at which only 5% of control cells are above, the proportion of cells responding by ratio elevations above this threshold vs time yields a presentation informative of the heterogeneity of the response (Figs. 3C and 3D). In some instances, however, plots of mean response and percentage responding cells versus time do not provide an adequate representation of the complexity of the data. For example, in Fig. 3, it can be seen from the "dot plot" and the "isometric" plot that the early response consists of a small population (~10%) of cells with an extremely high magnitude response of brief duration while the later response is comprised of ~50% of cells that have a low-magnitude response. Thus in this example, the time of the occurrence of the peak response is dissociated from the time of the maximal mean response (Figs. 3A and 3B vs 3C). Note that this heterogeneity of the pattern of response is not apparent in the displays of mean $[Ca^{2+}]_i$ vs time and the percentage responding cells vs time.

E. Simultaneous Analysis of $[Ca^{2+}]_i$ and Other Fluorescence Parameters

Although the broad spectrum of indo-1 fluorescence emission will likely preclude the simultaneous use of a second UV-excited dye, the use of two or more excitation sources allows additional information to be derived from visible light-excited dyes. Perhaps the most usual application will be determination of cellular immunophenotype simultaneously with the indo-1 assay, allowing alterations in $[Ca^{2+}]_i$ to be examined in, and correlated with, specific cell subsets.

Combining the use of FITC and PE-conjugated antibodies with indo-1 analysis allows determination of $[Ca^{2+}]_i$ in complex immunophenotypic subsets. On instruments without provision for analysis of four separate fluorescence wavelengths, detection of both FITC and the higher indo-1 wavelength with the same filter element may allow successful implementation of these experiments. Gating the analysis of indo-1 fluorescence upon windows of FITC vs PE fluorescence allows information relating to each identifiable cellular subset to be derived from a single sample.

Using other probes excited by visible light, it may be possible to analyze additional physiologic responses in cells simultaneously with $[Ca^{2+}]_i$. The simultaneous analysis of membrane potential and $[Ca^{2+}]_i$ has been accomplished by several groups (Lazzari et al., 1986; Ishida and Chused, 1988). Similarly, several investigators have measured pH_i and $[Ca^{2+}]_i$ simultaneously (Van Graft et al., 1993).

F. Sorting on the Basis of $[Ca^{2+}]_i$ Responses

The ability of the flow cytometric analysis with indo-1 to observe small proportions of cells with $[Ca^{2+}]_i$ responses different than the majority of cells suggests that the flow cytometer may be useful to identify and sort variants in the population for their subsequent biochemical analysis or growth. Results of artificial mixing experiments with Jurkat (T cell) and K562 (myeloid cell) leukemia lines indicated that subpopulations of cells with variant $[Ca^{2+}]_i$ comprising <1% of total cells could be accurately identified (Rabinovitch et al., 1986). Goldsmith and Weiss (1987, 1988) have reported the use of sorting on the basis of indo-1 fluorescence to identify mutant Jurkat cells that fail to mobilize $[Ca^{2+}]_i$ in response to CD3 stimulus, despite the expression of structurally normal CD3/Ti complexes. The basis for impaired signal transduction was found to be the absent expression of the lck protein tyrosine kinase (Straus and Weiss, 1992). These experiments suggest that sorting on the basis of indo-1 fluorescence can be an important tool for the selection and identification of genetic variants in the biochemical pathways leading to Ca^{2+} mobilization and cell growth and differentiation. The potential of this technique is not limited to lymphoid cells, as a specific calcium response was used to sort a rare subpopulation of gastrointestinal cells that secrete cholecystokinin (Liddle et al., 1992).

III. Use of Flow Cytometry and Fluo-3 to Measure $[Ca^{2+}]_i$

Fluo-3 is a fluorescein-based, calcium-sensitive probe developed by Minta and co-workers (1989). This was the first calcium indicator that did not require UV excitation, and, therefore, it could be used on all flow cytometers that have the capability to measure fluorescence emission from fluorescein. Fluo-3 may be less sensitive than indo-1 at detecting small changes in $[Ca^{2+}]_i$, in part because the K_d is higher. An advantage with fluo-3 is that it can be used with other probes such as caged calcium chelators that may themselves require UV excitation (Tsien, 1989). Because fluo-3 emission characteristics are virtually identical to those of FITC, the use of fluo-3 can be combined with almost all of the dyes used with multicolor analysis in addition to FITC, including simultaneous determination of cell cycle.

Fluo-3 has been successfully adapted for flow cytometry by several laboratories (Table IV). Loading of fluo-3 is performed similarly to the protocol described above for indo-1. Calibration must be performed either by use of spectrofluorimetry and nonratiometric equations as originally described for quin2 (Tsien et al., 1982) or by reference to a series of known calcium buffers (Chused et al., 1987). A direct comparison of fluo-3 to indo-1 was done by loading both dyes in T cells and analyzing the signals simultaneously (Rabinovitch and June, 1990). Both dyes showed a readily measurable response to agonist, although the change in indo-1 fluorescence ratio was greater than the increase in fluo-3 fluorescence intensity. Due to the heterogeneity in intracellular concentration of fluo-3 in loaded cells, the determination of $[Ca^{2+}]_i$ in any given cell is less accurate, and discrimination of heterogeneity in the cellular response to an agonist is less clear using fluo-3. These differences are not unexpected,

Table IV
Studies Using Fluo-3 or Fura–Red to Measure [Ca^{2+}]$_i$ by Flow Cytometry

Investigators	Study aims/conclusions	Comments
Vandenberghe and Ceuppens (1990)	Initial application of fluo-3 for flow cytometry. Human PBL assayed after fluo-3 loading using 488-nm excitation.	Leakage of dye was rapid, and this was a significant limitation, given absence of ratiometric properties with fluo-3.
Yee and Christou (1993)	Human PMNs assayed after fluo-3 loading using 488-nm excitation. Cells were primed with LPS and stimulated with FMLP.	Subpopulations of cells detected after FMLP stimulation.
Sei and Arora (1991)	Human PBL were stained with PE-conjugated antibodies and loaded with fluo-3.	Demonstration that simultaneous analysis of surface antigen expression and [Ca^{2+}]$_i$ is possible with single-beam excitation at 488 nm.
Van Graft et al. (1993)	Develop a method to measure [Ca^{2+}]$_i$ and pH$_i$ using fluo-3 and SNARF-1 in killer cell–target cell conjugates.	Conjugates were clearly distinguished from single cells. Killers and targets developed increased [Ca^{2+}]$_i$; only killer cells had changes in [pH]$_i$.
Rijkers et al. (1990)	Determine if simultaneous loading of cells with SNARF-1 and fluo-3 would improve [Ca^{2+}]$_i$ signal.	SNARF-1 emission collected at isobestic point. Ratio of fluo-3 to SNARF-1 reduces variation in fluorescence intensity caused by variable fluo-3 loading. Caveat: SNARF-1 can be protein bound.
Rabinovitch and June (1990)	Compare signals from T cells loaded with both indo-1 and fluo-3.	Resting CV smaller with indo-1; the quality of the agonist response was similar with fluo-3 and indo-1.
Nolan et al. (1992)	Determine if [Ca^{2+}]$_i$ can be measured in bull sperm with fluo-3.	Sperm exocytosis successfully studied.
Kozak and Yavin (1992)	PC12 cells studied with indo-1 and fluo-3 to determine if cell density affects [Ca^{2+}]$_i$.	Pronounced effects of cell density on [Ca^{2+}]$_i$ homeostasis suggested by both dyes.
Elsner et al. (1992)	Study characteristics of PMN response to fMLP, C5a, and IL-8.	Heterogeneity of [Ca^{2+}]$_i$ response suggests functional subsets of neutrophils.
Sanchez Margalet et al. (1992)	Determine if pancreastatin affects [Ca^{2+}]$_i$ in RINm5f cells.	Both pancreastatin and ATP caused a fourfold increase in [Ca^{2+}]$_i$.

given the nonratiometric properties of fluo-3. However, fluo-3 can be combined with SNARF-1 (Rijkers *et al.*, 1990), or, even better, with Fura-Red (see below), to obtain a ratiometric measurement that helps to overcome these difficulties.

IV. Use of Flow Cytometry and Fluo-3/Fura-Red for Ratiometric Analysis of $[Ca^{2+}]_i$

Indo-1 has been the ratiometric calcium indicator dye most commonly used in flow cytometry because of its shift in emission frequency when excited at a single wavelength. A significant drawback to the use of indo-1, however, is the requirement for UV excitation, which is not as widely available as 488-nm excitation. A ratiometric analysis with visible excitation is possible using two dyes simultaneously: Fura-Red (Haugland, 1992) and fluo-3, a fluorescein-based, calcium-sensitive probe (Minta *et al.*, 1989). As seen in Table I, both dyes are excited at 488 nm, but fluo-3 fluoresces in the green region with increasing intensity when bound to calcium while Fura-Red exhibits inverse behavior, fluorescing most intensely in the red region when not calcium bound. The spectral characteristics of the two dyes mixed together (Fig. 4) are reminiscent in shape of the UV-excited emission of indo-1 (Fig. 1). Fura-Red emissions are dimmer than emissions from cells loaded with indo-1 and fluo-3 at the same concentration, and approximately 2–3.5 times as much Fura-Red as fluo-3 may be required to produce emissions of optimal intensity; loading concentrations of 10 and 4 μM, respectively, were optimal for T cells (Novak and Rabinovitch, 1994). As Fig. 5 shows, the fluo-3/Fura-Red ratio provides a better resolved picture of the increase in $[Ca^{2+}]_i$ associated with cellular response to antibody

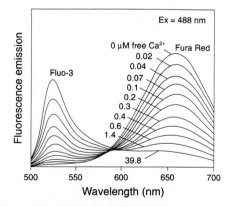

Fig. 4 Fluorescence emission spectra of a mixture of fluo-3 and Fura-Red in the presence of buffers containing various concentrations of free calcium, as noted. The fura red concentration is approximately 10 times that of fluo-3. (Courtesy of Molecular Probes, Inc., with copyright permission.)

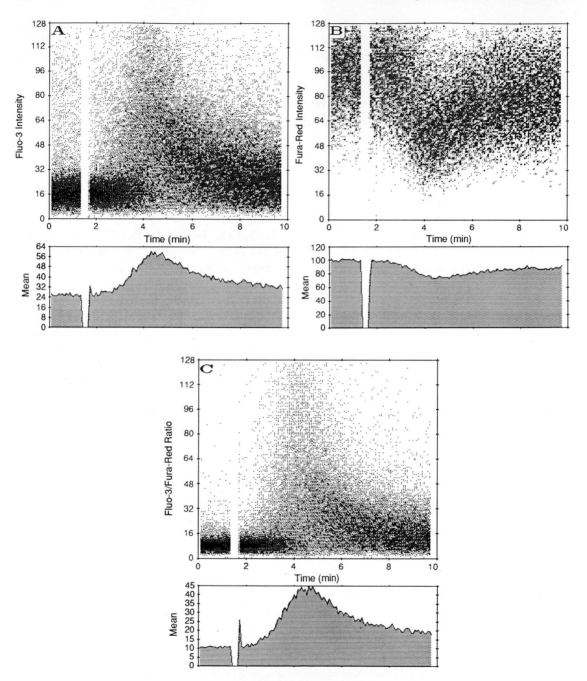

Fig. 5 Kinetic displays of fluorescence intensity/ratio and mean values of (A) fluo-3 intensity, (B) Fura-Red intensity, and (C) fluo-3/Fura-Red ratio of cells simultaneously loaded with 4 μM fluo-3 + 10 μM Fura-Red esters, gated for CD4$^+$ PE-labeled cells. Cells were stimulated by addition of 10 μg/ml anti-CD3 antibody at time 1 min. Data analysis and display performed using the software program MultiTime (Phoenix Flow Systems, San Diego, CA).

stimulus than does either fluo-3 or Fura-Red alone. Qualitatively, the fluo-3/ Fura-Red ratio (Fig. 5C) achieves results similar to those for indo-1. Compared to fluo-3 alone, there is a larger magnitude response, with less variability in measurements from different cells. The narrower distribution of resting ratios from nonstimulated cells can be observed graphically by comparing the prestimulated regions of Figs. 5A and 5B with the region in Fig. 5C. The fluo-3/ Fura-Red ratio can also be used simultaneously with analysis of PE-labeled immunofluorescence markers. The emission characteristics of this dye combination (Fig. 4) have a "window" between the fluo-3 and Fura-Red emission peaks that allows 488-nm-excited PE emission to be observed (Novak and Rabinovitch, 1994). When T lymphocytes, for example, are loaded with 10 μM Fura-Red, emission from Fura-Red is approximately 35% as intense as PE-CD4$^+$ emission in the region of 562–588 nm. Using spectral crossover compensation, this degree of overlap allows bright and moderately bright PE-labeled antibody probes to be used. If additional lasers are available, further label combinations can be used.

V. Cell Conjugate Assays Combined with Calcium Analysis and Flow Cytometry

The realization that many forms of signal transduction are initiated by intercellular contact combined with increasing interest in a number of cell adhesion molecules has led to the description of flow cytometric assays capable of measuring conjugate formation and calcium elevations. Abe and co-workers have devised a powerful technique to analyze adhesive and signaling interactions between bone marrow-derived cells (Fig. 6). This technique permitted the analysis of signal transduction initiated by viral "superantigens" that induce clonal deletion of self-reactive thymocytes (Abe *et al.*, 1992). Others have used a similar approach to measure calcium flux induced by tumor bearing cells from populations of T cells (Alexander *et al.*, 1992), permitting, in principle, the ability to isolate antigen-specific T cells. As was noted before, a conjugate assay was used to measure [Ca^{2+}]$_i$ and [pH]$_i$ simultaneously in natural killer cells bound to target cells (Van Graft *et al.*, 1993).

VI. Pitfalls and Critical Aspects

A. Instrumental Calibration and Display of Data

In some instruments there may be difficulty in displaying the ratio of indo-1 fluorescence so that increases in Ca^{2+} are depicted as increased ratio values (Breitmeyer *et al.*, 1987). In particular, the analog ratio circuits of older Coulter instruments were limited in their range of acceptable inputs; for example, that the "violet" signal never be greater than the "blue," yielding a ratio of greater

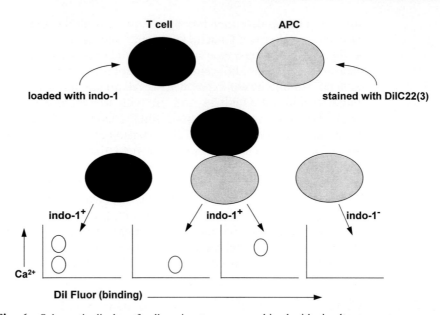

Fig. 6 Schematic display of cell conjugate assay combined with simultaneous measurement of $[Ca^{2+}]_i$. One population of cells (e.g., T cells) is loaded with indo-1, and another population [e.g., antigen presenting (APC) cells] with a cell marker dye such as DilC22(3). The cells are mixed, pelleted, and assayed. Indo-1 and DilC22(3) signals are detected by triggering on the indo-1 signal so that, in this case, either T cells or T cells conjugated to APC are analyzed. APC alone do not trigger analysis. T cells which have reacted with the APC and exhibit elevated $[Ca^{2+}]_i$ can be sorted for further analysis (for details see Abe *et al.,* 1992, and Alexander *et al.,* 1992).

than one. Rather than reversing the ratio (blue/violet) so that a rise in calcium results in a counterintuitive decline in ratio, the violet/blue ratio can be used as long as the signal gains are initially set such that subsequent rises in the ratio will not exceed the permitted value. Alternatively, several list-mode programs allow the ratio signal to be calculated with software subsequent to analysis.

Some instruments may have nonlinearity in signal amplification or introduce errors into the calculation of the indo-1 ratio. By either analog or digital calculation, it is important that no artifactual offset be introduced in the ratio; this can be quickly tested by altering the excitation power over a broad range of values in an analysis of a nonperturbed indo-1-loaded cell population—a correctly calculated ratio will not show any dependence upon excitation intensity. It can similarly be shown that loading of cells with a broad range of indo-1 concentrations results in a constant value of the violet/blue ratio (Rabinovitch *et al.,* 1986).

B. Difficulties with Loading Indo-1

Several problems may be encountered in cells loaded with indo-1. These include compartmentalization, leakage, secretion, quenching by heavy metals, and incomplete deesterification of indo-1 ester. The analysis of $[Ca^{2+}]_i$ using indo-1 is predicated upon achieving uniform distribution of the dye within the cytoplasm. In several cell types, the related dye Fura-2 has been reported to be compartmentalized within organelles (Di Virgilio *et al.*, 1989; Malgaroli *et al.*, 1987). In bovine aortic endothelial cells, Fura-2 has been reported to be localized to mitochondria; however, under those conditions, indo-1 remained diffusely cytoplasmic (Steinberg *et al.*, 1987). We have observed that indo-1 may become compartmentalized in cells. Some cell types, such as neutrophils and monocytes, and some cell lines rather than primary cells are more susceptible to compartmentalization. In addition, compartmentalization is enhanced by prolonged incubation of cells at 37°C. Thus, it is possible that there will be fewer problems with compartmentalization of indo-1 than with Fura-2; however, it is advisable to examine the cellular distribution of indo-1 microscopically, and in each new application to confirm the expected behavior of the dye. This is done by determining the ratio of R_{max} to R as a control for each experiment as described below. In addition, one should store indo-1-loaded cells at room temperature after loading and use the cells promptly after loading. In prolonged experiments, it is preferable to discard cells after several hours and to reload fresh batches of cells.

To reduce indicator quenching by heavy metals, the use of the membrane-permeant chelator diethylenetriaminepentaacetic acid (TPEN) for cell lines that contain increased amounts of heavy metals has been described (Arslan *et al.*, 1985). Probenecid, a blocker of organic anion transport may be useful in cells that actively secrete Fura-2, indo-1, or other indicators (Di Virgilio *et al.*, 1989). In some circumstances, the use of Pluronic F-127, a nonionic, high-molecular-weight surfactant, may aid in the loading of probe into cells that are otherwise difficult to load (Cohen *et al.*, 1974; Poenie *et al.*, 1986). Pluronic F-127 may be obtained from Molecular Probes (Eugene, OR).

If, for a particular cell type loaded with a calcium-sensitive dye, the magnitude of change between R and R_{max} are in good agreement with the values predicted from spectral curves of the probe in a cell-free buffer, then it would be unlikely that the dye is in a compartment inaccessible to cytoplasmic Ca^{2+}, in a form unresponsive to $[Ca^{2+}]_i$ (e.g., still esterified) or in a cytoplasmic environment in which the spectral properties of the dye were altered. With regard to the second condition, it has been proposed that since indo-1 fluorescence, but not that of the indo-1 ester, is quenched in the presence of millimolar concentrations of Mn^{2+}, Mn^{2+} in the presence of ionomycin can be used as a further test of complete hydrolysis of the indo-1 ester within cells (Luckhoff, 1986). It has been suggested that both Fura-2 and indo-1 may be incompletely deesterified within some cell types (Luckhoff, 1986; Scanlon *et al.*, 1987). Since the fluorescence of the ester has little spectral dependence upon changes in Ca^{2+}, the presence of this dye form could lead to false estimates of $[Ca^{2+}]_i$. Again, results of calibration experiments are helpful in excluding this possibility.

C. Unstable Baseline

Under typical conditions, the baseline indo-1 ratio should show little (<3%) variation from sample to sample. Some cell lines may have altered mean values of "resting" $[Ca^{2+}]_i$, which can often be ascribed to a subpopulation of cells with elevated $[Ca^{2+}]_i$. This may result from impaired viability of some cells or, presumably, may be due to cells traversing certain phases of the cell cycle. In circumstances where the baseline is not stable from sample to sample, the following situations should be considered. The cells should be equilibrated to 37°C for 3 to 5 min before analysis. Regulation of the temperature of the cell sample is essential, as transmembrane signaling and calcium mobilization are temperature dependent and active processes. Most applications will require analyses at 37°C. If cells are allowed to cool before they flow past the laser beam, calcium signals will often become impaired, so that either the sample input tubing should be warmed or narrow-gauge tubing and high flow rates should be used to keep transit times from warmed sample to flow cell minimized. As noted above, it is necessary to maintain the cells at room temperature and to warm the cells just prior to the assay. It is not clear as to what mechanism the variation in basal $[Ca^{2+}]_i$ with temperature variation may be ascribed. It is possible that the changes reported by indo-1 are real and reflect strict temperature requirements of the cell for the maintenance of calcium homeostasis. Alternatively, the changes of calcium reported by the indicator dye may in part be due to temperature-dependent changes in the effective dissociation constant of the dye for calcium.

Sometimes the baseline will start at a normal level and then rise with time. This may be due to the failure to completely remove an agonist from the sample lines from a previous experiment. The most common problem has been residual calcium ionophore; this can be efficiently removed by first washing the sample lines with dimethyl sulfoxide and then scavenging residual ionophore by washing with a buffer containing 2% bovine serum albumin.

D. Sample Buffer

The choice of medium in which the cell sample is suspended for analysis can be dictated primarily by the metabolic requirements of the cells, subject only to the presence of millimolar concentrations of calcium (to enable calcium agonist-stimulated calcium influx) and reasonable pH buffering. The use of phenol red as a pH indicator does not impair the flow cytometric detection of indo-1 fluorescence signals. Although the new generation of Ca^{2+} indicator dyes are not highly sensitive to small fluctuations of pH over the physiologic range (Grynkiewicz et al., 1985), unbuffered or bicarbonate-buffered solutions can impart large and uncontrolled pH shifts. Finally, if analysis of release of Ca^{2+} from intracellular stores is desired, independent of extracellular Ca^{2+} influx, addition of 5 mM EGTA to the cell suspension (final concentration) will reduce Ca^{2+} from several millimolars to <20 nM, thus abolishing the usual extracellular to intracellular gradient.

E. Poor Cellular Response

When one encounters cells that are poorly responsive to various treatments, it is necessary to first determine if there is difficulty with the cells or with the instrument. The cells should be stimulated with the calcium ionophore iono-mycin and the magnitude of R_{max} to R determined. If the ratio increases by the expected magnitude (approximately sixfold for indo-1), then the instrument is functioning properly. If the increase is less than expected, then one should obtain an independent preparation of cells, such as murine thymocytes or human peripheral blood lymphocytes. Aliquots of cryopreserved cells are convenient for this purpose. If these cells load properly and also respond poorly, then the instrument alignment should be checked. For ratiometric analyses, one of the two signals may not be properly focused, or perhaps there is interference from a second laser. This problem can be pinpointed by analyzing separately each of the two wavelength signals after ionophore treatment; for indo-1, the violet signal should increase at least threefold and the blue signal should decrease approximately twofold (Figs. 1 and 2).

If the instrument is functioning properly, then the problem may be in the cells. The cells must be loaded with sufficient dye to be easily detected. This should be checked independently with fluorescence microscopy. If the cells are too dim or excessively bright, or if the probe is compartmentalized, the ability to detect calcium signals will be impaired. For unknown reasons, the calcium signaling of B cells and not T cells is particularly sensitive to overloading with probes (Rabinovitch et $al.$, 1986; Chused et $al.$, 1987). The cells must be suspended in a medium that contains calcium; responses can appear blunted due to the inadvertent resuspension of cells in medium that contains no added calcium. In the simultaneous analysis of $[Ca^{2+}]_i$ and immunofluorescence, con-sider that the use of the antibody probe can itself alter the cellular $[Ca^{2+}]_i$. It is becoming increasingly clear that binding of monoclonal antibodies (mAbs) to cell-surface proteins can alter $[Ca^{2+}]_i$, even when these proteins are not previously recognized as part of a signal transducing pathway. For example, antibody binding to CD4 will reduce CD3-mediated $[Ca^{2+}]_i$ signals; if the anti-CD4 mAb is crosslinked to the CD3 complex, as with a goat–anti-mouse mAb, the CD3 signals are augmented (Ledbetter et $al.$, 1987,1988). Antibody binding to CD8 has similar effects.

As a consequence of these concerns, a reciprocal staining strategy should be used whenever possible, so that the cellular subpopulation of interest is unlabeled while undesired cell subsets are identified by mAb staining. The CD4$^+$ subset in PBL may be identified, for example, by staining with a combination of CD8, CD20, and CD11 mAbs (Rabinovitch et $al.$, 1986), and the CD5$^+$ subset can be identified by staining with CD16, CD20, and HLA-DR mAbs (June et

al., 1987). Finally, it is important when staining cells with mAbs for functional studies that the antibodies be azide free, in order that metabolic processes be uninhibited. Commercial antibody preparations may thus require dialysis before use.

As with many assays, artifacts can be encountered from diverse causes. When analyzing specimens of bone marrow, it is possible that fat droplets may be included in the sample preparation. This may pose a problem, as calcium probes can label fat droplets as well as cells (Bernstein *et al.*, 1989). Thimerosal, a commonly used preservative that is included in many drug preparations, can elevate calcium in cells (Gericke *et al.*, 1993). Finally, ethanol can inhibit the mitogen-induced initial increase in $[Ca^{2+}]_i$ in mouse splenocytes (Sei *et al.*, 1992), so that diluent controls must be performed diligently.

VII. Limitations

There are several limitations to the flow cytometric assay of cellular calcium concentration. First, certain problems attributable to the use of fluorescent indicators have been mentioned. In some cells indo-1 will not load uniformly into cells, or may not be uniformly hydrolyzed to the calcium-sensitive moiety. Quin2 at intracellular concentrations often used may block plasma membrane sodium–calcium transport (Allen and Baker, 1985), and quin2 may be quenched by heavy metals which are found in the cytoplasm of some cell lines (Arslan *et al.*, 1985). Similarly, quin2 has been reported to be mitogenic in certain cells and to alter certain cellular functions (Owen, 1988; Hesketh *et al.*, 1983), although this has not as yet been observed in indo-1-loaded cells (Chused *et al.*, 1987; Rabinovitch *et al.*, 1986). It is possible that indo-1 and newer dyes may have similar limitations, although this would be expected to be less of a problem due to the much lower concentrations of the probes that are required to attain a satisfactory fluorescent signal from cells. Second, there are limitations imposed by the nature of the flow cytometric assay system. Flow cytometry is unable to detect heterogeneity of cellular calcium concentrations within a single cell, and there are reports from assays using digital video microscopy that, in some situations, calcium transients may be compartmentalized (Poenie *et al.*, 1987). The use of the photoprotein aequorin may in some circumstances detect changes in cytosolic calcium not reported by indo-1 (Ware *et al.*, 1987), although use of the two indicators is complimentary because aequorin cannot measure Ca^{2+} in single cells (Cobbold and Rink, 1987). In addition, there is evidence that calcium elevations occurring after cellular stimulation may be oscillatory rather than sustained (Ambler *et al.*, 1988; Wilson *et al.*, 1987), thus raising the possibility that some cellular processes controlled by calcium may be frequency modulated as well as amplitude modulated. Since flow cytometry cannot measure the calcium concentration inside a single cell as a function of time, it is not possible to distinguish between a subpopulation of cells that is responding with an oscillatory response or, alternatively, two populations of cells, one that has sustained elevation of calcium concentrations and one that has basal levels.

Despite these limitations, determination of $[Ca^{2+}]_i$ in large numbers of single cells using flow cytometry offers great practical advantages and allows measurements not possible by alternative techniques that are currently available.

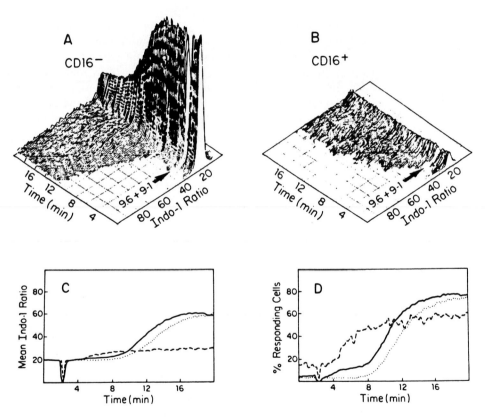

Fig. 7 Effects of CD2 stimulation on $[Ca^{2+}]_i$ of CD16$^-$ or CD16$^+$ lymphocytes. Indo-1-loaded peripheral blood T cells were stained with phycoerythrin-conjugated anti-CD16 antibody, which identifies a population of large granular lymphocytes that do not express the CD3 antigen, and stimulated with CD2 antibodies 9.6 plus 9-1. The cells were analyzed at 500 cells/sec and electronic "gating" was used to display the phycoerythrin-negative and -positive cells as isometric plots in panels A and B, respectively. The time course of the mean indo-1 ratio (C) and the percentage responding cells (D) are displayed (———, total cell population analyzed; ---, CD16$^+$ cells; ···, CD16$^-$ cells). Separate analysis showed that these cells were 99% CD2$^+$, 95% CD3$^+$, and 2% CD16$^+$. Reproduced from June *et al.* (1986), by copyright permission of the American Society for Clinical Investigation.

VIII. Results

The flow cytometric assay of cellular calcium concentration has already been applied to a wide variety of cells, providing interesting and sometimes unexpected results (Table IV). A small coefficient of variation (often <10%) was found in lymphocytes for the distribution of the basal calcium level; this value appears to reflect physiologic variation because the instrumental variation was only 4.5% (Fig. 2C). Surprisingly, there is heterogeneity in the response of lymphocytes to doses of calcium ionophores that are capable of stimulating physiologic responses in cells. Ishida and Chused (1988) found that the response of murine splenic B cells was lower than that of splenic or lymph node T cells. We have found a similar situation in human cells (Rabinovitch et al., 1986); large granular lymphocytes and B cells have lower magnitude responses than do peripheral blood T cells after treatment with ionomycin. After CD2 (T11) stimulation of peripheral blood lymphocytes, it was found that both CD3$^+$ T cells and CD16$^+$ CD3$^-$ large granular lymphocytes (LGL) mobilized calcium. Following CD3 (antigen-specific) stimulation, more than 90% of T cells responded, and, as was expected, LGL did not respond (June et al., 1986). The pattern of the calcium signal after CD2 stimulation differed in that calcium mobilization in LGL was early in onset and low in magnitude while, in T cells, the calcium signal was delayed and high in magnitude (Fig. 7). Thus, mechanisms of calcium homeostasis in T lymphocytes appear to differ among T cell subsets and in several respects from B cells. Infection with the HIV-1 retrovirus was found to impair signal transduction in CD4 cells (Linette et al., 1988). T cells from tumor bearing animals have been found to have abnormal calcium responses (Mizoguchi et al., 1992). There are many potentially exciting clinical applications of the flow cytometric assay of cellular calcium concentration (Rabinovitch et al., 1993). Demonstration of such heterogeneity in [Ca^{2+}]$_i$ signals would have been impossible to discern in conventional assays carried out in a fluorimeter where only the mean calcium response is recorded.

Acknowledgments

We thank Ryo Abe and Thomas Chused for the material adapted for Fig. 6, W. C. Gause and J. Bluestone for the reagents used for the experiment shown in Fig. 3, and R. J. Hartzman and N. Hensel for use of the spectrofluorimeter. This work was supported in part by National Institutes of Health Grant AG01751 and by the Naval Medical Research and Development Command, Research Task No. M0095.003-1402. The opinions and assertions expressed herein are those of the authors and are not to be construed as official or reflecting the views of the Navy Department or the naval service at large.

References

Abe, R., Ishida, Y., Yui, K., Katsumata, M., and Chused, T. M. (1992). *J. Exp. Med.* **176,** 459–468.
Alexander, R. B., Bolton, E. S., Koenig, S., Jones, G. M., Topalian, S. L., June, C. H., and Rosenberg, S. A. (1992). *J. Immunol. Methods* **148,** 131–141.

Allen, T. J., and Baker, P. F. (1985). *Nature (London)* **315,** 755–756.

Ambler, S. K., Poenie, M., Tsien, R. Y., and Taylor, P. (1988). *J. Biol. Chem.* **263,** 1952–1959.

Arslan, P., Di Virgilio, F., Beltrame, M., Tsien, R. Y., and Pozzan, T. (1985). *J. Biol. Chem.* **260,** 2719–2727.

Bernstein, R. L., Hyun, W. C., Davis, J. H., Fulwyler, M. J., and Pershadsingh, H. A. (1989). *Cytometry* **10,** 469–474.

Blinks, J. R., Wier, W. G., Hess, P., and Prendergast, F. G. (1982). *Prog. Biophys. Mol. Biol.* **40,** 1–114.

Breitmeyer, J. B., Daley, J. F., Levine, H. B., and Schlossman, S. F. (1987). *J. Immunol.* **139,** 2899–2905.

Chused, T. M., Wilson, H. A., Greenblatt, D., Ishida, Y., Edison, L. J., Tsien, R. Y., and Finkelman, F. D. (1987). *Cytometry* **8,** 396–404.

Cobbold, P. H., and Rink, T. J. (1987). *Biochem. J.* **248,** 313–328.

Cohen, L. B., Salzberg, B. M., Davila, H. V., Ross, W. N., Landowne, D., Waggoner, A. S., and Wang, C. H. (1974). *J. Membr. Biol.* **19,** 1–36.

Davies, T. A., Drotts, D., Weil, G. J., and Simons, E. R. (1988). *Cytometry* **9,** 138–142.

Di Virgilio, F., Steinberg, T. H., and Silverstein, S. C. (1989). *In* "Methods in Cell Biology (A. Tartakoff, ed.), Vol. 31, pp. 453–462. Academic Press, San Diego.

Elsner, J., Kaever, V., Emmendorffer, A., Breidenbach, T., Lohmann Matthes, M. L., and Roesler, J. (1992). *J. Leukocyte Biol.* **51,** 77–83.

Gericke, M., Droogmans, G., and Nilius, B. (1993). *Cell Calcium* **14,** 201–207.

Goldsmith, M. A., and Weiss, A. (1987). *Proc. Natl. Acad. Sci. U.S.A.* **84,** 6879–6883.

Goldsmith, M. A., and Weiss, A. (1988). *Science* **240,** 1029–1031.

Goller, B., and Kubbies, M. (1992). *J. Histochem. Cytochem.* **40,** 451–456.

Grynkiewicz, G., Poenie, M., and Tsien, R. Y. (1985). *J. Biol. Chem.* **260,** 3440–3450.

Harafuji, H., and Ogawa, Y. (1980). *J. Biochem. (Tokyo)* **87,** 1305–1312.

Haugland, R. P. (1992). *In* "Handbook of Fluorescent Probes and Research Chemicals" (K. D. Larison, ed.), pp. 117–118. Molecular Probes, Eugene, OR.

Hesketh, T. R., Smith, G. A., Moore, J. P., Taylor, M. V., and Metcalfe, J. C. (1983). *J. Biol. Chem.* **258,** 4876–4882.

Ishida, Y., and Chused, T. M. (1988). *J. Exp. Med.* **168,** 839–852.

June, C. H., Ledbetter, J. A., Rabinovitch, P. S., Martin, P. J., Beatty, P. G., and Hansen, J. A. (1986). *J. Clin. Invest.* **77,** 1224–1232.

June, C. H., Rabinovitch, P. S., and Ledbetter, J. A. (1987). *J. Immunol.* **138,** 2782–2792.

Kachel, V., Kempski, O., Peters, J., and Schodel, F. (1990). *Cytometry* **11,** 913–915.

Kozak, A., and Yavin, E. (1992). *J. Mol. Neurosci.* **3,** 203–212.

Kubbies, M., Schindler, D., Hoehn, H., and Rabinovitch, P. S. (1985). *Cell Tissue Kinet.* **18,** 551–562.

Lazzari, K. G., Proto, P. J., and Simons, E. R. (1986). *J. Biol. Chem.* **261,** 9710–9713.

Ledbetter, J. A., June, C. H., Grosmaire, L. S., and Rabinovitch, P. S. (1987). *Proc. Natl. Acad. Sci. U.S.A.* **84,** 1384–1388.

Ledbetter, J. A., June, C. H., Rabinovitch, P. S., Grossmann, A., Tsu, T. T., and Imboden, J. B. (1988). *Eur. J. Immunol.* **18,** 525–532.

Liddle, R. A., Misukonis, M. A., Pacy, L., and Balber, A. E. (1992). *Proc. Natl. Acad. Sci. U.S.A.* **89,** 5147–5151.

Linette, G. P., Hartzman, R. J., Ledbetter, J. A., and June, C. H. (1988). *Science* **241,** 573–576.

Luckhoff, A. (1986). *Cell Calcium* **7,** 233–248.

Malgaroli, A., Milani, D., Meldolesi, J., and Pozzan, T. (1987). *J. Cell Biol.* **105,** 2145–2155.

Miller, R. A., Jacobson, B., Weil, G., and Simons, E. R. (1987). *J. Cell. Physiol.* **132,** 337–342.

Minta, A., Kao, J. P., and Tsien, R. Y. (1989). *J. Biol. Chem.* **264,** 8171–8178.

Mizoguchi, H., O'Shea, J. J., Longo, D. L., Loeffler, C. M., McVicar, D. W., and Ochoa, A. C. (1992). *Science* **258,** 1795–1798.

Nolan, J. P., Graham, J. K., and Hammerstedt, R. H. (1992). *Arch. Biochem. Biophys.* **292,** 311–322.

Novak, E. J., and Rabinovitch, P. S. (1994). *Cytometry* (in press).

Owen, C. S. (1988). *Cell Calcium* **9,** 141–147.

Poenie, M., Alderton, J., Steinhardt, R., and Tsien, R. (1986). *Science* **233,** 886–889.

Poenie, M., Tsien, R. Y., and Schmitt-Verhulst, A. M. (1987). *EMBO J.* **6,** 2223–2232.

Rabinovitch, P. S., and June, C. H. (1990). *In* "Flow Cytometry and Sorting" (M. R. Melamed, T. Lindmo, and M. L. Mendelsohn, eds.), pp. 651–668. Wiley-Liss, New York.

Rabinovitch, P. S., June, C. H., Grossmann, A., and Ledbetter, J. A. (1986). *J. Immunol.* **137,** 952–961.

Rabinovitch, P. S., June, C. H., and Kavanagh, T. J. (1993). *Ann. N. Y. Acad. Sci.* **677,** 252–264.

Ransom, J. T., DiGiusto, D. L., and Cambier, J. C. (1986). *J. Immunol.* **136,** 54–57.

Ransom, J. T., DiGiusto, D. L., and Cambier, J. (1987). *In* "Methods in Enzymology" (P. Conn and A. Means, eds.), Vol. 141, pp. 53–63. Academic Press, Orlando, FL.

Rijkers, G. T., Justement, L. B., Griffioen, A. W., and Cambier, J. C. (1990). *Cytometry* **11,** 923–927.

Sanchez Margalet, V., Lucas, M., and Goberna, R. (1992). *Mol. Cell Endocrinol.* **88,** 129–133.

Scanlon, M., Williams, D. A., and Fay, F. S. (1987). *J. Biol. Chem.* **262,** 6308–6312.

Sei, Y., and Arora, P. K. (1991). *J. Immunol. Methods* **137,** 237–244.

Sei, Y., McIntyre, T., Skolnick, P., and Arora, P. K. (1992). *Life Sci.* **50,** 419–426.

Steinberg, S. F., Bilezikian, J. P., and Al-Awqati, Q. (1987). *Am. J. Physiol.* **253,** C744–C747.

Straus, D. B., and Weiss, A. (1992). *Cell (Cambridge, Mass.)* **70,** 585–593.

Tsien, R. Y. (1981). *Nature (London)* **290,** 527–528.

Tsien, R. Y. (1989). *In* "Methods in Cell Biology" (A. Tartakoff, ed.), Vol. 30, pp. 127–156. Academic Press, San Diego.

Tsien, R. Y., Pozzan, T., and Rink, T. J. (1982). *J. Cell Biol.* **94,** 325–334.

Vandenberghe, P. A., and Ceuppens, J. L. (1990). *J. Immunol. Methods* **127,** 197–205.

Van Graft, M., Kraan, Y. M., Segers, I. M., Radosevic, K., De Grooth, B. G., and Greve, J. (1993). *Cytometry* **14,** 257–264.

Ware, J. A., Smith, M., and Salzman, E. W. (1987). *J. Clin. Invest.* **80,** 267–271.

Wilson, H. A., Greenblatt, D., Poenie, M., Finkelman, F. D., and Tsien, R. Y. (1987). *J. Exp. Med.* **166,** 601–606.

Yee, J., and Christou, N. V. (1993). *J. Immunol.* **150,** 1988–1997.

CHAPTER 11

Cellular Protein Content Measurements

Harry A. Crissman and John A. Steinkamp

Division of Biomedical Sciences
Los Alamos National Laboratory
Los Alamos, New Mexico 87545

I. Introduction

Correlation of the synthesis and accumulation of protein, along with that of DNA and RNA, appears to play an important role in regulating cycle traverse capacity and cell division, growth, and size. Consistency in both the cycle generation time and the cellular protein content distribution, as reflected by the regularity of the volume distribution of the population, is controlled by transcriptional and translational processes that rigidly couple temporal metabolism of these macromolecules in cells. While DNA content increases only during S phase, protein and RNA syntheses are constant, and an increase in the cellular content of these macromolecules is linear and proportional across the cell cycle of exponentially growing mammalian populations. Jurt prior to mitosis, protein and DNA contents are doubled; however, cell division, which leads to a precise halving in DNA content, often results in an unequal proportioning in protein as well as RNA content in postmitotic G_1 cells. Flow cytometry (FCM) studies by Darzynkiewicz *et al.* (1982) have shown that following mitosis there is a significantly larger variability in the protein content distribution than observed in the mitotic subpopulation. This process increases the population heterogeneity in G_1 cells, with cells (G_{1A}) containing subthreshold levels of protein and RNA for entry into S phase and cells (G_{1B}) with threshold levels sufficient to proceed directly into S phase (Darzynkiewicz *et al.*, 1980). Flow cytometric bivariate distributions showed that G_{1A} cells are required to increase their protein content to the level of G_{1B} cells in order to synthesize DNA. When cells enter and traverse S phase the protein content variability is reduced. Although this process occurs through repeated cell cycles, heterogeneity within the population remains fairly constant. The mechanisms which control these processes are still poorly understood. However, a variety of cycle perturbing agents, including drugs, can induce a differential uncoupling of normal synthetic patterns causing a disproportionate accumulation of DNA, RNA, and protein (Crissman *et al.*, 1985). These conditions may lead to states of unbalanced growth, loss of long-term viability, and eventual cell death.

FCM provides a method of performing multiple biochemical measurements of both DNA and protein in single cells and allows for subsequent correlation of these metabolic parameters in studies on the relationship between cell growth and the cell division cycle. Simultaneous measurement of DNA and protein also allows assessment of the ratio of protein to DNA which serves as a sensitive gauge on the state of balanced growth of cells located at particular stages of the cell cycle under a variety of experimental conditions.

In this chapter we describe one-step staining procedures involving several dye combinations. These techniques have been applied to a variety of cell types. Cell staining involves the addition of staining solutions directly to ethanol-fixed or viable cells. Subsequent FCM analysis can be accomplished in the reagent solution after 10–20 min using a single- or dual-laser system (Steinkamp *et al.*, 1979) or a mercury arc excitation system.

II. Applications

Simultaneous FCM analyses of DNA and protein have been used for analysis of mouse squamous carcinoma cells and human vaginal samples (Steinkamp and Crissman, 1974); L1210 ascites (Crissman *et al.*, 1976, 1978); HeLa, L-929, and Chinese hamster (CHO) cells (Steinkamp *et al.*, 1976); Ehrlich ascites cells (Gohde *et al.*, 1970); rat hepatocytes; and hepatoma cells (Manske and Bohn, 1978) as well as yeast cells (Hutter and Eipel, 1979). In studies on samples from acute myeloid leukemia patients, Ffrench *et al.* (1985), using the propidium iodide–fluorescein isothiocyanate (PI–FITC) procedure, detected noncycling G1 cells with relatively low protein contents that had lost the capability to enter

S phase. In some studies (Crissman *et al.*, 1978; Steinkamp *et al.*, 1976) protein to DNA ratio analysis was used to characterize, sort, and identify cell types within heterogeneous cell populations. Also we have further demonstrated that simultaneous DNA–protein analysis methods could be applied for differentiating the cell proliferation patterns of normal cells from those of aneuploid myeloma cells in human bone marrow samples (Crissman *et al.*, 1981). In general, myeloma cells throughout the cell cycle had a greater protein content compared to normal cells. Roti-Roti *et al.* (1982) and Pollack (1990) have also used a combination of PI–FITC to examine and directly compare nuclear protein and DNA contents of cells in various phases of the cell cycle.

Sequential dual-laser excitation (i.e., UV and 488 nm) and Hoechst (HO) 33342-FITC staining has also been used for DNA and protein analysis. Using this method two subpopulations of cycling cells could be resolved in samples analyzed from non-Hodgkins lymphoma patients that had a bimodal protein content distribution (Crissman and Steinkamp, 1986). Bivariate analysis was required to detect the subpopulations since the two DNA distribution patterns were overlapping.

Many of the staining methods used in FCM involve a series of staining and rinsing steps with centrifugation between each step. This approach is not only time consuming, but it also tends to induce cell clumping and cell loss. For that reason we developed staining methods that achieve labeling of DNA and protein by combining all the reagents and adding them directly to ethanol-fixed or viable cells (Crissman and Steinkamp, 1982). By controlling the dye concentrations it is possible to minimize the dye–dye interations as well as the excessive background fluorescence. Also rinsing and centrifugations steps are eliminated.

Two-color staining of DNA and protein has most often been accomplished using a combination of either ethidium bromide (EB) and FITC (Gohde and Dittrich, 1970) or PI and FITC (Crissman and Steinkamp, 1973). However, another popular method developed by Stohr *et al.* (1978) involves staining with a combination of DAPI and sulfarhodamine 101 and only requires UV excitation for blue and red fluorescence analysis of DNA and protein content, respectively. The methods above require only a single excitation source. Other fluorescent labeling protocols incorporate modification of procedures for fluorochroming DNA with Hoechst 33342 (HO) (Arndt-Jovin and Jovin, 1977) or mithramycin (MI) (Crissman and Tobey, 1974) and applying these dyes in combination with either of the red protein stains, X-rhodamine isothiocyanate (X-RITC) or rhodamine 640 (R-640). Sequential, dual-laser excitation is required for FCM analysis. The combination of rhodamine-640 and HO 33342 has also been used to stain viable cells (Crissman and Steinkamp, 1982). These methods will be described and compared for their relative ease and utility in FCM. In general, using dual-laser excitation systems allows for combinations of dyes having less spectral overlap, and this improves the accuracy of the measurement. If one is limited to a single-laser FCM system, the staining procedures described in this chapter will help to alleviate the problems encountered due to the spectral properties of the dyes.

III. Materials

All stock solutions of reagents are prepared at 1.0 mg/ml and stored refrigerated for at least 1 month unless indicated otherwise. Stock solutions of propidium iodide (Molecular Probes, Inc., Eugene, OR) and mithramycin (Pfizer, Groton, CT) are prepared in PBS. Hoechst 33342 and DAPI (Molecular Probes, Inc.) are prepared in distilled water. Solutions of FITC (isomer 1, BBL, Division of Becton–Dickinson Co., Cockeysville, MD) are prepared in absolute ethanol just prior to use. (Exciton, Dayton, OH) and substituted RITC (Research Organics, Cleveland, OH) are dissolved in dimethyl sulfoxide (DMSO). Stock solutions of RNase (1.0 mg/ml) (Worthington, Freehold, NJ) are prepared in PBS.

IV. Cell Preparation and Staining

For ethanol fixation, cells are harvested from culture or from tissue dispersal solutions by centrifugation and then *thoroughly* resuspended, to prevent cell aggregation, in one part cold "saline GM" (g/liter: glucose, 1.1; NaCl, 8.0; KCl, 0.4; $Na_2HPO_4 \cdot 12\ H_2O$, 0.39; KH_2PO_4, 0.15) containing 0.5 mM EDTA for chelating free calcium and magnesium ions. Then three parts cold, 95% nondenatured ethanol are added to the cell suspension with mixing to produce a final ethanol concentration of about 70%. Viable cells are stained in culture medium as indicated below. Following ethanol fixation (at least 12 hr) cells are centrifuged and the fixative is removed prior to the application of the staining solutions listed below. The stained cell density is maintained at about 7.5 × 10^5 cells/ml. Unless otherwise indicated, staining is achieved at room temperature for at least 1 hr and in all cases analysis is performed on cells in the staining solution.

PI–FITC RNase: Solutions of PI (15 μg/ml) and FITC (0.05–0.10 μg/ml) prepared in PBS containing 50 μg/ml RNase A are added to the cell pellet.

MI–R-640 or MI–XRITC: Solutions of MI (50 μg/ml) and either R-640 or X-RITC (1.0 μg/ml) in PBS containing 15 mM $MgCl_2$ are added to the cell pellet.

HO 33342–R-640 or HO–XRITC: Ethanol-fixed cells are stained with HO 33342 (0.5–1.0 μg/ml) and either R-640 or X-RITC (1.0 μg/ml) in PBS. For viable cell staining HO 33342 (2–5 μg/ml) and R-640 (5.0 μg/ml) are added directly to cells in culture (37°C), for 1 hr prior to harvesting. Following staining, cells are harvested and the viable cells are then resuspended in culture medium containing the dyes for analysis.

DAPI-SR101: Ethanol-fixed cells are stained with DAPI (2 μg/ml) and SR101 (15-20 μg/ml) in PBS.

V. Critical Aspects of the Procedure

The dye concentrations of the DNA and protein stains may be adjusted to obtain an optimal color balance during analysis. The final dye concentrations selected will depend on the nuclear to cytoplasmic size of the cells. For example, human squamous cells, with a relatively large cytoplasmic volume will require a lower dye concentration for protein staining compared to lymphocytes. Adjustments will depend on the fluorescent intensities from both the DNA and protein signals. For sample-to-sample comparisons of both DNA and protein contents, the cell densities should be kept constant so that the ratio of dye to cell number remains somewhat consistent; otherwise variations in fluorescence intensities will result that do not truly reflect differences in cellular content but rather artifactual differences in stainability. To alleviate this problem, an accurate count of the number of viable cells should be made prior to fixation, and the final volume of fixative adjusted to yield 10^6 cells/ml. By staining the same volume of suspension of fixed cells for each sample, the relative cell number is easily kept constant. For the procedures described it should be noted that RNase at room temperature probably does not completely digest all the cellular RNA; however, the treatment adequately reduces double-stranded RNA to the single-stranded state and thereby prevents intercalation by dyes such as PI. Additionally, in two-color staining some energy transfer can occur such as from FITC to PI and from MI to either R-640 or X-RITC due to the overlap in the dye spectra. Little energy transfer should occur from HO 33342 to R-640 or to X-RITC.

VI. Controls and Standards

Preliminary studies utilizing a particular dye combination should be conducted with equal numbers of cells: (a) stained only with the DNA dye, (b) stained only with the protein dye, and (c) stained with both dyes at the appropriate concentrations. When the optimal dye concentrations and filter arrangements have been determined, signals from cells stained with just one fluorochrome only should be detected in the appropriate channel. By recording the data from cells stained with only one dye and comparing any change in the channel number and/or the coefficient of variation (CV) observed in cells stained with both dyes, it is possible to determine the extent of the dye–dye interactions. For example, an increase in the CV of the DNA histogram of cells stained with both dyes would indicate that protein fluorescence is leaking into the DNA channel. Decreasing the concentration of the protein dye would in this case alleviate the problem and increase the accuracy of the analyses.

A well-tested routine cell type, such as fixed CHO cells, is usually stained and analyzed to ensure that staining reagents have been properly prepared and that the instrument is adequately aligned and configured, with regard to wavelength excitations and filter combinations, for analysis. By comparison, the protein histogram usually resembles the electronic cell volume distribution obtained for unfixed cells.

VII. Instruments

A. PI–FITC RNase Cell Samples

Single-laser excitation at 488 nm is routinely used for analysis of the stained cell sample. The filter combination consists of a 600-nm long-pass dichroic filter, with 515 and 590-nm long-pass filters for detection of FITC and PI, respectively. Data routinely acquired includes forward angle light scatter (FLS), DNA content (red integrated signal), protein content (green fluorescence), protein to DNA (green to red signals) electronic ratio, and the red peak fluorescence signal. Subsequent bivariate analysis is used to correlate the individual parameters. The bivariate of the red integrated histogram vs the red peak histogram is used to discriminate doublets of G_1 cells in the G_2/M region, since the doublets have lower peak values compared to those of G_2/M cells. In some cases, fluorescence compensation may be necessary to adjust for overlap in the DNA and protein fluorescence.

B. MI–R-640 or MI–X-RITC Cell Samples

A dual-laser system, 457.9 nm (argon) for MI and 568 nm (krypton) for R-640 or X-RITC, was used for sequential excitation with the laser beams separated by approximately 200 μm. The filters used were a 600-nm long-pass dichroic, with 495 and 610-nm long-pass filters for analysis of MI and red (R-640 or X-RITC) fluorescence, respectively (see Steinkamp *et al.*, 1979, for details).

C. HO 33342–R-640 or X-RITC Cell Samples

A dual-laser system, 333–363 nm (combined lines) for HO 33342 and 568 nm (krypton) for R-640 or X-RITC, was used for sequential excitation. The filters used were a 500-nm long-pass dichroic, with 400 and 610-nm long-pass filters for analysis of HO and red (R-640 or X-RITC) fluorescence, respectively.

D. DAPI–SR101 Cell Samples

A single UV, laser, or mercury arc source can be used to excite both of these dyes simultaneously. The filter used would be the same as that used for the HO–R-640 or X-RITC analysis above.

VIII. Results and Discussion

Correlated analysis of DNA and protein provides information on the growth potential of cells at various phases of the cell cycle. In most instances, data are presented as bivariate contour plots (Fig. 1). Figure 1 shows the distribution of protein and DNA contents for exponentially growing CHO cells. The data are similar to those obtained for DNA and RNA in previous studies; however, the heterogeneity in protein content is much greater than that noted for RNA content (Darzynkiewicz *et al.*, 1982). The CV for the postmitotic population is about 15% greater than that for the mitotic cells, and the G1 cells can be divided into G_{1A} and G_{1B} subpopulations based on the protein content levels (Darzynkiewicz *et al.*, 1980). The heterogeneity in cell cycle, as reflected in part by the variability in protein and RNA content, has been suggested as an evolutionary survival mechanism that protects the population from cycle-specific, cytotoxic insult that could potentially destroy the entire population. The heterogeneity of the population ensures that cells are not at the same sensitive stage at any given time. This phenomenon also makes the design of effective chemotherapeutic agent much more difficult.

The DNA–protein measurements also provide information on the state of balanced growth of the cell population as determined by deviations from the normal patterns. Figure 2 shows the single-parameter DNA content and the DNA–protein profiles for exponentially growing (control) cells and for cells treated with 6 μg/ml adriamycin for 2 hr, rinsed and cultured in drug-free medium for 15 hr prior to analysis. Adriamycin treatment resulted in a G_2 block and the induction of a state of unbalanced growth as noted by the abnormally high protein content distribution for the G_2 cells, compared to control cells. Based on cell volume analysis and sorting of viable cells, the survival of the G_2-arrested cells ranged from about 2 to 4%, whereas G_1 cell, which have

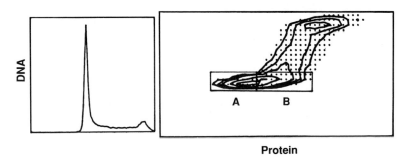

Protein

Fig. 1 DNA-protein contour distribution and a single-parameter DNA histogram for ethanol-fixed CHO cells stained with PI and FITC as described. The G_1 subpopulation in the contour distribution is subdivided to show G_{1A} and G_{1B} cells as described by Darzynkiewicz *et al.* (1980) in DNA–RNA contours.

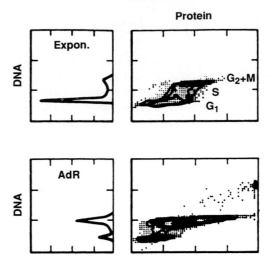

Fig. 2 Single-parameter DNA content histograms and bivariate contour distributions (DNA vs protein) for exponentially growing CHO control cells and for cells treated with adriamycin (6 μg/ml for 2 hr) (bottom). Analyses were performed on cells harvested and fixed 15 hr following an initial 2-hr drug treatment period.

protein content profiles similar to those of controls, had survival values near 27%, while the total population had a survival value of 14% (Crissman *et al.*, 1988). These analyses provide greater phase-specific information on the drug-treated population than does DNA content analysis alone.

Simultaneous analysis on DNA–protein contents have also been useful for making nuclear to cytoplasmic ratio determinations (Steinkamp and Crissman, 1974). Since the duration of the DNA and protein fluorescence signals is proportional to nuclear and cytoplasmic size, respectively, electronic ratio analyses of the signals can provide a comparison of these two parameters. Also gated analysis of the single-parameter DNA histogram can provide the protein content histograms and mean values for cells in various phases of the cell cycle (Crissman and Steinkamp, 1973). Considering the ease and simplicity of DNA–protein determinations and the increased amount of information available for gauging experimental effects, it would seem worthwhile in the design of a particular study to include the addition of protein content measurements along with the routine DNA cell-cycle analyses.

Acknowledgments

This work was supported by NIH Grant R24 RR06758, the Los Alamos National Flow Cytometry Resource (funded by the Division of Research Resources of NIH, Grant P41-RR01315), and the U.S. Department of Energy.

References

Arndt-Jovin, D. J., and Jovin, T. M. (1977). *J. Histochem. Cytochem.* **25,** 585–589.

Crissman, H. A., and Steinkamp, J. A. (1973). *J. Cell Biol.* **59,** 766–771.

Crissman, H. A., and Steinkamp, H. A. (1982). *Cytometry* **3,** 84–90.

Crissman, H. A., and Steinkamp, J. A. (1986). *In* "Techniques in Cell Cycle Analysis" (J. W. Gray and Z. Darzynkiewicz, eds.), pp. 163–206: Humana Press, Clifton, NJ.

Crissman, H. A., and Tobey, R. A. (1974). *Science* **184,** 1297–1298.

Crissman, H. A., Oka, M. S., and Steinkamp, J. A. (1976). *J. Histochem. Cytochem.* **24,** 64–71.

Crissman, H. A., Kissane, R. J., Wanek, P. L., Oka, M. S., and Steinkamp, J. A. (1978). *In* "Proceedings of the Third International Symposium on Detection and Prevention of Cancer" (H. E. Nieburgs, ed.), pp. 79–90. New York: Dekker.

Crissman, H. A., Von Egmond, J. V., Holdrinet, R. G., Pennings, A., and Haanen, C. (1981). *Cytometry* **2,** 59–62.

Crissman, H. A., Darzynkiewicz, Z., Tobey, R. A., and Steinkamp, J. A. (1985). *J. Cell Biol.* **101,** 141–147.

Crissman, H. A., Wilder, M. E., and Tobey, R. A. (1988). *Cancer Res.* **48,** 5742–5746.

Darzynkiewicz, Z., Traganos, F., and Melamed, M. R. (1980). *Cytometry* **1,** 98–108.

Darzynkiewicz, Z., Crissman, H. A., Traganos, F., and Steinkamp, J. (1982). *J. Cell Physiol.* **113,** 465–474.

Ffrench, M., Bryon, P. A., Fiere, D., Vu Van, H., Gentilhomme, O., Adeleine, P., and Viola, J. J. (1985). *Cytometry* **6,** 47–53.

Gohde, W., and Dittrich, W. (1970). *Z. Anal. Chem.* **252,** 328–330.

Gohde, W., Spies, I., Schumann, J., and Buchner, Th. (1978). *In* "Proceedings of the 2nd Symposium of Pulse-Cytophotometry" (W. Godhe, J. Schumann, and Th. Buchner, eds.), pp. 27–32. Ghent, Belgium: European Press.

Hutter, K. J., and Eipel, H. E. (1979). *Eur. J. Appl. Microbiol. Biotechnol.* **6,** 223–231.

Manske, W., and Bohn, B. (1978). *In* "Proceedings of the 2nd Symposium of Pulse-Cytophotometry" (W. Godhe, J. Schumann, and Th. Buchner, eds.), pp. 297–302. Ghent, Belgium: European Press.

Pollack, A. (1990). *In* "Flow Cytometry" (Z. Darzynkiewicz and H. A. Crissman, eds.), pp. 315–323. San Diego, CA: Academic Press.

Roti-Roti, J. L., Higashikubo, R., Blair, O. C., and Vygur, H. (1982). *Cytometry* **3,** 91–96.

Steinkamp, J. A., and Crissman, H. A. (1974). *J. Histochem. Cytochem.* **22,** 616–621.

Steinkamp, J. A., Hansen, K. M., and Crissman, H. A (1976). *J. Histochem. Cytochem.* **24,** 292–297.

Steinkamp, J. A., Orlicky, D. A., and Crissman, H. A. (1979). *J. Histochem. Cytochem.* **27,** 273–276.

Stohr, M., Vogt–Schaden, M., Knobloch, M., Vogel, R., and Futterman, G. (1978). *Stain Technol.* **53,** 205–215.

CHAPTER 12

Lysosomal Proton Pump Activity: Supravital Cell Staining with Acridine Orange Differentiates Leukocyte Subpopulations

Frank Traganos and Zbigniew Darzynkiewicz

Cancer Research Institute
New York Medical College
Valhalla, New York 10523

I. Introduction

Lysosomes are cellular organelles which store a variety of digestive enzymes, effectively segregating these potentially harmful hydrolytic constituents from the rest of the cellular cytoplasm. Lysosomes are known to be internally acidic and to require metabolic energy to maintain an intralysosomal pH of about 4.8 (Ohkuma and Poole, 1978). It appears that an ATP-dependent proton (H^+) pump is responsible for the internal acidic environment of lysosomes (Yamashiro *et al.*, 1983).

Many uncharged, lipophilic substances are able to cross biological membranes by an unselective permeation mechanism (Rundquist *et al.*, 1984). Some of these substances, such as the fluorescent dye acridine orange (AO), are weak bases. At acid pH, weak bases will accept a proton and be converted to a positively charged substance which is no longer capable of passing freely through cellular membranes (de Duve *et al.*, 1974). Thus, AO, like other lysosomotropic agents, will accumulate and become trapped within the lysosomes of living cells (Allison and Young, 1964; Robbins and Marcus, 1963; Robbins *et al.*, 1964; Rundquist *et al.*, 1984; Zelenin, 1966).

Generally, little is known of AO's intracellular targets in living, nonpermeabilized cells. Nevertheless, nucleic acids within living cells exposed to relatively low concentrations of AO ($<10^{-5}$ M) will bind the dye as monomers whereas lysosomes will concentrate the dye. The intralysosome dye concentration is predominately determined by the pH gradient across the membrane (de Duve *et al.*, 1974) which, in turn, is a reflection of the efficiency of the ATP-dependent proton pump (Yamashiro *et al.*, 1983). The granular, bright red luminescence of AO concentrated in lysosomes of living cells is thought to be the result of the high concentration of dye generated in lysosomal granules which may form stacked dye aggregates. Aggregates of AO are known to luminesce red (Kapuscinski and Darzynkiewicz, 1984) as compared to the normal green fluorescence associated with the monomer form of the dye that would be expected to predominate at low AO concentrations (Darzynkiewicz and Kapuscinski, 1990). Under appropriate staining conditions, the intensity of AO red luminescence is a function of both total lysosomal volume and the capacity of the lysosomal membrane to maintain a proton gradient (Rundquist *et al.*, 1984).

AO will also concentrate in the azurophilic granules of supravitally stained granulocytes (Abrams *et al.*, 1983). These granules are similar to lysosomes, concentrate AO, and luminesce red when excited with blue light. In both instances, however, loss of membrane integrity, either occurring naturally as cells die or as a result of fixation or following permeabilization with detergents, abolishes the metachromatic staining of these organelles (Darzynkiewicz and Kapuscinski, 1990).

II. Applications

A. Cell Viability

Supravital staining of cells with AO to detect lysosomal proton pump activity can be used to test cell viability following exposure to a variety of environmental insults and as a measure of general cellular integrity following treatment with toxic agents including chemotherapeutic drugs (e.g., Del Bino *et al.*, 1991).

B. White Blood Cell Differential

Staining of leukocyte granules has been used to provide automated white cell differentials of lymphocytes, monocytes, and granulocytes from whole blood (Melamed *et al.*, 1972a,b) and to assay neutrophil degranulation (Abrams *et al.*, 1983).

C. Marker of Cell Differentiation

Lysosomal activity has been observed to vary with the degree of cell differentiation. For instance, both in the case of some granulocytic leukemias (Melamed *et al.*, 1972b) and during acute bacterial infections (Melamed *et al.*, 1974), large numbers of immature granulocytes appear in the peripheral blood. Such immature granulocyte elements had characteristically increased red luminescence compared to more mature (segmented) granulocytes typically observed in peripheral blood (Melamed *et al.*, 1972b, 1974).

Lysosomal proton pump activity can also be used as a marker for the later stages of differentiation induced by various agents in some model systems. Thus, human myelogenous leukemic HL-60 cells differentiate into either granulocytes or monocyte/macrophages depending upon the type of inducer used. AO staining would be expected to differ depending upon which cell lineage was induced and the extent of differentiation within that lineage.

D. Functional Marker of Lysosomal Activity

Although largely unexplored, AO may be used to test the effects of various environmental stimuli on the activity of lysosomes. Thus, some drugs may affect lysosomal activity directly or indirectly. Lysosome activity is also likely to be affected by conditions which enhance or diminish cellular phagocytosis and/or pinocytosis. Some of these experiments have been carried out in isolated lysosomes in solution (Moriyama *et al.*, 1982).

III. Materials

A stock solution of AO should be made up in distilled water at a concentration of 1 mg/ml. Because AO can contain up to 50% impurities, it is important to utilize a chromatographically purified form as is available from Polysciences, Inc. (Warrington, PA; catalog No. 4539). The stock solution is stable for several months at 4°C.

IV. Staining Procedures

A. Lysosomes

AO staining for lysosomal protein pump activity should be done under conditions of optimum cell growth. AO should be added to cells in growth medium plus serum at a concentration of 1–2 μg/ml by direct dilution from the stock solution. Cells can then be incubated at 37°C for 30 min or longer. It is neither necessary nor desirable to rinse cells with AO-free medium or balanced salt solution prior to measurement.

B. Leukocyte Differential

While it is possible to obtain differential staining of leukocyte populations using buffy coat or density gradient-separated leukocytes, the original staining reaction was performed with whole-blood preparations and remains simple and reproducible within relatively wide ranges of cell concentration (Melamed *et al.*, 1972a).

A 1 mg/ml stock solution of AO is prepared as above. The dye is diluted to a concentration of 1 μg/ml in an isotonic buffer solution of pH 7.4 (Melamed *et al.*, 1972a). One part whole blood is then added to 25 parts AO solution and the mixture allowed to stand at room temperature for 6–8 min before flow cytometric analysis.

V. Critical Aspects of the Procedure

A. Cell Viability

Generally, at low concentrations of AO ($<10^{-5} M$), only live cells accumulate the dye in lysosomes or azurophilic granules (which luminesce red) and dead cells stain uniformly green. At that low AO concentration, the green component probably reflects dye interaction with DNA and RNA (Darzynkiewicz and Kapuscinski, 1990). However, it should be stressed that, at higher AO concentrations ($<10^{-4} M$), dead cells stain rapidly and uniformly red (nucleus and cytoplasm) while live cells still have red lysosomes but green nuclei and cytoplasm. However, with time, as the intracellular concentration of AO increases in living cells they also stain uniformly red. Caution should therefore be exercised in interpreting data on supravital staining with AO. Clearly, to ensure appropriate staining of lysosomes to detect viable cells, the appropriate (low) AO concentration is important.

Unfortunately, the question of cell viability does not fit neatly into a two-compartment model, i.e., live or dead. Cells exposed to stress, be it as a result of exposure to toxic chemicals, gases, or radiation or the withdrawal of necessary nutrients, may undergo a variety of physiological changes. Often these physiological changes represent a continuum ranging from no effect to immediate cell death. Cell death by necrosis means that the cell membranes have broken down. Such cells are no longer capable of concentrating dyes like AO in the lysosomal granules with the result that the remaining nuclei, with or without cytoplasmic tags attached, fluoresce green (see Section VIII, Results). In some instances, toxic agents (or removal of nutrients) will trigger a cascade of biochemical events leading toward specific morphological, biochemical, and molecular changes consistent with apoptosis (Darzynkiewicz *et al.*, 1992). Early stages of the apoptotic process are often marked by activation of an endogenous protease and endonuclease which results in digestion of internucleosomal stretches of nuclear DNA (Arends *et al.*, 1990). However, concomitant to nuclear DNA digestion, cellular membrane integrity, mitochondrial membrane potential, and lysosomal proton pump activity remain generally intact (Del Bino *et al.*, 1991). Therefore, cells undergoing apoptosis will retain their ability to concentrate AO into red luminescing lysosomal granules, at least in the early stages of the process. As a result, populations deemed viable by supravital AO staining of lysosomes may contain some portion of cells which have no reproductive capability.

B. Leukocyte Differential

Supravital staining of leukocyte subpopulations with AO require appropriate adjustments to ensure that cell density is within an acceptable range. Generally a 1 : 25 dilution of whole blood into the staining solution is optimum when the typical leukocyte count is within the range of $5-25 \times 10^6$ mm^3. However, in instances in which leukemic blood is being stained, it may be necessary to first dilute the blood with a balanced salt solution to bring the cell count within the range noted above. It is also necessary, as will be illustrated below, to wait until the staining reaches equilibrium, which is invariably between 6 and 8 min. After reaching the initial staining equilibrium, the staining pattern is quite stable for up to about 20 min.

Spectral overlap between the green and red emission of the dye should be considered and use of appropriate dichroic mirrors and long-pass filters for the "red" PMT chosen accordingly. Since the green fluorescence is fairly weak in this procedure, its spillover into the red PMT channel is not as much a problem as with other AO staining techniques. However, most flow cytometers use a 590-nm dichroic mirror and 600-nm long-pass filter for the red PMT which is optimized for propidium iodide (DNA) or phycoerythrin (immunofluorescence). A more appropriate filter assembly would contain a dichoic filter reflecting (or transmitting) at 610 nm and a long-pass filter transmitting above 640 nm for optimum measurement of AO red luminescence.

VI. Controls and Standards

Controls and standards for lysosomal proton pump activity can take several forms. Normally, treated, exponentially growing cells will provide a positive staining control for AO red luminescence (see Section VIII, Results). Generally, such a population should be set at midway to two-thirds maximum on the red luminescence scale. This will leave room for the possibility that certain treatments actually increase lysosomal proton pump activity over basal levels. Alternatively, the low end of the red luminescence scale can be established with appropriately chosen fluorescent beads, with peripheral blood lymphocytes which tend to have little or no lysosomal activity when inactivated, or by the examination of the cell system following permeabilization of cellular membranes with a detergent (e.g., Triton X-100). Care should be exercised in using detergent-treated cells as controls if it is important to retain the cytoplasm since detergents dramatically increase the fragility of the cytoplasmic membrane. Alternatively, isolated nuclei would also provide an excellent control for the lower limit of red luminescence of a particular cell system.

Whole blood from a healthy donor provides an optimal control for supravital AO staining of peripheral blood leukocytes. Granulocytes which stain the brightest and are most variable in their AO red luminescence should be positioned slightly higher than midway along the red luminescence axis since (see Section VIII, Results), in some granulocytic leukemias and in instances of bacterial infections (Melamed *et al.*, 1974), the red luminescence of the immature granulocyte population is skewed toward higher values.

VII. Instruments

Almost any flow cytometer with a light source capable of providing blue (~488 nm) excitation and able to simultaneously detect fluorescence emission at two separate wavelengths can be used for this assay, i.e., it has been successfully performed on Ortho Cytofluorographs with closed (nonsorting) channels, Ortho ICP 22s (utilizing a BG 38 filter to select the appropriate band of illumination from the mercury arc lamp), and a Becton–Dickinson FACscan. Generally, the shorter the distance traveled from the time the sample stream first comes into contact with the sheath flow and the intersection with the exciting light source, the easier and more straightforward the measurement. Often sorting channels because of their architecture tend to induce some interaction between sheath and sample streams which can adversely affect equilibrium staining with AO. Anything which affects (lowers) the dye concentration may result in too few dye molecules that, in turn, would inhibit formation of dye aggregates and/or dye-ligand condensation, diminishing or abolishing red luminescence.

VIII. Results

An example of the use of the AO supravital staining technique to assay lysosomal function is displayed in Fig. 1. The human lymphocytic leukemia cell line (MOLT-4) was incubated for 30 min, at 37°C, with 1 μg/ml AO in

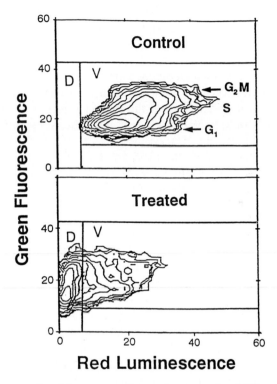

Fig. 1 The AO lysosomal staining pattern of control, untreated and H7-treated MOLT-4 cells. The cells were grown in RPMI 1640 medium supplemented with 10% fetal bovine serum as described previously (Traganos *et al.*, 1993). The treated culture received 200 μM H7 (an isoquinoline sulfonamide from Seikagaku Kogyo Co., Tokyo) for 24 hr. From a 1 mg/ml stock solution, AO was added to each culture to achieve a final concentration of 2 μg/ml, and the cultures were returned to the CO_2 incubator for 30 min at 37 °C. Aliquots of cells were removed from the cultures at the end of the incubation and run on an Ortho Cytofluorograf System 30 flow cytometer. An air-cooled argon ion laser was used to provide blue (488 nm) light for excitation of the AO. Filters and a dichroic mirror separated the resultant emission into green (510–530 nm) and red (>640 nm) luminescence which was amplified, digitized, and recorded using the Acqcyte software from Phoenix Flow Systems (San Diego, CA). The green and red luminescence of 1×10^4 cells was recorded and displayed as contour maps from control and H7-treated cultures. The approximate cell-cycle position of cells in control cultures is reflected in the green fluorescence whereas the red luminescence is proportional to the degree to which the cells were able to concentrate the stain in lysosomal granules. Treatment with H7 caused a decrease by more than 70% in the number of cells taking up AO intralysosomally.

complete tissue culture medium (RPMI 1640 plus 10% serum). The cells were removed directly from culture and their fluorescence was measured on an Ortho Cytofluorograf system 30. Dye excitation was provided by a 150-mW Omnichrome air-cooled argon ion laser. Green fluorescence and red luminescence of 2×10^5 cells were recorded using the Acqcyte program (Phoenix Flow Systems, San Diego, CA). The data were then plotted as contour maps of the correlated green fluorescence and red luminescence of the control (top) and treated (bottom) MOLT-4 cells. Treatment in this instance consisted of 24-hr exposure to 200 μM H7, a serine/threonine kinase inhibitor which has been demonstrated to inhibit cell growth at lower concentrations (Traganos et al., 1993).

The green fluorescence distribution of control cultures actually provides information on the cell-cycle phase distribution. However, under these staining conditions, AO great fluorescence is not expected to represent DNA content as would be the case when using the AO staining reactions designed for permeabilized cells (Darzynkiewicz et al., 1975). Nevertheless, G_0/G_1 cells are expected to have the least green fluorescence, and as cells increase in size their green fluorescence is expected to increase. The red luminescence of untreated MOLT-4 cells should be proportional to the volume of the cellular lysosomal granules and the gradient across the lysosomal membrane. Treatment with H7, which in this instance resulted in complete loss of proliferative potential and a decrease in viable cell number, also caused a shift of AO red luminescence in 70% of the cells from the viable (V) compartment to the dead (D) cell compartment (Fig. 1). The remaining 30% of the treated cells had red luminescence values toward the low end of the viable cell compartment. Note that there is one single, though heterogenous, population. This is presumably the result of the fact that some cells have entirely lost their membrane integrity, some have leaky membranes, and a few cells are still intact and capable of concentrating AO in lysosomes. Such a response is not uncommon when dealing with agents which induce apoptosis. Thus, 100 μM H7 caused cells to undergo apoptosis, which at 24 hr did not affect the ability of MOLT-4 cells to accumulate AO in lysosomes or exclude propidium iodide. It is not unusual that higher concentrations of agents which cause apoptosis induce necrosis in some or all cells depending on concentration and length of exposure. In this instance some 70–90% of the cells were nonviable based on AO and propidium iodide/rhodamine 123 cell staining, respectively (Traganos et al., 1993).

The use of the supravital AO staining technique to classify leukocyte populations from human peripheral blood is illustrated in Fig. 2. Whole blood from a healthy donor was stained as described above and the flourescence recorded after incubation times were increased using an Ortho Cytofluorograf System 30.

As early as 3 min following addition of the blood to the staining solution it is possible to differentiate between the three main populations: lymphocytes with little AO red luminescence consistent with the presence of few lysosomes,

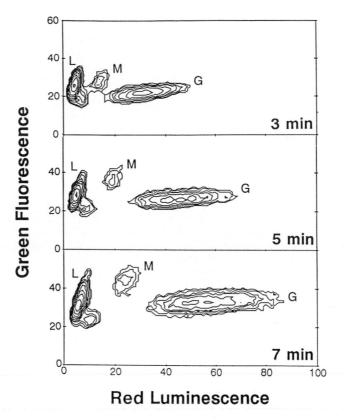

Fig. 2 Identification of white blood cell populations based on supravital staining of whole blood with AO. A 0.1-ml aliquot of whole heparanized blood obtained by venipuncture of a healthy control was admixed to 2.5 ml of AO (1 μg/ml) in a phosphate-buffered saline solution adjusted to pH 7.4. The specimen was introduced to the sample delivery system of an Ortho Cytofluorograf System 30 and the green and red luminescence of 1×10^4 cells recorded after 3, 5, and 7 min of incubation. The populations labeled L, M, and G represent lymphocytes, monocytes, and granulocytes as has been previously demonstrated (Melamed *et al.*, 1972a). A minor subpopulation of lymphocytes, characterized by increased AO uptake, is evident. Generally, stabilization of the staining pattern is achieved by 7 min of incubation providing maximum red luminescence and maximum separation of leukocyte populations.

monocytes with a moderate amount of red luminescence, and granulocytes which have the highest and most variable amount of red luminescence. As the incubation time increased to 5 min, total separation was achieved for all three populations (Fig. 2). In addition, a subpopulation associated with the lymphocytes became more evident. This subpopulation has not been sorted but represents lymphocytes with a slight increase in cytoplasmic red (AO) luminescence that is often observed early during lymphocyte activation (Darzynkiewicz and Kapuscinski, 1990). Finally, by 7 min the increase in red luminescence plateaued and the separation between populations reached a maximum (Fig. 2). This pattern remained relatively stable for the next 10–15 min.

Acknowledgments

This work was supported in part by U.S. Public Health Service NCI Grants R01 CA28704 and R37 CA23296, as well as the Carl Inserra Fund and the "This Close" Foundation.

References

Abrams, W. R., Diamond, L. W., and Kane, A. B. (1983). *J. Histochem. Cytochem.* **31,** 737–744.

Allison, A. C., and Young, M. R. (1964). *Life Sci.* **3,** 1407–1414.

Arends, M. J., Morris, R. G., and Wyllie, H. (1990). *Am. J. Pathol.* **136,** 593–608.

Darzynkiewicz, Z., and Kapuscinski, J. (1990). *In* "Flow Cytometry and Sorting" (M. R. Melamed, T. Lindmo, and M. L. Mendelsohn, eds.), 2nd ed., pp. 291–314. Wiley-Liss, New York.

Darzynkiewicz, Z., Traganos, F., Sharpless, T. K., and Melamed, M. R. (1975). *Proc. Natl. Acad. Sci. U.S.A.* **73,** 2881–2886.

Darzynkiewicz, Z., Bruno, S., Del Bino, G., Gorczyca, W., Hotz, M. A., Lassota, P., and Traganos, F. (1992). *Cytometry* **13,** 795–808.

de Duve, C., de Barsy, T., Poole, B., Trouet, A., Tulkens, R., and Van Hoof, F. (1974). *Biochem. Pharmacol.* **23,** 2495–2531.

Del Bino, G., Lassota, P., and Darzynkiewicz, Z. (1991). *Exp. Cell Res.* **193,** 27–35.

Kapuscinski, J., and Darzynkiewicz, Z. (1984). *Proc. Natl. Acad. Sci. U.S.A.* **81,** 7368–7372.

Melamed, M. R., Adams, L. R., Zimring, A., Murnick, J. G., and Mayer, K. (1972a). *Am. J. Clin. Pathol.* **57,** 95–102.

Melamed, M. R., Adams, L. R., Traganos, F., Zimring, A., and Kamentsky, L. A. (1972b). *Cancer (Philadelphia)* **29,** 1361–1368.

Melamed, M. R., Adams, L. R., Traganos, F., and Kamentsky, L. A. (1974). *J. Histochem. Cytochem.* **22,** 526–530.

Moriyama, T., Takano, T., and Ohkuma, S. (1982). *J. Biochem. (Tokyo)* **92,** 1333–1336.

Ohkuma, S., and Poole, B. (1978). *Proc. Natl. Acad. Sci. U.S.A.* **75,** 3327–3331.

Robbins, E., and Marcus, P. I. (1963). *J. Cell Biol.* **18,** 237–250.

Robbins, E., Marcus, P. I., and Gonatas, N. K. (1964). *J. Cell Biol.* **21,** 49–62.

Rundquist, I., Olsson, M., and Brunk, U. (1984). *Acta Pathol. Microbiol. Immunol. Scand.* **92,** 303–309.

Traganos, F., Knutti-Hotz, J., Hotz, M., Gorczyca, W., Ardelt, B., and Darzynkiewicz, Z. (1993). *Int. J. Oncol.* **2,** 47–59.

Yamashiro, D. J., Fluss, S. R., and Maxfield, F. R. (1983). *J. Cell Biol.* **97,** 929–934.

Zelenin, A. V. (1966). *Nature (London)* **212,** 425–426.

CHAPTER 13

Staining of DNA in Live and Fixed Cells

Harry A. Crissman* and Gregory T. Hirons†

* Division of Biomedical Sciences
Los Alamos National Laboratory
Los Alamos, New Mexico 87545
† University of California School of Medicine
Irvine, California 92715

I. Introduction

After almost 25 years, quantitative fluorescent staining and flow cytometric (FCM) analysis of cellular DNA remain the most rapid and reliable approach for assessing relative DNA contents of various cell types, as well as for obtaining cell-cycle frequency distributions and chromosome profiles of cell populations. The precision in these studies relies upon the specificity of the staining methods for tagging DNA in cells and chromosomes and the efficiency of flow instruments for quantitative fluorescence analysis. Early studies comparing computer-fit analysis of DNA content histograms to data obtained by conventional tritiated thymidine labeling and autoradiography confirmed the accuracy of the FCM technique for cell-cycle analysis. Those studies provided the credibility for firmly establishing FCM as an important technology, not only for DNA content, but also for the many other biological analyses that have subsequently been developed. Fortunately, the similarity in the design of FCM instruments and computer analysis programs has made it possible to compare results of DNA content studies conducted throughout the world.

In addition to the rapidity of both DNA staining and FCM analysis, there are other advantages, we have previously mentioned (Tobey and Crissman, 1975), such as the abilities to (a) monitor cell-cycle distributions in ongoing experiments and alter an experiment in progress; (b) localize cells in S phase and distinguish between early, mid, and late S phase; (c) additionally analyze other parameters such as RNA and protein contents of the cell populations; (d) monitor the various phases in populations composed of slowly progressing or arrested cells; (e) analyze populations devoid of S phase or mitotic cells; (f) analyze populations containing cells unable to transport, incorporate, or metabolize tritiated thymidine; and (g) detect abnormalities in progression through mitosis such as nondisjunction or polyploidization. DNA content analysis has also been useful for detecting aneuploid subclones within a cell population, and from a clinical standpoint such information is often of prognostic value. In addition, cells can be sorted based on DNA content, so there is not the need to synchronize cells for phase-specific biochemical analysis. There are, however, notable limitations on the technique.

Flow cytometric DNA content analysis describes the cycle position but provides no information on the cycling capacity and the rate at which cells will traverse the cell cycle. The cell-cycle frequency histogram is often compared to a "snapshot" photograph of individuals in a race. The picture shows the position of all runners at a given instant, but the speed and the capacity of each runner to successfully complete the race are indeterminant. For example, two exponentially growing cell populations may have very different doubling times [i.e., 15 hr for Chinese hamster ovary (CHO) cells versus 24 to 30 hr for HL-60 cells] but still yield similar DNA content distributions, if the duration of the individual phases of the cell cycle in proportion to the cycling time is similar.

Furthermore, under experimental conditions, it is difficult to ascertain the proportions of cycling and noncycling cells within the various phases, even though sequential sampling of the population over the course of an experiment may alleviate this problem to some extent.

In recent years flow cytometry has been used to elucidate other physiological parameters, which, in addition to DNA metabolism, regulate and control cell proliferation. Many current studies involve additionally labeling and measuring cellular constituents such as proteins and RNA simultaneously with DNA. Cellular levels of such descriptors and others are known to be important indicators of cell-cycle progression capacity, cell growth, and function. The sample preparation method, the fluorochrome probe combinations, and the flow instrument capabilities for analysis of the various cellular descriptors will determine the choice of DNA labeling procedure.

II. Applications with DNA-Specific Fluorochromes

Currently there is a significant number of fluorochromes available with different spectral properties and/or modes of binding to DNA. In this chapter we review and compare the various preparative methods and staining protocols employed with the different DNA-specific fluorochromes currently used for cell-cycle analysis by FCM. Each procedure has proven to be quite applicable to such studies, so the choice of a particular technique and/or fluorochrome depends primarily on the specific application.

A. Feulgen–DNA Reaction

The Feulgen procedure is rarely used today; however, it represents one of the early cytochemical procedures for labeling and quantitating cellular DNA content by FCM. Modifications of the procedure were used in earlier FCM studies employing the fluorochromes auromine 0 or acriflavine. The Feulgen method is time consuming and produces cell loss and nonspecific staining in many cell types. In addition, HCl hydrolysis, an essential step in the protocol, depurinates DNA, removes RNA and histones, and potentially damages other cellular components including membranes. Therefore, details on the procedure are omitted here but may be reviewed in Crissman *et al.*, 1979.

B. DNA-Reactive Hoechst Dyes, DAPI (4-6-Diamidino-2-phenylindole), DIPI [4-6-bis(2-Imidazolynyl-4H,5H)]-2-phenylindole], and LL 585

The Hoechst dyes are nonintercalating, benzimidazole derivatives that bind preferentially to A–T base regions and emit blue fluorescence when excited by UV light at about 350 nm. Latt (1973) showed that the fluorescence of Hoechst 33258 was quenched when bound to BrdUrd-substituted DNA and used the phenomenon to develop a technique for detecting regions of sister chromatid exchange in metaphase chromosomes. Arndt-Jovin and Jovin (1977) first demonstrated the use of both Hoechst 33258 and 33342 for quantitative DNA staining and FCM sorting of viable cells. The structural and spectral properties of DAPI and DIPI are similar and each can be used interchangeably with the Hoechst 33258 or 33342 derivatives for staining ethanol-fixed cells. However, neither DAPI or DIPI are as sensitive to BrdUrd as the Hoechst dyes.

Latt et al. (1984) have examined the DNA binding properties and cell staining characteristics of the fluorescent compound, LL 585 (Eastman-Kodak, Rochester, NY). The spectral characteristics of LL 585 in the visible range are similar to those of propidium iodide, but it binds preferentially to A–T sites and does not intercalate into double-stranded DNA. In contrast to HO 33342, LL 585 does not permeate viable cells but human lymphoblasts, permeabilized with 0.1% Triton X-100 and stained with LL 585 yielded typical cell-cycle distributions (coefficient of variation 2.9%) when analyzed by FCM using the 514-nm line of an argon ion laser beam.

C. Considerations for Viable Cell Staining with Hoechst 33342

Since synchronization protocol can potentially perturb the cell cycle, DNA-specific staining in viable cells, followed by FCM analysis and sorting provides an ideal alternative approach for selecting and recovering cells from various phases of the cell cycle. The sorted cells may be cultured and examined with regard to long-term viability, functional activity, as well as immunological and other physiological properties. These studies have significantly advanced the understanding and interpretation of cellular aspects which regulate and control cycle progression and cell proliferation.

Unfortunately, most DNA-specific fluorochromes are not easily transported into or retained by viable cells, so intracellular concentration levels sufficient for optimal staining and analysis are not often attained. Some dyes, such as DAPI, that can penetrate membranes of some cell types are very cytotoxic, so HO 33342 remains the preferred dye for viable cell staining even though use of this compound also has limitations.

Dyes such as HO 33342, which bind to DNA, can potentially interfere with cell replication and also impair viability. Fried et al. (1982) studied the effects of the dye on survival of various cell types and demonstrated that HeLa S-3 cells were highly resistant, but SK-DHL2 cells were very sensitive to the fluorochrome. Pallavicini et al. (1979) have shown that X-irradiated cells had greater sensitivity to HO 33342 than unirradiated cells, suggesting a possible synergistic cytotoxic effect. However, two recent studies showed that treatment of viable cells with membrane interacting agents may improve Hoechst uptake and possibly improve survival.

We found that the membrane potential, mitochondrial stain, DiO-C5-3, when applied to viable CHO cells in conjunction with Hoechst 33342, increased cellular uptake of the Hoechst dye twofold and provided coefficient of variation values of about 3.0% compared with 8.3% for cells treated with Hoechst alone (Crissman *et al.*, 1988). However, DiO-C5-3 treatment did not increase Hoechst uptake in L1210 cells or human skin fibroblasts, but these cell types routinely stain well with Hoechst compared with CHO cells. Krishan (1987) has also shown that calcium channel blocking agents, such as verapamil, can also increase Hoechst stainability in viable cells that are normally refractory to Hoechst uptake. Those studies indicated that rapid metabolic dye efflux could, in some cases, account for poor Hoechst stainability.

D. Mithramycin, Chromomycin, and Olivomycin

Mithramycin (MI) is a green-yellow fluorescent DNA-reactive antibiotic similar in structure and dye binding characteristics to chromomycin A3 and olivomycin. Complexes of these compounds with magnesium ions preferentially bind to G–C base regions by nonintercalating mechanisms. We studied mithramycin extensively with regard to its spectral characteristics and for quantitative DNA staining in ethanol-fixed cells examined by flow cytometry (Crissman *et al.*, 1979). Spectrofluorometric analysis of mithramycin or chromomycin A3–Mg complexes bound to DNA in PBS show two excitation peaks, a minor peak in the UV (about 320 nm) and a major peak at about 445 nm. A broad green-yellow emission spectrum is observed with a peak at 575 nm. Olivomycin by comparison has slightly lower wavelength excitation and emission characteristics.

Larsen *et al.* (1986) discriminated mitotic cells from interphase cells in FCM analysis of nuclear suspensions prepared with non-ionic detergent, fixed in formalin, and stained with MI, as well as propidium iodide (PI) or ethidium bromide (EB). Formaldehyde quenched the fluorescence of these dyes in interphase cell nuclei to a greater extent than mitotic nuclei which had MI fluorescence intensities 20–40% greater than but light scattering 30–60% lower than G_2 cell nuclei. Pierrez *et al.* (1987) obtained MI staining of peripheral blood and bone marrow samples following picric acid–alcohol fixation. In another application, van Kroonenburgh *et al.* (1985) distinguished and sorted eight different cell populations in MI-stained rat testis cell suspensions using bivariate analysis of the peak amplitude of the fluorescence signal versus the total fluorescence intensity integrated over time.

E. Propidium Iodide, Ethidium Bromide, and Hydroethidine

The red fluorochromes, propidium iodide and ethidium bromide, have similar chemical structures and both intercalate between base pairs of double-stranded DNA and RNA without base specificity. The dyes are used as DNA-specific stains following pretreatment of fixed cells with RNase. Dittrich and Gohde (1969) used EB for cell-cycle analysis studies by flow cytometry. PI was first used by Hudson *et al.* (1969) in a buoyant density procedure for isolating closed circular DNA, and later we introduced the dye in FCM in a double-staining technique with fluorescein isothiocyanate for analysis of DNA and protein (Crissman and Steinkamp, 1973). Spectral studies on PI bound to calf thymus DNA in PBS show two excitation peaks, a minor but substantial peak in the UV region (340 nm), and a major peak at about 540 nm. One emission peak was observed at 615 nm.

Neither PI nor EB penetrate viable cells, but cells with damaged membranes stain readily, so these dyes are often used for differentiating and quantitating viable and dead cells in a given population. Methods have been developed for rapid staining of unfixed cells with PI. Techniques vary with procedures used for permeabilizing cell membranes and/or tissue disaggregation (Krishan, 1975, 1990; Vindeløv and Christensen, 1990).

Mazzini and Giordano (1979) showed that the quantum efficiencies of PI and EB are significantly enhanced in solutions with deuterium oxide, and dye concentrations as low as 1.0 μg/ml adequately stained cells for FCM analysis. This technique requires that no significant mixing of the deuterium oxide in the cell stream occurs with the saline or water in the sheath fluid (Mazzini, personal communication).

Hydroethidine (HE) is a fluorescent compound produced by the reduction of EB. Gallop *et al.* (1984) showed that HE rapidly enters viable cells, where it is enzymatically dehydrogenated to ethidium which then intercalates into double-stranded DNA and RNA and fluoresces red. Nonreacted HE in the cytoplasm fluoresces blue when excited at 370 nm. Using 535-nm excitation only red (ethidium) is seen. Luce *et al.* (1985) have used HE in conjunction with sulfofluorescein diacetate to enumerate cytotoxic cell and target cell interactions by FCM.

F. Acridine Orange

Acridine orange is a metachromatic dye that fluoresces green while intercalated between DNA base pairs and red when stacked on RNA. The chemistry, binding properties, and flow cytometric applications of the dye have been carefully examined and experimentally exploited by Darzynkiewicz (1990).

G. TOTO-1, YOYO-1, TO-PRO-1, and YO-PRO-1

TOTO-1 and YOYO-1 are modified dimers of the dyes thiazole orange and oxazole yellow, respectively, with DNA and RNA specificity (Molecular Probes Inc., Eugene, OR). These fluorochromes bind to DNA by intercalation, and the relative fluorescence intensities of these dyes bound to the DNA bands in gels indicate that dye binding is proportional to DNA content rather than base composition. Haugland (Molecular Probes, Inc., Handbook 1992–1994) has reported quantum efficiencies (QE) of 0.34 for TOTO-1 and 0.52 for YOYO-

1. The ultrasensitivities of TOTO and YOYO, coupled with their excitation characteristics, 488 and 457 nm wavelengths, respectively, make these fluorochromes useful for FCM.

Previously we performed FCM studies, including DNA content analyses, of RNase-treated nuclei and RNase-treated fixed cells, stained with micromolar concentrations of TOTO and YOYO and their respective monomers, TO-PRO-1 and YO-PRO-1 (Hirons *et al.*, 1994). The DNA histograms obtained for nuclei were, in general, superior to those obtained with PI and MI, and by comparison YOYO-stained nuclei were over 1000 times more fluorescent than MI-stained nuclei, on an equimolar basis. The potential of TOTO and YOYO for staining human chromosome for flow karyotype analysis was also examined in our studies.

H. Additional DNA Binding Fluorochromes

A number of other DNA-specific dyes have been introduced into flow cytometry. 7-Amino-actinomycin D (7-Act D; Modest and Sengupta, 1974), a red fluorescent analogue of actinomycin D (excitation 540 nm), intercalates into G–C regions in DNA. Gill *et al.* (1975) studied the spectral properties of the dye, and Darzynkiewicz *et al.* (1984) examined the accessibility of DNA to 7-Act D following DMSO-induced differentiation of erythroid Friend leukemia cells. Evenson *et al.* (1986) studied changes in DNA accessibility during spermatogenesis of 7-Act D and several other fluorochromes, including another intercalating dye, ellipticine. Zelenin *et al.* (1984) performed a detailed study on the use of 7-Act D for cell staining and FCM analysis. Shapiro and Stephens (1986) examined DNA staining with the laser dyes, oxazine 750, LD700, and rhodamine 800. These dyes have advantages for FCM studies, since, they can be excited with a low-power, relatively inexpensive, helium-neon laser (633 nm).

I. Multiple Fluorochrome Labeling of DNA

Although DNA-specific stains are most often used as single-dye agents, several FCM studies have demonstrated advantages in combining some of these dyes. Van Dilla *et al.* (1983) used differences in DNA base composition to distinguish three bacterial strains stained with HO and CA3, and Dean *et al.* (1982) used this same dye combination to directly compare cell-cycle analysis results from cells stained with several DNA stains. Darzynkiewicz *et al.* (1992) reviewed procedures for staining with HO and PI for detecting apoptotic cells. Also, we have used a combination of HO, MI, and PI, for three-color staining and FCM analysis to detect modifications in chromatin structure in different phases of the cell cycle (Crissman and Steinkamp, 1993). Such structural modifications were reflected by the phase-specific redistribution in the relative proportion of sites on DNA available for binding by each of the three dyes.

III. Materials

The various DNA fluorochromes are available from several commercial companies including Molecular Probes, Inc. (Eugene, OR), Polysciences, Inc. (Warrington, PA), Calbiochem-Behring Corp. (La Jolla, CA), and Sigma Chemical Co. (St. Louis, MO). Mithramycin is available from Pfizer Co. (Groton, CT).

Stock solutions of most of the dyes are prepared in PBS at concentrations of 1.0 mg/ml; however, DAPI and the Hoechst dyes should be prepared in distilled water, since at relatively high concentrations these dyes tend to precipitate in PBS. TOTO and YOYO (Molecular Probes, Inc.) are available in DMSO at 1.0 mM concentrations. Stock solutions, refrigerated in dark-colored containers or wrapped in foil, have been used for at least 1 month without noticeable degradation.

IV. Cell Preparation and Fixation

Flow systems do not distinguish fluorescent cellular debris and cell clumps from properly stained single cells, so production of good quality, single-cell suspensions is a requirement for reliable analysis. Also, since DNA measurements are often coupled with analyses of other cellular constituents, cell preparation as well as fixation is an important consideration for ensuring that all the properties of interest are optimally preserved. For example, analysis of DNA content only requires intact nuclei, while analyses of DNA and cell membrane components require mild dispersal methods to preserve intact cells. Also, paraformaldehyde fixatives, which are preferable to ethanol for preserving membrane antigens, often interfere with DNA dye binding, which is best achieved with ethanol fixation. Therefore, the choice of preparation and fixation is determined by the biological sample, the constituent to be preserved, and the fixative that best preserves the cell constituents and subsequently provides for optimal fluorochrome labeling.

Membranes of viable cells that exclude many of the fluorochromes can be permeabilized by brief treatment with non-ionic detergents, or with hypotonic solutions, proteolytic enzymes, and/or for rapid DNA staining. However, membrane components and cytoplasmic constituents can be significantly lost from analysis by these treatments. Alternatively, ethanol fixation also perforates cell membranes, but preserves most cytoplasmic materials and does not appear to seriously affect dye binding to DNA. Detailed studies should be performed, however, to determine the extent of preservation of cellular constituents after fixation. Ethanol fixation involves harvesting cells from culture or from tissue dispersal solutions by centrifugation and then *thoroughly* resuspending the cells in one part cold "saline GM" (g/liter: glucose 1.1; NaCl, 8.0; KCl, 0.4; Na$_2$HPO$_4$·12 H$_2$O, 0.39; KH$_2$PO$_4$, 0.15) containing 0.5 mM EDTA for chelating free calcium and magnesium ions. Then three parts cold, 95% nondenatured ethanol is added to the cell suspension with mixing to produce a final ethanol concentration of about 70%.

Paraformaldehyde (1.0%) in cacodylate buffers (0.05–1.0 M) preserves cell membranes well and the electronic cell volume distributions of fixed cells are nearly identical to those obtained for viable cells at the same instrument gain setting. For DNA content by Hoechst 33342 (2.0 μg/ml) and membrane studies, cells can be fixed 2 hr in paraformaldehyde, rinsed, and stored for at least 1 month in Ca^{2+} and Mg/free phosphate-buffered saline.

Hedly *et al.* (1985) and Hedley (1990) devised methods for preparing nuclei from formalin-fixed, paraffin-embedded tissue. Following removal of paraffin with xylene or the less toxic commercial product "Histoclear," tissue samples are rehydrated, treated 30 min with aqueous 0.5% pepsin-HC1, and subsequently stained for DNA with DAPI or PI.

A protocol described by Gurley *et al.* (1973) has proven useful for preparing CHO and HL-60 nuclei. Cells are harvested from culture medium by centrifugation at 4°C for 8 min at 200g, and the supernatants removed by aspiration. Cells are then washed twice with 5.0 ml of cold saline GM (listed below) and 4.0 ml of RSB swelling solution, containing 0.01 M NaCl, 0.0015 M $MgCl_2$, 0.01 M Tris (pH 7.4), and 50 μg/ml RNase (Worthington Biochemical Corporation, Freehold, NJ), is added to each cell pellet, and the cell pellets are then vortexed vigorously for 30 sec and allowed to stand on ice for 5 min. After addition of 0.5 ml of 10% Nonidet P-40 (NP-40) detergent (Polysciences Inc.), the samples are vortexed again vigorously for 30 sec and put on ice for 15 min. In nuclear staining studies with TOTO and YOYO, 0.5 ml of 5% sodium deoxycholate (DOC) detergent (Mann Research Laboratories Inc., NY) was also added to the samples, after which they were vortexed for 30 sec and then put on ice for another 15 min. The final concentration of the nuclei was ~1.0 × 10^6/ml.

V. Cell Staining

Viable cell staining for DNA is usually performed by direct addition of HO 33342 (final concentration 2.0–5.0 μg/ml) to cells in culture medium (37°C) for incubation periods of 30–90 min depending on the cell type. Some cell types stained at 4°C yield poor DNA distributions, possibly indicating the requirement of metabolic conditions that favor active dye transport and retention. However, even with optimal staining conditions, dye uptake, cytotoxicity, DNA binding, and analytical resolution, as judged by coefficients of variation (CV) in the FCM-DNA histograms, are cell-type dependent. If the cell membrane is perforated with non-ionic detergents (i.e., NP-40, Triton X-100), staining in most cell types is rapid (i.e., 5–10 min) even at dye concentrations of 0.5 μg/ml. Untreated cells with damaged membranes also stain rapidly. For more details see Crissman *et al.* (1990).

Staining of ethanol-fixed cells can be performed in PBS containing dyes at the following concentrations: PI or EB, 15 μg/ml; MI or ChA3, 50 μg/ml with 20 mM MgCl$_2$; DAPI, 1–2 μg/ml; and Hoechst dyes, 0.5–1.0 μg/ml. RNase (50–100 μg/ml) is added to the stain solution for PI and EB. Dye concentrations required to obtain optimal results may vary slightly depending upon the flow instrument used. Staining time at room temperature is about 30 min–1 hr.

For staining fresh nuclear preparations, dye should be added directly to the preparation without centrifugation, which tends to cause clumping of the nuclei. The staining concentration for unfixed nuclei is 2.0 μg/ml with DAPI and 2–5 μg/ml with Hoechst depending on the cell type. Stock solutions of TOTO, YOYO, TO-PRO-1, and YO-PRO-1 (1.0 mM) are added directly to fresh nuclei in the isolation buffer, to obtain dye concentrations of 4.0×10^{-6} M for TOTO and YOYO and 4.0×10^{-5} M for TO-PRO-1 and YO-PRO-1. Fluorescence of nuclei or ethanol-fixed cells remains stable for at least 5–6 hr.

VI. Critical Aspects

In testing for DNA specificity of staining, ethanol-fixed cells are treated with DNase I (Worthington) in PBS containing 3 mM MgCl$_2$ for 1 hr at 37°C. Staining with the DNA dyes should be negative as determined by FCM and/or microscopic analysis.

Stained samples should be allowed to run in the FCM for about 1–2 min until the dye solution and the sheath fluid are equilibrated. The position of the G_1 peak will stabilize and the CV will generally decrease over this initial period prior to analysis and collection of data. The time required for stabilization will vary from instrument to instrument, depending on the hydrodynamics of the flow system.

Stoichiometry of DNA stains can be established from the ratio of the G_2/M and G_1 peak values in the DNA histograms. Ideally this ratio should be close to a value of 2.0. Also the computer fit of the DNA histogram should provide nearly the same percentage of G_1, S, and G_2/M phase cells as calculated by an independent method such as with tritiated thymidine. Coulson *et al.* (1977) tested the stoichiometry of various DNA staining reactions using cells from different animal species which varied in DNA content and found that PI staining of ethanol-fixed cells following RNase treatment provided the best results proportional to relative DNA content.

Experimental drugs which interact directly with DNA or interfere with DNA metabolism can have effects on subsequent staining. In *in vivo* studies, Alabaster *et al.* (1978) found differences in MI-DNA stainability of untreated (control) L1210 ascites cells and cells treated with a combination of cytosine arabinoside and adriamycin. Krishan *et al.* (1978) found that adriamycin diminished staining by propidium iodide.

All DNA-specific compounds should be considered potential carcinogens and therefore handled with some precautions. Solutions should be immediately washed from the skin if accidental contact is encountered. Aerosols created during cell sorting will contain some of the stain solution and some measures must be taken to avoid inhaling any of the dyes.

VII. Controls and Standards

The use of standards such a stained, nucleated trout or chicken red blood cells (RBC) has been proposed by Vindeløv *et al.* (1983). These cells provide internal markers for determining the DNA index (DI), which is calculated by dividing the G_1 peak channel of an aneuploid population by the G_1 peak of normal cells analyzed at the same instrument gain setting. The DI is often of prognostic value in certain types of cancer. Fluorescent microspheres can also be used to align the cell stream in the FCM system and also to calibrate the linearity of the instrument gain setting. When linearity is achieved the increase in the relative intensity of the microspheres should be proportional to the gain setting.

VIII. Instruments

Viable cells in suspension stained with HO are usually analyzed in equilibrium with the dye, using a UV laser beam (~350 nm) or a mercury arc lamp excitation source and analyzing emission above 400 nm. For sorting viable cells the laser excitation power should be reduced as low as possible to reduce potential phototoxicity to the HO-stained cells. The CV in the DNA histogram will increase somewhat at the lower power, but it is necessary to compromise resolution for cell survival. When using MI, chromomycin A3 or YOYO-1 staining, cells are excited at 457.9 nm and fluorescence is analyzed above 500 nm. Cells stained with PI, EB, TOTO-1, LL 585, 7-Act D, or HE are usually excited at 488 nm and emission is collected above 515 nm.

IX. Results and Discussion

As new fluorochromes become available their utility can best be judged on the basis of staining and analysis results compared to data obtained with conventional fluorochromes. DNA histograms for CHO nuclei stained with four new cyanine dyes, TOTO-1, YOYO-1, TO-PRO-1 and YO-PRO-1, are shown along with histograms obtained with PI and MI (Fig. 1A–1F), and Table I provides computer-fit analysis data for the respective DNA histograms (Hirons *et al.*, 1994). The DNA histogram for YOYO-stained nuclei analyzed with a

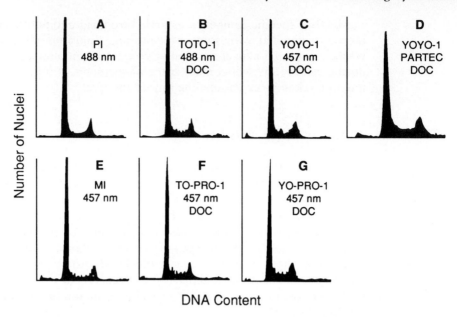

Fig. 1 Comparison of DNA histograms of CHO nuclei stained with (A) propidium iodide (PI), (E) mithramycin (MI), and (B–D and F, G) the cyanine fluorochromes. The histograms shown in A–C and E–G were obtained using the laser excitation wavelengths indicated. The histogram in D was obtained with a PARTEC flow cytometer using a mercury arc lamp source for excitation. In the Partec FCM nuclei were excited at wavelengths <500 nm, and fluorescence was collected at wavelengths >500 nm. Histograms obtained for nuclei prepared with the detergent sodium deoxycholate (DOC) are indicated. Statistical data, including coefficient of variation (CV) values for the histograms in A–C and E–G, are shown in Table I. From Hirons *et al.* (1994). Wiley–Liss © 1994.

Table I
Computer-Fit Analysis of DNA Histograms

Dye	%G_0/G_1	%S	%G_2/M	%CV	Ratio G_2/M to G_0/G_1
PI	55.1	37.3	7.6	4.0	2.05
TOTO	51.9	38.0	10.1	3.5	1.99
YOYO	53.0	37.5	9.5	4.0	2.00
MI	57.3	35.2	7.5	3.6	2.11
TO-PRO-1	50.3	41.5	8.2	3.1	1.96
YO-PRO-1	48.7	40.7	10.6	4.2	1.96

Partec flow cytometer using a mercury arc lamp excitation source (Fig. 1D) is similar to the histogram in Fig. 1C, indicating that low-power excitation can be used for analysis of the cyanine dyes. Preparation of nuclei in isolation buffers containing DOC detergent significantly improved the quality of histogram resolution, increased fluorescence intensity, and yielded lower CV values for both TOTO and YOYO (Figs. 1B and 1C), but only slightly improved these parameters for TO-PRO-1 and YO-PRO-1 (Figs. 1F and 1G).

The CVs, the percentages of cells in different phases of the cell cycle, and the G_2/M to G_0/G_1 modal channel ratios for samples stained with the cyanine dyes (Table I) compare well with data of samples stained with PI or MI. Although cell-cycle frequency distributions varied slightly among all of the dyes, the values were within the variations encountered during repetitive sample runs and computer fits of data. For all of the dyes, values for the modal channel ratios of the G_2/M peak to the G_0/G_1 peak are all close to the expected value of 2.0.

The relative intensities of nuclei stained with the cyanine dyes, compared to nuclei stained with either PI or MI, were also determined from the modal channel values of the G_1 peak of the respective DNA histogram obtained at the same electronic gain setting. At 488 nm excitation, TOTO-stained nuclei were found, on an equimolar dye basis, to be 43 times more intense than PI-stained nuclei, and at 457 nm excitation, YOYO-stained nuclei were 1058 times more intense than MI-stained nuclei.

The sensitivity of the cyanine dyes was examined by comparing the CVs of the MI- and YOYO-stained nuclei excited at 457 nm with decreased laser power. Using 25 mW, the CV for YOYO-stained nuclei remained nearly identical (CV = 3.8%) to the value shown in Table I using 320 mW, while the CV for MI-stained nuclei at 25 mW almost doubled (CV = 5.9%) from the value obtained using 320 mW (CV = 3.6%, Table I). These results and those obtained with the Partec instrument indicate that these cyanine dyes could be used in FCM instruments equipped with relatively low-power excitation sources.

Histograms for ethanol-fixed cells stained with all four cyanine dyes were of poor quality compared to those acquired from fixed cells stained with PI and MI. These results would seem to reflect the poor accessibility of these cyanine dyes to DNA in fixed cells where the DNA may be more tightly complexed with nucleoproteins compared to complexes in unfixed nuclei.

Based on these results and those obtained with human chromosomes (Hirons *et al.*, 1994), it appears that these new cyanine dyes will have applications to many FCM studies, and they should be especially useful in studies where high intensities are required for detection and analysis. The dyes have good specificity for DNA, following RNase treatment, but their utility for analysis of RNA content by FCM remains to be investigated. Also, studies are needed to improve DNA accessibility to these dyes in ethanol-fixed cells, since preparation of nuclei is not often convenient in studies where serial sampling of a cell population is required.

Acknowledgments

This work was supported by NIH Grant R24 RR06758, the Los Alamos National Flow Cytometry Resource (funded by the Division of Research Resources of NIH, Grant P41-RR01315), and the U.S. Department of Energy.

References

Alabaster, O., Tannenbaum, E., Habbersett, M. C., Magrath, I., and Herman C. (1978). *Cancer Res.* **38,** 1031–1035.

Arndt-Jovin, D. J., and Jovin, T. M. (1977). *J. Histochem. Cytochem.* **25,** 585–589.

Coulson, P. A., Bishop, A. O., and Lenarduzzi, R. (1977). *J. Histochem. Cytochem.* **25,** 1147–1153.

Crissman, H. A., and Steinkamp, J. A. (1973). *J. Cell Biol.* **59,** 766–771.

Crissman, H. A., and Steinkamp, J. A. (1993). *Eur. J. Histochem.* **37,** 129–138.

Crissman, H. A., Stevenson, A. P., Kissane, R. J., and Tobey, R. A. (1979). *In* "Flow Cytometry and Sorting" (M. R. Melamed, P. F. Mullaney, and M. L. Mendelsohn, eds.), pp. 243–261. New York: Wiley.

Crissman, H. A., Hofland, M. H., Stevenson, A. P., Wilder, M. E., and Tobey, R. A. (1988). *Exp. Cell Res.* **174,** 388–396.

Crissman, H. A., Hofland, M. H., Stevenson, A. P., Wilder, M. E., and Tobey, R. A. (1990). *In* "Flow Cytometry" (Z. Darzynkiewicz and H. A. Crissman, eds.), pp. 89–95. San Diego, CA: Academic Press.

Darzynkiewicz, Z. (1990). *In* "Flow Cytometry" (Z. Darzynkiewicz and H. A. Crissman, eds.), pp. 285–298. San Diego, CA: Academic Press.

Darzynkiewicz, Z., Traganos, F., Kapuscincki, J., Staino-Coico, L., and Melamed, M. R. (1984). *Cytometry* **5,** 355–363.

Darzynkiewicz, Z., Bruno, S., DelBino, G., Gorczyca, W., Hotz, M. A., Lassota, P., and Traganos, F. (1992). *Cytometry* **13,** 795–808.

Dean, P. N., Gray, J. W., and Dolbeare, F. A. (1982). *Cytometry* **3,** 188–1995.

Dittrich, W., and Gohde, W. (1969). *Z. Naturforsch.* **24B,** 360–361.

Evenson, D. E., Darzynkiewicz, Z., Jost, L., Janca, F., and Ballachey, B. (1986). *Cytometry* **7,** 45–53.

Fried, J., Doblin, J., Takamoto, S., Perez, A., Hansen, H., and Clarkson, B. (1982). *Cytometry* **3,** 42–47.

Gallop, P. M., Paz, M., Henson, S. A., and Latt, S. A. (1984). *Biotechniques* **1,** 32–36.

Gill, J. E., Jotz, M. M., Young, S. G., Modest, E. J., and Sengupta, S. K. (1975). *J. Histochem. Cytochem.* **23,** 793–799.

Gurley, L. R., Enger, M. D., and Walters, R. A. (1973). *Biochemistry* **12,** 237–245.

Hedley, D. W. (1990). *In* "Flow Cytometry" (Z. Darzynkiewicz and H. A. Crissman, eds.), pp. 139–147. San Diego, CA: Academic Press.

Hedley, D. W., Friedlander, M. L., and Taylor, I. W. (1985). *Cytometry* **6,** 327–333.

Hirons, G. T., Fawcett, J. J., and Crissman, H. A. (1994). *Cytometry* **15,** 129–140.

Hudson, B., Upholt, W. B., Deninny, J., and Vinograd, J. (1969). *Proc. Natl. Acad. Sci. U.S.A.* **62,** 813–820.

Krishan, A. (1975). *Cytometry* **8,** 642–645.

Krishan, A. (1990). *In* "Flow Cytometry" (Z. Darzynkiewicz and H. A. Crissman, eds.), pp. 121–125. San Diego, CA: Academic Press.

Krishan, A., Ganapathi, R., and Israel, M. (1978). *Cancer Res.* **38,** 3656–3662.

Larsen, J. K., Munch-Petersen, B., Christiansen, J., and Jorgensen, K. (1986). *Cytometry* **7,** 54–63.

Latt, S. A. (1973). *Proc. Natl. Acad. Sci. U.S.A.* **70,** 3395–3399.

Latt, S. A., Marino, M., and Lalande, M. (1984). *Cytometry* **5,** 339–347.

Luce, G. G., Sharrow, S. O., and Shaw, S. (1985). *Biotechniques* **3,** 270–272.

Mazzini, G., and Giordano, P. (1979). *In* "Flow Cytometry" (O. D. Laerum, T. Lindmo, and E. Thorud, eds.), pp. 74–77. Bergen, Norway: Universitetforlaget.

Modest, E. J., and Sengupta, S. K. (1974). *Cancer Chemother. Rep.* **58,** 35–48.

Pallavicini, M. G., Lalande, M. E., Miller, R. G., and Hill, R. P. (1979). *Cancer Res.* **39,** 1891–1897.

Pierrez, J., Guerci, A., and Guerci, O. (1987). *Cytometry* **8,** 529–533.

Shapiro, H., and Stephens, S. (1986). *Cytometry* **7,** 107–110.

Steinkamp, J. A., Stewart, C. C., and Crissman, H. A. (1982). *Cytometry* **2,** 226–231.

Tobey, R. A., and Crissman, H. A. (1975). *Exp. Cell Res.* **93,** 235–239.

Van Dilla, M. A., Langlois, R. G., and Pinkel, D. (1983). *Science* **220,** 620–622.

van Kroonenburgh, M. J., Beck, J. L., Scholtz, J. W., Hacker-Klom, V., and Herman, C. J. (1985). *Cytometry* **6,** 321–326.

Vindeløv, L., and Christensen, I. J. (1990). *In* "Flow Cytometry" (Z. Darzynkiewicz and H. A. Crissman, eds.), pp. 127–137. San Diego, CA: Academic Press.

Vindeløv, L. L., Christensen, I. J., Jensen, G., and Nissen, N. I. (1983). *Cytometry* **3,** 328–331.

Zelenin, A. V., Poletaev, A. I., Stephanova, N. G., Barsky, V. E., Kolesnikov, V. A., Nikitin, S. M., Zhuze, A. L., and Gnutchev, N. V. (1984). *Cytometry* **5,** 348–354.

CHAPTER 14

High-Resolution Analysis of Nuclear DNA Employing the Fluorochrome DAPI

Friedrich J. Otto

Fachklinik Hornheide
48157 Münster
Germany

I. Introduction

The quality and reproducibility of DNA analysis using flow cytometry (FCM) have increased over the years due to improvements in preparation and staining methods as well as in measuring techniques and instrumentation.

Nowadays protocols are available for high-resolution analysis of nuclear DNA which enable us to discriminate somatic cells from male and female donors and to characterize precisely the ploidy state of tumors.

The protocols presented here were developed for use in cells obtained from cell suspensions and solid tissues, respectively. The procedures include fixation with 70% ethanol. Thus, they are applicable in specimens that have to be stored before measurement.

In the case of solid tissues the procedure consists of a pretreatment with citric acid plus Tween 20, fixation in ethanol, and a second treatment with citric acid plus Tween 20 in order to produce a suspension of isolated nuclei, followed by addition of a solution of disodium hydrogen phosphate with 4′, 6-diamidino-2-phenylindole (DAPI), which raises the pH and stains the DNA.

In the case of cell suspensions the cells are fixed without pretreatment. After fixation they are treated with hydrochloric acid plus pepsin and stained by addition of a solution of trisodium citrate with DAPI.

The use of DAPI for flow cytometry was proposed by Göhde et al. (1978). Since then, DAPI has been used by many authors. It proved to be a specific, highly fluorescent stain, very well suited for FCM of DNA in whole cells, nuclei, and chromosomes.

The protocols presented here are recommended for accurate and reproducible measurements of the nuclear DNA, especially for the assessment of DNA content variations in cell populations affected by chemical or physical mutagens and for the reliable detection and characterization of aneuploid cell clones in tumors.

II. Applications

The described methods have been applied to a large variety of cell types. Flow cytometric DNA histograms with excellent resolution were obtained from fibroblasts, leukocytes, bone marrow, lung, liver, and skin cells from humans and experimental animals, from various human tumors, from chicken and trout erythrocytes, and from various cultured cell lines.

Using bone marrow of experimental animals it is possible to establish an *in vivo* mutagenicity test, taking the increased dispersion of the nuclear DNA content as a measure of the mutagenic effect (Otto et al., 1981, 1984; Otto and Oldiges, 1983).

In human solid tumors the methods proved to be highly sensitive for the detection of aneuploid cell populations (Otto et al., 1990; Otto and Schumann, 1991, 1992).

In testicular cells the method can be used for the assessment of spermatogenic and spermiogenic cell kinetics and for the detection of so-called diploid spermatozoa, which may be used as an indicator of genotoxic effects (Otto and Oldiges, 1986; Otto, 1987).

In mature mammalian spermatozoa the methods do not yield quantitative DNA staining because of the highly condensed chromatin. These cells need a pretreatment with strong decondensing agents such as dithioerythritol and papain (Otto et al., 1979; Otto and Arnstadt, 1988; Arnstadt and Otto, 1989; Lewalski et al., 1991).

It must be kept in mind that DAPI preferentially stains AT-rich DNA. Thus, the procedures described here are not suited to yield absolute DNA values or to compare cells with differing proportions of AT. Despite this reservation, the addition of chicken or trout erythrocytes as standard to one cell type permits the assessment of peak position and coefficient of variation (CV) of the mean cell fluorescence of that cell type.

III. Materials

The following solutions are used.

1. Detergent solution
    100 ml distilled water (dH$_2$O)
    2.1 g citric acid · H$_2$O (0.1 M)
    0.5 ml Tween 20 (SERVA, Heidelberg, Germany, Cat. No. 37470)
2. Pepsin solution
    10 ml 0.1 N hydrochloric acid
    25 mg pepsin (Riedel-de Haen, Seelze, Germany, Cat. No. 20821, 1200 units/g)
3. Staining solution (phosphate)
    100 ml dH$_2$O
    7.1 g Na$_2$HPO$_4$ · 2 H$_2$O
    0.2 mg DAPI (Partec, Münster, Germany, Cat. No. 06-5-4001)
4. Staining solution (citrate)
    100 ml dH$_2$O
    5.9 g citric acid trisodium salt · 2 H$_2$O
    0.2 mg DAPI.

The pepsin solution is not stable. Adequate amounts should be prepared immediately before use. The other solutions are stable at room temperature (RT) for 2 weeks. The DAPI staining solutions should be stored in the dark.

IV. Cell Separation and Fixation Method

Cells from suspension cultures and other suspensions like chicken and trout erythrocytes and human leukocytes are harvested without dispersing pretreatment and are fixed with 70% ethanol.

Cells from bone marrow and other suitable tissues are isolated by flushing with EDTA containing Ca- and Mg-free PBS. The cell suspension is then fixed with 70% ethanol.

Solid tissues are minced and incubated in the detergent solution for 20 min at RT. The resulting single-cell suspension is then fixed with 70% ethanol.

All fixed cells can be stored at 4°C or −20°C for as long as several months.

V. Staining Procedure

The fixed cells are centrifuged at $200g$ for 10 min and the fixative is removed completely.

Cells from suspension cultures, other suspensions, bone marrow, and similar tissues which have not been treated with the detergent solution before fixation are resuspended in one volume of the pepsin solution consisting of 0.1 N hydrochloric acid and 0.25% pepsin and incubated at RT for 15 min with gentle shaking. Subsequently, nine volumes of the staining solution containing 0.2 M citric acid trisodium salt and 5 μM DAPI are added to raise the pH and to stain the DNA.

Cells from solid tissues are resuspended in one volume of the detergent solution consisting of 0.1 M citric acid and 0.5% Tween 20 and incubated for 10 min at RT with gentle shaking. Then six volumes of the staining solution containing 0.4 M disodium hydrogen phosphate and 5 μM DAPI are added.

The stained cells are stable at RT for 24–48 hr. The best histograms are obtained at about 24 hr after staining.

VI. Instrument Setup

The excitation maximum of DAPI is in the near-ultraviolet (UV) range at about 360 nm and the emission maximum in the blue at 460 nm.

Using the PAS II flow cytometer (Partec GmbH, Münster, Germany) with a mercury lamp the following filters are recommended: KG 1, BG 38, and UG 1 for excitation; TK 420 as dichroic mirror; and GG 435 as barrier filter.

A quartz objective lens (Partec GmbH) proves to be suited to reduce the fluorescence background.

In order to guarantee a smooth and reproducible sample flow, use a syringe pump for the injection of the cell suspension.

VII. Standards

The use of standard cells is facilitated by this method. Chicken or trout erythrocytes, murine, or human lymphocytes can be stored in 70% ethanol for at least 6 months to be used as biological standards. They can be added before or after staining to be measured together with the cells of interest (additive standard). According to our experience, it is also possible to measure the standard cells in a separate run to generate a reference peak (external standard). The flow histogram of a mixture of three types of standard cells is shown in Fig. 1.

Due to the preferential staining of AT-rich DNA the peak positions do not provide quantitative information on the DNA content of cells from different

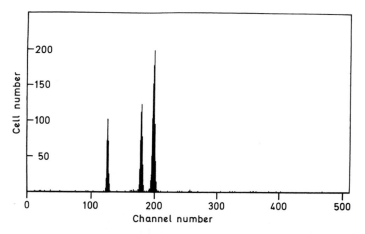

Fig. 1   Flow cytometric DNA histogram of a mixture of three biological standards (from left to right: trout erythrocytes, mouse spleen lymphocytes, human peripheral blood lymphocytes). The CVs of the peaks are 0.96, 0.84, and 0.84.

species. But the reference peaks obtained from those standard cells allow deviations in DNA content and CV in a given cell type to be detected.

VIII. Results and Discussion

The treatment of ethanol-fixed cells with hydrochloric acid plus pepsin or citric acid plus Tween 20 produces isolated nuclei free of cytoplasm and makes the DNA accessible for the dye. The addition of citrate or phosphate raises the pH and results in a buffer system in which DAPI is working as an effective and specific DNA stain.

The procedure yields FCM histograms of the nuclear DNA content with excellent resolution. Coefficients of variation in the range of 1% can be obtained in most specimens routinely. This allows the detection of small increases of DNA content variability induced by chemical or physical mutagens. The increased variability can be quantified by calculating the CV of the G_1 peak.

The protocols are suited for the discrimination of cell populations with small differences in DNA content such as the Y and X chromosome bearing round spermatids in mouse testicular tissue (Otto, 1990) or as the XY and XX chromosome bearing cells in a mixture of male and female lymphocytes (Fig. 2).

In human tumors the procedure is appropriate for the detection of aneuploid cell lines, even if the deviations from the diploid DNA content are very small, and for the precise characterization of the ploidy state by the DNA index (DI) (Fig. 3).

The DI correlates well with the chromosome count in the microscope. Because of the consequences for prognosis it is considered essential to use a high-resolution method when measuring the nuclear DNA content in tumors.

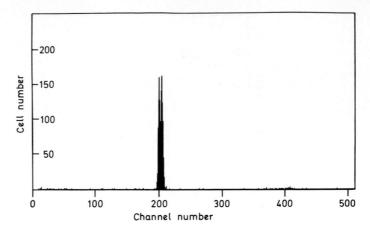

Fig. 2 Flow cytometric DNA histogram of a mixture of lymphocytes from a male and female donor.

During tumor progression the development of new aneuploid cell clones can be followed and the genetic instability of the tumor cells can be assessed (Otto and Schumann, 1992).

Using these protocols, double staining with DAPI and sulforhodamine 101 is feasible to yield two-parameter histograms of DNA and protein. In this case it must be noted that only nuclear proteins that are resistant to the acid-pepsin or acid-detergent treatment are measured.

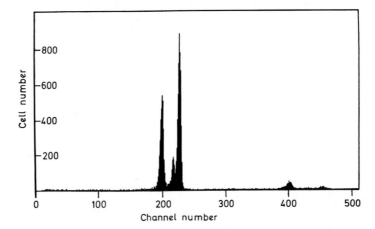

Fig. 3 Flow cytometric DNA histogram of a human malignant melanoma showing a diploid cell population at channel No. 200 and two aneuploid cell lines with DNA indices of 1.08 and 1.14.

References

Arnstadt, K.-I., and Otto, F. J. (1989). *Zuchthygiene* **24,** 146–147.

Göhde, W., Schumann, J., and Zante, J. (1978). *In* "Pulse Cytophotometry" (D. Lutz, ed.), pp. 229–232. European Press, Ghent.

Lewalski, H., Otto, F. J., Kranert, T., Glahn-Luft, B., and Wassmuth, R. (1991). *Genet. Sel. Evol.* **23,** Suppl. 1, 152s–156s.

Otto, F. J. (1987). *In* "Clinical Cytometry and Histometry" (G. Burger, J. S. Ploem, and K. Goerttler, eds.), pp. 297–299. Academic Press, London.

Otto, F. J. (1990). *In* "Methods in Cell Biology" (Z. Darzynkiewicz and H. A. Crissman, eds.), Vol. 33, pp. 105–110. Academic Press, San Diego.

Otto, F. J. (1992). *In* "Flow Cytometry and Cell Sorting" (A. Radbruch, ed.), pp. 65–68. Springer-Verlag, Berlin.

Otto, F. J., and Arnstadt, K.-I. (1988). *Cytometry, Suppl.* **2,** 80.

Otto, F. J., and Oldiges, H. (1983). *Wiss. Umwelt* **83,** 109–121.

Otto, F. J., and Oldiges, H. (1986). *Wiss. Umwelt* **86,** 15–30.

Otto, F. J., and Schumann, J. (1991). *Cytometry, Suppl.* **5,** 65.

Otto, F. J., and Schumann, J. (1992). *Anal. Cell. Pathol.* **4,** 209.

Otto, F. J., Hacker, U., Zante, J., Schumann, J., Göhde, W., and Meistrich, M. L. (1979). *Histochemistry* **61,** 249–254.

Otto, F. J., Oldiges, H., Göhde, W., and Jain, V. K. (1981). *Cytometry* **2,** 189–191.

Otto, F. J., Oldiges, H., and Jain, V. K. (1984). *In* "Biological Dosimetry" (W. Eisert and M. L. Mendelsohn, eds.), pp. 37–49. Springer-Verlag, Berlin.

Otto, F. J., Schumann, J., and Bartkowiak, D. (1990). *Cytometry, Suppl.* **4,** 55.

CHAPTER 15

Detergent and Proteolytic Enzyme-Based Techniques for Nuclear Isolation and DNA Content Analysis

Lars L. Vindeløv★ and Ib Jarle Christensen†

★ Department of Haematology and
† Finsen Laboratory
Rigshospitalet
DK-2100 Copenhagen, Denmark

I. Introduction

Performing flow cytometric DNA analysis on a suspension of nuclei has two major advantages. DNA nonspecific fluorescence from the cytoplasm is avoided, and fine-needle aspirates which contain a variable fraction of bare nuclei are well suited as starting material. The possibility of simultaneous analysis of cytoplasmic or plasma membrane-associated structures is thus sacrificed for easy and accurate determination of nuclear DNA in tissues and solid tumors.

A number of techniques have been published over the years. Some of them are listed in Table I. To choose a method for a particular purpose one should first consider the wavelengths available for excitation of the fluorochrome on the flow cytometer to be used. Second, one should assure, by testing, that high-resolution measurements can be obtained by preparing the cells in question with the method chosen. It should be remembered, in this context, that a properly aligned flow cytometer and a low flow rate are as necessary as stoichiometric staining of the DNA for obtaining good quality histograms. The latter are characterized by a minimal amount of debris and symmetrical G_1 peaks with low (1.5–2.5%) coefficients of variation (CV).

The amount of information that can be extracted from a DNA distribution depends not only on the technical quality of the analysis but also on the methods used for standardization and statistical analysis (deconvolution) of the histogram. The data reduction obtained by statistical analysis yields the desired endpoint results that will adequately and exhaustively describe the DNA distribution for most purposes. These are: (1) the number of subpopulations with different DNA content present in the sample and (2) for each subpopulation, (a) the relative size of the subpopulation, (b) the DNA index (DI), and (c) the fractions of cells in the cell-cycle phases G_1, S, and G_2 + M.

In the following we describe a set of integrated methods developed in our laboratories for sample acquisition and storage, standardization, fluorochrome staining, and statistical analysis.

Table I
Some Techniques for Nuclear Isolation and DNA Analysis

References	Nuclear isolation technique	Fluorochrome/excitation maximum
Krishan (1975, 1990)	Hypotonia	PI/536 nm
Vindeløv (1977); Vindeløv and Christensen (1990)	Nonidet-P40	PI/536 nm
Otto *et al.* (1979); Otto (1990)	Tween-20/citric acid	DAPI/350 nm
Taylor (1980); Palavicini *et al.* (1990)	Triton X-100	PI/536 nm DAPI/350 nm
Thornthwaite *et al.* (1980)	Nonidet-P40	PI/536 nm DAPI/350 nm
Vindeløv *et al.* (1983a–d); Vindeløv and Christensen (1990)	Nonidet-P40/trypsin	PI/536 nm

II. Basic Principles of the Methods

The integrated set of methods (Vindeløv *et al.*, 1983a–d; Vindeløv and Christensen, 1990) developed to solve the problems described above is outlined in Fig. 1. The analysis is performed on unfixed material. This is essential to avoid a potentially selective cell loss caused by centrifugation steps and keeps the cell requirement at a minimum. Clumping and staining artifacts caused by a fixative are avoided. Samples can be long-term stored by freezing in a citrate buffer with dimethyl sulfoxide (DMSO) (Vindeløv *et al.*, 1983a).

The key to accurate and reproducible DI determination, as well as statistical analysis of multiple overlapping populations, is adequate internal standardiza-

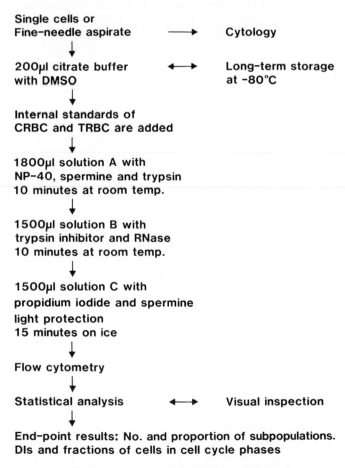

Fig. 1 Overview of methods and results.

tion. This is achieved by adding a mixture of chicken (CRBC) and trout erythrocytes (TRBC) to the sample before staining (Vindeløv *et al.*, 1983c). The peaks produced by the standards provide two points of reference at DIs approx 0.30 (CRBC) and 0.80 (TRBC) and allow deconvolution of the histogram and DI determination independent of zero-point shift.

The preparation consists of three steps (Vindeløv *et al.*, 1983b). Clean nuclei are obtained in the first step (solution A) by the combined action of the non-ionic detergent Nonidet-P40 (NP-40) and trypsin. In addition trypsinization increases the fluorescence of nuclei with dense chromatin such as granulocytes, presumably by splitting some chromosomal proteins. Spermine is essential for the stability of the unfixed nuclei during trypsinization. RNase A treatment in the second step (solution B) prevents dye binding to double-stranded RNA. Trypsin inhibitor is added, because trypsin activity after addition of propidium iodide (PI) results in unstable nuclei. In the final step (solution C) PI is added and the spermine concentration further increased for optimal stability. PI binds to double-stranded nucleic acid by intercalation (Crissman *et al.*, 1975; Krishan, 1975).

III. Applications

Flow cytometric DNA analysis has been established as a useful research tool for a number of years. A review of applications in our laboratory may be found in Vindeløv and Christensen (1989). The methods have been used for monitoring the stability of cell lines, for sensitivity testing, and for studying the action of anticancer drugs *in vitro*. Likewise the methods have been used to monitor the cell-cycle perturbations produced by radiation, chemotherapy, and hormonal substances in murine and human tumors in experimental animals and in malignant tumors in patients. Since the same methods can be applied at these different levels of complexity, the results are directly comparable and may serve as a link between exerimental and clinical results.

DNA analysis is currently assessed as a possible prognostic parameter in neoplastic disease (Merkel *et al.*, 1987). In a prospective study of 249 patients with bladder cancer, ploidy and S phase size were found to have an independent prognostic value in multivariate analysis of survival and response to treatment (Vindeløv *et al.*, 1993).

IV. Materials

A. Citrate Buffer

Sucrose (BDH)	85.50 g (250 mM)
Trisodium citrate, 2 H$_2$O (Merck)	11.76 g (40 mM)
Dissolve in distilled water	approx 800 ml
DMSO (Merck) is added	50 ml
Distilled water is added to a total volume of	1000 ml
pH is adjusted to	7.6

B. Stock Solution

Trisodium citrate, 2H$_2$O (Merck)	1000 mg (3.4 mM)
Nonidet-P40 (Shell)	1000 μl (0.1% v/v)
Spermine tetrahydrochloride (Serva, Cat. No. 35300)	522 mg (1.5 mM)
Tris (Sigma 7–9, Chemical Co., St. Louis. MO., Cat. No. T-1378)	61 mg (0.5 mM)
Distilled water is added to a total volume of	1000 ml

C. Solution A

Stock solution	1000 ml
Trypsin (Sigma, Cat. No. T-0134)	30 mg
pH is adjusted to	7.6

D. Solution B

Stock solution	1000 ml
Trypsin inhibitor (Sigma, Cat. No. T-9253)	500 mg
Ribonuclease A (Sigma, Cat. No. R-4875)	100 mg
pH is adjusted to	7.6

E. Solution C

Stock solution	1000 ml
Propidium iodide (Fluka Cat. No.)	416 mg
Spermine tetrahydrochloride (Serva, Cat. No. 35300)	1160 mg
pH is adjusted to	7.6

The citrate buffer is stored at 4°C. The staining solutions are stored in aliquots of 5 ml in capped plastic tubes at −80°C. The tubes with solution C are wrapped in aluminum foil for light protection of the PI. Before use the solutions are thawed in a water bath at 37°C, but not heated to 37°C. Solutions A and B are used at room temperature. Solution C is kept in an ice bath.

F. Internal Standards

Chicken blood can be obtained by heart puncture and collected in a tube containing 50 iu of heparin per ml of blood. The blood is subsequently diluted by citrate buffer. Rainbow trout blood can be obtained by caudal vein puncture after anesthesia with MS 222 (Sandoz, Basel, Switzerland) and should be mixed with citrate buffer immediately. The concentrations of CRBC and TRBC are determined by counting in a hemocytometer. The red cell concentrations are adjusted by dilution with citrate buffer to CRBC = 145 × 10^4 cells per ml and TRBC = 255 × 10^4 cells per ml. These suspensions are then mixed in equal volumes to obtain a final red blood cell concentration of 2 × 10^6 cells per ml and a ratio of CRBC:TRBC = 4:7, which will produce peaks of equal height in the histogram. The mixture of standards can be stored long term at −80°C in aliquots of, for instance, 100 μl (sufficient for three to five samples), after freezing as described in Section V,B.

V. Methods

A. Sample Acquisition

Fine-needle aspiration is used for initial mechanical disagregation of normal tissues, lymphomas, and solid tumors. It is a gentle and rapid procedure causing less debris than cutting with knives or scissors. It is therefore used even when a surgical biopsy is available. The tumor is secured by pinning it with injection needles to a plate of styrofoam. In the case of very small biopsies the tissue is wrapped in transparent plastic foil and secured during biopsy by pinning this to the plate as above. Biopsies as small as $2 \times 2 \times 2$ mm can be successfully aspirated in this way. A 0.5×25-mm (25G $\times$ 1) needle on a 20-ml disposable syringe fitted on a one-hand-operated handle (Cameco, Täby, Sweden) is adequate. Longer needles may be required for *in vivo* aspiration and thinner needles may give a better cell yield in fibrous tumors, such as some breast cancers. Suction is applied only when the point of the needle is within the tissue. The needle is moved back and forth in different directions within the tumor. The aspirate should stay within the needle during aspiration. The needle is flushed with citrate buffer (200 μl), and the cell yield is checked by counting in the hemocytometer. Aspiration is repeated until 10^6 cells have been obtained. The cells are stored by freezing (see Section V,B.) or stained and analyzed on the same day. Two additional aspirates are spread on slides for cytologic examination.

B. Storage

Samples and internal standards not stained and analyzed on the same day as obtained are stored at $-80°C$ after freezing (Vindeløv et al., 1983a). Samples have been kept in this way for up to 5 years, without change in the DNA histograms. Samples can be frozen either as aspirates in citrate buffer or as surgical biopsies in a dry tube. It is essential that each sample is frozen and thawed only once. The cell suspension in citrate buffer is frozen in polypropylene tubes (38 $\times$ 12.5-mm test tubes with screw cap, Nunc, Roskilde, Denmark) immersed in a mixture of dry ice and ethanol ($-80°C$). Before use the samples are thawed in a water bath at 37°C. The sample should only be thawed and not heated to 37°C.

C. Standardization and Staining

Staining with PI (Vindeløv et al., 1983b) is performed after addition of the internal standards (Vindeløv et al., 1983c) to 200 μl of sample cell suspension in citrate buffer. The standards are added in an amount equal to 20% of the cells in the sample. With the concentrations chosen, 20% of the sample cell count in 0.1 μl (the volume of the hemocytometer used in our laboratory) is equal to the number of microliters of standard to be added to 200 μl of sample. Staining is performed by stepwise addition of the staining solutions. A total of 1800 μl of solution A is added to 200 μl of tumor aspirate in citrate buffer and the tube is inverted to mix the contents gently. After 10 min at room temperature, during which the tube is inverted two to three times, 1500 μl of solution B is added. The solutions are again mixed by inversion of the tube and after 10 min at room temperature 1500 μl ice-cold solution C is added. The solutions are mixed and the sample is filtered through a 25-μm nylon mesh into tubes wrapped in aluminum foil for light protection of the PI. The samples are kept in an ice-bath until analysis which should take place between 15 min and 3 hr after addition of solution C (Vindeløv et al., 1983b). If cells are scarce the volumes of staining solution can be halved to increase the cell concentration.

There are few critical aspects of this simple procedure. Pure, analytic grade reagents should be used for the solutions. Accurate weighing of the reagents and accurate pipetting of the solutions is important. The samples should be handled gently throughout, since agitation will increase debris and clumping.

D. Flow Cytometric Analysis

The FACScan flow cytometer (Becton–Dickinson, Mountain View, CA) is currently used in our laboratory. Cells are analyzed at a rate of 20–40 cells/sec, with a flow rate of 12 μl/sec. This low rate of measurement is chosen to ensure a thin stream of sample, with the cells intersecting the laser beam in the same path. The blue 488-nm line of the argon ion laser is used for excitation. The instrument is equipped with a doublet discriminator module that allows measurement of pulse area (FL2-A), detected through a 585-nm band pass filter.

E. Statistical Analysis

The end-point results mentioned in Section I are determined by statistical analysis of the DNA distribution (Vindeløv and Christensen, 1990). Visual inspection of the histogram plays a role in determining the quality of the analysis and the number of subpopulations present. Figure 2A shows an example of a DNA histogram. Statistical analysis or deconvolution, described briefly, involves the following steps:

1. Normal distributions are fitted to the peaks of the standards. Debris is subtracted by fitting a truncated exponential between the CRBC and TRBC peaks, and the zero point is corrected, based on the means of the standards.

2. A model describing the G_1 and G_2 + M peaks with normal distributions, and the S phase with an exponential function of a polynomial of a given degree,

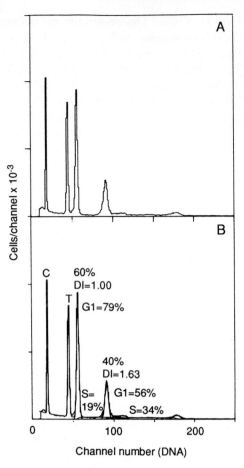

Fig. 2 (A) A DNA distribution from breast cancer. Fine-needle aspiration of a surgical biopsy stored by freezing. Visual inspection indicates a good quality histogram with two subpopulations. (B) The histogram after statistical analysis. The peaks produced by the CRBC (C) and TRBC (T) are indicated. The first subpopulation is strictly diploid (DI = 1.00) and constitutes 60% of the cells. The second subpopulation is hypertriploid (DI = 1.63) and constitutes 40% of the cells. The percentages of cells in the cell-cycle phases are indicated on the figure. The CVs of the G_1 peaks are 2.3 and 2.2%.

is fitted to the histogram by maximum likelihood. For histograms with more than one subpopulation, a mixture of this density is fitted (Fig. 2).

3. The DIs of the G_1 peaks are calculated by comparison with known values for normal cells (Vindeløv *et al.,* 1983d), and the areas under the curves are estimates of the fractions of cells in the cell-cycle phases G_1, S, and G_2 + M. Further details regarding deconvolution of DNA content frequency histograms are presented in Chapter 18 of this volume.

Fig. 2B shows the deconvoluted histogram and the estimated values. Important features are:

1. The very accurate zero-point determination obtained by the standards.

2. A point of reference (TRBC) close to the diploid peak, with makes DI determination more accurate.

3. A slight nonlinearity of the relationship fluorescence/DNA, which has been determined experimentally and corrected for in the computer program. This allows prediction of the G_2 + M mean location and thereby deconvolution of heterogeneous tumors with overlapping populations, as shown in Fig. 2.

4. Histograms with a treatment-induced perturbation of the cell cycle can be deconvoluted by increasing the degree of the polynomial used for fitting the S phase. An example is shown in Fig. 3. The S phase distribution is estimated as an additional end-point result.

VI. Results

These methods were developed with emphasis on obtaining optimal results in a wide range of cells and tissues, particularly solid tumors. The staining method was developed in 1978 as a modification of an initial attempt (Vindeløv, 1977). In the past 14 years we have analyzed approx 20,000 samples from

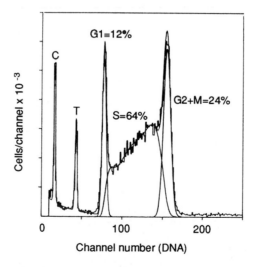

Fig. 3 An example of deconvolution of a histogram from a drug-perturbed cell population. The small cell lung cancer cell line NCI-N592 was exposed to the alkylating agent BCNU. The S phase was fitted by an eighth degree polynomial. The standards and the percentages of cells in the cell-cycle phases are indicated on the figure.

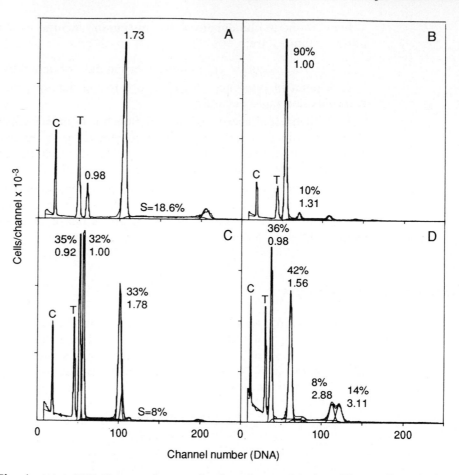

Fig. 4 (A) A DNA histogram from small cell carcinoma of the lung. Fine-needle aspiration *in vivo* of a lymph node metastasis. A single aneuploid (DI = 1.73) population is present. The peak with DI = 0.98 could represent normal cells and possibly some tumor cells. The CVs of the G_1 peaks are 1.9 and 1.7% (B) A distribution from carcinoma of the oral cavity. Fine-needle aspiration of a small surgical biopsy. Ninety percent of the cells are strictly diploid. A small subpopulation with DI = 1.31 is present. The S phases are confounded and cannot be estimated reliably. The CV of the diploid G_1 peak is 2.2%. (C) A distribution from a non-Hodgkin lymphoma. Fine-needle aspiration *in vivo*. Three subpopulations of nearly the same size are present. The separation of the diploid (DI = 1.00) and the hypodiploid (DI = 0.92) peaks illustrate the need for a high resolution. The CVs range from 1.6 to 1.8%. The S phases of the diploid and hypodiploid subpopulations are confounded. An average value can be calculated. (D) A distribution from breast cancer. Fine-needle aspiration of surgical biopsy. Four subpopulations are present. Their sizes and DIs are indicated on the figure. Only one or two of the S phases can be estimated. The CVs vary from 2.1 to 2.8.

clinical and experimental studies. Satisfactory results have been obtained in all types of cells and tissues examined with the exception of sperm. The samples analyzed include (1) normal tissues: human lymphocytes, granulocytes and spleen and mouse lymphocytes, bone marrow, spleen, liver, kidney, and thymus; (2) human neoplasms: lung cancer, breast cancer, lymphoma, leukemia, bladder cancer, and cancer of the oral cavity; (3) human tumors in nude mice: breast cancer, lung cancer, melanoma, and colon cancer; and (4) mouse ascites tumors: JB-1, L1210, Ehrlich, and P388. Some examples are shown in Fig. 4. CVs of around 2% are obtained routinely. The theoretical and practical problems of resolving minor DNA differences as well as long-term reproducibility and comparability of results obtained by the methods have been examined in some detail (Vindeløv *et al.*, 1983a,d).

Acknowledgments

The work was supported by grants from The Danish Cancer Society, The Danish Medical Research Council, and The Lundbeck Foundation.

References

Crissman, H. A., Mullaney, P. F., and Steinkamp, J. A. (1975). *In* "Methods in Cell Biology" (D. M. Prescott, ed.), Vol. 9, pp. 179–246. Academic Press, New York.

Krishan, A. (1975). *J. Cell. Biol.* **66,** 188–193.

Krishan, A. (1990). *In* "Methods in Cell Biology" (Z. Darzynkiewicz and H. Crissman, eds.), Vol. 33, pp. 121–125. Academic Press, San Diego.

Merkel, D. E., Dressler, L. G., and McGuire, W. L. (1987). *J. Clin. Oncol.* **8,** 1690–1703.

Otto, F. J. (1990). *In* "Methods in Cell Biology" (Z. Darzynkiewicz and H. Crissman, eds.), Vol. 33, pp. 105–110. Academic Press, San Diego.

Otto, F. J., Hacker, U., Zante, J., Schumann, J., Göhde, W., and Meistrich, M. L. (1979). *Histochemistry* **61,** 249–254.

Palavicini, M. G., Taylor, I. W., and Vindeløv, L. L. (1990). *In* "Flow Cytometry and Sorting" (M. R. Melamed, T. Lindmo, and M. L. Mendelsohn, eds.), 2nd ed., pp. 187–194. Wiley-Liss, New York.

Taylor, I. W. (1980). *J. Histochem. Cytochem.* **28,** 1021–1024.

Thornthwaite, J. T., and Thomas, R. A. (1990). *In* "Methods in Cell Biology" (Z. Darzynkiewicz and H. Crissman, eds.), Vol. 33, pp. 111–119. Academic Press, San Diego.

Thornthwaite, J. T., Sugarbaker, E. V., and Temple, W. J. (1980). *Cytometry* **3,** 229–237.

Vindeløv, L. L. (1977). *Virchows Arch. B* **24,** 227–242.

Vindeløv, L. L., and Christensen, I. J., (1989). *Eur. J. Haematol Suppl.* **42**(48), 69–76.

Vindeløv, L. L., and Christensen, I. J. (1990). *Cytometry* **11,** 753–770.

Vindeløv, L. L., Christensen, I. J., Keiding, N., Spang-Thomsen, M., and Nissen, N. I. (1983a). *Cytometry* **3,** 317–322.

Vindeløv, L. L., Christensen, I. J., and Nissen, N. I. (1983b). *Cytometry* **3,** 323–327.

Vindeløv, L. L., Christensen, I. J., and Nissen, N. I. (1983c). *Cytometry* **3,** 328–331.

Vindeløv, L. L., Christensen, I. J., Jensen, G., and Nissen, N. I. (1983d). *Cytometry* **3,** 332–339.

Vindeløv, L. L., Christensen, I. J., Engelholm, S. A., Guldhammer, B. H., Højgaard, K., Sørensen, B. L., and Wolf, H. (1993). *Cytometry* (submitted for publication).

CHAPTER 16

DNA Analysis from Paraffin-Embedded Blocks

David W. Hedley

Departments of Medicine and Pathology
Ontario Cancer Institute/Princess Margaret Hospital
Toronto, Ontario, Canada M4X 1K9

I. Introduction

Routine pathological examination of tumors involves fixation, usually with formaldehyde, followed by dehydration and embedding in paraffin wax. Microtome sections can then be cut from these paraffin "blocks" for histological examination. Preserved in this way tumor tissue is remarkably durable, and because of the occasional need to cut further tissue sections for pathological review, the standard practice is to archive paraffin blocks for up to several decades. This material can be used for flow cytometric DNA analysis.

II. Application

Compared to fresh material, sample preparation from paraffin blocks is more time consuming, and the DNA histograms are generally of poorer quality in terms of CVs and debris. As originally conceived, the main advantage of the method was that it allowed retrospective analysis of material from cohorts of patients whose clinical outcome was already known, thus allowing the prognostic significance of DNA index and S phase in the various tumor types and stages to be determined. It has however become widely used for the prospective evaluation of individual cancer patients when fresh material is not available, or with very small tumors where the entire sample requires processing for diagnostic purposes. In these latter cases, microscopic examination of a parallel tissue section can allow selective sampling from an area of the block that contains tumor, minimizing the proportion of stromal cells in the DNA histogram.

Since its original publication (Hedley *et al.*, 1983), many modifications of the method have been described, adapting it for use either in routine pathology laboratories or for analysis of particular tumor types (Hedley, 1989; Hitchcock and Ensley, 1993). Unfortunately, there does not appear to be one method that is ideal for all situations, and laboratories using the technique should be prepared to be creative. The purpose of this chapter is to guide users through the basic method and to review the commonly encountered problems and their possible solutions, rather than to review all of the published variations. Comprehensive and critical evaluations of the technique are discussed by Hitchcock and Ensley (1993) and by Heiden *et al.* (1991), and these are recommended for further reading.

III. Methods

The basic steps of the method are selection of a block that contains an adequate and representative population of tumor cells, cutting of thick sections using a microtome, dewaxing, and rehydration. This material is then subjected to enzymatic digestion, and the resulting nuclear suspension processed for DNA analysis.

A. Selection of Blocks

Although the method has been successfully applied to blocks that are several decades old (Toikkanen *et al.*, 1989), poorly preserved material is a major reason for failure to obtain an interpretable DNA histogram. This problem is further discussed in Section IV, Critical Aspects of the Technique. Because %S phase is increasingly used as a prognostic determinant, and because reliable estimates require that admixture with normal stromal elements be kept to a minimum, parallel tissue sections should be examined by light microscopy and blocks selected which contain a substantial proportion of tumor cells. Methods for enriching tumor cells are discussed below.

B. Section Cutting

Sections are usually cut using a microtome and should be as thick as possible, since there is a clear relationship between the thickness of a microtome section and the extent and distribution of debris due to partially sectioned nuclei. Fifty micrometers or thicker is advised, which is greater than can be cut with some microtomes. Sections of this thickness curl up as they come off the microtome, and there is no need to uncurl them as it is more convenient to handle them in this form. Depending on tumor type and cellularity, between one and four sections are needed for DNA analysis.

C. Dewaxing

A variety of organic solvents can be used for dissolving the paraffin wax. Xylene, which was described in the original publication, is still widely used but is toxic and can be substituted for by more environmentally friendly agents such as Histoclear (National Diagnostics, Summerville, NJ). Place thick sections in a tube that is solvent compatible, add 3 ml of solvent, let stand at room temperature for 10 min, aspirate, and repeat this process, finally reaspirating the solvent.

D. Rehydration

Add 3 ml of 100% ethanol, let stand at room temperature for 10 min, and aspirate. Repeat this with 95, 70, and 50% ethanol for 10 min each, and finally add 100% water. During this rehydration process the tissue will become soft and friable, and care is required to prevent aspirating of pieces of tissue. If this is a problem, use either centrifugation or process tissue in nylon mesh bags (see below). Aspirate water.

E. Pepsin Digestion

Prepare a 0.5% solution of pepsin in 0.9% saline, and adjust pH to 1.5 by adding 2 N HCl. Note that inferior grades of pepsin may have low activity or contain other digestive enzymes as contaminants. Sigma product number P7012 is of high purity and catalytic activity and gives satisfactory results. Add 1 ml to tissue, and place in a 37°C water bath for 30 min. Agitate or briefly vortex mix periodically. During this procedure the sample often becomes visibly turbid, and microscopic examination will show bare nuclei or intact cells resembling those seen in the parallel thin section. The enzyme digestion step is one of the most critical in the assay, and optimum conditions may differ between tumor types, or depend on the original fixation procedure. Following digestion, the nuclei should be washed once in buffered medium, filtered, and counted.

F. Cell Counting

Optimum use of equilibrium dyes such as the DNA stains requires that the ratio of dye molecules to DNA be kept within fairly narrow limits, and for this reason nuclei/cells should be counted and the final concentration adjusted. For most instruments, a concentration of 1×10^6/ml is satisfactory. Although use of an appropriate dye:DNA ratio holds for all flow cytometric DNA analysis techniques, it is often technically more difficult to obtain reliable counts of nuclei obtained from paraffin-embedded material than with fresh tissue, due to the increased amount of debris. Counts can be made either manually with a hemocytometer or using a Coulter counter.

G. Staining

1. DAPI

The original method of Hedley *et al.* (1983) used the DNA-specific dye 4′,6′-diamidino-2-phenylindole dihydrochloride (DAPI), which binds to the minor groove of the double helix. It is obtainable from Boehringer-Mannheim and made up as a 500 μg/ml stock solution in water. This is stable at 4°C for several months. Stain at a final concentration of 1 μg/ml at room temperature for 30 min and run on a flow cytometer, using UV excitation. DAPI has a broad blue emission spectrum, and band pass filters centered between 400 and 500 nm are suitable.

2. Propidium Iodide

Because most clinical flow cytometers use air-cooled argon lasers, propidium iodide (PI) is used more often than DAPI. Propidium iodide intercalates into the DNA double helix, and its binding is influenced by nuclear proteins to a greater extent than is DAPI. It is probably also more susceptible to the DNA denaturation that occurs during the procedure, and the CVs obtained with PI tend to be wider than those obtained with DAPI. Use a final propidium iodide concentration of 50 μg/ml for 30 min or longer, and the addition of RNase is recommended, as for propidium iodide staining of fresh material.

H. Data Analysis

For many tumor types %S phase is a more powerful prognostic factor than DNA index, but reliable estimates of S phase are much more dependent of the extent of cell debris. Debris is a particular problem using paraffin-embedded material because of the harsh conditions involved in sample preparation and because sectioning with a microtome produces sliced nuclei. The latter show a characteristic distribution, with a disproportionate degree of debris to the left of the G_1 peak. Conventional debris subtraction overcompensates for this, giving an erroneously low (or even negative!) %S phase. Current versions of both MultiCycle and ModFit have debris subtraction that models for sliced nuclei (Fig. 1), and this has been shown to significantly improve the prognostic power of S phase (Kallioniemi *et al.*, 1991). Use of these modeling programs is therefore strongly recommended for routine clinical samples.

I. Use of Internal DNA Standards

Unfortunately, there are no reliable internal standards of cellular DNA content that can be used with this method. The basic problem is that the initial fixation with formaldehyde causes crosslinkages between DNA and other macromolecules, which are then broken down during the enzyme digestion step. The degree of crosslinking varies considerably between samples, and this critically and unpredictably influences the availability of DNA for staining, rendering comparisons with internal or external standards unreliable. The least unsatisfactory standard is a piece cut from the thick section which is microscopically free of tumor; but because formalin penetrates tissue at a slow rate, the extent of fixation will vary within large blocks and may affect the stainability with DNA dyes. It is not therefore possible to identify with certainty hypodiploid tumors, which are rare but might carry a poorer prognosis than near-hyperdiploid tumors. By convention the G_1 peak with the lowest DNA content is considered normal diploid, and DNA index calculated by comparing other G_1 peaks to its modal position. This gives a close correlation with the DNA index obtained for the same tumor using unfixed material, but hypodiploid tumors are erroneously assigned at a DNA index of >1.0.

IV. Critical Aspects of Technique

Compared to DNA analysis using fresh tissue this method is technically more demanding, commonly encountered problems being failure to obtain a DNA

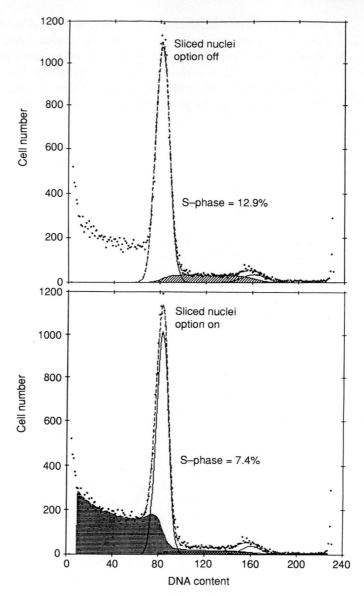

Fig. 1 Analysis of %S phase using the "sliced nuclei option" of the MultiCycle program (Phoenix Flow Systems, San Diego). Note the distribution of debris is not exponential because a major component is derived from partially sectioned nuclei. Since most cells are in the G_1 peak, debris falls off abruptly beyond this position. The program makes allowance for this during debris subtraction (right panel, shaded), and the resulting S phase estimate is a more reliable guide to prognosis (Kallioniemi *et al.*, 1991). A similar option is available in ModFit (Verity Software).

histogram, excessive amounts of debris, or unacceptably high CVs. Very often wide CVs and large amounts of debris go together. These problems can usually be traced to either the initial fixation of the tissue or to the enzyme digestion step.

A. Tissue Fixation

The quality of the histograms obtained is critically dependent on the initial fixation of the tissue. Formaldehyde-based fixatives are generally satisfactory, although the rate of penetration into tissue is slow, resulting in uneven fixation of large pieces of tissue and the possible generation of false aneuploid peaks. More exotic fixatives, such as those based on mercury or Bouin's fixative, frequently give very poor quality DNA histograms, and the pedigree of a sample should therefore be obtained if there are problems with analysis. Anecdotally, results with material fixed in Bouin's fluid are said to be improved by modifying the rehydration step to include several changes in 50% ethanol and then allowing to stand overnight in 50% ethanol.

Because the flow cytometry laboratory usually has no control over the quality of tissue fixation, it is strongly recommended that the method be worked up using material that is known to give a good quality DNA histogram, possibly obtained from another laboratory experienced with the method. It is also a good idea to include a sample of this material in each staining run, as a quality control check.

B. Enzyme Digestion

If you are sure that the material being examined is well fixed and capable of giving a good quality DNA histogram, but are getting wide CVs or excessive debris, the problem probably lies in the enzyme digestion step. Enzyme solutions are commonly described in terms of percentage by weight of dry powder, but note that most commercial sources originate in the local abattoir, rather than the biotechnology industry and that their purity and catalytic activity varies widely. Furthermore, do not assume that the potency of an enzyme written on the label is necessarily accurate, or that preparations have an indefinite shelf life. In other words, be prepared to try other enzyme preparations and to vary the strength, pH, or incubation time.

The original method using 0.5% pepsin at pH 1.5 for 30 min at 37°C usually gives satisfactory results, although these conditions may be too harsh for some tumor types, such as lymphomas or testis cancer, and produce nuclear fragmentation. In such cases a reduction in pepsin concentration or incubation time would be appropriate. It should be noted that this enzymatic step was optimized for use with the minor groove binding stain DAPI and that it is not necessarily ideal for use with intercalating dyes such as propidium iodide, which are probably more susceptible to DNA degradation. There is still no completely satisfactory published review of alternative methods for enzymatic digestion, but it is possible that trypsin may give lower CVs when propidium iodide is used as the DNA stain (Tagawa *et al.*, 1993), while Heiden *et al.* (1991) have reported that the use subtilisin Carlsberg (pronase) gives fewer cell clumps than pepsin. Details of these enzyme procedures are given below in Section V, Alternative Methods for Sample Preparation.

C. Other Factors

In the dewaxing and rehydration sequence, probably the only really critical stage is ensuring complete removal of paraffin wax. Unless some unusual embedding material has been used, this should be achieved easily by using the procedure described above, but consider altering these conditions if consistently poor results are obtained despite adequate fixation and enzyme digestion. The rehydration procedure was originally taken verbatim from a local immunohisto-chemistry lab, and although it work there is no reason to suppose that it needs detailed adherence.

V. Alternative Methods for Sample Preparation

A. Selecting Areas of Interest within a Block

Microscopic examination of a parallel thin section frequently shows that only a small area contains tumor cells, and in these cases detection of aneuploid peaks or estimates of %S phase will be compromised by the large admixture of stromal elements. The sample processed for flow cytometry can be enriched for tumor cells using the thin section to map out the required region of the block. Spread the thick sections flat, and trim away unwanted areas with a scalpel. Alternatively, use a scalpel blade to score around the area of interest on the surface of the block, and this will detach when the thick section is cut.

B. Dewaxing and Rehydration

Although these are readily done on a small scale using a row of test tubes, it is possible to automate the procedure using a tissue processor (Babiak and Poppema, 1991), such as that produced by Shandon-Elliot and used in routine histology laboratories. Prepare bags of 90-μm mesh nylon gauze, about 1×1 cm, and place thick sections in these (Heiden *et al.*, 1991). Bags may be weighted with 1-mm diameter glass balls to make them sink. These bags are then placed in biopsy cassettes, which are loaded into a tissue processor programmed to run through a sequence of dewaxing followed by rehydration through graded ethanol to distilled water. The bags can then be placed in enzyme solution. It has been shown that centrifugation is a major cause of cell aggregation in this procedure and that this can be avoided by shaking the bags at the end of the digestion phase and adding DNA stain, buffered to neutralize the pepsin solution, directly to the tube containing the bag (Heiden *et al.*, 1991). The supernatant is then run on the flow cytometer.

C. Alternative Enzymatic Procedures

Enzyme digestion is required to release nuclei from tissue. In addition, experiments using cells in suspension show that formalin fixation decreases stainability with DNA dyes, which is restored with pepsin treatment, usually with an improvement in the CV. The relative effects of formalin and pepsin on DNA staining vary between cell types and probably reflect differences in the formation of crosslinks involving DNA and their dissolution by the enzyme and/or acidic conditions. Although the original procedure has proved to be generally applicable, there is no reason to suppose that it is the best possible method because there has been no definitive study of all the possible enzymes in all the permutations of concentration, time, tumor type, etc. A report (Tagawa *et al.*, 1993) suggests that a trypsin method originally described by Schutte *et al.* (1985) gives smaller CVs and less debris than pepsin, when used in association with propidium iodide staining, while pronase (subtilisin Carlsberg) has been reported to produce fewer aggregates (Heiden *et al.*, 1991). Methods for these enzymes, taken from the original publications, are as follows:

1. Trypsin

Dewax and rehydrate as for the pepsin method. Samples are incubated in 0.25% trypsin (Difco) in citrate buffer (3 mM trisodium citrate, 0.1% v/v Nonidet-P40, 1.5 mM spermine tetrachloride, 0.5 mM Tris, pH 7.6) overnight at 37°C. Vortex, filter, and stain.

2. Pronase

Prepare a 0.1% solution of pronase (subtilisin Carlsberg, Sigma protease XXIV) in 0.01 M Tris, 0.07 M NaCl, pH 7.2. Add 1 ml to rehydrated tissue and incubate at 37°C for 30 min to 2 hr, depending on the type of tissue. At the end of the digestion the sample should be shaken, and the stain is added directly to the enzyme mix without a centrifugation step. Run on flow cytometer.

D. Multiparametric Analysis

Despite the harsh conditions used in this method, several nuclear antigens are sufficiently preserved for dual-parameter analysis of DNA content versus fluorescent antibody binding. Examples include the p105 proliferation antigen, originally described by Clevenger *et al.* (1985), and several oncogene products, including c-*myc* (Watson *et al.*, 1985). More recently, the proliferation-dependent Ki-S1 antibody has also been shown by Camplejohn *et al.* (1993) to give good staining in paraffin-embedded samples of breast cancer. Laboratories intending to develop these techniques should be prepared to vary the enzyme digestion procedure and to accept compromises between preserving sufficient nuclear protein for detection while obtaining an adequate DNA histogram.

Acknowledgment

I thank Marijka Koekebakker for her critical review of this chapter and helpful suggestions.

References

Babiak, J., and Poppema, S. (1991). *Am. J. Clin. Pathol.* **96,** 64–69.

Camplejohn, R. S., Brock, A., Barnes, D. M., Gillett, C., Raikundalia, B., Kreipe, H., and Parwaresch, M. R. (1993). *Br. J. Cancer* **67,** 657–662.

Clevenger, C. V., Bauer, K. D., and Epstein, A. L. (1985). *Cytometry* **6,** 208–214.

Hedley, D. W. (1989). *Cytometry* **10,** 229–241.

Hedley, D. W., Friedlander, M. L., Taylor, I. W., Rugg, C. A., and Musgrove, E. A. (1983). *J. Histochem. Cytochem.* **31,** 1333–1335.

Heiden, T., Wang, N., and Tribukait, B. (1991). *Cytometry* **12,** 614–621.

Hitchcock, C. L., and Ensley, J. F. (1993). *In* "Clinical Flow Cytometry: Principles and Application" (K. D. Bauer, R. E. Duque, and T. V. Shankey, eds.), pp. 93–110. Williams & Wilkins, Baltimore, MD.

Kallioniemi, O. P., Visakorpi, T., Holli, K., Heikkinen, A., Isola, J., and Koivula, T. (1991). *Cytometry* **12,** 413–421.

Schutte, B., Reynders, M. M., Bosman, F. T., and Blijham, G. H. (1985). *Cytometry* **6,** 26–30.

Tagawa, Y., Nakazaki, T., Yasutake, T., Matsuo, S., and Tomita, M. (1993). *Cytometry* **14.**

Toikkanen, S., Joensuu, H., and Klemi, P. (1989). *Br. J. Cancer* **60,** 693–700.

Watson, J. V., Sikora, K., and Evan, G. I. (1985). *J. Immunol. Methods* **83,** 179–192.

CHAPTER 17

Controls, Standards, and Histogram Interpretation in DNA Flow Cytometry

Lynn G. Dressler* and Larry C. Seamer†

* University of North Carolina–Chapel Hill
Lineberger Comprehensive Cancer Center
School of Medicine
Chapel Hill, North Carolina 27599

† Flow Cytometry Facility
School of Medicine
University of New Mexico Cancer Center
Albuquerque, New Mexico 87151

I. Introduction

The routine use of flow cytometric techniques to evaluate DNA content and cell-cycle parameters in clinical and research specimens requires consistency and standardization in preparation of sample, acquisition of histogram data, and interpretation of the data obtained. This chapter focuses on the use of controls and standards that can provide quality control and quality assurance in the acquisition and interpretation of DNA histogram data.

There are two critical issues one must address prior to data acquisition or interpretation: one is quality assurance that a representative tissue sample is used for assay and the other is the use of a DNA reference standard to identify diploid position. Quality assurance most frequently relates to tumor tissue and requires that an aliquot or section of the exact material used for flow cytometry is reviewed by cytology or histology for presence of sufficient tumor cell nuclei. DNA reference standards are required to interpret the histogram ploidy status and thereby allow for appropriate cell-cycle analysis. DNA reference standards can be diploid or hypodiploid, but must reproducibly and accurately define the diploid position in the histogram.

Additional issues include standardization of cell/nuclear concentration in each sample and evaluation of the fluorochrome stain using light or fluorescence microscopy, respectively. Standardization of these factors provides quality control for fluorochrome dye saturation and consistency in flow rate which will influence the precision and accuracy of the DNA fluorescence measurements.

In cell-cycle analysis a small deviation in quality can lead to serious erroneous results. Therefore, instrument quality control, standardization, and maintenance are also essential to accurate DNA histogram acquisition and data interpretation and, ultimately, the quality of cell-cycle analysis.

Quality control in this context is defined as maximizing the precision and accuracy of single-cell DNA measurements. Precision is the measure of the reproducibility of a given measurement and is usually expressed as a coefficient of variation (C.V.). Accuracy reflects deviation of a measurement from reality. An analysis technique may have great precision but due to a systematic error may not be accurate. In flow cytometric cell-cycle analysis these considerations are reflected in the ability to correctly quantify the amount of fluorescent dye associated with the DNA of a cell and do that reproducibly.

Precision of a flow cytometer is affected by several factors, including optical alignment, stability of the stream, excitation light intensity, and electronics. It is essential that the cells or nuclei flow through the focal point of the laser as well as the detection optics at a constant speed. Partial nozzle plugs or pneumatic instability will cause variation in illumination and therefore fluorescence emission variations. System variations are detected as an increase in the CV.

Accuracy can be compromised by a variety of factors. Incomplete illumination of the cells/nuclei due to poor optical alignment results in a reduced signal to the photomultiplier tubes (PMTs) and an incorrectly lowered DNA content. Partial nozzle plugs can cause the cells/nuclei to slow as they pass through the excitation light resulting in an increased signal to the PMT and an apparently increased DNA content. Alternately, a nozzle plug can cause the cells/nuclei to miss the optical alignment point resulting in a reduced signal to the PMT and elevated coefficient of variation (CV) (Fig. 1). Also, any one of the elements which can cause nonlinearity, from nonstoichiometric staining to nonlinear analog-to-digital converters (ADCs), results in inaccuracies in DNA measurements.

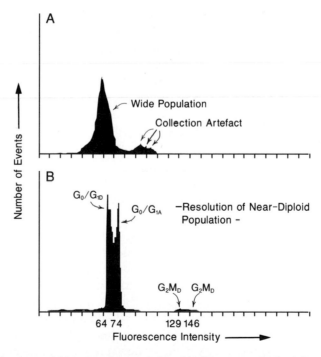

Fig. 1 Nozzle plug and histogram interpretation. (A) Histogram obtained from a sample run through a partially clogged nozzle tip. We observe a major wide single asymmetric population. (B) Histogram obtained from the same sample after flushing the nozzle to remove the clog. We observe resolution of a near-diploid population and the corresponding G_2M regions.

Evaluation of DNA histograms involves the incorporation of defined parameters to allow consistency and quality control of data interpretation. A defined set of evaluation criteria is required that includes acceptable ranges for the CV of the measured cell population, relative DNA content or DNA index (DI), and specific definitions for DNA diploid versus DNA aneuploid and subgroups of DNA aneuploid (hypodiploid, hyperdiploid, near-diploid, near-tetraploid, tetraploid, hypertetraploid, and multiploid). Criteria for evaluability of cell-cycle analysis should also be defined.

In this chapter, technical details for controls, standards, and criteria used to acquire and interpret DNA FCM data that have proved efficient and reproducible are presented. These guidelines should prove useful to investigators setting up DNA FCM methods in their laboratories and hopefully will stimulate suggestions, improvements, and eventually standardization.

II. Application

A. Quality Assurance of Representative Tissue

The first step in a good quality assay is to ensure that the data obtained are representative of the tissue sample. When dealing with tumor tissue it is essential to evaluate the proportion of malignant to nonmalignant cells to ensure sufficient tumor cell nuclei comprise the sample used in the assay. We have found that a minimum preparation of 15–20% tumor nuclei is required to meet these conditions (Dressler and Bartow, 1989). This evaluation is usually performed by a pathologist with the aid of a micrometer grid fitted in one of the oculars of the microscope or by visual estimation of the entire specimen.

1. Fresh or Frozen Tissue

Cytologic controls are practical in fresh and frozen tissue specimens, in which an aliquot (usually 50–100 μl) of the sample is obtained following tissue disaggregation or, in the case of body fluids, upon receipt of the specimen (see Section V, Methods). Obtaining the morphologic control following dissociation is highly recommended. Cytocentrifugation of the aliquot usually results in the best quality slide. Cytologic smears have also been prepared but are sometimes more difficult to evaluate. Slides are stained using the laboratory's standard procedure [Giemsa, Hematoxylin and eosin (H&E), DiffQuik, etc.]. Morphologic controls can also be obtained prior to dissociation as an imprint, touch preparation, or frozen section (provided sufficient tissue is submitted) but will not control for any cell selection that might result from the dissociation technique used. Cytocentrifugation of an aliquot following dissociation is routinely used in our laboratory with good results. (This morphologic control can also be applied to alcohol fixed specimens.)

2. Formalin-Fixed Paraffin-Embedded Tissue

Samples received fixed and embedded in paraffin wax can be easily evaluated by histologic review of the 5-μm sections immediately preceding and following the thick sections cut for assay. Review of these "top" and "bottom" H&E-stained sections allows for assessment of the proportion of tumor nuclei relative to nontumor nuclei throughout the sample cut for assay and ensures that the appropriate block was selected for assay. This review also allows an opportunity to maximize tumor tissue nuclei recovery by microdissection of localized tumor areas from nontumor (necrotic, fibrotic, hemorrhagic) areas (see Section V, Methods).

B. DNA Reference Standards

Several different external and internal DNA standards have been described in the literature (Vindeløv et al., 1983; Iverson and Laerum, 1987; Jakobsen, 1983). An external standard is one that is prepared in a separate tube from the sample and run in parallel with the sample. An internal standard is one that is actually added to the sample tube prior to fluorochrome staining. Ideally, a sample should be run in duplicate, where one sample tube contains the internal standard and the other does not. This allows for quality control of staining variability. However, because a large quantity of available sample is needed, running duplicate samples is not always feasible or practical. Therefore, some investigators have chosen to use either an internal or an external standard and others use both kinds of standards to identify the DNA diploid position. There are two strategies for using DNA reference standards: one is to use a DNA hypodiploid reference(s) to identify where the human DNA diploid G_0/G_1 peak should be located; this can be accomplished by using the reference as an internal or external control. The other strategy is to use a normal DNA diploid reference that will mark the DNA diploid channel and will superimpose with the DNA diploid peak of the sample; this again has been performed as an internal as well as an external control. There are advantages and disadvantages to both strategies. Use of DNA hypodiploid internal controls can mask the appearance of unknown hypodiploid populations in the sample with a DNA index of approximately 0.80 and, more importantly, have the potential to "overwhelm" small aneuploid populations found in the sample. Therefore, careful cell counts must be made to determine the appropriate proportion of reference cells that should be added to the sample tube. For this reason, the use of human peripheral blood lymphocytes (hPBLs) as a diploid reference is generally only used as an external control. If, however, two sample tubes can be set up, one tube can contain the hPBL as an internal control and the other tube would contain sample alone.

1. Fresh or Frozen Tissue

Vindeløv and colleagues (1983) described a system using the ratio of two hypodiploid references, chicken red blood cell nuclei (CRBC) and trout red blood cell nuclei (TRBC), as internal controls, to help identify the position of the human diploid G_0/G_1 population. Nuclei of CRBC and TRBC have different DNA contents, which are both less than a normal human lymphocyte (33 and 80%, respectively). Aliquots of both TRBC and CRBC are added to the tumor specimen prior to staining, and the ratio of the peak positions of the trout G_0/G_1 to the chicken G_0/G_1 are used to normalize the histogram and identify the human diploid G_0/G_1 position.

Normal diploid controls, such as unstimulated hPBLs or nonmalignant tissue from the same organ as the tumor sample, are also used as DNA diploid standards. Human PBLs are the most widely used DNA reference standard and are generally used as an external standard. Nonmalignant tissue from the same organ, however, is the ideal, yet not always practical, control.

CRBC are also used as an external standard for instrument alignment, setting high voltages and gains, setting gates and threshold for light scatter, size or fluorescence intensity on the flow cytometer (see Section V, Methods, for details). Because the peak position of CRBC can be variable in the low channels, it is not recommended that CRBC alone be used to estimate the DNA diploid peak position.

The authors use unstimulated hPBL as an external DNA reference standard. The hPBL are first run on the flow cytometer and adjustments in gain and/or high voltage are made to position the peak or mean height of the G_0/G_1 population at channel 60 (on a 256 scale). This channel number allows for a tetraploid tumor (4N, G_0/G_1, and 8N, G_2M) to be on scale. We routinely collect 5000 events for the hPBL standard. We then run the tumor sample and collect 25,000–50,000 events (25,000 if collected gated; 50,000 if collected ungated). To confirm the position of the DNA diploid G_0/G_1 peak, an aliquot of hPBL nuclei is added to the sample tube to a final concentration of 20%. The "mix" of DNA reference standard and sample is then rerun on the flow cytometer and 25,000–50,000 events are collected in the mix as was collected for the sample (Dressler and Bartow, 1989). In the case of an aneuploid tumor, the diploid peak height will increase and the aneuploid peak height will correspondingly decrease in the "mixed" relative to the "unmixed" sample (Fig. 2). In the case of a diploid tumor, one should obtain a superimposition of the G_0/G_1 peaks and a proportional decrease in the height of the G_2M peak.[1] In addition, unfixed PI-stained CRBC nuclei in phosphate-buffered saline (PBS) can be used to set the threshold for fluorescence intensity on the flow cytometer. Events having fluorescence intensity less than that of the CRBC nuclei are considered debris and do not trigger the instrument to record these events. This allows collection of total events less likely to represent debris. After the threshold is set, collection of ungated data still allows the user to estimate and model the debris in the histogram.

---

[1] Several users have indicated differences in fluorescent intensity obtained between male and female reference standards. There have also been reports of differences in DNA supercoiling or DNA compactness resulting in differential uptake of an intercalating dye.

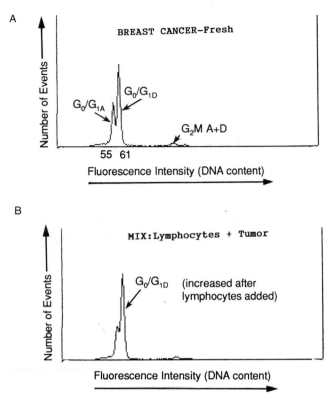

Fig. 2 Use of external control to identify human DNA diploid G_0/G_1 peak. (A) Histogram of a breast tumor obtained fresh from surgery. Two distinct peaks are observed at channels 55 and 61, respectively. To confirm which of these two peaks represents the normal DNA diploid population, an aliquot of unstimulated normal human peripheral blood lymphocytes (hPBL) was mixed with the tumor sample (to a final concentration of 20%). (B) The peak corresponding to channel 61 is seen to have increased after the addition of hPBL and confirms this peak as representing the normal DNA diploid G_0/G_1 (G_0-G_{1D}) population. The DNA aneuploid tumor was assigned an interpretation of "DNA hypodiploid" and a DI of 0.90 was calculated.

2. Paraffin Blocks

An optimal normal DNA reference standard for paraffin block samples would be nonmalignant tissue from the same organ that was fixed and processed in parallel with the tumor sample; ideally nonmalignant tissue would be from the same paraffin block as the sample. If nonmalignant tissue is to be used as a DNA reference, it must have a highly cellular epithelial component to allow enough cells to be obtained for assay. An alternative DNA reference standard, which may be more practical in certain tissue types, is an uninvolved lymph node from the cancer dissection. For biopsy specimens or when lymph nodes or corresponding nontumor tissue is not available, a normal lymph node or tonsil block from another patient, whose tissue was processed in the same laboratory, at about the same time, and by the same methods as the tumor specimen, can be used.

The use of control tissue in the paraffin block procedure, however, does not provide a DNA reference standard as consistent as the trout/chick ratio or hPBL used for fresh or frozen tissue. Differences in fixation and/or processing as well as differences in DNA structure of the control tissue as compared to the tumor tissue can affect the binding and stoichiometry of dyes, specifically the intercalating dyes, resulting in different fluorescence intensity of diploid nuclei (Fig. 3). In our laboratory, we have experienced this difficulty in <6% of all paraffin block cases. Listed here are several suggestions for minimizing difficulty with the paraffin block technique.

1. Any tissue, especially lymph nodes, should be uniformly well fixed.

2. If fixation of the lymph node or any other block is inconsistent or poor, an alternate block should be selected.

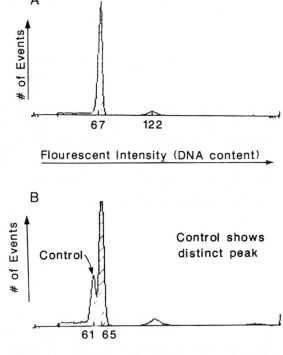

Fig. 3 Problems with paraffin block controls. (A) Histogram obtained from a formalin-fixed paraffin-embedded breast tumor. A major single symmetric population is observed in channel 67 with a corresponding G_2M in channel 122. (B) Histogram obtained after mixing formalin-fixed paraffin-embedded control lymph node nuclei ($\sim$ 20%) with the tumor sample. We observe a distinct population at channel 61 representing control nuclei. These differences in DNA fluorescence between control and sample have been observed in less than 6% of over 5000 paraffin samples.

3. Prescreening of corresponding histologic slides stained with hematoxylin and eosin prior to assay is very helpful in assessing the quality of fixation and tumor composition in the selected block.

4. In cases where blocks contain tumor only in specific areas, the technician can dissect out the indicated areas of the block to maximize recovery of tumor tissue. This selection technique is also useful for obtaining both normal and neoplastic tissue from the same block and for excluding areas of necrosis or host inflammatory cells from the tumor specimen.

5. CRBC nuclei can be quite useful in the paraffin-embedded procedure to help determine whether a shift in fluorescence is instrument related or sample related.

C. Standardization of Cell Concentration

Standard practices in our laboratory require that cell/nuclear counts are performed to ensure that a concentration of approximately 2 million cells/ml is consistently run on the flow cytometer. Standard concentrations in all tubes allow for more consistent flow rates on the flow cytometer and ensure a saturating concentration of the fluorochrome for all samples. These parameters can impact on the CV of both the diploid and aneuploid populations (Fig. 4). An aliquot of cells/nuclei is taken prior to fluorochrome staining and cell counts are easily estimated using a hemocytometer.

D. DNA Fluorescence Controls

Evaluation of fluorescence of stained cells or nuclei can be quickly and easily performed using a fluorescence microscope and gives the user much information about the sample preparation before running on the flow cytometer. This evaluation serves to ensure that both DNA reference and sample have incorporated the dye into the nucleus with minimal, if any, cytoplasmic staining. If cytoplasmic staining is observed the user should carefully review their staining procedure and modify accordingly for appropriate concentrations and incubation times for RNase and non-ionic detergent digestion. In addition, the user can assess the percentage of nuclei that have an intact vs a broken nuclear membrane, the degree of aggregation/clumping present, and the degree of debris material that may cause nonspecific trapping of the DNA fluorochrome.

III. Instrument Quality Control and Standardization

A. Instrument Maintenance

Instrument quality control begins with an effective and well-documented preventive maintenance plan. The preventive maintenance documentation should include a specific procedure manual as well as a schedule to record the

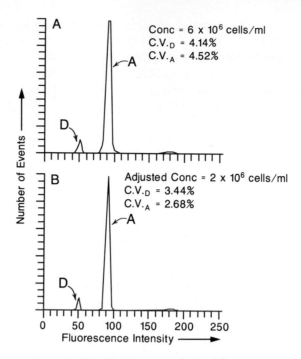

Fig. 4 Cell concentration and CV. (A) Histogram obtained from a tumor sample whose cell concentration was approximately 6×10^6 cells/ml. Coefficient of variation (CV) of the G_0/G_1 populations for the diploid and aneuploid population was 4.14 and 4.52%, respectively. An aliquot of this sample was diluted to a concentration of 2×10^6 cells/ml (B). The histogram obtained in B shows a lower CV for both the diploid and aneuploid population as compared to the more concentrated sample in A.

name of the individual performing the maintenance and the date performed. Instruments vary in their requirements for routine maintenance. Therefore, the manufacturer's procedure manual should be consulted when developing a preventive maintenance schedule. However, there are a few procedures that should be included in any maintenance schedule. An effective quality control plan will include routine cleaning of the sample line, sheath system, and waste system. The sheath filter should also be replaced regularly. It is also important to keep a problem log, recording the date a problem is detected and describing the nature of the problem and all steps taken to correct it.

Maintaining instrument linearity is critical when performing DNA flow cytometry. Nonlinearity can result from a variety of sources, many of which are due to poor instrument calibration. Therefore, linearity checks should be performed as part of an overall preventive maintenance plan. If your instrument allows for electronic ADC linearity calibration, that procedure should be included in the preventive maintenance schedule. There is some debate concerning the required frequency of these checks (quarterly vs weekly vs daily). A linearity check should be performed after any component of the signal processing electronics is replaced. These signal processing components include photomultiplier tubes, amplifiers, and ADCs.

B. Daily Instrument Setup Procedure

To ensure run-to-run reproducibility, instrument setup must also be monitored daily. An instrument performance log should be maintained by the operator and checked at regular intervals by the laboratory supervisor. At the start of each day's run, certain laser and system fluidic parameters should be noted and logged (Table I): (A) Laser parameters, wavelength (if adjustable), output power, current necessary to achieve output, and aperture setting (if available). (B) Systemic fluidics: gas tank pressure or system input pressure, sheath pressure (voltage), sample pressure (voltage), differential pressure necessary to maintain 100 particle/sec flow rate (if available).

C. Daily Quality Control

After an appropriate instrument warm-up period, set the PMT voltage and amplifier gains to a standard configuration. A standard particle should then be run, monitored, and logged for mean fluorescence level and coefficient of variation (Table II). This control verifies proper instrument setup, appropriate and

Table I
Sample Daily Instrument Setup Logs

A. Laser					
Date	Wavelength	Power	Current	Aperture	Initials
1/11/93	488	600 mw	28 amps	6	LS

B. System fluidics					
Date	Gas	Sheath	Sample	Diff	Initials
1/11/93	50 psi	13 psi	na	9 psi	LS

Table II
Sample Daily QC Log

Red fluorescence PMT #2				
Date	Bead lot	Mean 110–135	CV < 2.0	Initials
1/11/93	12345	125	1.65	LS

clean optical filters, flow stability, laser stability, and proper optical alignment. It is important to select a particle that has an internal CV of less than 2.0% in fluorescence intensity when acquired in the detector channel where the DNA fluorochrome is measured. A Levey-Jennings plot of mean channel fluorescence is helpful in detecting day-to-day instrument drift or other "out of control" situations. However, it is not appropriate to use this kind of plot for CV since there is no situation where the CV can be too low. Therefore, acceptability of CV is defined by an upper threshold level, above which an unacceptably unstable or misaligned instrument is suggested. Each new lot of beads should be run in parallel with the old lot to determine an acceptable range of the mean fluorescence for the new bead while continuing to monitor the previous lot.

Finally, to establish the position of the normal DNA diploid G_0/G_1 peak and to detect instrument errors which may occur only at the instrument setting used for sample acquisition, it is necessay to run normal human cells stained with the fluorochrome buffer used on test samples. The PMT voltage and amplifier gain necessary to place the diploid G_0/G_1 peak in the desired channel should be recorded along with the CV and the position of the apparent G_2M peak to check system linearity.

D. Histogram Acquisition

Quality control during sample acquisition also significantly affects the accuracy of the final results. Due to the dynamic nature of a flow cytometer, many problems can occur during sample acquisition.

The flow rate is perhaps the most critical component affecting histogram quality. Flow cytometers must maintain a constant particle speed through the excitation light. Therefore, to increase the flow rate, the instrument increases the volume of sample flowing while maintaining a constant particle speed. Increased sample volume widens the sample portion of the stream. Therefore, cells have a greater probability of passing through suboptimal excitation light which diminishes their fluorescence intensity. The result is, the faster the sample flow rate, the higher the CV and the less accurate the resultant cell-cycle analysis will be. We recommend a flow rate of approximately 100–200 cells/sec on instruments that can be controlled. On other instruments, we recommend using the slowest acquisition setting that approaches this level.

As a final monitor of system linearity, it is necessary to calculate a G_2/G_1 ratio on each population noted during all sample acquisitions. In a few older flow cytometers, nonlinear acquisition will occur only on large particles. This phenomenon will often be seen on aneuploid nuclei because of their size.

Because of the long acquisition times necessary to collect DNA cell-cycle data, momentary fluctuations and gradual, progressive changes can cause unnoticed artifacts in the data. These artifacts will appear as shoulders on existing peaks or as extra peaks which could be misidentified as DNA aneuploidy. To monitor for drift errors it is sometimes recommended that a time vs fluorescence dot plot be acquired with every sample (Muirhead, 1993). Instability will appear as a shift of the peak position over time. Alternately, for instruments that are unable to acquire time, analysis software is available which can display list-mode data sequentially in a pseudo-time display or can calculate mean fluorescence on sequential batches of cells. Sample drift can then be detected during cell-cycle analysis.

IV. DNA Histogram Interpretation

Numerous papers have been published using DNA FCM measurements to characterize a variety of solid tumors. However, standard quality control parameters are not universally in place in clinical or research laboratories at present. Before standardization can occur, definitions for ploidy status, criteria for evaluability, and modeling parameters for S phase need to be understood and well described. The following guidelines have proven efficient and practical and provide consistency and quality control in DNA histogram interpretation.

A. DNA Ploidy Status

In order to distinguish results obtained by DNA flow cytometry from data obtained by cytogenetic techniques, it was recommended by the Convention on Nomenclature in 1984, that the terms "normal" and "abnormal stemline" be used instead of diploidy and aneuploidy, respectively (Hiddeman *et al.*, 1984). The term "DNA aneuploid" was agreed on as being a synonym for "abnormal stemline." In our laboratory, the terms "DNA diploid" and "DNA aneuploid" are used for normal and abnormal stemlines, respectively.[2]

Definition of tumor ploidy status is based on the amount of DNA relative to normal. The DNA index (DI) is a value given to express the amount of DNA content relative to normal and is calculated by the following equation.

$$DI^{3,4} = \frac{\text{Mean or modal channel No. of DNA aneuploid } G_0G_1 \text{ peak.}}{\text{Mean or modal channel No. of DNA diploid } G_0G_1 \text{ peak}}$$

A DNA diploid population (G_0/G_1) is given a DI of 1.00 by definition. DNA aneuploid or "nondiploid" tumors can have less than the 2N amount of DNA and are termed "DNA hypodiploid" or, more frequently, have greater than the 2N amount of DNA and are termed "DNA hyperdiploid." Those tumors that have a 4N amount of DNA are termed "DNA tetraploid," while those

[2] Other investigators have chosen to refer to the abnormal population as "nondiploid."

[3] Most frequently the modal channel number is used when manual calculation of the DNA index is performed, whereas mean channel number is used in software modeling algorithms. DNA index value obtained from modal channel may be different compared to mean channel number depending on the symmetry (or asymmetry) of the G_0/G_1 peaks.

[4] DNA index can only be reported when the position of the DNA diploid population is confirmed.

that are "near" 4N are sometimes called "DNA near tetraploid," and those greater than 4N are "DNA hypertetraploid." Tumors may also have more than one abnormal population, in which case they are referred to as multiple aneuploid or "DNA multiploid" (Dressler and Bartow, 1989). Table III shows the DNA index values that define the aneuploid subgroups in our laboratory. It is important for each laboratory to establish consistent definitions for ploidy subgroups and publish these definitions to allow for comparison of data and eventual standardization.

There are situations where the interpretation of DNA diploid versus DNA aneuploid is not straightforward. For example, a fresh tumor may show a single peak with a slight left shoulder. Addition of control lymphocytes causes a single peak to be observed, shifted to the left by one channel, and no shoulder is visible. Because two distinct peaks cannot be resolved, it cannot be determined by FCM whether there are indeed two distinct populations. Some investigators have interpreted this type of histogram as "diploid-like," "near-diploid," or "questionable ploidy." Other investigators define a DNA index range for DNA diploidy, based on resolution of the flow cytometer. For example, if the resolution of the flow cytometer is 3%, the diploid range would be defined as 0.97–1.03.

It should be emphasized that only a fraction of nuclear DNA is stainable (accessible to dyes) and that the extent of DNA accessibility varies depending on the dye used, chromatin structure, and staining procedure (Darzynkiewicz et al., 1984). The terms "DNA ploidy" or "DNA index" have been considered operational, and actual DNA content of the studied cells may be different from that established using "normal" cells as the DNA reference standard. However, several studies have reported that DNA index is a good estimate of total chromosome number in solid tumors.[5] (Remvikos et al., 1988; Siegfried et al., 1991; Dressler et al., 1993; Smeets et al., 1987).

Table III
DNA Index and Ploidy Category

DNA Index (DI)	Ploidy category
= 1.00	Diploid
≠ 1.00	Aneuploid
< 1.00	Hypodiploid
> 1.00 ≤ 1.90[a]	Simple hyperdiploid
> 1.90; ≤ 2.20[a]	Tetraploid/near tetraploid
> 2.20[a]	Hypertetraploid
2 DI ≠ 1.00	Multiploid

[a] These DNA index limits may vary from laboratory to laboratory. Reprinted with permission (Dressler and Bartow, 1989).

[5] A crude estimate of expected total chromosome number can be obtained by multiplying DNA index value by 46.

B. Criteria for Evaluation

The use of data-derived criteria help the investigator to evaluate and interpret a DNA histogram with consistency and accuracy. It should be understood that these criteria may not be applicable to all tumor types, and individual sets of criteria may be necessary for different disease groups and specific studies, especially if the specimen has been obtained from a patient undergoing treatment. The following discussion of interpretation criteria is based on the authors' experience with a variety of solid tumors with an emphasis on breast cancer.

One of the most important criteria for determining if a histogram is evaluable for ploidy status is CV of the mean DNA content of the G_0/G_1 subpopulation. Its value corresponds to the ability to resolve two different DNA content populations in the same sample and is influenced by sample preparation and instrument alignment. The larger the CV, the wider the peak and the less resolution, i.e., the greater the chance of missing an abnormal population that lies close to 2 N. In addition, the wider the CV, the less accuracy and reproducibility will be achieved in trying to estimate the S phase region of the histogram. Therefore, criteria for CV are often set to allow for distinction between which histograms are evaluable for ploidy status as well as cell-cycle analysis (S or S + G_2M) and which are not. It is important to define how the CV is determined and it is essential to report CV values and ranges in all published studies.[6] Histograms with CV values of <5% are generally acceptable to most investigators (Dressler et al., 1987)[7].

Another important yet controversial criterion for evaluating histograms is the definition of a DNA aneuploid peak. Many investigators do not report their definitions. Those investigators who have reported definitions for DNA aneuploid generally only consider an abnormal stemline present when two distinct G_0/G_1 populations are observed (Bauer et al., 1986; Dressler and Bartow, 1989). In our laboratory, a peak is defined as being DNA aneuploid if at least 10% of the total events collected in the histogram are found in the G_0/G_1 peak and a corresponding G_2M can be identified.[8] There are examples, however, of discrete, symmetrical nondiploid G_0/G_1 peaks being observed that comprise

---

[6] CV values can be obtained from a variety of equations. Statistically CV = SD/mean; however, many programs estimate CV as the value at half the maximum height. CV values have been reported for the raw data or processed data. It is important to be consistent in reporting CV values to establish criteria for evaluability. If histogram data is subjected to cell-cycle analysis, values obtained from the processed (software-generated) data should be used.

[7] In our laboratory, a CV cutoff for fresh/frozen samples of 5% was derived from inter- and intraarray variability studies in which 10 aliquots of the same tumor were run on 10 different days (interassay) or on the same day (intraassay). We observed that tumors with a single peak on histograms that had a CV >5% showed an increased frequency of being resolved as two distinct populations on subsequent runs. Those single-peak tumors with CV <5% showed only one peak on subsequent runs. For paraffin block samples, a CV of >7% establishes the cutoff (Dressler, 1987).

[8] One can set cursors around the peak of interest and assess the total number of events within the cursor region relative to total number of events collected.

<10% of the total events collected.[9] When this observation is reproduced in a duplicate sample, it is sufficient criteria for our laboratory to indicate "the presence of a small, nondiploid population."

An exception to this definition of DNA aneuploidy, in our laboratory, is the case of a tetraploid tumor. Because one cannot distinguish DNA diploid G_2M cells from DNA tetraploid G_0/G_1 cells in the 4N peak, a DNA tetraploid tumor is defined as having at least 15–20% of the total events collected in the 4N peak. Other investigators have defined a DNA tetraploid tumor as having 10% of cells in the 4N peak on the histogram (Baildam *et al.*, 1987). Still others do not include DNA tetraploid tumors as being among the DNA aneuploid group (Ewers *et al.*, 1984), but rather refer to any even multiple of 2N as being euploid (diploid, tetraploid, octaploid). Essential to the interpretation of DNA tetraploid is: (1) fluorescent microscopic examination of the stained specimen at the time of FCM analysis to quantify aggregates/clumping; (2) doublet discrimination by hardware or aggregate estimation by software programs. These parameters can help discriminate clumps of diploid G_0/G_1 nuclei vs tetraploid G_0/G_1 nuclei.

Overall, we use four classifications to describe the DNA histogram: DNA diploid, DNA aneuploid, nondiploid, and uninterpretable. Nondiploid includes (1) inability to resolve a shoulder on a population in the diploid range, (2) a small discrete peak <10% of the total collected events, (3) a single peak with wide CV (>5.00 for fresh–frozen; >7.00 for paraffin), or (4) a single G_0/G_1 peak, with a broadened G_2M (4N) region. "Uninterpretable" refers to (1) a sample that is all debris, (2) one that does not yield a sufficient number of cells for analysis, (3) one in which very poor resolution was obtained, or (4) one that does not meet our criteria for quality assurance of tumor tissue. In these instances, we either try to repeat the assay, if there is sufficient material available, or request another specimen (e.g., another paraffin block) to be submitted for assay. Specific criteria for evaluability that have been useful in our laboratory to provide consistency in interpretation are in Section VI.[10]

V. Methods

A. Cytospin Protocol for Cytologic Evaluation

1. Following tissue dissociation, place ~100,000 cells in suspension (50–100 μl) in a 12 x 75-mm tube.

[9] This criterion can be confirmed when comparing primary lesions with corresponding lymph node metastases; however, a comment is added to the DNA histogram report indicating that the "clinical significance" of this observation is not confirmed at this time. Quite often in these cases, the G_2M population may not be identifiable. Likewise, in nonbreast cancer solid tumors, the G_2M population may not be easily detectable even though the G_0/G_1 population comprises 10% of the total events collected.

[10] The reader is also referred to a special report of the DNA Cytometry Consensus Conference published in *Cytometry* **14**, 471–500. (Shankey *et al.* 1993). This report covers seven articles in this issue of *Cytometry* which includes guidelines for specific disease sites: bladder, breast, colorectal, hematopathology, and prostate cancer, subsequently published to submission of this chapter.

   2. Add an equal volume of 95% ethanol to fix samples.

   3. Gently vortex the fixed sample and add 150–200 μl to the cytospin funnel.

   4. Centrifuge for 5 min at 350g.

   5. Let air dry and follow routine staining procedure for cytologic specimens (hematoxylin and eosin, Giemsa, Wright, DiffQuik, etc.).

B. Histologic Protocol

   1. Cut one 5-μm section immediately preceding and directly following thick sections. Stain according to routine method.

   2. Evaluate "top" and "bottom" sections for percentage of tumor versus nontumor. At least 15% of tissue section should be composed of tumor cells for quality assurance.

   3. Evaluate for necrosis, hemorrhage, and degree of fibrosis.

   4. If tumor is localized to one area, or both tumor and nontumor are present in separate areas on the same block, mark block with razor blade to outline specific areas of interest.

   5. When microtoming, each "outlined" part of the section can be placed in separate tubes and processed independently.

C. Fluorochrome Control

   1. Immediately before running sample on flow cytometer, aliquot 20 μl of sample and apply to glass slide as a wet preparation.

   2. Add a small coverslip and seal edges with nail polish to prevent sample from drying out.

   3. View under fluorescence microscope and record percentage of cells/nuclei staining, intensity of stain (0–3; 3 highest), integrity of cells and nuclei (percentage broken or intact), degree of noncellular material staining (0–3, 3 highest), percentage of cells/nuclei in aggregates versus single-cell dispersion, and size of particles (small, medium, large).

D. DNA Reference Standards

   To obtain a large number of reference aliquots, the authors routinely use unstimulated hPBL, obtained as whole blood from the local blood bank prepared as follows:

   1. Dilute blood 1 : 1 with PBS containing 1% heparin (sodium salts; 1000 units/ml) in a sterile glass bottle.

   2. Layer over Ficoll-Paque (four parts diluted blood to one part Ficoll-Paque), and centrifuge for 30 min at 750g.

3. Carefully pipet off lymphocyte layer and wash cells in sterile Hanks' balanced salt solution (HBSS).

4. Centrifuge at 250*g* for 5 min. Resuspend in 1 ml HBSS and count cells.

5. Resuspend cells in appropriate volume of freezing medium (Stone *et al.*, 1985) to allow a concentration of 5×10^6 cells/ml to be frozen in 1.0-ml aliquots.

E. Freezing Medium

NaCl: 1.765 g

Dimethyl sulfoxide (DMSO): 14.940 ml

RPMI (or MEM, Medium 199) + 20% serum: qs to 100 ml

1. Resuspend cells in cryopreservation medium. Concentration should not exceed 5×10^6 cells/ml for ease of FCM preparation.

2. Place vials containing cells in $-70°C$ freezer.

3. Keep sample at $-70°C$ until day of assay.

F. System Linearity Check Procedure

The procedure described here checks most components of the flow cytometer's signal processing electronics. Two sets of microspheres (or stained cells) with slightly different fluorescence intensity are required (Bagwell *et al.*, 1989). The amplifier should be set at the value typically used when acquiring DNA fluorescence histograms and particles should be sought with fluorescence intensities that are close to each other and in the range of stained cells.

1. Adjust the PMT voltage so that the two microsphere peaks are in the lowest channels that just barely separate them.

2. Record the difference in channels of the mean or the mode of the two peaks.

3. Increase the PMT voltage until the first peak is in a channel that is approximately 5% (of the total histogram range) greater than its original position. If a histogram has a 256-channel maximum range, 5% would be approximately 12 channels.

4. Repeat steps 2 and 3 until the highest intensity peak falls off scale on the right.

5. Create a graph of the differences in the peaks on the *y*-axis versus the peak position of the lowest intensity bead on the *x*-axis.

6. Calculate correlation coefficient on the points of the graph using the following formula:

$$r = \frac{\Sigma xy}{\sqrt{\Sigma x^2 + \Sigma y^2}}.$$

7. Although each laboratory should establish their own cutoff level, an *r* value <0.95 should be viewed with some concern.

G. Daily Instrument Setup Procedures

1. Align instrument with fluorescent beads (or glutaraldehyde-fixed CRBC). Measure and record integrated red fluorescence (IFL) and forward angle light scatter (FALS). Coefficient of variation on both should be <2.00%.

2. Run external control (PBL nuclei for fresh or frozen samples). Adjust high voltage and/or gain to position G_0/G_1 peak in channel 60. On a linear scale (0–255) this will allow a tetraploid tumor (4N) and corresponding G_2M (8N) to be on scale. Ensure that CV is <3.5% (this is usually the case when the beads have been used for proper alignment in step 1).

3. Set threshold intensity (for fresh or frozen samples) with unfixed, stained CRBC nuclei. (CRBC should be found at approximately channel 20.)

4. Run external diploid control (hPBL for fresh or frozen samples) and collect 5000 events. For control and samples, record channel numbers for G_0/G_1 and G_2M peak(s), corresponding CV values, high-voltage setting, and ratio of G_2M peak position to the G_0/G_1 peak position (for both DNA diploid and DNA aneuploid populations). (This ratio gives a control for the stoichiometric relationship of the stain system and should be fairly consistent throughout the run using the same staining system.)

5. Following stabilization of sample flow rate, run the tumor specimen and collect 25,000–50,000 events, if possible. During collection of first 10,000 events, carefully monitor for the following:

   a. Shoulders on peaks (debris vs off-center alignment).[11]

   b. Potential G_0/G_1 peaks observed in channels higher than 120 (G_2M events will be off scale).[12]

VI. Summary of Criteria for Interpretation of DNA Histograms[10]

A. Quality Assurance of Representative Tissue

1. Five-μm sections (top and bottom) or cytocentrifuged aliquots must contain ≥15% tumor cells; percentage of tumor vs nontumor, degree of nec-

---

[11] If shoulders are observed, we stop data acquisition and then reacquire to ensure that we are not observing artifact due to baseline drift or off-center alignment. External standards or beads are run again to ensure proper alignment, if necessary. Acquiring a two-parameter histogram of fluorescence versus time in list mode can also assess possible drifting and hydraulic stability.

[12] If the specimen shows a potential aneuploid G_0/G_1 peak whose position is greater than channel 120, only 10,000 events are initially collected at this voltage. The voltage and/or gain are then lowered to adjust the position of the abnormal peak to channel 120 or lower and 25,000 events are collected at this lowered voltage. This adjustment allows the corresponding G_2M to be on scale and makes possible cell-cycle analysis.

rosis, fibrosis, or presence of hemorrhage should be recorded on paraffin blocks.

2. Control should be confirmed as being uninvolved with tumor.

B. Evaluation of DNA Fluorescent Stained Cells

The following characteristics are noted for each sample:

1. Percentage of cells/nuclei in a single-cell suspension versus that in aggregates
2. Integrity of cells/nuclei (i.e., percentage broken vs intact)
3. Intensity of stain (1–3; where 3 is brightest)
4. Percentage of debris in the sample.

C. Histogram QC Checks

1. At least 10,000 total events have been collected (bladder irrigations may be 5,000)
2. Sample concentration at 2×10^6 cells/ml for optimal flow rate
3. Threshold is set to reduce debris contamination in total events.

D. Ploidy Status

1. DNA Diploid

a. Histogram shows a single G_0/G_1 peak with CV $\leq 7.0\%$ ($\leq 5.0\%$ for fresh/frozen specimens), and a corresponding G_2M peak.
b. Diploid position is confirmed by mix of control with tumor specimen.[13]
c. Histologic/cytologic evaluation confirms sufficient representative tumor nuclei are present.

2. DNA Aneuploid

a. Histogram shows more than one discrete G_0/G_1 peak.
b. Percentage of events in the abnormal G_0/G_1 peak is $\geq 10\%$ of the total number of events collected.
c. The CV of the DNA aneuploid G_0/G_1 peak is $\leq 7.00\%$ ($\leq 5.0\%$ for fresh/frozen specimens) for S phase analysis.

[13] When diploid position cannot be confirmed using an external reference, careful evaluation of the histologic section can confirm that sufficient nonmalignant cells are present to allow for an internal control.

3. Special Cases of DNA Aneuploid

a. DNA Hypodiploid

1. Histogram shows a discrete abnormal G_0/G_1 population to the left of the DNA diploid population.
2. The position of the DNA diploid population is confirmed using the DNA reference sample, i.e., a mix of reference and sample is performed.
3. The abnormal G_0/G_1 peak should contain 5–10% of all events collected.

b. DNA Tetraploid

1. Histogram shows a population at 4N comprised of at least 15–20% of all events collected.[14]
2. A corresponding G_2M (at 8N) should be identified.
3. Presence of cell aggregates has been rigorously estimated and minimized by evaluation of fluorescent cells and one of the following:
   a. Doublet discrimination by hardware gate.
   b. Aggregate subtraction by software.

4. Non Diploid

a. Single G_0/G_1 population with wide CV (CV $>7.00\%$; >5.00 for fresh – frozen sample)
b. Single G_0/G_1 population with right or left shoulder.
c. Small discrete population comprising less than 10% of the total collected events
d. Single G_0/G_1 population with a corresponding broad G_2M (4N) region (CV >7.00), comprising less than 15% of events collected.

5. Uninterpretable

a. Poor quality histogram obtained due to excessive debris, poor resolution, insufficient number of events collected ($<10,000$).
b. Insufficient tumor cell nuclei are observed on histologic/cytologic section.

Acknowledgments

The authors thank Dr. James Cervin and David Pena for their review of this chapter and helpful discussions.

References

Bagwell, C. B., Baker, D., Whetstone, S., Munson, M., Hitchcox, S., Ault, K. A., and Lovett, E. J. (1989). *Cytometry* **10**, 689–694.

[14] It should be recognized that this value is somewhat arbitrary in that normal tissues including liver, bladder, thyroid, and testes contain significant fractions of tetraploid cells.

Baildam, A. D., Zaloudik, T., Howell, A., Barnes, D. M., Turnbull, L., Swindell, R., Moore, H., and Sellwood, R. A. (1987). *Br. J. Cancer* **55,** 553–559.

Bauer, K. D., Merkel, D. E., Winter, J. N., Harder, R. J., Hauck, W. W., Wallemart, C. B., Williams T. J., and Variakojis, D. (1986). *Cancer Res.* **46,** 3173–3178.

Darzynkiewicz, Z., Traganos, F., Kapuscinski, J., Staiano-Coico, L., and Melamed, M. R. (1984). *Cytometry* **5,** 355–363.

Dressler, L. G., and Bartow, S. A. (1989). *Semin. Diagn. Pathol.* **6,** 55–82.

Dressler, L. G., Seamer, L., Owens, M. A., Clark, G. N., and McGuire, W. L. (1987). *Cancer Res.* **47,** 5294–5302.

Dressler, L. G., Varsa, E., Duncan, M., Bartow, S., Allgood, G., Sandoval, J., Seamer, L., Baca, P., Woodward, L., and McConnell, T. (1993). *Cancer* **72,** 2033–2041.

Ewers, S. B., Langstorm, E., Baldetorp, B., and Killander, D. (1984). *Cytometry* **5,** 408–419.

Hiddeman, W., Schumann, J., Andreeff, M., Barlogie, B., Herman, C. J., Leif, R. C., Mayall, B. H., Murphy, R. F., and Sandberg, A. A. (1984). *Cytometry* **5,** 445–446.

Iverson, O. E., and Laerum, O. D. (1987). *Cytometry* **8,** 190–196.

Jakobsen, A. (1983). *Cytometry* **4,** 161–165.

Muirhead K. A. (1993). *In* "Clinical Flow Cytometry: Principles and Application" (K. Bauer, R. Duque, and T. V. Shankey, eds.), pp. 177–199. Wiliams & Wilkins, Baltimore, MD.

Remvikos, Y., Muleris, M., Vielh, P. H., Salmon, R. J., and Dutrillaux, B. (1988). *Int. J. Cancer* **42,** 539–543.

Shankey, T. V., Rabinovitch, P. S., Bagwell, B., Bauer, K. D., Duque, R. E., Hedley, D. W., Mayall, B. H., and Wheeless, L. (1993). *Cytometry* **14,** 472–477.

Siegfried, T. M., Ellison D. T., and Resau, T. H. (1991). *Cancer Res.* **51,** 3257–3273.

Smeets, A. G., Pauwels, R. E., Beck, T. M., Geraedts, J. M., Debruyne, F., and Laarakkers, L. (1987). *Int. J. Cancer* **39,** 304–310.

Stone, K. R., Craig, B. R., Palmer, J. O., Rivkin, S. E., and McDivit, R. (1985). *Cytometry* **6,** 357–361.

Vindeløv, L. L., Christensen, I. J., and Nissen, N. I. (1983). *Cytometry* **3,** 323–327.

CHAPTER 18

DNA Content Histogram and Cell-Cycle Analysis

Peter S. Rabinovitch

Department of Pathology
University of Washington
Seattle, Washington 98195

I. Introduction

The analysis of cells stained with DNA-specific fluorochromes was one of the first applications of flow cytometry, and it continues to be one of the most common uses of this technique. A primary reason for this is that this procedure can rapidly determine both DNA ploidy and cell-cycle measurements. Recent interest in the clinical application of such information has accentuated the need for greater care and accuracy in the analysis of DNA content histograms. At the same time, this interest has stimulated improvements in the computational models used to extract ploidy and cell-cycle information from DNA content histograms. This chapter reviews guidelines for DNA content histogram analysis and the principles and advances in methods of analysis.

II. DNA Content Histogram Basic Principles

The DNA content of each cell in an organism is generally highly uniform. In the resting (G_1) phase of the cell cycle there are exactly 23 chromosomes per human somatic cell, and a DNA content of approximately 7 pg/cell. This *diploid* DNA content is designated in flow cytometry by DNA index (DI) 1.0 (Hiddeman *et al.*, 1984). When diploid cells which have been stained with a dye that stochiometrically binds to DNA are analyzed by flow cytometry, a "narrow" distribution of fluorescent intensities is obtained. This is displayed as a histogram of fluorescence intensity (*x*-axis) vs number of cells with each observed intensity. Since all G_1 cells have the same DNA content, the same fluorescence should (in theory) be detected, and only a single channel of the histogram should be filled (Fig. 1). In practice, however, instrumental errors and biological variability in DNA dye binding result in a Gaussian (normally distributed) fluorescence distribution from G_1 cells (Fig. 1). Greater variation in measurement results in broader DNA content peaks, and the term coefficient of variation (CV) is used to describe the width of the peak: CV $= 100 \times$ SD/mean of the peak.

When beginning the process of replication, cells enter DNA synthesis, or S phase. Initially their DNA content is imperceptibly greater than the G_1 DNA

Fig. 1 The difference between a histogram from a "perfect" flow cytometer with no errors in measurement (A) and the Gaussian broadening of the histogram that is encountered in all real analyses (B). In B, actual data points are displayed as small diamonds, solid lines indicate the Gaussian G_1 and G_2 phase components and the S phase distribution, as fit with the Dean and Jett (1974) polynomial S phase model. The dashed line shows the overall fit of the model to the data. (C) The same model fitting, but to a histogram that has overlapping diploid and aneuploid cell cycles. (D) A similar histogram of diploid and aneuploid cells, but with the addition of debris resulting from extraction of nuclei from paraffin, together with aggregates of cells with cells and cells with debris. A solid line shows the combined distribution of background aggregates and debris.

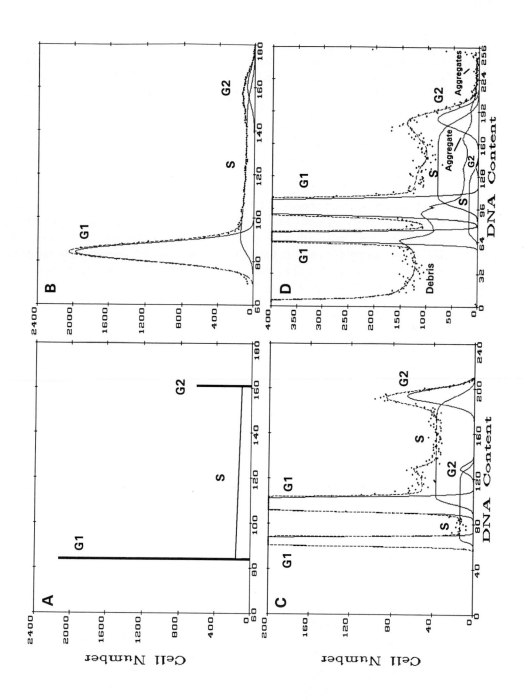

content; as DNA synthesis proceeds, cellular DNA content progressively increases until, with complete DNA replication, cells enter the G_2 phase with a DNA content twice that of G_1. When DNA damage is repaired and chromosomes are organized, cells enter mitosis. Cell division returns the two daughter cells to the G_1 DNA content. In a growing cell population, the distribution of cells in various stages of S phase results in a broad distribution of DNA contents between G_1 and G_2. In the theoretical distribution, these S phase cells are easily identified between G_1 and G_2 DNA contents (Fig. 1A). In actual histograms, however, the uncertainty in measurements and Gaussian broadening of G_1, S, and G_2 phase distributions results in considerable overlap between G_1 cells and early S phase cells and G_2 cells and late S phase cells (Fig. 1B).

When cells with an abnormal DNA content are present in tissue, a second G_1, S, and G_2 phase is present (the aneuploid cells are almost always accompanied by a component of cells with normal, diploid DNA content—for example stromal fibroblasts, capillary endothelial cells, lymphocytes). The overlap between the diploid and aneuploid cell cycles can be variable, but adds to the complexity of the histogram analysis (Fig. 1C).

Finally, actual DNA content measurements, especially those performed from tissue specimens, frequently are complicated by the presence of debris (fragments of nuclei) and aggregation of nuclei with each other or with debris (Fig. 1D). It is the collective deviation from the theoretical simplicity of the DNA content analysis that gives importance to the computer modeling of DNA content histograms.

III. Cell–Cycle Analysis of DNA Content Histograms

DNA content histograms require mathematical analysis in order to extract the underlying G_1, S, and G_2 phase distributions; methods for this analysis have been developed and refined over the past 2 decades. Methods to derive cell-cycle parameters from DNA content histograms range from simple graphical approaches to more complex deconvolution methods using curve fitting. For detailed descriptions of the basic methods, the reader is referred to one of several more extensive reviews (Dean, 1985, 1990; Bagwell, 1993).

All of the simpler methods are based upon the assumption that the G_1 and G_2 phase fractions may be approximated by examining the portions of the histogram where the G_1 or G_2 phases have less overlap with S phase. There are two such approaches. The first is to calculate the area under the left half of the G_1 curve and the right half of the G_2 curve, and multiply each by two (i.e., reflecting these about the peak mean); what remains is S phase. The second approach is to use only the center-most portion of the S phase distribution and extrapolate this leftward to the G_1 mean and rightward to the G_2 mean. What remains on the left is G_1 and on the right is G_2. These methods can be reasonably accurate when one cell cycle is present and the histogram is optimal in shape. Both methods assume that the G_1 and G_2 peaks are symmetrical (DNA staining variability in tissues does not always provide this) and that the midpoint (mean) of each peak can be precisely identified. Because of the overlap of G_1 and G_2 peaks with the S phase, the mean of these peaks is not always at their maximal height (mode), especially for the G_2. If a second overlapping cell cycle is also present, then the overlap of the two cell cycles usually precludes safe use of these methods. In addition, modeling of debris and aggregates is usually not a part of these simpler graphical approaches.

The most flexible and accurate methods of cell-cycle analysis are based upon building a mathematical model of the DNA content distribution and then fitting this model to the data using curve fitting methods. The most well-established model, proposed by Dean and Jett (1974), is based upon the prediction that the cell-cycle histogram is a result of the Gaussian broadening of the theoretically perfect distribution (Fig. 1A). The underlying distribution can be recovered or "deconvoluted" by fitting the G_1 and G_2 peaks as Gaussian curves and the S phase distribution as a Gaussian-broadened distribution. As originally proposed, the shape of this broadened S phase distribution is modeled as a smooth second-order polynomial curve (a portion of a parabola). The model can be simplified by using a first-order polynomial curve (a broadened trapezoid) or a zero-order curve (a broadened rectangle). When the quality of the histogram is less than ideal, especially if G_1 or G_2 peaks are non-Gaussian (broadened bases, skewed, or having shoulders), then the simplified models may give results that are less affected by artifacts that increase the overlap of G_1, S, and G_2 peaks. This often is the case in analysis of clinical samples, as described in a subsequent section.

Some experimentally derived S phase distributions (usually from cultured cells) are more complex, and several alternative schemes have been proposed to model such distributions. The most flexible models are those of fitting S phase by the sum of Gaussians (Fried, 1976), in which the S phase is fit by a series of overlapping Gaussian curves, and the sum of broadened rectangles (Bagwell, 1979), in which the S phase is fit with a series of 5–10 broadened rectangles. In these models each of the Gaussian or broadened rectangle curves can be of any height. Therefore, the shape of the S phase is extremely flexible, and these models can fit S phase distributions that have complex shapes. This is also a primary drawback in practical use of these models, however. The very flexible S phase shape allows accurate fitting of any artifacts in the data and allows increased ambiguity in fitting the region of S near G_1 and G_2 (i.e., the areas of greatest overlap of G_1 and S and S and G_2). A generally successful compromise was suggested by Fox (1980), who added one additional Gaussian curve to Dean and Jett's polynomial S phase model. Fox's model provides a more flexible S phase shape, but still retains the smoothness of the S phase that is characteristic of the Dean and Jett model. It is especially suited to cell-cycle analysis of populations highly perturbed or synchronized by drug treatments.

Curve fitting models are almost universally fit to the histogram data by use of least square fitting. The fitting model is used to generate a mathematical expression, or function, for the predicted histogram distribution. The function has a number of parameters (usually between 7 and 22) that must be adjusted to give the optimum concordance between the fitting model and the observed data. Since the fitting function used by the model is not a simple linear equation, nonlinear least squares analysis is utilized. An excellent description of methods of nonlinear least squares analysis, and sample computer subroutines, is contained in Bevington (1969). The most commonly used technique of nonlinear least squares analysis in these applications is that described by Marquardt (1963). All of the nonlinear least square fitting techniques are *iterative:* successive approximations are made, in which the parameters in the fitting model equations are revised and the fit to the data is successively improved. When no further improvement is obtained, the fit has *converged* and is theoretically optimal. Goodness of fit is usually quantified by the χ^2 statistic

$$\sum \frac{(y_{\mathrm{fit}_i} - y_{\mathrm{data}_i})^2}{\sigma_i^2}$$

or the reduced χ^2 statistic,

$$\chi_v^2 = \frac{\chi^2}{\text{degrees of freedom}},$$

which measure the deviation of the fitting function from the data. The speed of the least square fitting is determined by the efficiency in searching for and finding the optimum combination of fitting parameter values. The Marquardt algorithm uses an optimized strategy for searching for the lowest χ^2 value along the n-dimensional "surface" defined in the space of the χ^2 vs n fitting variables.

An advantage of the least square fitting methods is that the models can be directly extended to analysis of two or even three overlapping cell cycles. The overlapping model components are mathematically deconvoluted to yield individual cell-cycle estimates. An additional advantage of curve fitting methods is that they tend to be less dependent upon the initial or "starting parameters" used to begin the fitting process. Such parameters include initial estimates of peak means and CVs, as well as the limits of the region of the histogram included in the fit. When the cell cycle and debris model is most accurate in fitting the data, the result is least dependent on starting values, and interoperator variation in results is reduced (Kallioniemi *et al.*, 1991a).

It has been important to recognize that DNA content histograms from tumor tissue are often far from optimal (broad CVs, high debris, and aggregation) or complex (multiple overlapping peaks and cell cycles) and frequently contain artifactual departures from expected shapes (e.g., skewed and non-Gaussian peak shapes). This is even more true when analyses are derived from formalin-fixed specimens. *When a skewed G_1 peak or a peak with a "tail" on the right side extends visibly into the S phase, S phase estimates should be used with extreme caution* (Shankey *et al.*, 1993a).

An important aspect of the analysis of imperfect histograms is the ability to reduce the model's complexity by using simplifying assumptions to reduce the number of model parameters being fit. This may reduce the ability of the model to fit the finer details of a histogram, but it also reduces the possibility of incorrect fitting of the data. As described above, some models may assume that a skew or broad base in G_0 or G_1 peaks is part of the S phase, which can lead to an overestimation of the true S phase. More conservative models may be more accurate in situations where CVs are wide or peaks are not well resolved, when multiple peaks are extensively overlapping, or when background debris and aggregation are high. These situations are more fully described in later sections. The Dean and Jett algorithm may be used with a zero-order (broadened rectangle) or first-order S phase polynomial (broadened trapezoid), instead of the more flexible, but error-prone second-order polynomial. Additional constraints can be imposed to require that the CVs of the G_2 and G_1 peaks be equal (they are usually very similar), or the CVs of DNA diploid and aneuploid peaks can be made equivalent, or the G_2/G_1 ratios can be constrained to have a user-supplied value, based upon past laboratory experience.

IV. Critical Aspects of DNA Content and Cell–Cycle Analysis

This section details seven critical aspects of DNA histogram analysis:

A. Cell number
B. CV and detection of near-diploid aneuploidy
C. Number of histogram channels, histogram range, and histogram linearity
D. DNA content standards
E. Debris modeling
F. Modeling of cell or nuclear aggregation
G. Quantitation of aggregates and debris.

Guidelines for each of these aspects have been described by the DNA Cytometry Consensus Conference (Shankey *et al.*, 1993a). The following descriptions will expand upon the rationale for each of these guidelines. In this and subsequent sections, recommendations of the Consensus Conference are noted in italics.

A. Cell Number

One of the principle advantages of flow cytometry is that large numbers of cells may be analyzed in a short time. The object of acquiring larger numbers of cells is to reduce statistical fluctuations in the histogram, most apparent in areas of fewer cell numbers, and particularly in the region of S phase. Nevertheless, in order to speed the analysis, or if the tissue sample is small, there is a tendency to acquire the minimum number of events necessary. Figure 2 illustrates the effect of different cell numbers in two typical histogram types. Multiple histograms were acquired at each cell number, and the variation in the S phase measurement for each is shown. Above 10,000 events per histogram, S phase measurements are highly reproducible. Below this number, especially with fewer than 1000 events, accuracy in the S phase measurement in an individual histogram deteriorates substantially. A *minimum of 10,000 events in DNA content cell-cycle analysis has been recommended by the DNA Cytometry Consensus Conference Guidelines, although detection of DNA aneuploid populations may be possible from histograms with fewer cells or nuclei.* Note that this refers to the number of cells in the cell cycle; if a substantial proportion of events is from debris or aggregates, the total number of events acquired must be correspondingly higher in order to assure the required minimum number of

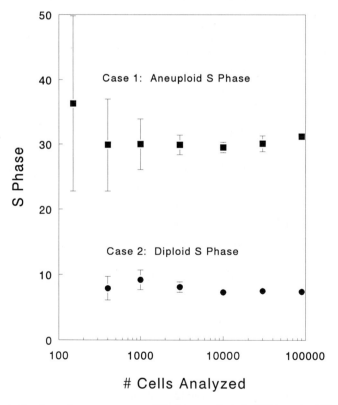

Fig. 2 Effect of cell number on accuracy of S phase analysis in DNA content histograms. Two human tumor specimens (case 1, squares, DNA aneuploid, aneuploid S phase analyzed, and case 2, circles, DNA diploid) were analyzed by acquisition of histograms with from 150 to 100,000 cells each, in triplicate. For each cell number, the mean and standard deviation of the S phase measurement was plotted.

intact single cells or nuclei. While ploidy information may be sometimes accurately derived from fewer cells, accurate S phase analysis may sometimes require more than this number. This is especially true if the proportion of aneuploid cells is low; again, in order to collect a minimum acceptable number of cells for the aneuploid cell-cycle analysis, the total cell number must be increased.

B. CV and Detection of Near-Diploid Aneuploidy

Distinguishing two populations with different ploidy becomes increasingly difficult as they become closer in DNA content. There is a consensus that near-diploid DNA contents cannot be reliably diagnosed in the clinical laboratory unless the histogram is bimodal, i.e., there is a depression or trough between two separate peaks (Hiddeman et al., 1984; Shankey et al., 1993a). How close to diploid a population may be, and still be detected, depends upon both the CV of the analysis and the relative proportions of diploid and near-diploid cells. Both sample preparation and instrument performance can have a significant impact on the CV. Using the presence of a 10% dip between DNA diploid and aneuploid curves as a criterion for the detection of bimodality, the relationship between the minimum detectable DNA index, CV, and the proportion of aneuploid cells is shown in Fig. 3. Detection is optimal when the diploid and near-diploid cells are in equal proportions (Fig. 3, solid line). When the proportions of DNA diploid and aneuploid cells are unequal, then aneuploidy detection requires a lower CV or a larger ploidy difference. Note, for example, that with a CV of 6, a DI of 1.13 can be detected if the aneuploid cells are 50% of the total. If the aneuploid cells are only 5% of the total, only a DI of 1.23 can be detected with the same CV. A CV of 3.5 would be required to detect a 5% subpopulation of DI 1.13 cells. The values shown in Fig. 3 are minimum estimates, since problems with real data, such as non-Gaussian-shaped peaks or low numbers of cells in the histogram, will adversely affect peak discrimination.

Note that when fitting near-diploid DNA aneuploid peaks, best results are often obtained using a software option to constrain the CVs of the diploid and aneuploid peaks to be equal. Examine this option if the fit without such constraint produces CVs for DNA diploid and aneuploid peaks that are very dissimilar.

C. Number of Histogram Channels, Histogram Range, and Histogram Linearity

The number of channels into which the DNA content histogram is digitized can have an important effect upon the data. A wide spectrum of histogram "resolutions" is in common use, ranging from 64 to 1024 channels or "bins." Too few channels may not provide the resolution needed to preserve the accuracy of the original analog signal. As an extreme example, note that if a peak were placed in channel 10, then a DI 1.05 ploidy would never be detected

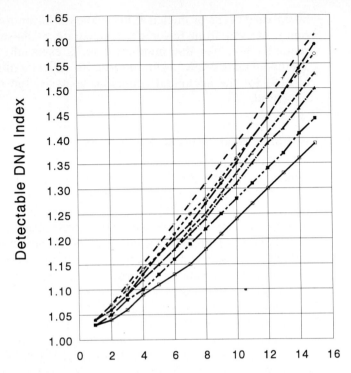

Fig. 3 Minimum detectable DNA index. The effect of CV and proportion of aneuploid cells on the ability to detect a near-diploid aneuploid peak is shown. The criterion for detectability is the presence of bimodality, with a 10% dip between diploid and aneuploid peaks. Gaussian peaks with different CV and separation were created by mathematical modeling and examined at different percentages of aneuploid cells: 5% (+), 10% (O), 25% (▲), 50% (□), 75% (■), 90% (△), 95% (●). For any CV, as the proportion of aneuploid cells increases or decreases away from 50%, the capacity to detect near-diploid aneuploidy is diminished, and only a higher DNA index is detectable.

because the difference between channels 10 and 11 is 10%. This is more than a theoretical concern since channel numbers as low as 64 may be used in some bivariate cytograms (for example, when one axis is DNA content and the other immunofluorescence). Conversely, if a very large number of channels is used, then unless very large numbers of cells are analyzed, statistical fluctuations in the number of events per channel will be greater. This results in a less satisfactory appearance and can produce greater uncertainty in data analysis.

Figure 4 shows practical examples of the effect of channel number on the result of histogram analysis. Variability and errors in CV, S phase fraction (SPF), and G_2/G_1 ratios (as well as calculations of DI, not shown) all increase when the mean of the G_1 peak is placed below channel 50. CVs always rise if the G_1 peak is place in channels below 50, whereas S phase and G_2/G_1 ratio discrepancies are erratic. The magnitude of the S phase errors can be substantial, whereas the G_2/G_1 ratio variation is small (below 1%). The current recommenda-

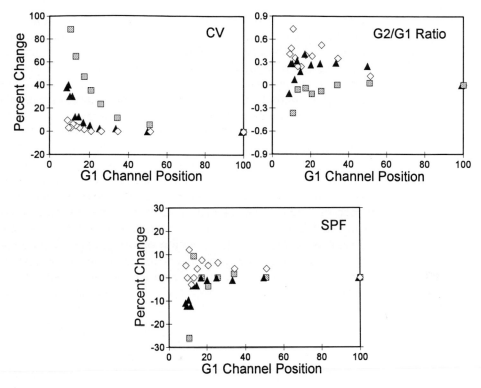

Fig. 4 The effect of reducing the number of channels used in the cell-cycle histogram. The histogram was initially analyzed with the G_1 peak in channel 100, and the G_1 channel position was subsequently reduced. The mean channel position of the G_1 peak in the histogram is shown on the abscissa vs the change in CV of the G_1 peak relative to the initial analysis (top left), the G_2/G_1 ratio (top right), and the percent S phase estimate (bottom center). The three different initial histograms had a G_1 peak CV of approximately 2 (squares, fresh tissue analysis), 4 (triangles, paraffin-embedded tissue), and 6 (diamonds, paraffin-embedded tissue). Cell-cycle analysis was performed by the method of Dean and Jett using a zero-order (rectangular) S phase polynomial. Debris was analyzed with the sliced nucleus model (see text).

tion is that *the lowest G_1 population should be accumulated in channels greater than 30, and probably above 50* (Shankey *et al.*, 1993a).

The range of DNA content values that is collected is another very important consideration. Valuable information can be lost if an adequate range of data to the left and right of the cell cycle(s) is not contained in the histogram. The left portion of the histogram contains most of the information describing the shape of the debris distribution, and these data are essential to proper application of debris models (see below). *Sufficient debris should be collected in the histogram to enable sound judgment of specimen quality and to allow software programs sufficient data to construct a model for debris compensation* (Shankey *et al.*, 1993a). *Setting a lower limit of data acquisition at a channel that corresponds to DI 0.1 is recommended* (Shankey *et al.*, 1993a).

Similarly, the hypertetraploid region must be examined. Hypertetraploid peaks may not be detected if these "high" data channels are discarded or are accumulated in the last "overflow" channel. In addition, data above the G_2 of the population with highest ploidy contain much of the information relating to the degree of aggregation present in the sample, and, if software aggregation modeling is applied, these data are essential to proper use of such modeling (see below). As a minimum, DNA contents up to at least 50% above the highest G_2 peak should be sampled. As an even more stringent rule, the *DNA Cytometry Consensus Guidelines recommend collecting data up to DI 6.0, or even to DI 10 if aneuploid with DI > 2.0 is present* (Shankey *et al.*, 1993a). The requirements that the diploid G_1 peak be positioned above channel 50 and that channels up to DI 6–10 be observed require that 300–500 channels are present in the histogram. Thus, while sufficient resolution is almost always achieved with 512 channels, 128-channel histograms often have insufficient resolution.

Visual observation of the position of triplets (DNA index = 3 for diploid triplets) using an expanded vertical histogram scale allows approximation of the extent of aggregation at the time of data acquisition. Dissociation or trituration of the sample to minimize aggregates is very important. The use of software aggregation modeling subsequent to data acquisition, as discussed below, can further compensate for the effects of aggregation if adequate hypertetraploid data are contained within the histogram.

Lack of histogram linearity is a common problem in many instruments. Departures from linearity can produce nonstandard G_2/G_1 ratios, altered DNA indices, and potential difficulty in computer modeling of aggregation. One common source of nonlinearity is an incorrect "zero" setting in the analog-to-digital converter—the channel in which a signal of zero intensity is placed. Problems with nonlinearity from this and other sources are most commonly manifest in the lowest and highest ends of the histogram; this can have the greatest effect on the use of a DNA content standard in the lower channels or on the evaluation of G_2 peaks and aggregates in the higher channels. *Instrument linearity should be determined on a regular basis, using standard particles and/or suitable methods* (Bagwell *et al.*, 1989), *and appropriate corrections made* (Shankey *et al.*, 1993a).

D. DNA Content Standards

Some authors have proposed that aneuploid peaks overlapping with diploid peaks might be detected by the use of external DNA content standards such as lymphocytes or nucleated red blood cells. Relative to the fluorescent standard, a shift in the position of the diploid/near-diploid composite peak away from the expected diploid position might be taken as evidence of DNA aneuploidy. However, differences in DNA staining between different cell types can result from variations in DNA dye binding (Darzynkiewicz *et al.*, 1984; Bertuzzi *et al.*, 1990; Kubbies, 1992; Wolley *et al.*, 1982; Iverson and Laerum, 1987; Evenson *et al.*, 1986; Klein and White, 1988; Wersto *et al.*,1991; Heiden *et al.*, 1990) and this *can produce ambiguity in the correct diploid DNA content* (Shankey *et al.*, 1993a). Nonstoichiometric dye binding is also observed in dead or dying cells, apoptotic cells, or cells with DNA damage (Kubbies, 1990; Stokke *et al.*, 1991; Telford *et al.*, 1991; Roti Roto *et al.*, 1985; Nicoletti *et al.*, 1991; Alanen *et al.*, 1989; Joensuu *et al.*, 1990) and cells in different cell-cycle state and cell differentiation (Bruno *et al.*, 1991; Darzynkiewicz *et al.*, 1977). Thus, small differences in staining intensity relative to "standards" cannot be interpreted as evidence of DNA aneuploidy. Using an external reference standard to establish a range of positions or CVs to define as "diploid" results in overdiagnosis of DNA aneuploidy (Wersto *et al.*, 1991; Heiden *et al.*, 1990). *The best DNA content standard is the normal tissue component that represents the normal counterpart of the neoplastic cells* (Shankey *et al.*, 1993a). Malignant tissue almost always has at least a small component of normal diploid elements. If two peaks are present (from non-formalin-fixed tissue), but it is unclear which is aneuploid, a DNA content standard may be very useful. If human lymphocytes are added as a standard, the diploid peak position should be elevated in magnitude. Alternatively, if the standard has a DNA content that is much less than that of diploid human cells (e.g., chicken or trout red blood cells), then the standard will appear as a distinct peak at the left of the histogram that does not overlap with the diploid human cells. Software analysis of this peak will provide a ratio to diploid for evaluation relative to the expected range. Analysis of two standards simultaneously has been suggested (Vindelov *et al.*, 1983); in our experience this rarely adds significant advantage in the clinical laboratory if histogram linearity is satisfactory (Koch *et al.*, 1984).

In the case of formalin-fixed tissue, variability in fixation and DNA dye accessibility (Larsen *et al.*, 1986; Becker and Mikel, 1990) prohibits any reproducibility in the position of the standard peak, and use of a standard is unfeasible. Even diploid cells from a paraffin block may not be a reliable standard, as there is substantial variability in DNA stainability from block to block and even within different portions of the same block (Price and Herman, 1990). The criteria that distinctly bimodal peaks be present to diagnose DNA aneuploidy must be used, and *it is recommended that the left-most peak from paraffin-embedded material be assumed to represent the DNA diploid population* (Shankey *et al.*, 1993a).

E. Debris Modeling

Damaged or fragmented cells or nuclei are almost always present in samples prepared for DNA flow cytometry; only the relative amount and origin of the debris vary. The debris produces events that are most visible on the left side of the histogram. More significantly, but often less obviously, the debris also extends into the cell-cycle region of the histogram. Since the S phase is the lowest and broadest cell-cycle compartment in the histogram, S phase calculations are most affected by the presence of debris. Thus, it is important to include modeling of the debris distribution in the histogram analysis in order to subtract the effects of the underlying debris from the cell-cycle fitting.

The earliest concept in debris fitting was that a steadily declining background debris curve could be fit by an exponential function (e^{-kx}). There are two primary reasons why a simple exponential curve does not usually provide an accurate fit. First, it is common to observe a debris component that rapidly declines with increasing DNA content combined with a portion that declines more slowly or plateaus. This latter portion has a much greater effect upon the cell-cycle fitting than would be predicted by an exponential curve. Second, debris is produced by degradation, fragmentation, or actual cutting of nuclei. Created in this manner, the fragments are always smaller than the DNA content of the nucleus from which they are derived. Thus, they are present only leftward of each DNA content position from which they are derived. In modeling this debris, each DNA content in the histogram must be considered as a separate source of debris, and, thus, the shape of the debris curve is dependent upon where the peaks in the DNA histogram are. Debris models of this kind are termed ''histogram dependent.''

Figure 5 illustrates the difference between the classical and histogram-dependent exponential debris. Figure 5A shows fitting of a simple exponential curve to the debris region left of the G_1 peak. This model does not take into consideration that most of the debris is created by fragmentation of G_1 nuclei; thus, the curve predicts too much background over the S and G_2 phases. Cell-cycle analysis with this model yields a 0% SPF. Application of the histogram-dependent exponential model is shown in Fig. 5B. The background debris curve drops rapidly from the left side of the G_1 peak to the right side of the G_1 peak. Fitting with this model yields an 11.2% SPF.

Figures 5C and 5D illustrate the shape of debris that is often observed in histograms derived from paraffin-embedded tissue. This debris is obviously not exponential in shape, as it has a much flatter distribution left of the G_1. Since the analysis of DNA histograms from cells preserved in paraffin blocks has become an increasingly important part of DNA flow cytometry, using a model which is consistent with the shape of this debris is of considerable practical importance for accurate cell-cycle analysis.

The examples in Figs. 5C and 5D have been fit with a model which recognizes the effect of cutting nuclei with a knife. As part of process of extraction of nuclei from paraffin blocks, sections are usually cut with a microtome at a thickness near 50 μm, and sectioning or slicing or nuclei is an unavoidable consequence. Nuclei in the path of the knife are cut randomly into two portions. If the nuclei were considered in a simple model to be identical cubes randomly cut perpendicularly to one face, then the volume of each randomly cut portion would be from near-zero to nearly full volume. In such a model, the debris distribution would be a flat plateau to the left of the whole-cell G_1 peak. This simplified model (Rabinovitch, 1988) was subsequently revised to be consistent

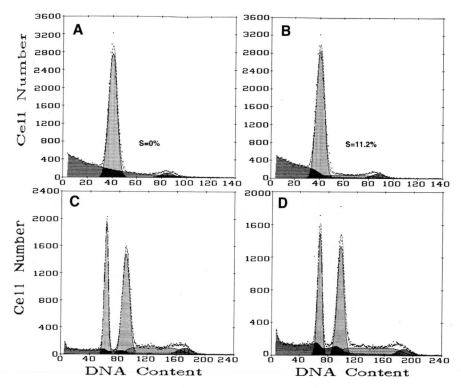

Fig. 5 Simple exponential debris model (A) and histogram-dependent exponential debris model (B) applied to a DNA diploid histogram from a fresh tissue sample containing degenerating cells. The S phase fractions resulting from the cell-cycle analysis are 0 and 11.2%, respectively. (C and D) Sliced nucleus debris modeling is applied to a DNA aneuploid tumor extracted from paraffin with microtome sectioning at 50 μm (C) or 20 μm (D). In all panels, the debris component of the fitting model is shown with horizontal hatching. Cell-cycle analyses in this and other figures were performed using the MultiCycle program written by the author (Phoenix Flow Systems, San Diego, CA).

with spherical and ellipsoidal nuclei (Bagwell *et al.*, 1991). As illustrated in Fig. 6, there tends to be a slightly greater fraction of small and large fragments produced in this model, which results from the small volume of the "crescents" produced by cutting the rounded ends of the nuclei. The distribution of sliced fragments thus exhibits a concave rather than a flat distribution. The histogram-dependent implementation of this model considers each channel of the distribution to be a discrete population of DNA contents, of which a certain proportion are cut and therefore form a flat-concave curve to the left of that channel. The probability of a nucleus being cut is proportional to its radius. The radius is proportional to the cubed-root of the volume, which is presumed to be proportional to DNA content (for example, S and G_2 phase nuclei are larger than G_1 nuclei). The process of least squares fitting is used to determine the probability

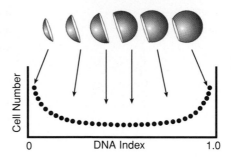

Fig. 6 The DNA content distribution of fragments of nuclei created by random sectioning. Sectioning at the "ends" of the nucleus results in larger numbers of very small and large fragments, resulting in a "smile" shaped distribution. DNA index = 1.0 corresponds to the DNA content of an intact nucleus.

of nuclear cutting that yields the best fit to the data. Figures 5C and 5D illustrate the close fit of this model to histograms from paraffin-embedded cells.

As illustrated in Fig. 7, in addition to the flat-concave sliced nucleus debris, many clinical specimens from paraffin also have an additional component of more rapidly declining debris originating from degenerating cells and fragments other than those caused by cutting with the microtome. This same shape may be seen in fresh specimens that are minced with a scalpel, forced through mesh, or otherwise cut. To fit this combined effect of degradation and cutting, histogram-dependent exponential debris can be combined with sliced nucleus debris modeling. The relative contribution of each is determined by the least squares fitting. This combined model shows the greatest flexibility in modeling diverse types of histograms (Figs. 7C and F). The ability to fit the entire spectrum of the noise distribution has the additional benefit that this combined model is relatively insensitive to the end points chosen for the fitting region. This, in turn reduces interoperator variation in results (Kallioniemei *et al.*, 1991). Figure 7 illustrates this behavior: the simple exponential debris shows the greatest dependence on the end points chosen for fitting (Figs. 7A and 7D), histogram-dependent exponential debris shows less variation (Figs. 7B and 7E), and the sliced nucleus model combined with histogram-dependent exponential debris (hereafter referred to simply as the sliced nucleus model) shows the least

Fig. 7 Fitting of a histogram derived from paraffin-embedded diploid cells using a simple exponential background debris curve (A), histogram-dependent exponential debris (B), and the sliced nucleus debris model (C). The S phase fraction of the cell-cycle analysis is different in each case, as indicated. Simple exponential background debris applied with a left end point of the region of fitting that is closer to the G_1 peak is shown in (D), resulting in a very different S phase measurement than that in (A). The histogram-dependent exponential debris applied with the narrower fitting region is illustrated in (E), showing a 22% reduction in S phase compared to B. In contrast, the sliced nucleus model (F) is very insensitive to the change in fitting region.

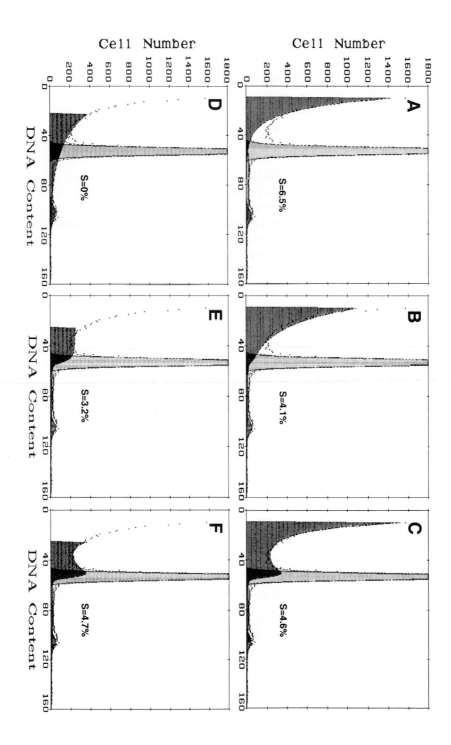

variability because it accurately fits the entire debris distribution (Figs. 7C and 7F).

The accuracy of the sliced nucleus modeling was evaluated by applying it to a test system comprised of PHA-stimulated diploid human lymphocytes and aneuploid Hela cells (derived from an adenocarcinoma) and mixtures of these cell types. The cells were analyzed both fresh and after formalin fixation, paraffin embedding, and extraction from paraffin. Figures 5C and 5D show examples of the mixture of lymphocytes and Hela cells in paraffin sectioned at 50 and 20 μm, respectively. The results of cell-cycle analysis (Fig. 8) show that, for the fresh cells, the effects of the sliced nucleus debris modeling are small, except for the estimate of the lymphocyte S phase in the mixture with Hela cells; in this case, Hela nuclear fragments overlap the lymphocyte S phase, giving rise to a 3% overestimation of S phase without sliced nuclei debris modeling, but a satisfactory correction when using the sliced nucleus model.

For the paraffin-derived lymphocytes and Hela cells examined independently (unmixed) there is an overestimation of S phase, which increases progressively as the section thickness decreases; this is almost completely corrected by debris modeling. When the diploid and aneuploid cells are mixed, many of the sliced Hela nuclei overlap the lymphocyte cell-cycle distribution, resulting in an artifi-

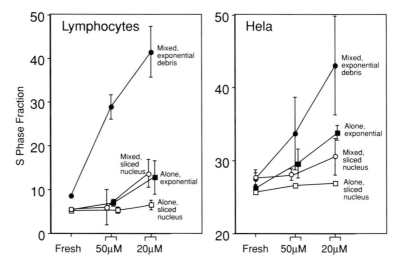

Fig. 8 S phase fraction measurements obtained with sliced nucleus debris modeling (open symbols) vs simple exponential debris (closed symbols, fit as in Fig. 7A) in cell-cycle analysis of lymphocytes alone, Hela cells alone, and mixtures of these two cell types. The abscissa indicates results performed on fresh cells and paraffin-embedded cells sectioned at 50 or 20 μm. While each of the models yields similar results when applied to fresh cells that have little debris, when cells are mixed to simulate samples with DNA diploid and aneuploid peaks, and especially when cells are derived from paraffin, the sliced nucleus model gives S phase fractions much closer to those obtained from fresh unmixed cells.

cial elevation of the lymphocyte S phase (note in Figs. 5C and 5D that the majority of events between diploid and aneuploid G_1 peaks are modeled as debris). Cell-cycle fitting using the sliced nucleus model produces S phase estimates that are very close to those of the fresh cells in 50-μm sections and nearly complete correction for 20-μm sections (the latter contain over twice the proportion of sliced nucleus debris as the former, see Fig. 5). Figure 8 also shows that the reproducibility of S phase estimates is better when sliced nucleus debris modeling is applied than when it is not. The comparison of 50 vs 20-μm sections indicates, however, that, even with an improved model, greater amounts of debris reduce the accuracy of S phase estimates. Because of this, quantitation of the proportion of debris in histograms is an important part of the assessment of cell-cycle analysis reliability.

Finally, it must be stressed that when debris modeling is used to improve the accuracy of cell-cycle measurements (as is currently recommended; see below), it is necessary that the histogram contain all of the necessary data pertaining to the debris. The requirement for an adequate range of data to the left and right of the cell cycle(s) has been described above. Similarly, it is imperative to avoid the use of light scatter gates. Although the histogram "appearance" may be improved by gating out the smaller-sized fragments and debris, this gating destroys the mathematical relationships that correlate the extent of low DNA content debris with the amount of larger debris that overlaps the cell-cycle distribution.

F. Modeling of Cell or Nuclear Aggregation

Although analysis of the cell cycle(s) together with debris is sometimes sufficient to model the DNA content histogram, careful inspection of many histograms will reveal, in addition, evidence of cell aggregation. An aggregate of two G_1 cells (a doublet) will have the same DNA content as a G_2 cell and may be overlooked; however, diploid triplets will be seen at DI 3.0, quadruplets at DI 4.0, etc. In addition, S and G_2 cells and nuclear fragments (debris) also can aggregate with G_1 cells and with each other. If aneuploid cells are present, they may aggregate with diploid cells and debris, as well as with each other. The net effect upon the histogram can be complex. *Aggregates can affect detection of DNA aneuploid peaks and may cause major errors in S phase, especially for DI 2.0 tumors* (Shankey et al., 1993a).

In the past, the primary approach to detection of aggregates has been to distinguish the altered pulse shape that they may produce when analyzed by a flow cytometer using a focused laser beam. Although this method may be successful, especially when examining uniformly spherical whole cells, with some cell preparations the method has less success (Rabinovitch, 1993a). If doublets of two G_1 cells pass through the laser beam aligned parallel to the laser beam, rather than perpendicular to it, then the fluorescence profile cannot be distinguished from that of a G_2 cell. In addition, an oblong G_2 cell, for example, cannot be easily distinguished from a G_1 doublet on the basis of pulse shape. Aggregates of three or more cells or nuclei may form a "spheroid" without a longer axis, and these also may not be distinguishable from a single large cell. Aggregated nuclei have no intervening cytoplasm, and their altered pulse shape may be less discernable than those of aggregated whole cells.

The pulse shape analysis is most commonly performed by plotting fluorescence peak vs area (or less often time-of-flight or peak width vs area) signals from each cell. A diagonal line is then drawn through the origin, making the assumption that aggregates will fall below the line (i.e., their pulse peak value will be lower than those of nonaggregates for a given pulse area). Placement of the diagonal line is subject to user interpretation. For oblong nuclei, such as from many epithelial cells, the G_1 and G_2 peak fluorescence is variable, and a diagonal that appears adequate to exclude aggregates may also exclude G_1 or G_2 cells or nuclei (Rabinovitch, 1993a).

A DNA content histogram with aggregates may show additional peaks corresponding to doublets or triplets. In the past, software modeling of these peaks has been attempted by adding an extra peak to the cell-cycle model to fit the triplet peak position or by predicting doublets on the basis of the frequency of triplets (Beck, 1980). This approach will not, however, fit the more complicated patterns of aggregation discussed above, and aggregates may not, in fact, produce distinctly visible peaks, especially in more complex histograms with both diploid and aneuploid DNA contents. As an extension of the "histogram-dependent" modeling approach, a computer model can be applied that allows a generalized approach to the fitting of aggregation in DNA histograms (Rabinovitch, 1990; 1993a). The basis of this model is the simple assumption that any two particles, i.e., elements of the histogram, have a certain probability of aggregating with each other. Thus, doublets form with a probability, p. Triplets are assumed to form by association of a doublet with a singlet; the singlet can "attach to" either of the two cells in the doublet, with a net probability of $2p^2$. Quadruplets can form in two ways: two doublets can aggregate with each other with a probability of $4p^3$ (there are four ways the two doublets can attach to each other, or $4p$ times p^2), or a triplet can combine with a singlet with a probability of $6p^3$ (there are three ways to combine the triplet with the singlet, or $3p$ times $2p^2$). The constants 2, 4, and 6 can be derived in this simplest fashion, or it is possible to modify these based on alternative models, which changes the final aggregate distribution slightly (Rabinovitch, P. S., unpublished data; Bagwell, 1993). The net aggregate distribution has a shape that is formed from the composite of all possible aggregate combinations. Expressed mathematically:

The doublet distribution: $$D(i) = p \cdot \sum_{j=1}^{i} \sum_{k=1}^{i} Y(j) \cdot Y(k)$$
$$\text{(for all } j + k = i),$$

where $Y(i)$ is the cell distribution without aggregation.

The triplet distribution: $T(i) = 2p^2 \cdot \sum_{j=1}^{i} \sum_{k=1}^{i} D(j) \cdot Y(k)$

(for all $j + k = i$).

The quadruplet distribution: $Q(i) = 4p^3 \cdot \sum_{j=1}^{i} \sum_{k=1}^{i} D(j) \cdot D(k)$

$+ 6p^3 \cdot \sum_{j=1}^{i} \sum_{k=1}^{i} T(j) \cdot Y(k)$ (for all $j + k = i$).

The net aggregate distribution: Aggregates$(i) = D(i) + T(i) + Q(i)$.

Higher order aggregates can be added to the aggregate equation; in practice, however, these have little observable effect in most histograms and are usually not calculated. The aggregate distribution can be added to the cell-cycle and debris models and the combined model is fit to the observed data by using least squares fitting. This will determine the value of the variable, p; this single variable determines the amount of aggregation present in the DNA histogram, but does not change the shape of the aggregate distribution, which is determined only by $Y(i)$, the number of cells in each channel of the histogram.

An example of this fitting is shown in Figure 9. There are several aggregate peaks in the region of the DNA diploid and aneuploid cell cycles, and there are additional aggregates that are not obvious as peaks. Both the DNA diploid and DNA aneuploid S and G$_2$ phase measurements obtained with aggregation modeling are lower than those obtained without aggregation modeling. Comparison of S and G$_2$ phase measurements obtained in samples with various degrees of aggregation demonstrate that in analysis of epithelial tumors, software aggregation compensation may give more accurate S and G$_2$ phase measurements than hardware doublet discrimination (Rabinovitch, 1993a). Since gating from peak/area analysis affects the aggregate distribution in the histogram, it may alter the "expected" aggregate relationships. Therefore, pending future detailed study of these interactions, *use of software aggregation modeling should be restricted to data collected without hardware gating* (Shankey et al., 1993a). As for debris modeling, this also means avoiding the use of light scatter gates. Although there is not yet a consensus on appropriate use of hardware gating vs software compensation for aggregate correction, it is hoped that further study will address this issue. At present, the most practical demonstration of the effectiveness of software aggregation modeling is obtained by the correlation of proliferative measurements with clinical outcome, as described subsequently.

G. Quantitation of Aggregates and Debris

The relative proportion of events analyzed by the flow cytometer that consist of cell or nuclear debris or aggregates is highly variable. The debris is generally

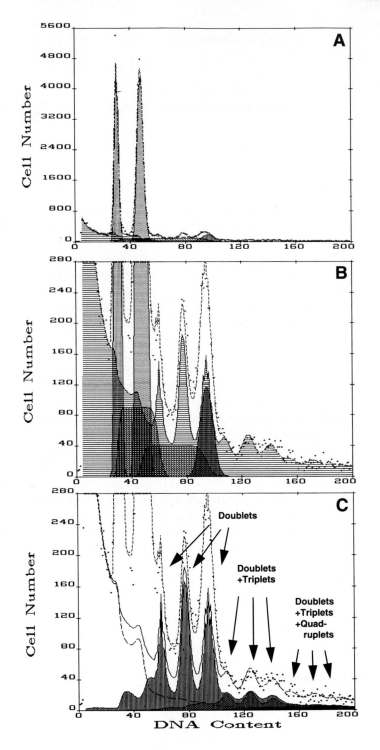

higher in paraffin-processed tissue, due to nuclear slicing, and in degenerating or necrotic tissue, but these magnitudes are difficult to predict. To address the need for a quantitative measure of aggregates and debris, the DNA Cytometry Consensus Conference defined a parameter termed background aggregates and debris (BAD), defined as the proportion of the histogram events between the leftmost G_1 and the rightmost G_2 that is modeled as debris or aggregates. The reason that this parameter is defined in this manner, rather than as the total percentage debris and aggregates in the entire histogram, is that left and right end points of a histogram are variable and arbitrary, depending on instrument settings. The proportion of debris in the histogram is especially sensitive to variation in the left limit of data acquistion. The BAD is unaffected by histogram end points. It is, however, very much dependent on the choice of histogram modeling. For greatest accuracy and interlaboratory comparison, it is suggested that histogram-dependent models of debris and aggregation be utilized.

V. Interpretation of DNA Content and Cell–Cycle Histograms

Choices that the operator must make, both in the cell-cycle modeling and in the interpretation of the fitted histogram, can lead to substantial variability in results. This variability has been documented in interlaboratory studies that have compared analysis and interpretation of histograms derived from replicate cells or tissues (Wheeless *et al.*, 1989, 1991; Hitchcock, 1991; Kallioniemi *et al.*, 1990; Coon *et al.*, 1989). Variation in histogram analysis may be seen both between different software programs analyzing the same data and between laboratories using the same software (Hitchcock, 1991). Some of this variability may potentially be reduced by use of fully automated "semi-intelligent" peak detection and cell-cycle analysis software (Rabinovitch and Kallioniemi, 1991; Kallioniemi *et al.*, 1994). Reproducibility in analysis within the laboratory and

Fig. 9 Application of aggregation modeling to a DNA aneuploid histogram from a carcinoma of the breast. (A) Cell-cycle analysis without aggregation modeling. A peak is present near channel 80 that might be mistaken for a second DNA aneuploid population. (B) $(20 \times$ *y*-axis scale) shows the same histogram analyzed with aggregation modeling added to the sliced nucleus debris (horizontal hatching). Diploid and aneuploid S phases are shown by diagonal hatching, and Gaussian G_1 and G_2 peaks are shown by stippling. The total fit is indicated by the short-dashed line. (C) $(20 \times$ *y*-axis scale) shows the individual components of the background fit. At the left of the histogram is the sliced nucleus debris (long-dashed line). The debris plus aggregates are shown as a solid line. The doublet distribution (vertical stripes) is complex in shape, reflecting the fact that all histogram components (debris, diploid G_1, S, and G_2 and aneuploid G_1, S, and G_2) aggregate with each other. The triplet distribution (diagonal stripes) has higher DNA content overall than the doublets, but there is extensive overlap. The quadruplet distribution (solid) is even higher in DNA content, but overlaps the triplet distribution to a large extent.

between laboratories is, however, most dependent on the consistent application of sensible rules for histogram interpretation.

A. Diagnosis of Aneuploidy

Both the convention on nomenclature for DNA cytometry in 1984 (Hiddeman *et al.*, 1984) and the guidelines of the DNA Cytometry Consensus (Shankey *et al.*, 1993a) indicate that *DNA aneuploidy can be reliably diagnosed only when two or more distinct (bimodal) peaks are present.* The ability to detect near-diploid DNA contents is directly related to the CV of the diploid and aneuploid peaks, as described previously. Detection of aneuploid populations of any DI also becomes more uncertain when the aneuploid peak is small, relative to the total number of cells in the histogram (Cusick *et al.*, 1990). *Ordinarily, for univariate DNA histogram analysis, the DNA aneuploid population should comprise at least 5–10% of the cell or nuclear events, after correction for aggregates* (Shankey *et al.*, 1993a). The number of total cells in the histogram can also be important in the identification of aneuploidy (see above). In samples with low cell number, statistical fluctuations should not be confused with an aneuploid peak; *if this is ambiguous, additional tissue samples should be used to confirm the presence or absence of DNA aneuploidy* (Shankey *et al.*, 1993a).

The requirement for the presence of bimodality in ploidy diagnosis is conservative, since in theory an aneuploid peak overlapping with the diploid peak might be detected by the use of DNA content standards. As described previously, a shift in the position of a peak with respect to standards cannot reliably be taken as evidence of DNA aneuploidy; otherwise, overdiagnosis of near-diploid aneuploidy may occur. In the past, differing criteria of diagnosis of DNA aneuploidy utilized by different investigators have complicated the interpretation and comparison of published data.

Finally, it should be noted that differentiation of DNA diploid from near-diploid cells may be more important in some tumors than others. For example, in breast cancer, aneuploid tumors with DI up to 1.3 show an improved survival compared to tumors with greater degrees of aneuploidy (Hedley *et al.*, 1993), and it has been suggested that these should be grouped together with DNA diploid cells (Toikkanen *et al.*, 1990). In contrast, the presence of near-diploid aneuploidy in some lymphomas may be clinically significant (Duque *et al.*, 1993).

B. DNA Index 2.0 Peaks: G_2 vs DNA Tetraploidy

The difference between G_2 vs DNA tetraploid peaks cannot always be distinguished by DNA content analysis. By convention, *a definition of tetraploid peaks is DI values from 1.9 to 2.1 with proportions of cells greater than the G_2/M fraction of normal control tissue samples, after correction for aggregates* (Shankey *et al.*, 1993a). In many published studies arbitrary cutoff values, most commonly >15% total cells, have been utilized. However, some studies have correlated elevated DI 2.0 peaks with adverse clinical outcome, even when they comprise less than 15% of cells. Examples are prostatic adenocarcinoma (Nativ *et al.*, 1989; Jones *et al.*, 1990; Stephenson *et al.*, 1987) and Barrett's esophagus (Reid *et al.*, 1992). Because of this, careful assessment of the normal "control" range of G_2 values should be made in each organ system and laboratory. Hopefully, future studies will better validate the definition of tetraploidy in each organ system.

The following considerations should also be kept in mind:

1. When the diploid S phase is high, the diploid G_2 may be expected to be higher as well. In some cases of rapid proliferation a G_2 peak may be in excess of 15%.

2. The DNA index of a G_2 peak should be within the range expected for G_2 cells. Careful recording of the actual G_2/G_1 ratios for a given cell or tumor type will help to establish the laboratory confidence interval for true G_2 cells. Peaks outside this range may be aneuploid, even if they are smaller than the control G_2 level.

3. If there is evidence of S and G_2 phases for the DI 2.0 population, then there is a stronger argument that the DI 2.0 peak may be a G_1 near-tetraploid population. Note that software aggregation modeling, as described above, is often very useful in demonstrating that the G_2 of the near-tetraploid population is greater than can be accounted for by the effects of aggregation alone.

4. When a tetraploid or near-tetraploid population is present, it will, by definition, overlap with the G_2 of the "diploid" population. When the overlap is close, there is no way to accurately determine how much of the near-tetraploid peak is diploid G_2 and how much is near-tetraploid G_1. Cell-cycle analysis software should, however, be prevented from making the G_2 an unreasonably large proportion of the total near-tetraploid peak [by assuming, for example, that the G_2 of the diploid (or lower ploidy) population is proportional to the S phase fraction of the same population].

C. Which S Phase Is Most Relevant When DNA Diploid and Aneuploid Populations Are Both Present?

More sophisticated cell-cycle modeling does not require that S phases be nonoverlapping in order to give independent estimates of each. If at least some part of the S and G_2 phase distributions is nonoverlapping, then the relative components of each can be evaluated. Exactly how close two ploidy populations can be before their S phases can no longer be independently evaluated depends mainly upon the CV of the analysis. If there is no region of the histogram in which the two cell cycles are largely nonoverlapping, then the individual diploid and aneuploid S phases cannot be reliably established; *in this circumstance it is recommended at present that the estimate of the combined S average phase of the two populations should be reported* (Shankey *et al.,* 1993a). This recommendation will, hopefully, be subject to additional future experimental confirmation. When the ploidy value and quality of the histogram allow individual S phase estimates to be derived, the current recommendation is that *the S phase of the aneuploid population should be used for clinical interpretation* (Shankey *et al.,* 1993a). This interpretation is based, in part, upon the argument that the DNA aneuploid cells, when present, represent the malignant cells, and it is assumed that the proliferative behavior of the malignant cells is the most relevant to the biological aggressiveness of the tumor. In the past, published reports have frequently failed to specify the origin of the S phase calculation. The study of node-negative breast cancer summarized in Table I demonstrated that utilizing the DNA aneuploid S phase (when present) provided superior prognostic value compared to the use of the average of the DNA diploid and aneuploid S phase. In contrast, if only the S phase of the DNA diploid cells was utilized, little prognostic information was obtained. Additional studies will hopefully also address this issue.

D. Optimal Choices of Fitting and Debris Models for Histogram Analysis

As described above, there have been a number of recent developments in cell-cycle analysis modeling, particularly in debris and aggregation compensation. It is generally assumed that these newer modeling techniques will lead to improved accuracy of S and G_2 phase measurements (Shankey *et al.,* 1993a). At the present time, however, only a limited amount of direct evidence is available to confirm this improvement in the clinical setting. Several reports indicate that the use of sliced nucleus modeling of debris, especially in samples derived from paraffin, improves the predictive strength of S phase measurements in breast (Kallioniemi *et al.,* 1991a; Clark *et al.,* 1992) and prostate (Kallioniemi *et al.,* 1991a) cancer. Similar results from the data set used by Kallioniemi *et al.* (1991a) are shown in Table I and illustrate the improvement obtained by sliced

Table I
The Prognostic Value of Different S Phase Estimates[a]

Method	P value[b]
Diploid S for diploid tumors, aneuploid S for aneuploid tumors, average S phase for near-diploid tumors. Classical exponential debris.	0.004
As above but with sliced nuclei debris model.	0.0005
Average of diploid and aneuploid S phase utilized.	0.029
Diploid S phase only utilized.	0.13

[a] Node-negative breast cancer. Modified after Kallioniemi *et al.* (1994).
[b] Difference (Wilcoxon-Breslow analysis) between 5-year survival with above or below median S phase.

nucleus modeling compared to classical exponential debris modeling. Based on theoretical considerations and limited published study, it is the consensus at present that *simple exponential debris modeling is not considered reliable, and histogram-dependent debris models should be used* (Shankey *et al.*, 1993a).

Comparison of sliced nucleus debris with and without software aggregation modeling is shown in Table II. The addition of aggregation modeling improves the predictive strength of S phase in this data set, supporting the use of this approach. As mentioned previously, the presence of aggregates can obscure the detection of an aneuploid peak, or an aggregate peak can be falsely interpreted as an aneuploid peak. Aggregates can especially compound the difficulty of identifying DNA ''tetraploid tumors'' (Shankey *et al.*, 1993a).

The use of aggregation modeling generally results in a reduction in the average values of both S and G_2 phase measurements. In diploid breast cancers, the reduction in S phase estimates averaged only 0.89 percentage points; however, aneuploid S phase estimates were reduced an average of 2.5 percentage points (Rabinovitch, 1993a). S phase estimates in bladder and prostate cancer (which are frequently tetraploid) may be reduced by 20–50% when software aggregation correction is applied (Rabinovitch, 1993a; Shankey *et al.*, 1993b). As a rule, aggregates of >5% cells can cause significant overestimation of ''tetraploid'' tumor S phases (Shankey *et al.*, 1993a).

Most often, S phase is the proliferative parameter evaluated in flow cytometric literature, although occasionally S + G_2 is utilized. Although the DNA Cytometry Consensus Guidelines recommend that *S phase, rather than S + G_2/M, be utilized for tumor proliferation estimates*, the relative utility of S vs S + G_2/M is in need of further study. Table II illustrates that, in that data set, S + G_2/M was of greater prognostic strength than S phase alone and that this strength was appreciably enhanced with the use of aggregation modeling. Because the G_2/M calculation is especially affected by the presence of doublets, it is possible that the use of improved aggregation compensation will lead to a general improvement in the prognostic strength of the S + G_2/M measurement.

Simplification of the cell-cycle model used in analysis of clinical specimens has been suggested in a preceding discussion, because histograms from clinical

Table II

Relative Risk (Top/Bottom Tertile) and Significance (P value) of S or S + G_2/M Phase Fraction in Cox Multivariate Analysis[a]

Model	SPF	S + G_2/M
Sliced nucleus debris, zero-order S	1.75 (0.01)	2.00 (0.001)
Sliced nucleus debris, zero-order S + Aggregation	2.16 (0.0002)	2.52 (0.00001)
Sliced nucleus debris + first-order S	1.69 (0.02)	1.84 (0.002)
Sliced nucleus debris + zero-order S + CVs fixed	1.73 (0.01)	1.96 (0.002)

[a] Breast cancer, all stages, surgical treatment only. Modified after Barlow *et al.* (1994).

tissues often contain artifacts which are not consistent with the fitting model. An example is artifactual skewing of the G_1 peak which can be falsely confused with S phase. The use of a zero-order (rectangular) S phase model was compared to a first-order (trapezoidal) S phase model, as shown in Table II. The simpler zero-order S phase model provided a slightly better prognostic strength for both S and S + G_2. Simplifying the fitting model by constraining the CVs of diploid and aneuploid G_1 and G_2 peaks to the same value had no beneficial effect.

E. Prognostic Categories Based on Ploidy and Cell-Cycle Results

As several interlaboratory comparisons have described (Hitchcock, 1991; Kallioniemi *et al.*, 1990), the values of S and/or G_2/M derived by different laboratories vary considerably, due to lack of standardized techniques and use of different methods of S and G_2/M calculation and histogram interpretation. Thus, *use of published values for numerical cut points for survival or disease course should be avoided* (Shankey *et al.*, 1993a). At present *each laboratory must define its own range of values (high, intermediate, and low S phase for DNA diploid and DNA aneuploid tumors) based upon their own analysis of individual types of tumor* (Shankey *et al.*, 1993a). Even this method is subject to the caveat that the patient population examined locally may differ from that investigated in a published study.

Most often, clinical survival or relapse-free interval has been compared for patients with above or below mean or median S phase, although some authors have determined the S phase cutoff that yields the maximum intergroup difference in clinical outcome (Clark *et al.*, 1989). Although a single cutoff point yields a binary result—good or bad prognosis, for example—this division does not make good biological or statistical sense: cell-cycle estimates that differ by insignificant amounts may be placed in opposite prognostic categories. *A single cut point should be avoided and tertiles such as low, intermediate, and high are preferable* (Shankey *et al.*, 1993a).

At present, the DNA Cytometry Consensus Conference Guideline recommendation is that high, intermediate, and low S phase ranges for DNA diploid and DNA aneuploid tumors be calculated separately (see above). Although survival rates for diploid and aneuploid SPF ranges have occasionally been reported (Clark *et al.*, 1992), they have only rarely (Kallioniemi *et al.*, 1988) been calculated by tertiles. Use of the recommended tertiles within ploidy class may require some adjustment in current perceptions and may cause some confusion until it is more widely utilized. While the clinical hazards estimated from separate diploid and aneuploid tertile cut points (Table III) lead to improved prognostic strength, the cut points are very different from the more conventional cut points derived from "all" ploidies, and the resultant hazards reflect the combined influence of ploidy and proliferative rate. Note that in the separate analysis of diploid and aneuploid cases, diploid tumor S phase values are overall much

Table III

Estimated Hazards for a Cox Model Analysis of Breast Cancer Survival[a]

Factor	S phase range	Hazard
Size 2–5 cm		1.69
Size >5 cm		3.24
Positive nodes		2.24
Metastases		3.30
S phase tertiles, all tumors		
Low S	0–1.9	1.0
Intermediate S	1.9–5.4	1.54
High S	>5.4	2.10
S phase tertiles, diploid and aneuploid tumors separately		
Diploid low S	0–0.8	1.0
Diploid intermediate S	0.8–2.1	1.39
Diploid high S	>2.1	2.17
Aneuploid low S	0–4.1	1.50
Aneuploid intermediate S	4.1–9.1	2.35
Aneuploid high S	>9.1	2.40

[a] Breast cancer, all stages, surgical treatment only. Aneuploid S phase utilized when separately calculateable; sliced nucleus debris and aggregation modeling done with the MultiCycle program (Rabinovitch, 1993b). Modified after Barlow *et al.* (1994).

lower than aneuploid tumor S phase values. This is a consistent finding in large numbers of published reports and may be due both to intrinsically lower proliferative rates of diploid malignant cells and to the unavoidable admixture of diploid tumor cells and slower growing stromal cells in the univariate DNA histogram. Differences between the "all" vs separate prognostic categories can be appreciable in some cases. For example, diploid tumors with S phase values within the diploid-specific "intermediate" range of 0.8–1.9 or the diploid "high" range of >2.1 are associated with higher calculated risk when analyzed separately by ploidy than when all ploidies are lumped together (Table III). Similarly, aneuploid tumors with low proliferative rates of <1.9 have higher risk than the "all" low tertile would suggest (aneuploid tumors with low SPF have higher risk than diploid tumors with low SPF). Publication of analyses of additional datasets using separate diploid and aneuploid tertiles would be very valuable. As the Consensus Guidelines indicate, the specific cutoff values shown in Table III should not be directly extrapolated into other laboratories. In particular, values will be higher if sliced nucleus and aggregation modeling are not utilized.

F. Histogram Reliability and Confidence Estimation

After performing a cell-cycle analysis, how the resulting parameters are used often hinges upon the assessment of the accuracy and reliability of the DNA content and cell-cycle estimates. *Not all histograms are adequate for identification of DNA aneuploid populations, or for the estimation of S phase . . . when the histogram is inadequate, due to high CVs, debris, or aggregates, it should be reported as inadequate* (Shankey *et al.,* 1993a). Criteria for the ability to detect near-diploid aneuploidy based on CV have been described previously in Fig. 3. *In general the CV of normal diploid cells in a histogram should be less than 8%* (Shankey *et al.,* 1993a). As demonstrated above, debris and aggregates can appreciably affect S phase estimation, although newer models help to compensate for these effects. Quantitating the effect of CV, debris, and aggregates on the accuracy of S phase measurement is, however, complex.

The DNA Consensus Conference Guidelines recommend that *if a sample has a high percentage of aggregates (>10% as determined by manual counting), it should be further disaggregated by mechanical means or rejected for analysis. The percentage aggregates and debris should be evaluated for each histogram . . . it is recommended that >20% histogram background aggregates and debris are unsatisfactory for S phase analysis.*

The DNA Consensus Conference Guidelines further recommend that *accurate calculation of S phase generally requires that the proportion of tumor cells is greater than 15–20% of the total cells, although samples with lower proportions may be useful for DNA ploidy analysis.* When aneuploid cells are present, accurate S phase estimation requires that they should be greater than 15–20% of the total cells. This is because (1) overlap of cells from the a large diploid population into the S phase region of the aneuploid population can impair the accuracy of estimation of the aneuploid S phase, and (2) even when the two populations do not overlap extensively, debris and aggregates from the more abundant population may still overlap the cell cycle of the rarer population. A common clinical setting in which low proportions of tumor cells may be found is in the presence of an abundant lymphocytic infiltrate (Eckhardt *et al.,* 1989).

Guidelines such as the above for rejection of histograms based on "poor" quality are based upon a consensus of experience, but there has been minimal publication of objective data. Existing data bases of DNA content histograms and the associated clinical outcome can be used to objectively define improved reporting methods, including tests of criteria for histogram rejection. In such an analysis, "poor" histogram quality would be assumed to be associated with reduced prognostic strength of ploidy and cell-cycle results. One such analysis of predictors of poor reliability of S phase measurements is illustrated in Table IV. Consistent with the DNA Cytometry Consensus Conference Guidelines, rejection of cases with low proportions of aneuploid cells and high BAD was beneficial, although the exact rejection criteria were slightly different (more stringent for percent aneuploid cells, less stringent for BAD). A predictor of S phase reliability that is based upon a statistical error analysis of the least squares fitting itself, the "intramodel error" (Rabinovitch, 1993a), was also beneficial. In addition, comparison of the range of S phase values produced by alterations

Table IV

Predictors of S Phase Reliability: Prognostic Strength of the Remaining Cases after Rejection of Subsets of Histograms[a]

Cases removed	Percent cases removed	Hazard (relative risk)	Significance (P)
None	0	2.10	0.0005
% Aneuploid cells < 30%	14	2.54	0.00002
% Aneuploid cells < 40%	21	2.82	0.000005
BAD > 35%	10	2.40	0.00004
BAD > 28%	20	2.54	0.00002
Intramodel error			
80th percentile	20	2.50	0.0002
70th percentile	30	2.64	0.0002
Intermodel error			
90th percentile	10	2.18	0.0004
80th percentile	20	2.32	0.0002
CV > 8.6	10	2.00	0.002
CV > 7.5	20	1.94	0.006
$\chi^2 > 2.0$	10	1.93	0.005
$\chi^2 > 1.7$	20	2.05	0.003

[a] Breast cancer, all stages, surgical treatment only. Cox multivariate analysis. Modified after Barlow *et al.* 1994.

in the fitting model (e.g., with and without aggregation modeling, zero-order vs first-order S phase models, with vs without G_2/G_1 ratio constraints), the "intermodel error" (Rabinovitch, 1993a), was a moderately useful predictor of reliability. These four predictors can be combined into an overall estimate of histogram reliability (Rabinovitch, 1993a,b). In contrast, the analysis shown in Table IV failed to demonstrate that either elevated CV or higher χ^2 of the least squares fit was useful as a criterion for histogram rejection. The absence of utility of CV is surprising, but may indicate that other measures of histogram quality are more important. The χ^2 is affected by a large number of variables, not all related to goodness of the fit; these include the number of cells acquired in the histogram and the end points of the analysis region used within the histogram.

VI. Conclusion

The careful and more uniform use of guidelines for DNA content and cell-cycle analysis should lead to greater accuracy and reproducibility in performing and reporting ploidy and cell-cycle measurements. If this is done, then this simple methodology can be applied with greater confidence over a wider range of clinical settings. The future of DNA cytometry also includes greater use of multiparameter DNA analyses, especially the use of cell-type and proliferation-specific antibodies. Although the added information that is derived from the multiparameter analysis will help improve accuracy, most of these analyses will still involve cell-cycle analysis of gated histograms or bivariate histogram regions. Thus, DNA content and cell-cycle analysis will retain importance well into the future.

References

Alanen, K. A., Joensuu, H., and Klein, P. J. (1989). *Cytometry* **10**, 417–425.

Bagwell, C. B. (1979). Ph.D. Thesis, University of Miami, Coral Gables, FL.

Bagwell, C. B. (1993). *In* "Clinical Flow Cytometry: Principles and Applications" (K. D. Bauer, R. E. Duque, and T. V. Shankey, eds.), pp. 41–62. Williams & Wilkins, Baltimore, MD.

Bagwell, C. B. Baker, D. Whetstone, S., Munson, M., Hitchcox, S., Autl, K. A., and Lovett, E. J. (1989). *Cytometry* **10**, 689–694.

Bagwell, C. B., Mayo, S. W., Whetstone, S. D., Hitchcox, S. A., Baker, D. R., Herbert, D. J., Weaver, D. L., Jones, M. A., and Lovett, E. J. (1991). *Cytometry* **12**, 107–118.

Barlow, W., Kallioniemi, O.-P., Visakorpi T., Isola, J., Rabinovitch, P. S. (1994). Submitted for publication.

Beck, H. P. (1980). *Cell Tissue Kinet.* **13**, 173–181.

Becker, R. L., Jr., and Mikel, U. V. (1990). *Anal. Quant. Cytol. Histol.* **12**, 333–341.

Bertuzzi, A., D'Agnano, I., Gandolfi, A., Graziano, A., Star, G., and Ubezio, P. (1990). *Cell Biophys.* **17**, 257–267.

Bevington, P. R. (1969). "Data Reduction and Error Analysis for the Physical Sciences," pp. 153–160. McGraw-Hill, New York.

Bruno, S., Crissman, H. A., Bauer, K. D., and Darzynkiewicz, Z. (1991). *Exp. Cell Res.* **196**, 99–106.

Clark, G. M., Dressler, L. G., Owens, M. A., Pounds, G., Oldaker, P., and McGuire, W. L. (1989). *N. Engl. J. Med.* **320**, 627–633.

Clark, G. M., Mathiew, M. C., Owens, M. A., Dressler, L. G., Eudey, L., Tormey, D. C., Osborne, C. K., Gilchrist, K. W., Mansour, E. G., Abeloff, M. D., *et al.* (1992). *J. Clin. Oncol.* **10**, 428–432.

Coon, J. S., Deitch, A. D., de Vere White, R. W., Koss, L. G., Melamed, M. R., Reeder, J. E., Weinstein, R. S., Wersto, R. P., and Wheeless, L. L. (1989). *Cancer* (*Philadelphia*) **63**, 1592–1600.

Cusick, E. L., Milton, J. I., and Ewen, S. W. B. (1990). *Anal. Cell. Pathol.* **2**, 139–148.

Darzynkiewicz, Z., Traganos, F., Sharpless, T. K., and Melamed, M. R. (1977). *Cancer Res.* **37**, 4635–4640.

Darzynkiewicz, Z., Traganos, F., Kapuscinski, J., Staino-Coico, L., and Melamed, M. R. (1984). *Cytometry* **5**, 355–363.

Dean, P. N. (1985). *In* "Flow Cytometry: Instrumentation and Data Analysis" (M. A. Van Dilla, P. N. Dean, O. D. Laerum, and M. R. Melamed, eds.), pp. 195–221. Academic Press, New York.

Dean, P. N. (1990). *In* "Flow Cytometry and Sorting" (M. R. Melamed, T. Lindmo, and M. L. Mendelsohn, eds.), 2nd ed., pp. 415–444. Wiley-Liss, New York.

Dean, P. N., and Jett, J. (1974). *J. Cell Biol.* **60**, 523.

Duque, R. E., Andreeff, M., Braylan, R. C., Diamond, L. W., and Peiper, S. C. (1993). *Cytometry* **14**, 492–496.

Eckhardt, R., Feichter, G. E., and Goerttler, K. (1989). *Anal. Quant. Cytol. Histol.* **11**, 384–390.

Evenson, D., Darzynkiewicz, Z., Jost, L., and Ballachey, B. (1986). *Cytometry* **7**, 45–53.

Fox, M. H. (1980). *Cytometry* **1**, 71.

Fried, J. (1976). *Comp. Biomed. Res.* **9**, 263–276.

Hedley, D. W., Clark, G. M., Cornelisse, C. J., Killander, D., Kute, T., and Merkel, D. (1993). *Cytometry* **14,** 482–485.

Heiden, T., Strang, P., Stendahl, U., and Tribukait, B. (1990). *Anticancer Res.* **10,** 49–54.

Hiddeman, W., Schumann, J., Andreeff, M., Barlogie, B., Herman, C. J., Leif, R. C., Mayall, B. H., Murphy, R. F., and Sandberg, A. A. (1984). *Cytometry* **5,** 445–446.

Hitchcock, C. L. (1991). *Cytometry, Suppl.* **5,** 46.

Iverson, O. E., and Laerum, O. D. (1987). *Cytometry* **8,** 190–196.

Joensuu, H., Alanen, K., Klemi, P., and Aine, R. (1990). *Cytometry* **11,** 431–437.

Jones, E. C., McNeal, J., and Bruchovsky, N. (1990). *Cancer (Philadelphia)* **66,** 752–757.

Kallioniemi, O.-P., Blanco, G., Alavaikko, M., Hietanen, T., Mattila, J., Lauslahti, K., Lehtinen, M., and Koivula, T. (1988). *Cancer (Philadelphia)* **62,** 2183–2190.

Kallioniemi, O.-P., Joensuu, H., Klemi, P., and Koivula, T. (1990). *Breast Cancer Res. Treat.* **17,** 59–61.

Kallioniemi, O.-P., Visakorpi, T., Holli, K., and Heikkinen, A., Isola, J., and Koivula, T. (1991). *Cytometry* **12,** 413–421.

Kallioniemi, O.-P., Visakorpi, T., Holli, K., Isola, J. J., and Rabinovitch, P. S. (1994). *Cytometry* **16,** 250–255.

Klein, F. A., and White, K. H. (1988). *J. Urol.* **139,** 275–278.

Koch, H., Bettecken, T., Kubbies, M., Salk, D., Smith, J. W., and Rabinovitch, P. S. (1984). *Cytometry* **5,** 118–123.

Kubbies, M. (1990). *Cytometry* **11,** 386–394.

Kubbies, M. (1992). *J. Pathol.* **167,** 413–419.

Larsen, J. K., Munch-Peterson, B., Christiansen, J., and Jorgensen, J. (1986). *Cytometry* **7,** 54–63.

Marquardt, F. W. (1963). *J. Soc. Ind. Appl. Math.* **11,** 431–441.

Nativ, O., Winkler, H. Z., Raz, Y., Therneau, T. M., Farrow, G. M., Myers, R. P., Zincke, H., and Lieber, M. M. (1989). *Mayo Clin. Proc.* **64,** 911–919.

Nicoletti, I., Migliorati, G., Pagliacci, M. C., Grignani, F., and Riccardi, C. (1991). *J. Immunol. Methods* **139,** 271–279.

Price, J., and Herman, C. J. (1990). *Cytometry* **11,** 845.

Rabinovitch, P. S. (1988). "MultiCycle Program." Phoenix Flow Systems, San Diego.

Rabinovitch, P. S. (1990). *Cytometry, Suppl.* **4,** 27.

Rabinovitch, P. S. (1993a). *In* "Clinical Flow Cytometry: Principles and Applications" (K. D. Bauer, R. E. Duque, and T. V. Shankey, eds.), pp. 117–142. Williams & Wilkins, Baltimore, MD.

Rabinovitch, P. S. (1993b). "MultiCycle Program: Advanced Version." Phoenix Flow Systems, San Diego.

Rabinovitch, P. S., and Kallioniemi, O.-P. (1991). *Cytometry, Suppl.* **5,** 138.

Reid, B. J., Blount, P. L., Rubin, C. E., Levine, D. S., Haggitt, R. C., and Rabinovitch, P. S. (1992). *Gastroenterology* **102,** 1212–1219.

Roti Roto, J. L., Wright, W. D., Higashikubo, R., and Dethlefsen, L. A. (1985). *Cytometry* **6,** 101–108.

Shankey, T. V., Rabinovitch, P. S., Bagwell, C. B., Bauer, K. D., Duque, R. E., Hedley, D. W., Mayall, B. H., and Wheeless, L. (1993a). *Cytometry* **14,** 472–477.

Shankey, T. V., Dougherty, S., Manion, S., and Flanigan, R. C. (1993b). *Cytometry, Suppl.* **6,** 83.

Stephenson, R. A., James, B. C., Gay, H., Fair, W. R., Whitmore, W. F., and Melamed, M. R. (1987). *Cancer Res.* **47,** 2504–2509.

Stokke, T., Holte, H., Erikstein, B., Davies, C. L., Funderud, S., and Steen, H. B. (1991). *Cytometry* **12,** 172–178.

Telford, W. G., King, L. E., and Fraker, P. J. (1991). *Cell Proliferation* **24,** 447–459.

Toikkanen, S., Joensuu, H., and Klemi, P. (1990). *Am. J. Clin. Pathol.* **93**(4); 471–479.

Vindelov, I. I., Christenson, I. J., and Nissen, N. I. (1983). *Cytometry* **3,** 328–331.

Wersto, R. P., Liblit, and R. A., and Koss, L. G. (1991). *Hum. Pathol.* **22,** 1085–1098.

Wheeless, L. L., Coon, J. S., Cox, C., Deitch, A. D., deVere White, R. W., Koss, L. G., Melamed, M. R., O'Connell, M. J., Reeder, J. E., Weinstein, R. S., and Wersto, R. P. (1989). *Cytometry* **10,** 731–738.

Wheeless, L. L., Coon, J. S., Cox, C., Deitch, A. D., deVere White, R. W., Fradet, Y., Koss, L. G., Melamed, M. R., O'Connell, M. J., Reeder, J. E., Weinstein, R. S., and Wersto, R. P. (1991). *Cytometry* **12,** 405–412.

Wolley, R. C., Herz, F., and Koss, L. G. (1982). *Cytometry* **2,** 370–373.

CHAPTER 19

Immunochemical Quantitation of Bromodeoxyuridine: Application to Cell–Cycle Kinetics

Frank Dolbeare* and Jules R. Selden†

*Biology and Biotechnology Program
Lawrence Livermore National Laboratory
Livermore, California 94550
†Department of Safety Assessment
Merck Research Laboratories
West Point, Pennsylvania 19486

I. Introduction

The immunological BrdUrd techniques developed for flow cytometry and
the microscope have almost replaced autoradiographic techniques with tritiated
thymidine in cell kinetic studies. The immunochemical evaluations of BrdUrd
are being used largely in clinical situations to measure labeling index (LI), S
phase duration (Ts), potential doubling time (Tpot), growth fraction, and drug
resistance (Riccardi *et al.*, 1988, Waldman *et al.*, 1988; Suzuki, 1988; Danova
et al. 1990; Duprez *et al.*, 1990; Hardonk and Harms, 1990; Ito *et al.*, 1990;
Raza and Preisler, 1990; Garin *et al.*, 1991; Giaretti, 1991; Raza *et al.*, 1991a,
1992; San-Galli *et al.*, 1991; Ward *et al.*, 1991; Yu *et al.*, 1992). BrdUrd can be
used in the same way as [^{3}H]TdR to provide detailed cell kinetics measurements.
Continuous exposure to BrdUrd provides an estimate of the fraction of noncycling cells, and pulse-chase experiments provide quantitative measures of cell
progression. For use with human material, where only limited sampling is
practical, the pulse-chase approach has been refined to a single time point
method (Begg *et al.*, 1985). The introduction of halogen-selective antibodies
(e.g., anti-BrdUrd selective and anti-IdUrd selective) (Dolbeare *et al.*, 1988;
Shibui *et al.*, 1989; Raza *et al.*, 1991b) allows double-labeling experiments like
those using ^{3}H- and ^{14}C-labeled nucleotides.

The incorporation of BrdUrd into cells offers a more accurate estimation
of the fraction of cells in S phase than does DNA measurement alone, and
the optimal method is the simultaneous measurement of both DNA content
and incorporated BrdUrd in each cell (Dolbeare *et al.*, 1983). Fluorescein-
conjugated anti-BrdUrd antibodies provide a green fluorescent signal that is
proportional to incorporated BrdUrd when plotted on a logarithmic *y*-axis.
Propidium iodide intercalation provides a red fluorescent signal that is proportional to DNA content when plotted on a linear *x*-axis. G_1 and G_2 cells are low
in green fluorescence but are well resolved on the basis of red fluorescence. S
phase cells have intermediate DNA contents, but are readily distinguished by
their intense green fluorescence.

II. Applications

The primary applications of the technique are the quantification of cell-cycle
phase fractions, phase durations, doubling time, labeling index, and growth
fractions. Specific effects of agents which stimulate cell proliferation, block

specific phases of the cell cycle, or slow the progression through the cell cycle can be quantified. The combination of one or more nucleoside analogues with varying amounts of time between cell exposure and sample collection can be used to determine the total cell-cycle time and the duration of the various cell-cycle compartments.

The quantitative equations for determining phase durations, e.g., time duration of G_1 (Tg1), Ts, and cell-cycle time (Tc), were derived by Takahashi (1966). Although the mathematics provided by this model are beyond the scope of this chapter, a simplified mathematical equation can describe the movement of cells from compartment to compartment (Gray et al., 1990). For example, the rate of change in the number of cells in compartment i at time t, $N_i(t)$ might be described as

$$dN_i(t)/dt = l_{i-1}N_{i-1}(t) - l_iN_i(t),$$

where l_i defines the rate at which cells leave compartment $i + 1$. For compartment 1, the equation becomes

$$dN_1(t)/dt = 2l_kN_k(t) - l_1N_1(t),$$

where k is the number of the last compartment in the cycle and the factor 2 takes account of the fact that the cell number doubles as cells move from G_2M to G_1 phase.

Nonkinetic applications of BrdUrd or IdUrd incorporation include (1) testing of drug resistance or sensitivity (Waldman et al., 1988; Lacombe et al., 1992), (2) analogues incorporated into DNA during the repair of DNA damage (unscheduled DNA synthesis) (Beisker and Hittleman, 1988; Selden et al., 1993), (3) the incorporation of BrdUrd into DNA for demonstrating sister chromatid exchanges (SCEs) by fluorescent antibody labeling (Pinkel et al., 1985), (4) Replication patterns in chromatin and nuclei during DNA synthesis (Allison et al., 1985; Nakamura et al., 1986), (5) the isolation of nascently replicated DNA using immunoaffinity columns directed against incorporated BrdUrd (Leadon, 1986), (6) BrdUrd-labeled probes for in situ hybridization (Frommer et al., 1988), (7) analysis of proliferation during differentiation (Gratzner et al., 1985; Kaufman and Robert-Nicoud, 1985).

III. Materials

A. General Purpose Reagents

1. Bromodeoxyuridine or iododeoxyuridine: Generally, anti-BrdUrd antibodies are more efficient at stoichiometric binding of BrdUrd when the level of BrdUrd substitution is at about 25% of the total thymidines. At substitutions higher than this amount, steric factors limit binding, while below about 1% substitution the possibilities of multivalent binding and the absolute signal

strength are limited. In the most rigorous experiments for *in vitro* labeling of cultured cells, BrdUrd is diluted in culture media along with TdR at a defined molar ratio, and 5-fluorouracil (5-FU) is included to block endogenous thymidine synthesis. This mixture of BrdUrd + dThd = 10 μM, 5 μM 5-FU, and 5 μM dCytd ensures incorporation at a specified substitution level. Less rigorous incorporation protocols use either BrdUrd or IdUrd between 0.1 and 10 mM, assuring that at this level the analogue is present in excess of intracellular dThd pools.

2. Anti-halopyrimidine antibody: We describe here procedures using monoclonal antibodies against either BrdUrd or IdUrd. Animal antisera derived against one of the halopyrimidines may also be used, but historically these have lacked the needed selectivity. Table I is a list of commercially available anti-bromodeoxyuridine antibodies. The final sensitivity of the measurement depends on the purity, affinity, and specificity of the particular antibody. High-affinity antibodies permit quantification of low levels of BrdUrd incorporation (e.g., <0.1% substitution) (Beisker *et al.*, 1987). Antibodies directly conjugated to fluorophores may be used, or the anti-BrdUrd antibodies may be detected using indirect immunochemical methods. For methods involving the simultaneous detection of cell-surface antigens and BrdUrd, direct conjugates are required. The monoclonal antibody pair of IU-4 and Br-3 (Caltag, South San Francisco, CA) allows double-labeling with IdUrd and BrdUrd, since Br-3 shows halogen selectivity in binding, preferring BrdUrd to IdUrd (Dolbeare *et al.*, 1988; Shibui *et al.*, 1989).

3. Phosphate-buffered saline (PBS); 0.05 M sodium phosphate, pH 7.2 (no Ca^{2+} or Mg^{2+}), and 0.15 M NaCl.

4. Antibody diluting buffer: PBS as described above containing 2 × salt sodium citrate (SSC) (SSC = 0.15 M NaCl + 0.015 M sodium citrate), 0.5% Tween 20, (Sigma Chemical Company, St. Louis, MO), and a blocking protein to limit nonspecific sticking of the antibody. The blocking protein may be 1% bovine serum albumin, 1% gelatin, or 2% dry nonfat milk protein.

5. Ribonuclease A (RNase) stock solution: Ribonuclease A (Sigma Chemical Company) at 0.5 mg/ml in PBS. This solution can be stored refrigerated with 0.1 mg/ml sodium azide for up to 6 months.

6. Paraformaldehyde solution (E.M. grade, Polyscience, Inc., Warrington, PA): Stock solution is 0.25 or 1% paraformaldehyde in PBS, pH 7.2.

7. 0.1 M HCl plus 0.5% Triton X-100: Five grams of Triton X-100 (Sigma Chemical Company) in 1 liter of 0.1 M HCl.

8. Wash buffer: Five grams of Tween 20 (Sigma Chemical Company) in 1 liter of PBS.

9. Goat anti-mouse IgG–fluorescein conjugate: This antibody may be obtained from a number of sources (see "Linscott's Directory of Immunological and Biological Reagents," 7th Ed.). Depending on the particular needs of the

Table I
Commercial Sources of Anti–BrdUrd Antibodies[a]

Clone	Ig type	State	Source
>1 clone	IgG1	Purified	Bioclone Australia
3D9	IgG1	Purified	Oncogen Science, Inc.
3D9	IgG1	Purified-FITC	Oncogene Science, Inc.
76-7	IgG1	Purified	Amac, Inc.
			Biodesign Invc
			Serotec, Ltd.
			Sigma Chemical
B44	IgG1	Purified	Becton–Dickinson
B44	IgG1	Purified-FITC	Becton–Dickinson
BMC9318	IgG1	Purified	Boehringer-Mannheim
		Purified-FITC	Boehringer-Mannheim
Br3	IgG1	Purified	Caltag
Br3	IgG1	Purified biotin	Caltag
Br3		Purified-FITC	Caltag
BU5.1	IgG2a	Purified purified-FITC	Cymbus Bioscience
			IBL Research Products
			Paesel&Lorei GmbH
			Progen Biotechnik GmbH
BU1-75	IgG Rat	Ascites	Sera-Labs, Ltd.
		Purified	Accurate Chemical
		Suprnatant	
Bu6-4	IgG1	Purified	Pierce Chemical
Bu20a	IgG1	Suprnatant	Dako
BU-33	IgG1	Purified	Sigma
MBU	IgG1	Ascites	Medscand USA
IU4	IgG1	Purified	Caltag
SB18	IgG1	Ascites	Accurate Chemical
			Medica, Inc.
			Sanbio BV
ZBU30	IgG1	Purified	Zymed Labs
		Purified Alkphos	Zymed Labs
		Purified biotin	Zymed Labs
		Purified-FITC	Zymed Labs
		Purified rox	Zymed Labs
		Purified phyco	Zymed Labs
Anti-BrdUrd	IgG1	Acites	Chemicon
Anti-BrdUrd	??		Janssen Biochimica
			Accurate Chemical

[a] Most of the sources are listed in Linscott's Directory of Immunological and Biological Reagents, Eighth Edition, obtainable from Linscott's Directory, Santa Rosa, CA 95404. Complete addresses of all of the above sources and their international distributors are given in the directory.

experiment (e.g., membrane antigen labeling in addition to BrdUrd/DNA analysis) one may require an alternative fluorophore as a conjugate, such as the blue emitting fluorophore, aminomethylcoumarin acetic acid (AMCA), or a red emitting fluorophores Texas red, Princeton red, or phycoerythrin (Molecular Probes, Eugene, Oregon).

10. Propidium iodide (Sigma Chemical Company, St. Louis, MO): The working solution is 10 μg/ml in PBS, pH 7.2. A stock solution of 1 mg/ml PI in 70% ethanol stored in the refrigerator is stable for at least a year. Propidium iodide is a suspected carcinogen and should be handled with proper caution.

11. Exonuclease III and *Eco*RI (Bethesda Research Laboratories, Gaithersberg, MD).

12. 0.1 M Citric acid/0.5% Triton X-100.

13. *Eco*R I buffer (see recommendations of manufacturer for specific endonucleases): 0.1 M Tris–HCl, pH 7.5, containing 50 mM NaCl and 10 mM MgCl$_2$.

14. Exonuclease III buffer: 50 mM Tris–HCl, pH 8.0, 10 mM 2-mercaptoethanol, 5 mM MgCl$_2$.

15. 40-μm nylon mesh (Small Parts, Inc., Miami Lakes, FL).

B. Reagents for Labeling Surface Antigens plus BrdUrd

This method follows that of Carayon and Bord (1992).

1. A biotin-conjugated antibody to cell-surface antigen, such as biotin-coupled anti-CD4.
2. Streptavidin–phycoerythrin (Becton–Dickinson, San Jose, CA).
3. 1% paraformaldehyde with 0.01% Tween 20.
4. DNase I(Sigma Chemical Company): 50 Kunitz unit/ml in PBS containing Ca^{2+} and Mg^{2+}.

C. Reagents for Staining Tissue Sections

While there are many suitable techniques for immunohistochemical staining of paraffin-embedded histologic sections, we have chosen the method of Shibuya *et al.* (1992) as illustrative of the methods. This study describes methods for dual labeling of both incorporated BrdUrd and IdUrd, using the halogen-selective monoclonal antibody pair Br-3 and IU-4 and a combination of alkaline phosphatase enzyme immunohistochemistry and immunogold staining. Simpler methods can be developed by the reader, using parts of this comprehensive method.

1. 6 μM histologic sections mounted on glass slides from tissues exposed to BrdUrd and/or IdUrd, fixed in 70% ethanol, paraffin embedded, and sectioned.
2. 4 N HCl.

3. 2.5% glutaraldehyde.

4. Gold-conjugated goat anti-mouse IgG and a silver enhancement kit (IntenSE M, Amersham Corp., Arlington Heights, IL).

5. 5% Acetic acid in water.

6. Tris-buffered saline (50 mM Tris, pH 7.6, and 0.15 M NaCl).

7. Alkaline phosphate-conjugated rabbit anti-mouse immunoglobulins (Dako Corp., Santa Barbara, CA).

8. Alkaline phosphatase substrate kit I (Vector Immunochemicals, Burlingame, CA).

9. 5% Gills No. 1 hematoxylin (Sigma).

10. Histo-Clear (National Diagnostics, Manville, NJ).

11. Permount (Fisher Chemical Company, Fair Lawn, NJ).

IV. Procedures

A. Immunochemical Labeling Followed by Flow Cytometric Detection

1. Thermal Denaturation Method (Beisker *et al.*, 1987; Dolbeare *et al.*, 1990)

a. Use 1×10^6 to 5×10^6 cells previously labeled with BrdUrd or IdUrd and fixed in cold 50% ethanol or methanol/acetic acid (3/1 v/v). Centrifuge at 500g for 2 min. (Note: avoid over centrifugation which can lead to serious cell clumping.) Pour off supernatant. Suspend cells by gentle vortexing.

b. Add 1.5 ml of RNase stock solution and incubate for 10 min at 37°C.

c. Centrifuge at 500g for 2 min, pour off supernatant, vortex to loosen pellet, and suspend cells in 2 ml of 0.25–1.0% paraformaldehyde solution for 30 min at room temperature.

d. Centrifuge, decant, and vortex pellet. Wash with 2 ml PBS. Centrifuge, decant, and vortex pellet.

e. Suspend cells in 1.5 ml 0.1 M HCl/Triton X-100 for 10 min on ice.

f. Add 5 ml of PBS and centrifuge for 2 min at 500g. Drain pellets well before vortexing.

g. Suspend cells in 1.5 ml of distilled H$_2$O and place tubes in a water bath at 95°C for 10 min. This is the DNA denaturing step and often must be adjusted depending on the cell type to maximize denaturation while avoiding extensive cell loss.

h. Remove samples from hot water bath and place in an ice/water mixture until the suspensions are cold. Then add 3 ml of PBS, and centrifuge 500g for 2 min. Drain tubes and vortex gently to loosen pellet. At this point some clumping may be observed. Disperse clumps by pipetting or syringing through a 25-gauge needle. Failure to disperse the clumps completely may prevent antibody access to BrdUrd labeled cells. These cells may appear later as apparently unlabeled, resulting in a measurement error.

Note: Steps 2 through 8 may be deleted if HCl denaturation is used (Gratzner, 1982; Dolbeare *et al.*, 1983). Cells are incubated at room temperature with 200 μl of 1.5 to 4 *M* HCl for 20 min and then washed twice with 2.5 ml of borate or phosphate buffer to restore pH to neutral before antibody treatment.

i. Suspend cells in 100 μl of diluted anti-BrdUrd antibody for 30 min at room temperature. Antibody should be diluted in antibody buffer to 0.2–0.5 μg/ml. We have observed that staining is better at 25°C than at 4°C for the short incubation time.

j. Add 5 ml of wash buffer, and centrifuge at 500*g* for 2 min. Drain well, and vortex pellet.

k. Add 100 μl of diluted second antibody (goat anti-mouse IgG FITC conjugate, typically diluted 1:50 to 1:500 in antibody buffer) for 20 min at room temperature. This step may be omitted if a direct conjugate anti-BrdUrd is used.

l. Add 5 ml of wash buffer, centrifuge, drain well, and vortex pellet.

m. Suspend cells in 1.5 ml of PI working solution.

n. Filter by running sample through a 40-μm nylon mesh and analyze with a flow cytometer (optical filters for flow cytometer are described in Section VII, below).

2. Restriction Enzyme/Exonuclease III Method (Dolbeare and Gray, 1988)

This method avoids the use of heat for denaturation, thus preserves many other antigens, and minimizes cell loss.

a. Following treatment with RNase, wash the cells and incubate for 30 min in 1 ml 1% paraformaldehyde at room temperature.

b. Wash cells with 2 ml of PBS and incubate cells 1 ml with cold 0.1 *M* citric acid containing 0.5% Triton X-100 for 10 min.

c. Wash cells with 2.5 ml 0.1 *M* Tris–HCl, pH 7.5, and incubate in 100 μl of 0.1 *M* Tris–HCl containing 50 m*M* NaCl and 10 m*M* MgCl$_2$ and 10 units of *Eco*RI for 30 min at 37°C.

d. Wash cells with 1 ml 0.1 *M* Tris–HCl, pH 7.5, and resuspend in 100 μl of exonuclease III buffer containing 30 units of exonuclease III.

e. Wash cells with 2 ml PBS and continue with thermal protocol at step i (incubation with anti-BrdUrd antibodies).

3. Double Pyrimidine Label Method (Dolbeare *et al.*, 1988; Shibui *et al.*, 1989; Bakker *et al.*, 1991)

This method permits a combination of pulse label, e.g., with BrdUrd, and extended labeling with, e.g., IdUrd, making possible calculation of growth fraction in addition to cell kinetic parameters (phase duration, labeling index, and cell-cycle time). This method depends on halogen-selective monoclonal antibodies. A suitable pair of monoclonal antibodies are Br-3, which binds preferentially to BrdUrd over IdUrd, and IU-4, which does not show halogen selectivity.

a. Following the denaturation step, incubate cells in the presence of 100 μl Br-3 for 20 min at 25°C.

b. Without removing the Br-3, add 100 μl of the second antibody (with affinity IdUrd > BrdUrd, e.g., IU-4) at low concentration, e.g., 1:5000 dilution in antibody diluting buffer. Direct conjugate antibodies simplify the technique since both Br-3 and IU-4 are mouse monoclonal IgG1 subtypes. Simultaneous double indirect antibody staining can be done if one of the antibodies is a mouse monoclonal and the second is a rat monoclonal, a rabbit polyclonal antibody, or a mouse monoclonal of a different IgG subtype.

c. For single-laser flow cytometric analysis of the double-label, FITC and phycoerythrin antibody conjugates can be used. 7-Aminoactinomycin D can also be used as the DNA stain (Bakker *et al.*, 1991).

4. Combined BrdUrd/DNA/Cell–Surface Antigen (Method of Carayon and Bord, 1992)

This method permits detection of cell-surface antigens along with proliferation status, allowing determination of proliferation in each population in a mixture of cells.

a. Wash cells with 2 ml of PBS following BrdUrd incorporation.

b. Resuspend cells in 1 ml PBS containing phycoerythrin-labeled antibody to cell-surface antigen (anti-CD4) and incubate for 30 min at 4°C.

c. Wash cells once with 2 ml PBS and resuspend in 1 ml 1% paraformaldehyde/0.01% Tween 20 overnight at 4°C. This fixes the first antibody onto its antigen. Centrifuge, aspirate supernatant, and loosen pellet.

d. Wash cells in 2 ml PBS to remove paraformaldehyde. Incubate in PBS containing Mg^{2+} and Ca^{2+} and 50 Kunitz units DNase I, 30 min at 37°C.

e. Wash and resuspend cells in 150 μl of PBS containing 10% bovine serum albumin and 0.5% Tween 20 and 20 μl FITC-coupled anti-BrdUrd for 45 min at room temperature.

f. Wash cells with 2 ml PBS and resuspend in 1 ml PBS containing propidium iodide (10 μg/ml) and analyze on a flow cytometer using three-color fluorescence detection.

5. Washless Technique (Larsen, 1990, and also Chapter 24 of this volume)

This technique permits histochemical staining of cells without centrifugation. Reagents are added in a stepwise manner so that cell loss is minimized.

B. Immunochemical Labeling Followed by Microscopic Detection

1. Fluorescent Staining Method

a. Cells centrifuged on slides by cytospin techniques or grown on slides may be stained using the above protocols with times being reduced 50% for washing, denaturation, and incubation with antibodies.

b. After fluorescent or colorimetric staining with anti-BrdUrd, slides may be washed with PBS and then distilled water. Fluorescence or light microscopy may be used directly for determining the labeling index. Labeling index is defined here as the number of BrdUrd-labeled cells/total cells observed in the field. For bright field microscopy after color staining, counterstain cells with 100 μl 0.05% Giemsa in 0.05 M phosphate buffer (pH 6.5).

2. Histochemical Staining of Tissue Sections (Shibuya *et al.*, 1993)

This method uses a combination of colloidal gold and alkaline phosphatase enzyme immunohistochemical methods to detect both BrdUrd and IdUrd incorporated into DNA in histologic sections. Nuclei are counterstained with hematoxylin, and sections are examined by microscopy.

a. Sections are deparaffinized by immersion in xylene for 5 min followed by immersion 2 min each in 100, 70, and 50% ethanol. DNA is denatured by immersion of slides in 4 N HCI for 10 min at room temperature.

b. Incubate for 30 min with 100 μl Br-3 diluted 1:20,000 in PBS with 5% normal goat serum as a blocking protein.

c. Rinse slides with 5 ml PBS.

d. Incubate slides for 30 min with 100 μl gold-conjugated goat anti-mouse IgG diluted 1:50 in PBS.

e. Wash slides with 5 ml PBS and fix with 100 μl 2.5% glutaraldehyde for 10 min. Rinse slides with 10 ml deionized water.

f. Silver precipitate for 20 to 30 min with silver enhancement kit as described in the protocol provided by the manufacturer. Result is a black staining of the nuclei containing BrdUrd.

g. Immerse slides in 5% acetic acid for at least 1 hr. Shibuya *et al.* (1993) note that slides may be left overnight in this solution. Rinse slides with TBS following incubation.

h. Incubate 30 min with 100 μl IU-4 dilutes 1:800 in TBS with 1% normal rabbit serum as a blocking protein. Rinse slides with 5 ml TBS following incubation.

i. Incubate 30 min with 100 μl alkaline phosphatase-conjugated rabbit anti-mouse immunoglobulins diluted 1:50 in TBS. Rinse slides with 5 ml TBS following incubation.

j. Incubate with 100 μl suitable alkaline phosphatase substrate, e.g., 0.2 mM naphthol ASMX phospate plus 0.2 mM Fast red TR salt in 0.05 M Tris buffer, pH 8.2, containing 5 mM Mg^{2+}. Wash with 5 ml distilled water. Result is a red staining of the nuclei containing IdUrd. BrdUrd containing nuclei also stain, but this is marginal against their previous black staining by Br-3. Counterstain nuclei with 100 μl of 5% Gills No. 1 hematoxylin, giving a light blue color to nuclei not containing either BrdUrd or IdUrd. Wash with 5 ml distilled water.

k. Air dry the slides, clear with HistoClear, and mount glass coverslip with Permount.

C. Cell Kinetics Measurements

1. Generally the window technique is used for FCM analysis (Dean *et al.*, 1984; Pallavicini *et al.*, 1985). Commercial software for these computations is available from Phoenix Flow Systems, Verity Software House, Cell Pro, and others. An arbitrary electronic window can be selected for determining events in a particular part of the bivariate BrdUrd/DNA histogram. The number of events appearing in that window as a function of time after the BrdUrd pulse can be used to derive cell-cycle parameters, e.g., G_1, S, and Tc. If cells are continuously pulsed with BrdUrd for a period equal to or greater than Tc, it is possible to evaluate the growth fraction of the tumor.

2. The alternate method proposed by Begg *et al.* (1985) permits determination of Tpot with a single sample several times following a BrdUrd pulse (see Fig. 2, the bivariate distribution at 3 hr). A window may be drawn around the S phase portion and a value for relative movement, Rm, can be determined. The relationship defined by Begg *et al.* (1985) shows that Rm could be derived with the expression

$$Rm = F_L - F_{G_1}/F_{GM} - F_{G_1},$$

where F_L is the mean red fluorescence of the green-labeled cells and F_{G_1} and F_{GM} are the mean red fluorescence of the G_1 and $G_{2+}M$ cells, respectively. Rigorous mathematical derivation of this expression was published by White and Meistrich (1986) and subsequently modified by Terry *et al.* (1991).

3. Using double halopyrimidine labels makes it possible to determine Ts and Tpot with a single biopsy sample (see Shibui *et al.*, 1989; Raza *et al.*, 1991a). One of the analogues (e.g., IdUrd) is given several hours before surgery, the second analogue given just prior to surgery. Either thin-section histochemistry or flow cytometry can be used to derive quantitative terms for Ts and Tpot.

V. Critical Aspects of the Procedure

A. DNA Denaturation

All antibodies to halogenated nucleosides suitable for cell kinetics studies reported to date recognize the incorporated analogue only in the context of single-stranded DNA. Thus denaturation of the DNA is required before antibody staining. Extensive denaturation is neither required nor desirable, however, since denaturation leads to a loss of the ability to stain with DNA intercalating dyes, causes unacceptably broad coefficients of variation in DNA staining, and increases cell loss. As such, there must be a balance between the need to denature the DNA for antibody recognition of the BrdUrd and the need to retain intact double-stranded DNA for propidium iodide binding. Limited denaturation by either heating or with HCl or partial DNA digestion by nucleases is the preferred method. Carefully controlled denaturation may also allow resolution of tetraploid G_2 and M cells into distinct G_2 and M peaks in the bivariate histogram by exploiting the differences in denaturation of DNA in mitotic and G_2-phase cells (Nüsse et al., 1989).

B. Paraformaldehyde Fixation

Paraformaldehyde fixation will lower the sensitivity of the staining reaction probably by reducing the denaturability of the DNA and crosslinking of chromatin proteins. Even with 1% paraformaldehyde fixation, however, quantification of a 10 nM BrdUrd 30-min pulse is attainable. The paraformaldehyde treatment also helps to prevent cell loss, especially of lymphoid cells, during the staining procedure. We have found also that the fixation step improves the quality of the DNA histogram with lower CVs for the G_1 peak.

C. Nonspecific Fluorescence

Nonspecific fluorescence is due primarily to nonspecifically bound antibody (anti-BrdUrd or fluorescein-conjugated second antibody). Additional washes after antibody treatment or incorporation of 1–5% blocking protein in the wash buffer can reduce nonspecific binding of antibody. Since most nonspecific antibody binding localizes in the cytoplasm and on the cytoplasmic membrane, using nuclei rather than whole cells can also greatly reduce nonspecific fluorescence (Landberg and Roos, 1992).

D. Cell Loss

Cell loss generally results from clumping and cell adherence to the walls of the test tube being used to process the cells. Lymphoid cell loss is generally greater from this procedure but may be reduced greatly by using lower centrifuge speeds during the pelleting of cells. Limit centrifugation to between 400 and 800g for no more than 4 min. Clumping may depend also on the nature of the fixative used. Frequently large clumps appear after the thermal denaturation step. Clumps may be disaggregated by a combination of mild vortexing and syringing the suspension gently through a size 25 needle. Cell adherence to the centrifuge tube can be decreased by either siliconizing the tubes or by using microfuge tubes to reduce tube surface area.

E. Double-Labeling with BrdUrd and IdUrd

When staining cells that have double halopyrimidine labels, e.g., with BrdUrd pulse and continuous IdUrd label, one should be concerned with the differences in specificities and affinities of the specific antibodies. If very specific antibodies are used, there may not be a problem. As an example of halogen specificity, IU-4 has affinity for the halopyrimidines in the following order IdUrd > BrdUrd > CldUrd, whereas Br-3 has the following specificity: BrdUrd = CldUrd ≫ IdUrd. If some crossreactivity occurs with one of the antibodies (e.g., IU-4 will react with both IdUrd and at lower affinity with BrdUrd), then add that antibody at a much lower concentration after the specific antibody Br-3 has incubated with the cells for 20 to 30 min. Then continue the incubation for an additional 30 min. In this way the Br-3 will saturate BrdUrd sites but not react with the IdUrd sites. Adding the IU-4 then will preferentially bind only to the exposed IdUrd sites. Using a lower concentration of this antibody will prevent displacement of the Br-3 from the BrdUrd sites.

F. Optical Filters for FCM

Using correct optical filters during the flow cytometric analysis will prevent the crosstalk between the photomultipliers detecting the fluorescein and propidium iodide signals. Propidium iodide exhibits a broad band of fluorescence ranging from 530 to 700 nm. If a 550-nm short-pass filter coupled with a 500-nm long-pass filter is used for the green fluorescence, then some PI fluorescence will be observed in the green fluorescence channels causing a skewing of the BrdUrd histogram. While commercial instruments have capabilities for compensating for some excess crosstalk, they cannot correct for a large excess of one or the other fluorophore in the sample. As such, the absolute fluorescence intensity of the separate staining dyes should be approximately equal.

VI. Controls and Standards

A. Negative Controls

To some extent, the presence of G_1 cells in the cell population serve as a negative control cell population. G_1 cells should have minimal green fluorescence from antibodies to BrdUrd, since they do not contain BrdUrd. Elevated G_1 staining indicates nonspecific antibody binding or autofluorescence. The latter may result from inappropriate fixation techniques, particularly those using aldehyde fixatives. The ratio of green fluorescence intensity of the mid S phase to G_1 cells should be at least 10 and under ideal staining conditions may be several hundred. If the protocol used continuous BrdUrd labeling so unlabeled cells are not part of the sample, then a specific negative control sample where BrdUrd has been omitted should be used.

B. Crosstalk

To determine whether crosstalk is occurring between fluorescence channels (i.e., whether PI fluorescence is being detected by the fluorescein detector) add PI but no antibodies after the thermal denaturation step. If the green fluorescence is above background then crosstalk is present. Correct this by either reducing the concentration of PI used or by selecting more appropriate filters. We recommend a final PI concentration of about 10 μg/ml.

VII. Instruments

A. Flow Cytometer

Any flow cytometer equipped with a single argon ion laser with two photomultiplier tubes is adequate. The instrument should also be equipped with a log amplifier to accommodate the large range of fluorescence signal generated by the anti-BrdUrd fluorescence. Both fluorescein and PI can be excited at 488 nm. Use a 514-nm band pass filter for the fluorescein fluorescence (BrdUrd content) and a 600-nm long-pass filter for the PI fluorescence (DNA content). If AMCA is used as the fluorophore for an antibody, then use an excitation of 363 nm and a 450-nm band pass filter to collect fluorescence. The incorporation of a doublet eliminator will prevent accumulation of doublet G_1 and early S phase cells in windows that should contain either G_2 or late S phase cells.

B. Microscope

A bright field or fluorescent microscope can be used to determine labeling indices of cells stained on slides.

VIII. Results

Figure 1A shows a bivariate DNA/BrdUrd contour histogram generated by CHO cells stained according to the above protocol. This is the kind of histogram generated after a 30-min pulse of 1 mM BrdUrd. G_1 and G_2 + M populations should have only background green fluorescence. S phase cells have green fluorescence and produce the horseshoe-shaped pattern with mid S phase cells having a 64-fold higher fluorescence than G_1 cells. The fraction of cells with S

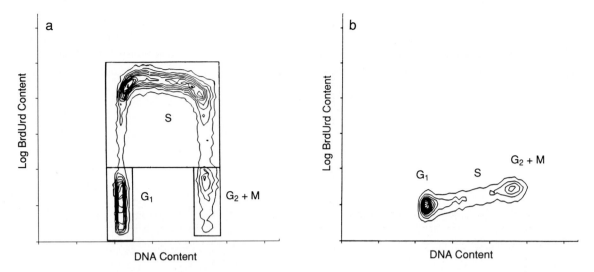

Fig. 1 (a) Bivariate histogram showing DNA content (*x*-axis = PI fluorescence) and BrdUrd content (*y*-axis = log fluorescein fluorescence) following a 30-min pulse of 1 m*M* BrdUrd. (b) Bivariate histogram showing DNA content and BrdUrd content of CHO not receiving BrdUrd pulse.

phase fluorescence divided by the total population will give the labeling index. Cells that have not been incubated with BrdUrd but stained by the same protocol will exhibit only background fluorescence (Fig. 1B).

The cell kinetic applicability of the method is demonstrated by Fig. 2 where a single BrdUrd pulse was given at $t = 0$ followed by a thymidine pulse chase after 30 min. Samples were taken at 30 min (0 hr) and at $t = 3,6,9,$ and 12 hr, fixed in 50% ethanol, and stained by the thermal denaturation protocol. S phase cells that incorporated BrdUrd show the typical green fluorescence after 30 min. This same cohort of S phase cells progress through the cell cycle with the fluorescence appearing in $G_2 + M$ and daughter G_1 cells at 3 and 6 hr and progressing further into G_1 and back into S phase at 9 hr with most of the label reappearing in S phase at approximately one cell-cycle time, i.e., 12 hr after the initial pulse of BrdUrd.

The window method for determining fraction of cells in each compartment is shown in Fig. 3. A BrdUrd/DNA bivariate histogram was derived from an *in vivo* pulse label of BrdUrd in the brown Norway rat injected 17 days previously with the myeloid leukemic tumor cells (Kuo *et al.*, 1993). The windows in 3a are drawn so that events in G_1 or mid S can be evaluated. When these counts are collected as a function of time after the BrdUrd pulse, then the kinetic parameters, T_{G_1}, Ts, and Tc can be determined. The S phase window in Fig. 3b permits the calculation of LI, i.e., the fraction of cells in S phase. Figure 4a shows the quantitative analysis of the cytokinetic properties of asyn-

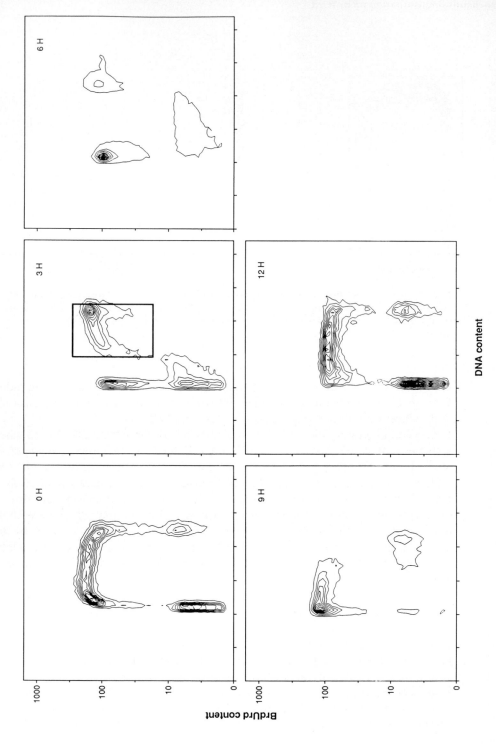

Fig. 2 A series of bivariate DNA/BrdUrd distributions taken after a 30-min pulse of 10 μM BrdUrd followed by a thymidine chase and subsequent sampling at 3-hr intervals.

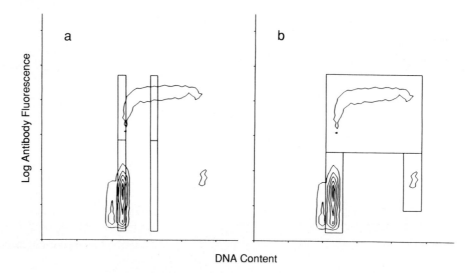

Fig. 3 Graphical presentation of windows set up to calculate the fraction of labeled cells in G_1 phase (left boxes in a) or mid S phase (right boxes in a) and the labeling index (b). Fraction of labeled cells in G_1 phase = cells in labeled G_1 box(upper G_1 window)/total cells in G_1 box (both labeled and unlabeled G_1 cells). Fraction of labeled cells in mid S phase = cells in labeled mid S box/cells in S box including both labeled and unlabeled mid S box. Labeling index = cells in labeled, high FITC fluorescence box/total cells in distribution. From Kuo *et al.* (1993).

chronous BNML spleen cells. Fractions of labeled cells in mid S phase (a) and in G_1 phase (b) are derived from pulse-chase bivariate distributions. Ts is the time for cells to go from one maximum or minimum in label to a second, e.g., the Ts is the time for the cells to go from the first arrow in Fig. 4a to the second, i.e., from the first minimum S phase label to the second. Growth fractions were obtained from bivariate distributions after continuous labeling. The labeling index increases with time until the entire growing fraction is labeled. Thus, further continuous labeling results in very little change in the fraction of labeled cells. An extrapolation of this limit (Fig. 4b) leads to the derivation of the growth fraction (0.76). Solid lines represent the best-fitted mathematical model of the experimental data points (●) derived from a computer analysis of the three sets of data. Table II is a summary of the kinetic data for BNML cells derived from the curves in Fig. 4.

IX. Summary

We have described several laboratory procedures for the immunochemical staining of the halopyrimidines, BrdUrd and IdUrd, in cell suspensions for flow cytometry and a method for staining histological sections on slides. Halogenated

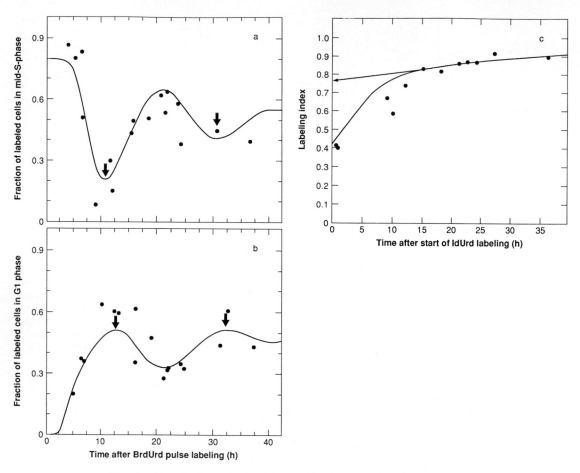

Fig. 4 Quantitative analysis of the cytokinetic properties of asynchronous BNML spleen cells. Fractions of labeled cells in mid S phase (a) and in G_1 phase (b) are derived from pulse-chase bivariate distributions. Growth fractions were derived from bivariate distributions after continuous labeling with boxes described in Fig. 3c. Solid lines represent the best fitted mathematical model of the experimental data points (●) derived from a computer analyses of the three sets of data. From Kuo *et al.* (1993).

pyrimidine quantitation allows cell-cycle parameters, including total cell-cycle time, phase durations, and growth fraction to be determined. We have presented some flow cytometric data to demonstrate the use of these methods in determining bivariate BrdUrd/DNA histograms with CHO cells and in kinetic studies with the brown Norway rat myeloid leukemia model.

Table II
BNML Cytokinetic Parameters[a]

Tg1	5.4 hr
Ts	12.5 hr
Tg2m	2.6 hr
Tc	20.5 hr
GF	0.76

[a] These parameters are derived from the curve fitting, quantitative analyses shown in Fig. 3. Tg1, Ts, Tg2m and Tc represent the time for G_1, S, G_2M phases and cell-cycling time, respectively. GF, growth fraction.

Acknowledgment

Part of this work was performed under the auspices of the U.S. Department of Energy at Lawrence Livermore National Laboratory under contract No.W-7405-Eng-48.

References

Allison, L., Arndt-Jovin, D. J., Gratzner, H., Ternynck, T., and Robert-Nicoud, M. (1985). *Cytometry* **6**, 584–590.

Bakker, P. J. M., Tukker, C. J., van Ofen, C. H., and Aten, J. (1991). *Cytometry* **12**, 366–372.

Begg, A. C., McNally, N. J., Shrieve, D. C., and Karcher, H. A. (1985). *Cytometry* **6**, 620–626.

Beisker, W., and Hittleman, W. (1988). *Exp. Cell Res.* **174**, 156–161.

Beisker, W., Dolbeare, F., and Gray, J. W. (1987). *Cytometry* **8**, 235–239.

Carayon, P., and Bord, A. (1992). *J. Immunol. Methods* **147**, 225.

Dean, P. N., Dolbeare, F., Gratzner, H., Rice, G. C., and Gray, J. W. (1984). *Cell Tissue Kinet.* **17**, 427–436.

Dolbeare, F. (1993). *Histochem. J.* (in press).

Dolbeare, F., and Gray, J. W. (1988). *Cytometry* **9**, 631.

Dolbeare, F., Gratzner, H., Pallavicini, M., and Gray, J. W. (1983). *Proc. Natl. Acad. Sci. U.S.A.* **80**, 5573–5577.

Dolbeare, F., Kuo, W. L., Vanderlaan, M., and Gray, J. W. (1988). *Proc. Am Assoc. Cancer Res.* **29**, 1896.

Dolbeare, F., Kuo, W. L., Beisker, W., Vanderlaan, M., and Gray, J. W. (1990). *In* "Methods in Cell Biology" (Z. Darzynkiewicz and H. Crissman, eds.), Vol. 33. Academic Press, San Diego.

Duprez, A., Barat, J. L., Girard, A., Hoffmann, M., and Hepner, H. (1990). *Neurochirurgie* **36**, 157–166.

Frommer, M., Paul, C., and Vincent, P. C. (1988). *Chromosoma* **97**, 11–18.

Garin, L., Barona, R., O'Connor, E., Armengot, M., and Basterra, J. (1991). *An. Otorrinolaringol. Ibero* **18**, 567–574.

Giaretti, W. (1991). *Tumori* **77**, 403–419.

Gratzner, H. G. (1982). *Science* **218**, 474.

Gratzner, H. G., Ahmad, P. M., Stein, J., and Ahmad, F. (1985). *Cytometry* **6**, 563–569.

Gray, J. W., Dolbeare, F., and Pallavicini, M. (1990). *In* "Flow Cytometry and Sorting" (M. R. Melamed, T. Lindmo, and M. L. Mendelsohn, eds.), 2nd ed., 93–137. Wiley-Liss, New York.

Hardonk, M. J., and Harms, G. (1990). *Acta Histochem., Suppl.* **39**, 99–108.

Ito, H., Kamiryo, T., Kajiwara, K., Nishizaki, T., and Oshita, N. (1990). *No Shinkei Geka* **18,** 595–599.

Kaufmann, S. J., and Robert-Nicoud, M. (1985). *Cytometry* **6,** 570–577.

Kuo, W. L., Dolbeare, F., Vanderlaan, M., and Gray, J. W. (1993). *In* "DNA Cytometric Analysis" (A. Sampedro, ed.) 219–238. Oviedo Univ. Press.

Lacombe, F., Belloc, F., Dumain, P., Puntous, M., Lopez, F., Bernard, P., Boisseau, M. R., and Reiffers, J. (1992). *Cytometry* **13,** 730–738.

Landberg, G., and Roos, G. (1992). *Cytometry* **13,** 230.

Larsen, J. (1990). In "Methods in Cell Biology" (Z. Darzynkiewicz and H. Crissman, eds.), Vol. 33, p. 227. Academic Press, San Diego.

Leadon, S. A. (1986). *Nucleic Acids Res.* **14,** 8979–8995.

Nakamura, H., Morita, T., and Sato, C. (1986). *Exp. Cell Res.* **165,** 291–297.

Nüsse, M., Julch, M., Geido, E., Bruno, S., DiVinci, A., Giaretti, W., and Russo, K. (1989). *Cytometry* **10,** 312–319.

Pallavicini, M. G., Summers, L. J., Dolbeare, F., and Gray, J. W. (1985). *Cytometry* **6,** 602–610.

Pinkel, D., Thompson, L., Gray, J., and Vanderlaan, M. (1985). *Cancer Res.* **45,** 5795–5798.

Raza, A. G., and Preisler, H. D. (1990). *CRC, Crit. Rev. Oncol.* **1,** 373–378.

Raza, A. G., Miller, M., Mazewski, C., Sheikh, Y., Lampkin, B., Sawaya, R., Crone, K., Berger, T., Reisling, J., Gray, J., Khan, S., and Preisler, H. D. (1991a). *Cell Proliferation* **24,** 113–126.

Raza, A. G., Bokhari, J., Yousuf, N., Medhi, A., Mazewski, C., Khan, S., Baker, V., and Lampkin, B. (1991b). *Arch. Pathol. Lab. Med.* **115,** 873–879.

Raza, A. G., Yousuf, N., Bokhari, A., Masterson, M., Lampkin, B., Yanik, G., Mazewski, G., Khan, S., and Preisler, H. D. (1992). *Cancer* (*Philadelphia*) **69,** Suppl., 1557–1566.

Riccardi, A., Donavo, M., and Ascari, E. (1988). *Haematologica* **73,** 423–430.

San-Galli, F., Maire, J. P., and Guerin, J. (1991). *Neurochirurgie* **37,** 3–11.

Selden, J., Dolbeare, F., Clair, J., Nichols, W., Miller, J., Kleemeyer, K., Hyland, R., and DeLuca, J. (1993). *Cytometry* **14,** 154–167.

Shibui, S., Hoshino, T., Vanderlaan, M., and Gray, J. (1989). *J. Histochem. Cytochem.* **37,** 1007–1011.

Shibuya, M., Ito, S., Davis, R. L., and Hoshino, T. (1993). *Biotech. Histochem.* **67** (in press).

Suzuki, H. (1988). *Jpn. J. Surg.* **18,** 483–486.

Takahashi, M. (1966). *J. Theor. Biol.* **13,** 195–202.

Terry, N. H., White, R. A., Meistrich, M. L., and Calkins, D. P. (1991). *Cytometry* **12,** 234–241.

Vanderlaan, M., Watkins, B., Thomas, C., Dolbeare, F., and Stanker, L. (1986). *Cytometry* **7,** 499.

Waldman, F., Dolbeare, F., and Gray, J. W. (1988). *Cytometry, Suppl.* **3,** 65–72.

Ward, J. M., Wedghorst, C. M., Diwan, B. A., Konishi, N., Lubet, R. A., Henneman, J. R., and Devor, D. E. (1991). *Prog. Clin. Biol. Res.* **369,** 369–388.

White, R., and Meistrich, M. (1986). *Cytometry* **7,** 486–492.

Yu, C. C., Woods, A. L., and Levinson, D. A. (1992). *Histochem. J.* **24,** 121–131.

CHAPTER 20

Application and Detection of IdUrd and CldUrd as Two Independent Cell-Cycle Markers

Jacob A. Aten,★ Jan Stap,★ Ron Hoebe,★ and Piet J. M. Bakker†

★ Laboratory for Radiobiology and † Division of Medical Oncology
University of Amsterdam
1105 AZ Amsterdam
The Netherlands

I. Introduction

The introduction in 1982 of immunological reagents for the detection of BrdUrd (Gratzner, 1982; Dolbeare *et al.*, 1983) has initiated the recent impressive advance in the field of cell-cycle analysis. The BrdUrd method is highly suitable for application in both cell biological studies (Nakamura *et al.,* 1986) and clinical studies (Begg *et al.,* 1985; Wilson *et al.,* 1988) as it is virtually nontoxic as well as sensitive, fast, and easy to perform.

However, for a detailed analysis of a large range of cell kinetics processes, in particular those involving time-dependent phenomena occurring during changes in the dynamics of cell proliferation or during DNA synthesis activity at the subcellular level, the BrdUrd method is not sufficiently versatile. Studies of that type can only be carried out effectively by using two or more independent DNA replication labels. One way of approaching this problem is by application of tritiated thymidine in combination with BrdUrd, but that requires the reintroduction of autoradiography techniques. An additional disadvantage of this approach, in particular for studies at the subcellular level, is the limited spatial resolution of autoradiography. These handicaps can be avoided by the introduction of a double-labeling procedure based on the application of a combination of two nonradioactive DNA precursors that are incorporated by cells *in vivo* (Shibui *et al.,* 1989).

To establish a reliable DNA double-labeling and staining procedure using two replication markers, we tested different combinations of halogenated deoxyuridines. After a series of experiments we were able to select a pair of antibodies that very effectively distinguishes IdUrd from CldUrd and vice versa. With this set of antibodies the IdUrd and CldUrd labels can be detected with little crosstalk. The methods for DNA double-labeling and for staining of nuclei are an extension of the methods for the immunocytochemical detection of BrdUrd described by Dolbeare and Selden in Chapter 19 of this volume. The two halogenated nucleotides are detected using two monoclonal antibodies in combination with Texas red-labeled and fluorescein-conjugated polyclonal antibodies (Bakker *et al.,* 1991; Aten *et al.,* 1992). DAPI was used for staining of the nuclear DNA. As an alternative phycoerythrin and fluorescein can be used for staining IdUrd and CldUrd, in combination with propidium iodide as DNA stain (Pollack *et al.,* 1993).

II. Application

Using the procedure reported here, all four classes of cells from double-labeled populations can be distinguished: (1) cells labeled with IdUrd only, (2) cells labeled with CldUrd only, (3) cells with both labels, and (4) unlabeled cells. This double-labeling method is highly suitable for flow cytometry studies in tumor and cell biology. It is required for the evaluation of complex cell-cycle parameters such as the fraction of noncycling cells or the rate of recruitment of resting cells into the cell cycle (Aten *et al.,* 1992; Bakker *et al.,* 1993). In combination with fluorescence microscopy techniques the spatial and temporal development of replication patterns in interphase nuclei and in metaphase chromosomes can be analyzed in detail (Manders *et al.*, 1992). The staining method can be applied to cells in suspension as well as to cells and chromosomes fixed on slides, although the details given here mainly refer to application in flow cytometry. Double-labeling experiments with IdUrd and CldUrd have been performed with cultured cells, lymphocytes, and solid tumours.

III. Materials

A. Antibodies

1. Mouse anti-BrdUrd (Cat. No. 7580, Becton–Dickinson, Mountain View, CA), is used to detect IdUrd.

2. Fluorescein-conjugated mouse anti-BrdUrd (Cat. No. 7583, Becton–Dickinson) may be used as an alternative.

3. Texas red-conjugated goat anti-mouse IgG (Cat. No. 115-075-100, Jackson, West Grove, PA) is used as a second step antibody in the staining procedure, in combination with the unconjugated Becton–Dickinson antibody.

4. Rat anti-BrdUrd, MAS 250C Clone Bu/75 (obtained from Sera-lab, Crawley Down, Sussex, UK) is used to detect CldUrd.

5. Fluorescein-conjugated goat anti-rat IgG (Cat. No. 112-015-102, Jackson) is used as a second step antibody in the staining procedure. It is used in combination with the Sera-lab antibody.

6. Phycoerythrin-conjugated goat anti-rat IgG (Cat. No. R400 4-1, Caltag, S. San Francisco, CA) may be used as an alternative.

B. Blocking Agents

1. Bovine serum albumin (BSA) (Cat. No. 42904, Organon Technika, Boxtel, the Netherlands) is diluted to a working solution of 10% in phosphate-buffered saline (PBS). It is used to decrease crossreaction of the first step monoclonal antibodies.

2. Normal goat serum, cat. no. X907, Dako A/S, Glostrup, Denmark. The serum is used at a concentration of 1 mg/ml in PBS to decrease cross-reaction of the second step antibodies.

C. Enzymes

Pepsin (Cat. No. 108057, Boehringer-Mannheim GmbH, Mannheim FRG): This enzyme is used at a concentration of 0.4 mg/ml in 0.1 M HCl to improve accessibility of the incorporated IdUrd and CldUrd for the antibodies.

D. Buffers

1. PBS: (8.75 g NaCl + 2.86 g Na_2HPO_4 * 12 H_2O + 0.21 g KH_2PO_4)/l, at pH 7.4. PBS is used at various stages in the protocol.

2. PBT consists of PBS supplemented with 0.05% Tween 20 (Cat. No. P1379, Sigma, St. Louis, MO). This buffer is also used at several stages in the protocol.

3. 20× SSC: (175.3 g NaCl + 100.3 g sodium citrate)/liter. A total of 2× or 4× SSC may be used for diluting the monoclonal antibodies to decrease background and/or cross-reactivity.

4. Tris "high salt buffer": (29.22 g NaCl + 4.44 g Tris–HCl + 2.65 g Tris base)/liter + 0.5% Tween 20, at pH 8.0. This is a washing buffer used after incubating the cells with the Becton–Dickinson anti-BrdUrd antibody.

5. Borax buffer: 0.1 M sodium tetra borate, at pH 8.5. This buffer is used to neutralize the HCl after DNA denaturation.

E. DNA Fluorochromes

1. DAPI (Cat. No. D1388, Sigma) is used to stain the nuclear DNA, in combination with Texas red and FITC for immunofluorescence staining of IdUrd and CldUrd.

2. Propidium iodide (Cat. No. P4170, Sigma) is used to stain the nuclear DNA, in combination with phycoerythrin and FITC for immunofluorescence staining of IdUrd and CldUrd.

IV. Methods

A. Cell Labeling with Halogenated Deoxyuridines

1. Incubate cells in culture for a period of 2 to 30 min with 10 μM IdUrd.

2. Remove the medium and wash the cells with prewarmed medium with normal pH. To secure a more complete depletion of the first label, cells may be washed more than once. During the first washing step, thymidine may be added to a concentration of 100 μM.

3. After culturing in normal medium for a selected time interval, pulse label the cells with 10 μM CldUrd, also for a period of 2 to 30 min. The order in which the labels IdUrd and CldUrd are applied can be reversed.

4. Double-labeling of rodent tumours: Host animals are injected intraperitoneally with the nucleosides (50 mg/kg body wt for mice and rats).

B. Cell Harvesting and Fixation and DNA Denaturation

1. After the labeling procedure, harvest the cells by trypsinization, centrifuge, resuspend in 2 ml PBS, and fix the cells by adding 6 ml of cold ethanol (final concentration ± 70%). Fixed cells can be stored at 4°C for up to 3 weeks.

2. The staining procedure can be performed with smaller numbers of cells, but it is convenient to use a sample of at least 1 × 10⁶ ethanol-fixed cells.

Centrifuge the suspension, aspirate and resuspend in the remaining drop of liquid, and incubate in 1 ml of pepsin solution for 30 min at room temperature. Add 3 ml PBS to terminate the enzymatic treatment and centrifuge again.

3. Partial denaturation is carried out by incubating in 1 ml 2 M HCl for 30 min at 37°C. Add 3 ml of borax buffer to neutralize the HCl, centrifuge the cells, and resuspend in 1 ml PBT supplemented with BSA (1 mg/ml). The cell nuclei are now ready for the immunofluorescence staining procedure.

C. Immunofluorescent Staining of Cell Nuclei

1. Centrifuge the cells and incubate for 30 min in 100 μl PBS containing BD mouse anti-BrdUrd antibody, diluted between 1:2 and 1:4, and BSA (1 mg/ml).

2. Centrifuge and incubate for 30 min in 5 ml Tris buffer.

3. Centrifuge and wash cells in 1 ml PBT containing normal goat serum (1 mg/ml). Centrifuge again and incubate for 30 min in 100 μl Texas red-conjugated goat anti-mouse IgG (1:100 in PBS).

4. Centrifuge and wash in 1 ml PBT supplemented with 1 mg/ml BSA. Sediment again and incubate for 30 min in 100 μl PBS containing BSA and the Sera-lab anti-BrdUrd antibody diluted about 1:600.

5. Centrifuge and incubate for 30 min in 5 ml Tris buffer.

6. Centrifuge cells and wash in PBT with normal goat serum (1 mg/ml). Sediment again and incubate for 30 min in 100 μl PBS with normal goat serum and fluorescein-conjugated goat anti-rat IgB, 1:100.

7. Finally centrifuge and resuspend in 0.5 ml PBS (volume depending on the number of cells) containing DAPI (1 μg/ml).

8. Stained specimens can be stored for up to 2 weeks at -12°C after glycerol has been added to a concentration of 30%. Before flow cytometric analysis syringe the cells through a 21-gauge needle to disaggregate cell clumps.

9. Analyze in an FCM instrument with two laser beams. The FITC (= incorporated CldUrd) and the Texas red (= incorporated IdUrd) can be excited at 514 nm. Use a 545-nm band pass filter for the FITC fluorescence and a 630 long-pass filter for the Texas red fluorescence. These signals can be separated by a 590-nm dichroic mirror. For the DAPI excitation use a UV laser beam. The DAPI signal can be detected through a 450-nm band pass filter.

V. Critical Aspects of the Procedure

1. Several important critical aspects with respect to the application of mono-clonal antibodies in BrdUrd DNA analysis have been described elsewhere in this volume by Dolbeare and Selden (Chapter 19). The same remarks apply to the

procedure for IdUrd/CldUrd double-labeling described in this chapter. These remarks will not be repeated here but they certainly merit attention.

2. We have demonstrated that the presence of IdUrd and CldUrd in the cell culture medium, has, at the low concentrations used in these experiments, no effect on the cell cycle. However, an important aspect of experiments with two replication markers is that the labels, IdUrd and CldUrd, are not given simultaneously but separated by a time interval of several hours. The medium changes required in these double-labeling experiments may significantly perturb the progression of the cells through the cell cycle. Results may vary with the cell type used. It is essential that the media added to the cells have the right temperature and pH. You should check this! For sensitive cells a 1 : 1 mixture of fresh medium and conditioned medium may be used for washing the cells and for culturing them after the washing procedure. This type of cell-cycle perturbation does not occur during double-labeling experiment *in vivo*. Halogenated nucleosides administered to test animals are eliminated within 30 min by the organism itself.

3. All centrifugation was carried out at $500g$ for 1 min, to minimize cell loss. We used a Hettich Rotixa/KS-type 7250 centrifuge. Other centrifuges may give different results, depending on the time it takes the rotor to come to a standstill. Results may also vary with the cell type used.

4. Incubations involving fluorescent antibodies should be performed in the dark.

5. Despite the high specific affinity of the Sera-lab and Becton–Dickinson antibodies for CldUrd and IdUrd, respectively, some cross-reactivity and a specific binding may occur. Always wash in Tris buffer with a high salt concentration after incubating the cell nuclei with the Becton–Dickinson antibody to reduce cross-reactivity. If further reduction is necessary dilute the monoclonal antibodies in $2 \times$ SSC or $4 \times$ SSC for incubation instead of PBS. Small variations between different batches of antibodies and aging of the antibodies may require adjustments of the antibody concentrations used.

6. When using a different combination of fluorochromes, it is possible to analyze samples with one instead of two laser beams. IdUrd and CldUrd are then stained with FITC and phycoerythrin. These fluorochromes are used in combination with propidium iodide as a DNA stain. The antibodies used are the FITC-conjugated mouse anti-BrdUrd (Becton–Dickinson), the rat anti-BrdUrd (Sera-lab), and the phycoerythrin (PE)-conjugated goat anti-rat (Caltag). The laser is tuned to 488 nm. For the detection of the fluorescence signals the following filter combination can be used: a 600-nm dichroic mirror separating the PI fluorescence from the PE fluorescence; a 630-nm long-pass filter for detection of the PI signal; a 550-nm dichroic mirror separating the FITC fluorescence from the PE fluorescence; 520- and 575-nm band pass filters for detection of the FITC and PE signals, respectively. One should be aware that considerable crosstalk between these three fluorescence signals precludes the analysis of

weak labels. Moreover, this combination of fluorochromes renders the procedure for the quantitative analysis of the three-parameter flow cytometry data more difficult.

7. During the flow cytometry analysis, doublet detection should be applied to prevent contamination of the data by signals from cell aggregates.

VI. Controls and Standards

Two controls should be run.

1. The first control is a sample of cells labeled with IdUrd but without CldUrd label. The cells are treated by the normal staining protocol.

2. The second control is a sample of cells labeled with CldUrd but without IdUrd label. These cells are also treated by the normal staining protocol.

These controls allow you to evaluate the quality of the cell staining. They can be used also to assess the amount of crosstalk between the signals. The

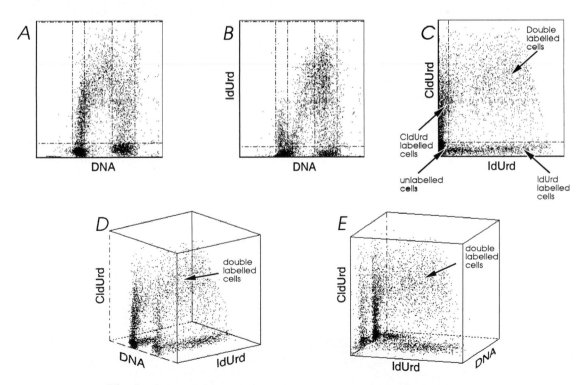

Fig. 1 Two- and three-parameter distributions of a cell population pulse labeled twice: first with IdUrd and 3 hr later with CldUrd.

observed crosstalk is a combination of staining effects, of spectral overlap between the fluorochromes, and of the spectral properties of the optical filters. This crosstalk can be partially corrected for, as described under Results.

VII. Results

The data in Fig. 1 represent a proliferating cell population pulse labeled twice; first with IdUrd and 3 hr later with CldUrd. The cells were harvested at the end of the second labeling period and stained according to the protocol described. Figure 1A shows the CldUrd signal as a function of the DNA content of the cells. This distribution resembles the horseshoe-shaped pattern well known from BrdUrd labeling experiments. Figure 1B shows the IdUrd signal as a function of the DNA content of the cells. This distribution shows a horseshoe pattern shifted toward higher DNA content values. A small fraction of the IdUrd-labeled cells has divided. These cells have migrated to a position just above the G_1 cells. The shift of the labeled cells toward higher DNA content values corresponds with the time elapsed between the labeling of the cells with IdUrd and the harvesting of the cells. Figure 1C shows the correlation between the two cell-cycle markers IdUrd and CldUrd. By adding the DNA content information, a 3D distribution is obtained. This distribution is displayed in Figs. 1D and 1E, as viewed from different angles. The lines in Figs. 1A–1C are used for sectoring the 3D distribution. A computer program, operating under Windows, is used to correct for crosstalk and to derive cell-cycle data. Values for Tpot, Ts, cell recruitment, and other parameters can be obtained in this way.

The quality of the cell staining and the amount of crosstalk between the fluorescence signals can be assessed easily from single-label controls. 3D distributions for a CldUrd-labeled sample and for an IdUrd-labeled sample, corrected for crosstalk, are displayed in the Figs. 2A and 2B.

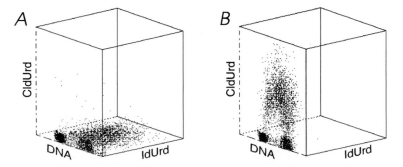

Fig. 2 Three-parameter distributions of control samples. (A) Cells labeled with IdUrd only. (B) Cells labeled with CldUrd only.

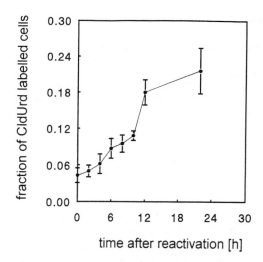

Fig. 3 Analysis of cell recruitment. The fraction of cells that is labeled IdUrd negative and CldUrd positive is displayed as a function of the time after reactivation of cell growth.

One of the applications of the double-labeling technique is the analysis of cell recruitment. Figure 3 shows results of an experiment in which resting cells were recruited into the cell cycle by a change of medium. In cultures with resting as well as proliferating V79 cells, all proliferating cells were labeled. This was done by culturing the cells in the presence of IdUrd during a period equal to the cell-cycle time. At the end of that period the medium, and the label, were washed away after which the cells were allowed to continue their growth in fresh medium. The presence of fresh medium reactivated the resting cells. Recruitment into the cell cycle was then monitored by pulse labeling the cultures with CldUrd; each culture was pulse labeled at a different time. The cells of interest were resting during the first labeling period but cycling during the second labeling period. A fraction of these resting (IdUrd-negative) cells took up the CldUrd label after their recruitment. This fraction is displayed in Fig. 3. The data indicate that the fraction of recruited cells increased as a function of the time after reactivation of cell growth.

References

Aten, J. A., Bakker, P. J. M., Stap, J., Boschman, G. A., and Veenhof, C. H. N. (1992). *Histochem. J.* **24**, 251–259.
Bakker, P. J. M., Stap, J., Tukker, C. J., van Oven, C. H., Veenhof, C. H. N., and Aten, J. A. (1991). *Cytometry* **12**, 366–372.
Bakker, P. J. M., de Vries, R. J. L., Tukker, C. J., Hoebe, R. A., and Barendsen, G. W. (1993). *Cell Prolif.* **26**, 89–100.
Begg, A. C., McNally, N. J., Schrieve, D. C., and Kärcher, G. (1985). *Cytometry* **6**, 620–626.

Dolbeare, F., Gratzner, H., Pallavicini, M. G., and Gray, J. W. (1983). *Proc. Natl. Acad. Sci. U.S.A.* **80,** 5573–5577.

Gratzner, H. (1982). *Science* **218,** 474–475.

Manders, E. M. M., Stap, J., Brakenhoff, G. J., vanDriel, R., and Aten, J. A. (1992). *J. Cell Sci.* **103,** 857–862.

Nakamura, H., Morita, T., and Sata, C. (1986). *Exp. Cell Res.* **165,** 291–297.

Pollack, A., Terry, N. H. A., Van Nguyen, T., and Meistrich, M. L. (1993). *Cytometry* **14,** 168–172.

Shibui, S., Hoshima, T., Vanderlaan, M., and Gray, J. W. (1989). *J. Histochem. Cytochem.* **37,** 1007–1011.

Wilson, G. D., McNally, N. J., and Dische, S. (1988). *Br. J. Cancer* **58,** 423–431.

CHAPTER 21

Cell-Cycle Analysis Using Continuous Bromodeoxyuridine Labeling and Hoechst 33358–Ethidium Bromide Bivariate Flow Cytometry

**Martin Poot,* Holger Hoehn,* Manfred Kubbies,†
Angelika Grossmann,‡ Yuchyau Chen,§
and Peter S. Rabinovitch§**

*Department of Human Genetics
University of Würzburg
Biocenter / Am Hubland
97074 Würzburg
Germany

†Boehringer–Mannheim Research Center
8122 Penzberg
Germany

‡Department of Comparative Medicine
§Department of Pathology
School of Medicine
University of Washington
Seattle, Washington 98195

I. Introduction

A limitation shared by current cell-cycle techniques is that continued growth beyond a single cell cycle cannot be discriminated. In conventional, univariate DNA flow cytometry (FCM) this is because the G_1 product of mitosis is indistinguishable from the G_0 cell which has never entered the S phase. The newer bivariate techniques using acridine orange (Darzynkiewicz *et al.*, 1976) or 5'-bromodeoxyuridine (BrdUrd) antibody staining (Gray and Mayal, 1985) distinguish replicating from resting cells, but so far they cannot enumerate the times a given cell has replicated within a given observation period.

Since the mid-1980s, our laboratories have designed modifications of the BrdUrd-Hoechst (BrdUrd-HO) quenching technique that was originally developed by Latt for mitotic cells (Latt, 1973) and applied to interphase cells and FCM by Latt (1977), Böhmer (1979), Nüsse (1981), Beck (1981), and others. These modifications consisted of careful titrations of BrdUrd and HO concentrations in relation to cell types and cell densities and in the definition of staining conditions that would yield highly reproducible bivariate flow cytograms of a variety of human and murine cell cultures, both primary and permanent. The unique feature of these cytograms is that they permit the tracing of dividing cells throughout multiple cell cycles. Moreover, a number of kinetic parameters that characterize replicative heterogeneity and cell-cycle traverse can be derived by analyzing temporally spaced samples of a growing cell culture.

II. Application

Applications to date have ranged over several cell types, examining questions in both cell biology and clinical medicine. The method was initially applied to human diploid fibroblast (hDF) and lymphocyte cultures (Rabinovitch, 1983; Kubbies and Rabinovitch, 1983). With hDF cultures, it was shown that their *in vitro* growth rate is regulated by both a noncycling cell fraction and by the probability by which resting cells are recruited into the cell cycle (Rabinovitch, 1983). The mitogen response of human peripheral blood lymphocytes (hPBL) was studied as a function of cell culture conditions and donor age (Kubbies *et al.*, 1985a,b); contrary to previous reports, striking differences that relate to donor age were found only between prepubertal and postpubertal donors (Schindler *et al.*, 1988). The CD4 subset of T lymphocytes appears to be the most functionally defective in older lymphocyte donors (Grossmann *et al.*, 1989). Currently, a number of highly specific cell-cycle lesions are being defined in human genetic disorders whose common denominator is a disturbance of cellular proliferation. For example, the cell kinetic lesion in Fanconi's anemia (FA) consists of accumulations of cells within the G_2 compartments of the first and second cell cycles after activation (Schindler *et al.*, 1985; Poot and Hoehn, 1993). Strongly elevated G_2 phase accumulations are also displayed by Ataxia telangiectasia (AT) cells in response to X-irradiation (Seyschab *et al.*, 1992). Under both conditions, the quantitative assessment of cell blockage in the G_2 phase compartment via the BrdUrd-HO technique has proved useful as a clinical test (Schindler *et al.*, 1987). Characteristic cell-cycle alterations have also been found in Bloom's syndrome (Poot *et al.*, 1989; Poot and Hoehn, 1993). In addition to the definition of specific cell-cycle defects, the continuous BrdUrd labeling method permits a highly sensitive *in vitro* monitoring of growth factor effects (Kubbies *et al.*, 1987), cytotoxic agents (Vogel *et al.*, 1986; Poot *et al.*, 1988a,b, 1990a,b, 1991; Seyschab *et al.*, 1993), cellular differentiation (Giaretti *et al.*, 1988; Seyschab *et al.*, 1989; Poot *et al.*, 1990c), and human metastatic melanoma cells (Weilbach et al., 1990; Poot *et al.*, 1992).

III. Materials

A. Cell Cultures

The diverse cell types that are amenable to the assay are illustrated by the following examples. Human peripheral blood mononuclear cells are obtained by Ficoll-Hypaque centrifugation from anticoagulated venous blood. Murine splenic B lymphocytes are separated through a Percoll gradient from finely minced spleens. The source of DF-like cells were skin biopsies or second trimester amniotic fluids. Mouse primary and spontaneously transformed mesenchymal cells were from explants of lung and kidney. NIH-3T3 cells, the murine lymphoid CTLL line, and human Epstein–Barr virus (EBV)-transformed cell lines were obtained from public cell banks. With the exception of PBL and murine splenic B lymphocytes (which are naturally quiescent), all other cell types need to be rendered quiescent prior to exposure to mitogens and BrdUrd. This step is only mandatory if full exit kinetic data are to be derived from an initially quiescent cell culture. In case rendering cells quiescent turns out to be difficult, it is possible to obtain usefull cell kinetic information from asynchronous cell cultures (Ormerod and Kubbies, 1992). Table I lists the commonly used cell types, their culture conditions, and the protocols used to induce proliferative quiescence. Also shown in Table I are the conditions under which growth was induced in a particular culture (i.e., plating density, types of mitogens, BrdUrd concentration). These conditions must be carefully observed. It is recommended that all cultures be grown in commercial tissue culture grade plastic ware; the cultures must be shielded from light by tight

Table I

Cell Culture and Staining Conditions Employed for Bivariate BrdUrd-HO-EB Flow Cytometric Analysis of Six Prototype Cell Types[a]

Condition	Cell type					
	hPBL	hDF	EBV-LCL	NIH 3T3	M-BCL	M-CTLL
Cell culture media	RPMI 1640, 16% FBS, 20 μM αTG	MEM, 10% FBS	RPMI 1640, 10% FBS	MEM/HAM-F10, 50:50, 10% FBS	RPMI 1640, 16% FBS	RPMI 1640, 10% FBS, 20 U/ml IL-2
Quiescence	Natural	48 hr 0.1% FBS	4 days unfed	24 hr 0.5% FBS	Natural	10% FBS, 0.2 U/ml IL-2
Plating density	10^4 cells/ml	2.5×10^3 cells/cm^2	5×10^4 cells/ml	2.5×10^3 cells/cm^2	10^4 cells/ml	10^4 cells/ml
Mitogen	PHA, CD3	10% FBS	10% FBS	10% FBS	LPS + anti-μ	IL-2
BrdUrd	100 μM	65 μM	100 μM	100 μM	100 μM	65 μM
dCyt	—	65 μM	100 μM	100 μM	—	
HO258	1.2 μg/ml	1.2 μg/ml	1.2 μg/ml	1.2 μg/ml	1.2 μg/ml	1.2 μg/ml
EB	1.5 μg/ml	2.0 μg/ml	2.0 μg/ml	2.0 μg/ml	1.5 μg/ml	1.5 μg/ml
Staining density	4×10^5 cells/ml	5×10^5 cells/ml	10×10^5 cells/ml	5×10^5 cells/ml	4×10^5 cells/ml	4×10^5 cells/ml

[a] hPBL, human peripheral blood lymphocytes; hDF, human diploid fibroblastlike cells; EBV-LCL, human Epstein–Barr Virus-transformed lymphoblastoid cell lines; NIH 3T3, murine 3T3 cells (BIH strain); M-BCL, murine splenic B lymphocytes; M-CTLL, murine interleukin 2-dependent T lymphocyte cell line; FBS, fetal bovine serum; αTG, α-thioglycerol; IL-2, interleukin 2; PHA, phytohemagglutinin; LPS, lipopolysaccharide; anti-μ, anti-B cell differentiation antigen.

aluminium foil wrappings; incubation is at 37.5°C in humidified incubators gassed with 5% CO_2 and air.

B. Cell Culture Harvest

At the desired times after exposure to mitogens in the presence of BrdUrd (usually beginning at 20 to 24 hr and ending between 72 and 96 hr) adherent cell cultures are harvested by standard trypsinization. Suspended cells are pipetted briefly, transferred to 15-ml centrifuge tubes, pelleted at 400g for 10 min, and resuspended in 10 ml culture medium containing 10% FBS and 10% DMSO. The samples are then stored at -20 to -40°C in the dark until analysis. Storage times of up to 2 years have not noticebly affected the quality of the FCM analysis.

IV. Procedures

A. Staining Solution

The staining solution consists of (final concentration)

100 mM Tris, pH 7.4
154 mM NaCl
1 mM $CaCl_2$
0.5 mM $MgCl_2$
0.1% Nonidet-P40
0.2% Bovine serum albumin (BSA)
1.2 μg/ml Hoechst 33258.

Batches of 100 ml of this solution are prepared weekly from 100$\times$ and 10$\times$ concentrated stock solutions in deionized water. This working solution is stored in dark glass bottles at 4°C and can be filtered immediately prior to use if necessary.

B. Staining Procedure

The staining procedure consists of two parts. First, the thawed cells are pelleted at 400g for 5 min, after which they are resuspended in ice-cold staining solution and incubated during 15 min at 4°C in the dark. The cell concentration during staining is very important; it may range from 4×10^5 to 8×10^5 cells/ml depending on cell type (see Table I). Second, from a 100$\times$ concentrated stock solution, ethidium bromide is added to a final concentration of 1.5 or 2.0 μg/ml, depending on cell type (see Table I). After a further 15 min in the dark at 4°C, samples are ready for analysis by flow cytometry; typical flow rates are 200 to 500 cells/sec.

C. Critical Aspects

1. A very critical aspect of the continuous BrdUrd labeling assay is the cell culture methodology. Cells must be in good condition. Otherwise, cellular debris with liberated proteases and nucleases will cause cell clumping and confusing bivariate cytogram binding. Poorly controlled cell culture conditions and use of suboptimal-quality mitogens (such as growth factors, serum, and PHA) are the most likely reasons for poor proliferation.

2. At least as critical as optimal cell culture is the avoidance of short-wavelength light from the time of first exposure to the halogenated base analogue. Careful attention to this requirement must be paid throughout the entire harvesting, freezing, and staining steps. These procedures must be carried out in the dark with minimal lighting by a red darkroom lamp. BrdUrd-substituted DNA is extremely sensitive to shortwave $\updownarrow$ length χ light, and any such exposure will lead to suboptimal dye binding, chromatin damage, and nuclear decay.

3. A number of factors bear on the degree of the desired quenching of the HO258 fluorochrome (Kubbies and Rabinovitch, 1983): (1) The density at which cells are grown and exposed to the base analogue is critical. Plating densities of 2.5×10^3 cells/cm^2 for adherent cultures and of 2×10^5 cells/ml for suspension cultures should not be exceeded (see Table I). Higher seeding densities, particularly in the case of fast growing cells, will lead to rapid depletion of BrdUrd. (2) presumably due to variable amounts of nucleotide precursors, different batches of sera may affect the efficacy of HO dye quenching. (3) The final HO and EB dye concentrations given in Table I must be carefully observed; we also recommend that fluorochromes from different commercial sources be tested side by side in order to screen out inferior dye batches. (4) As one would expect from the AT base pair affinity of the HO dye, both the AT/GC base pair ratio and the interspersion pattern will affect the quenching efficiency of BrdUrd-substituted chromatin. AT-rich genomes produce a greater quenching effect than GC-rich genomes (Kubbies and Friedl, 1985).

4. BrdUrd cytotoxicity. The continuous exposure to halogenated nucleoside analogues may entail a number of adverse effects on cellular functions (e.g., metabolic changes due to alterations in the balance of nucleotide pools, direct DNA damage, alterations in DNA–protein interaction). Directly or indirectly, these effects could impair proliferation. Our experience shows that proliferation is not measurably affected, at BrdUrd concentrations in the order of 65 to 100 μM, in two frequently used cell types: PBL and NIH 3T3 cells. Provided that they do not belong to a subline deficient in thymidine kinase, NIH 3T3 cells were found to tolerate BrdUrd concentrations up to 300 μM before they

show signs of growth inhibition. On the other hand, human lymphoblastoid cells lines and hDF are sensitive to BrdUrd. Fibroblasts experience a 10 to 15% reduction of 72-hr cell counts when exposed to 10 μM BrdUrd, but no further reduction occurs in the interval between 10 and 65 μM (Poot *et al.,* 1988a). The BrdUrd sensitivity of hDF and of EBV-transformed lymphoblastoid cells is expressed as arrest in the G_1 compartments of the second and third cell cycles after activation, but not in the G_0 to S phase transition. Using hypoxic (5% v/v) rather than atmospheric oxygen cell culture conditions, these arrests can be minimized. Incorporation of BrdUrd apparently sensitizes cultured fibroblasts toward the growth inhibitory effects of ambient and elevated oxygen (Poot *et al.,* 1988a). This may also hold for cultures of metastatic melanoma cells (Poot *et al.,* 1992).

V. Controls and Standards

Given the complex type of information yielded, and the high level of resolution required by BrdUrd-HO FCM, several standards have to be included to avoid artifacts of various types. First, regarding a possible cytotoxic effect of BrdUrd incorporation, the growth of cells under analysis has to be tested with a concentration series of the halogenated pyrimidine analogue. Analysis can best be done with conventional univariate DNA FCM using DAPI (4′,6-diamidino-2-phenylindole) as DNA stainig fluorophore concomitant with cell counting. In case accumulation of cells in the G_2 phase of the cell cycle or a severe BrdUrd-dependent reduction in cell growth is encountered, lowering the BrdUrd concentration and/or the addition of an equimolar amount of deoxycytidine (dCyt) can be considered. Second, the culture medium, the batch of FBS, and possible mitogens to be used have to be analyzed with BrdUrd-HO FCM using a cell type of known growth characteristics. Third, the staining procedure should be tested with a cell type of known AT content in its DNA, in order to assure optimal staining and FCM resolution. Fourth, the flow cytometer in use has to be carefully adjusted and optimized. In our experience chicken red blood cells (CRBC) are well suited to this purpose, but also other cell types, such as NIH 3T3 or nonstimulated hPBL can be used. Instrument adjustment can also be performed with a sample of the quiescent cells frozen immediately after initiation of a given experiment. This allows checking of the procedure for rendering cells quiescent. The cluster of cells in G_0/G_1 should be focused such that a coefficient of variation (CV) of 3–5% is achieved with respect to the HO258 and to the EB axis.

VI. Instruments

Most of the published work with BrdUrd-HO FCM has been performed on an ICP-22 epiillumination system, but also the laser-powered Ortho Cytofluorograph 50H has been used. The ICP-22 instrument (PHYWE, Göttingen, West Germany) and likewise the PAS II (Partec, Münster, Germany) uses for excitation an Osram HBO 100W/2 mercury arc lamp combined with an FT 450 dichroic mirror (Ditric Optics), a UG1, and a BG38 glass filter (Schott, Mainz, West Germany). HO258 fluorescence is collected with a FT 450 dichroic mirror (Ditric Optics) and a K45 glass filter (Schott), and fluorescence from EB is selected with a K65 glass filter (Schott). The Ortho cytofluorograph 50H and other cell sorters use 351- to 364-nm excitation from an argon ion laser, or a laser type with equivalent excitation wavelength. The filter combinations used to select the fluorescence from the two DNA dyes are similar to those employed with the ICP-22.

VII. Results

Figure 1 illustrates the response of hPBL (obtained from a 52-year-old donor) to polyclonal activation by PHA. After initiation of the culture, cell aliquots were harvested at the times indicated. The most prominent signal cluster in the 35-h panel represents G_0/G_1 cells. As activated cells enter the S phase, their EB fluorescence increase; because of the quenching of HO258 fluorescence by BrdUrd-substituted chromatin, there is a concomitant shift of replicating cells to lower fluorescence intensities on the HO axis. At 45 hr, the first cells have divided and arrive in the G_1 compartment of the second cell cycle (G_1'). During the second round of replication in the presence of BrdUrd, S phase cells increase both their EB and HO258 fluorescence, which results in a signal distribution that is a "mirror image" of the first cycle distribution. Further divisions result in signal distributions that run parallel to those of the second cycle; because of overlapping distributions, cells beyond the G_1 phase of the fourth cell cycle (G_1''') can no longer be distinguished. In the sequence shown in Fig. 1, *de novo* recruitment still occurs at 80 hr after stimulation (note the presence of first-cycle fluorescence signals). At 80 and 96 hr, signals are seen that indicate nuclear lysis or degradation in third- and fourth-cycle populations (the signals connecting the G_1' and G_1'' signal clusters to the origin of the cytogram).

Figure 2 illustrates the activation and cell-cycle progression of serum-deprived hDF-like cells by exposure to 10% serum. DNA synthesis starts between 18 and 22 hr (not shown) and reaches its maximum at 30 hr after stimulation. Most of the cells have arrived in the second cell cycle by 42 hr; a portion of these enters the third cell cycle (54- to 90-hr panels). However, there is a preponderance of G_1 over S and G_2 phase cells at these later times. This indicates that hDF-like cells are subject to increasing cell-cycle arrest as they enter the G_1 phases of the second and third cell cycles.

In Fig. 3, a sequence of bivariate cytograms obtained with mouse NIH 3T3 cells is shown. DNA synthesis starts as early as 10 hr after release from quiescence (not shown), and cells are well into the first cycle with some already in

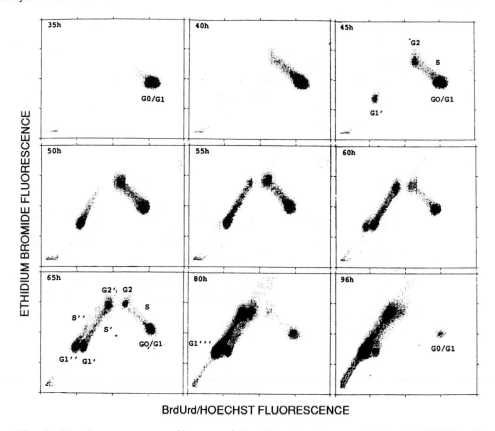

Fig. 1 Bivariate scattergrams of human peripheral blood lymphocytes stimulated with PHA and cultured for various time intervals.

the G_1' compartment after 18 hr of serum stimulation. At 22 hr most cells are in the G_2 and in the G_1' compartment, and very few reside in G_0/G_1. Four hours later, the majority of cells is in the second and third cycle, whereas a few are still retained in the first cycle. The striking degree of synchrony with which cells traverse the various cell-cycle compartments is evident from the bivariate cytograms. At 30 hr of serum stimulation, almost all cells are in the second and third cell cycle, and only a minute fraction of cells is still in the G_0/G_1 and in the G_2 compartment. The latter two compartments are devoid of cells after 34 hr, and cells continue to traverse the second and third cycle. The G_1 of the fourth cycle (G_1''') is reached at 42 hr, whereas the second cycle is at this time point almost entirely empty. Beyond 42 hr of culture, analysis becomes difficult because of the less satisfactory separtion between the third and the fourth cycle. NIH 3T3 cells typically show no perceptible arrest in any cell cycle compartment under these culture conditions (Weller *et al.*, 1993).

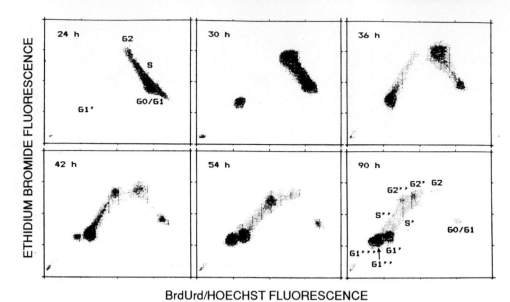

Fig. 2 Bivariate scattergrams of human primary diploid fibroblasts grown in MEM with 10% FBS for various time intervals.

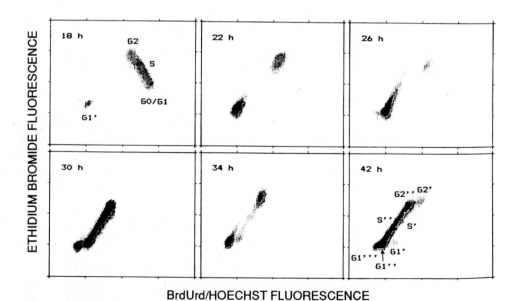

Fig. 3 Bivariate scattergrams of mouse NIH 3T3 cells grown in DMEM/Ham's F-10 with 10% FBS for various time intervals.

Figures 4A and 4B illustrate the procedure by which the distribution of cells within individual cell cycles in the cytogram is determined. The signal distributions in the first, second, and third cycle are electronically framed, rotated, and projected onto a single axis. The resulting univariate fluorescence histograms can then be processed by conventional curve fitting (bottom row of panels in 4A and 4B) in order to determine the fraction of cells in the G_1, S, and G_2/M phases of the cell cycle. These actual fractions ("real data") are listed at the bottom of each figure in the sequence (from left to right): G_1, S, G_2/M. By taking into account the number of times a given population has divided, these "real" data are converted to "original" data. This is done by dividing the "real" number of cells in the second cycle by two, and those within the third cycle by four, and so on. By this simple mathematical procedure, one obtains a quantitative measure of the accurate proliferative history of a given cell culture. Note that Fig. 4B, which represents the cytogram of a 72-hr harvest from a patient with Fanconi anemia, shows much greater accumulations of cells in the G_2/M phases of the first and second cell cycles than Fig. 4A which shows the cytogram from a healthy lymphocyte donor.

Figure 5 illustrates the conversion of the results of sequential, temporally spaced cell-cycle distribution analyses into a kinetic analysis. This conversion is realized by plotting the cell-cycle fraction data of each time point (obtained by the procedure explained in Fig. 4) on a semilogarithmic scale. In such plots the fraction of cells in the G_1, S, and G_2/M compartments of the first, second, and third cell cycle versus time results in a series of curves which can be fitted to a number of cell-cycle models. The curves shown in Fig. 5 are the results of computer fits to a modified version of the Smith-Martin transition probability model (Smith and Martin, 1973; Rabinovitch, 1983; Kubbies et al., 1985a,b). Each such line thus represents the boundary for the time of transition from one cell-cycle compartment to the next. The following cell kinetic information can be derived from the mathematical analysis of these curves: duration of the initial lag phase (that is the time between induction and first entrance into the S phase), minimal duration of each cell-cycle compartment (given by the x-axis intercepts of two successive exit curves), cell-cycle compartment specific transition probabilities (from the slopes of the exit curves), and cell-cycle compartment specific arrest fraction (from the differences of the extrapolated plateau phase levels between successive exit curves). The examples of exit curves shown in Fig. 5 illustrate the changes of these kinetic parameters as a function of donor age for PHA-activated 96-hr cultures of human blood lymphocytes. Other applications of this comprehensive cell-cycle analysis are presented in recent reviews (Rabinovitch et al., 1988; Poot and Hoehn, 1993; Seyschab et al., 1993).

Acknowledgments

We are indebted to Miss J. Köhler (Würzburg) for the preparation of the figures. The authors are supported by Deutsche Forschungsgemeinschaft Grant Nos. HO 849-2-1 (M. P., H. H.) and SFB 172 Porject A7 (M. P. and H. H.), and by NIH Grant No. AG 01751 (A. G., Y. C. C., P. S. R.).

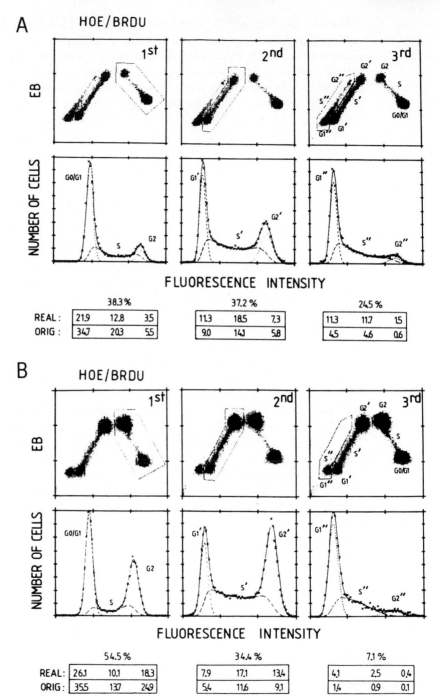

A HOE/BRDU

EB

1st   2nd   3rd

NUMBER OF CELLS

FLUORESCENCE INTENSITY

	38.3 %			37.2 %			24.5 %		
REAL:	21.9	12.8	3.5	11.3	18.5	7.3	11.3	11.7	1.5
ORIG:	34.7	20.3	5.5	9.0	14.1	5.8	4.5	4.6	0.6

B HOE/BRDU

EB

1st   2nd   3rd

NUMBER OF CELLS

FLUORESCENCE INTENSITY

	54.5 %			34.4 %			7.1 %		
REAL:	26.1	10.1	18.3	7.9	17.1	13.4	4.1	2.5	0.4
ORIG:	35.5	13.7	24.9	5.4	11.6	9.1	1.4	0.9	0.1

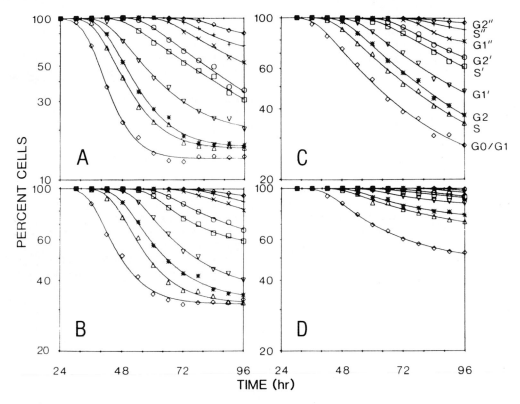

Fig. 5 Exit kinetics of peripheral blood lymphocytes from four donors of different ages. (A) Newborn; (B) 15 years of age; (C) 35 years of age; (D) 75 years of age. Ordinate: percentage of cells that have not excited beyond a particular cell cycle compartment. Abscissa: time (hr) after PHA stimulation. The respective cell cycle compartments are indicated to the right of (C).

Fig. 4 Example of computer analysis of bivariate scattergrams. The top row indicates electronic framing of each cell cycle component of the scattergram: the second row shows univariate cell distributions obtained via projection of the framed data onto a single axis parallel to each respective cell cycle. (A) Peripheral blood lymphocytes from a normal donor 72 hr after PHA stimulation. (B) Peripheral blood lymphocytes from a donor with Fanconi's anemia 72 hr after PHA stimulation.

References

Beck, H. P. (1981). *Cytometry* **2**, 146–158.
Böhmer, R. M. (1979). *Cell Tissue Kinet.* **12**, 101–114.
Darzynkiewicz, Z., Traganos, F., Sharpless, T., and Melamed, M. R. (1976). *Proc. Natl. Acad. Sci. U.S.A.* **73**, 2881–2884.
Giaretti, W., Moro, G., Quarto, R., Bruno, S., DiVinci, A., Geido, E., and Cancedda, R. (1988). *Cytometry* **9**, 281–290.

Gray, J. W., and Mayal, R. H., eds. (1985). *Cytometry* **6**(6).

Grossmann, A., Ledbetter, J. A., and Rabinovitch, P. S. (1989). *Exp. Cell Res.* **180,** 367–382.

Kubbies, M., and Friedl, R. (1985). *Histochemistry* **83,** 133–137.

Kubbies, M., and Rabinovitch, P. S. (1983). *Cytometry* **3,** 276–281.

Kubbies, M., Schindler, D., Hoehn, H., and Rabinovitch, P. S. (1985a). *Cell Tissue Kinet.* **18,** 551–562.

Kubbies, M., Schindler, D., Hoehn, H., and Rabinovitch, P. S. (1985b). *J. Cell. Physiol.* **125,** 229–234.

Kubbies, M., Schindler, D., Hoehn, H., Schinzel, A., and Rabinovitch, P. S. (1985c). *Am. J. Hum. Genet.* **37,** 1022–1030.

Kubbies, M., Hoehn, H., and Rabinovitch, P. S. (1987). *Exp. Cell Biol.* **55,** 225–236.

Latt, S. A. (1973). *Proc. Natl. Acad. Sci. U.S.A.* **70,** 3395–3402.

Latt, S. A. (1977). *J. Histochem. Cytochem.* **25,** 913–922.

Nüsse, M. (1981). *Cytometry* **2,** 70–77.

Ormerod, M. G., and Kubbies, M. (1992). *Cytometry* **13,** 678–685.

Poot, M., and Hoehn, H. (1993). *Toxicol. Lett.* **67,** 297–308.

Poot, M., Schindler, D., Kubbies, M., Hoehn, H., and Rabinovitch, P. S. (1988a). *Cytometry* **9,** 332–339.

Poot, M., Esterbauer, H., Rabinovitch, P. S., and Hoehn, H. (1988b). *J. Cell. Physiol.* **137,** 421–429.

Poot, M., Nicotera, T., Rüdiger, H. W., and Hoehn, H. (1989). *Free Radical Res. Commun.* **7,** 179–187.

Poot, M., Köhler, J., Rabinovitch, P. S., Hoehn, H., and Priest, J. H. (1990a). *Hum. Genet.* **84,** 258–262.

Poot, M., Kausch, K., Köhler, J., Haaf, T., and Hoehn, H. (1990b). *Cell Struct. Funct.* **15,** 151–157.

Poot, M., Rizk-Rabin, M., Hoehn, H., and Pavlovitch, J. H. (1990c). *J. Cell. Physiol.* **143,** 279–286.

Poot, M. Schuster, A., and Hoehn, H. (1991). *Biochem. Pharmacol.* **41,** 1903–1909.

Poot, M., Hoehn, H., Bogdahn, U., and Otto, F. (1992). *Melanoma Res.* **2,** 241–247.

Rabinovitch, P. S. (1983). *Proc. Natl. Acad. Sci. U.S.A.* **80,** 2951–2960.

Rabinovitch, P. S., Kubbies, M., Chen, Y. C., Schindler, D., and Hoehn, H. (1998). *Exp. Cell Res.* **174,** 319–318.

Schindler, D., Kubbies, M., Hoehn, H., Schinzel, A., and Rabinovitch, P. S. (1985). *Lancet* **2,** 937.

Schindler, D., Seyschab, H., Poot, M., Hoehn, H., Schinzel, A., Fryns, J. P., Tommerup, N., and Rabinovitch, P. S. (1987). *Lancet* **2,** 1398.

Schindler, D., Kubbies, M., Priest, R. E., Hoehn, H., and Rabinovitch, P. S. (1988). *Mech. Ageing Dev.* **44,** 253–263.

Seyschab, H., Friedl, R., Schindler, D., Hoehn, H., Rabinovitch, P. S., and Chen, U. (1989). *Eur. J. Immunol.* **19,** 1605–1612.

Seyschab, H., Schindler, D., Friedl, R., Barbi, G., Boltshauser, E., and Hoehn, H. (1992). *Eur. J. Pediat.* **151,** 756–760.

Seyschab, H., Sun, Y., Friedl, R., Schindler, D., and Hoehn, H. (1993). *Hum. Genet.* **92,** 61–68.

Smith, J. A., and Martin, L. (1973). *Proc. Natl. Acad. Sci. U.S.A.* **70,** 1263–1267.

Vogel, D. G., Rabinovitch, P. S., and Mottet, N. K. (1986). *Cell Tissue Kinet.* **19,** 227–236.

Weilbach, F. X., Bogdahn, U., Poot, M., Apfel, R., Behl, C., Drenkard, D., Martin, R., and Hoehn, H. (1990). *Cancer Res.* **50,** 6981–6986.

Weller, E. M., Poot, M., and Hoehn, H. (1993). *Cell Proliferation* **26,** 45–54.

CHAPTER 22

Detection of BrdUrd-Labeled Cells by Differential Fluorescence Analysis of DNA Fluorochromes: Pulse–Chase and Continuous Labeling Methods

Harry A. Crissman,★ Noboru Oishi,† and Robert Habbersett★

★Division of Biological Sciences
Los Alamos National Laboratory
Los Alamos, New Mexico 87545
†San Antonio Cancer Institute
San Antonio, Texas 78229

I. Introduction

Analysis of DNA content by flow cytometry (FCM) provides a convenient method for obtaining the frequency distribution of cells in various phases of the cell cycle, and when such determinations are coupled with cellular 5-bromo-deoxyuridine (BrdUrd) incorporation data, kinetic information is obtained that allows further subdivision of the cell cycle based on cycle traverse capacity.

Not only is the cycle position of cells obtained, but the rate and the percentage of cells proceeding from one compartment of the cell cycle to the next can be quantified under a wide range of experimental conditions. Information from such studies has been useful for studying the differences in the kinetic patterns of normal and tumor cells and also for investigating the effects of cycle pertubing agents on cell-cycle progression.

Several FCM techniques have been developed for DNA-BrdUrd measurements, but probably the most widely used is the assay devised by Dolbeare *et al.* (1983) that combines propidium iodide (PI) DNA staining with immunofluorescent labeling of BrdUrd as described by Gratzner (1982). This method requires partial denaturation of cellular DNA by heat or acid treatment to expose the incorporated BrdUrd to the antibody; however, the technique has the advantage that it requires only a single laser (488 nm) for excitation of both PI and fluorescein-labeled BrdUrd. Using conventional two-color, FCM analysis, cells containing BrdUrd are detected and quantitated and their cell-cycle position is readily assessed (see Dolbeare and Selden, Chapter 19).

Most of the other FCM procedures for DNA-BrdUrd analysis utilize the cytochemical method first described by Latt (1973) who demonstrated that the fluorescence of Hoechst 33258 (HO), an A-T base binding fluorochrome, is quenched, quantitatively, when bound to A-BrdUrd regions in double-stranded DNA. In current FCM studies, a second DNA fluorochrome that is not significantly affected by BrdUrd, usually ethidium bromide (EB), is combined with Hoechst for cell staining, and analysis performed using a single UV excitation source. A number of investigators (Bohmer, 1979; Bohmer and Ellwart, 1981a,b; Kubbies and Rabinovich, 1983; Ellwart and Dohmer, 1985; Poot *et al.,* 1990; and Poot *et al.,* Chapter 21) have used this approach. Usually BrdUrd labeling periods of 4–6 hr or longer are required for cells to incorporate sufficient amounts of the base analogue in order to be detected by HO quenching. These FCM techniques, therefore, do not provide resolution and sensitivity comparable to the immunofluorescent assay, and their application has been primarily for continuous BrdUrd labeling studies.

In this chapter, we describe FCM methods for the analysis of populations of cells following pulse-chase and continuous BrdUrd labeling using a two-color technique that can potentially detect BrdUrd-labeled DNA in cells treated for 30 min or less (Crissman and Steinkamp, 1987). The procedure, which is similar to one described previously (Crissman and Steinkamp, 1990), involves staining with two nonintercalating, DNA fluorochromes: HO and G-C binding mithramycin (MI) (Crissman and Tobey, 1974; Crissman *et al.,* 1979), a dye whose stoichiometry for DNA content is not affected by BrdUrd. Using dual-laser excitation, and performing differential fluorescence analysis on the two fluorochromes, a measure of the cellular BrdUrd content is obtained. The technique requires only one-step staining; it is mild, and, therefore, cell loss and loss of other important cellular markers, such as RNA, and proteins, including cellular antigens, are minimal.

II. Applications

In pulse-chase studies, the duration of the BrdUrd exposure period is brief and adjusted to only label cells in S phase. The cells are rinsed, cultured in BrdUrd-free medium, and harvested at time intervals thereafter. The analysis yields information on the cycle progression capacity of cells that were in S phase during treatment with various agents. In some instances, the data are used to calculate the rate of DNA synthesis (Begg *et al.*, 1985).

Continuous BrdUrd labeling studies are designed to, potentially, determine DNA synthetic capacity of all the cells in the population. In these studies, cells are continuously labeled with BrdUrd in culture and then harvested at desired time intervals and analyzed. Calculations of the percentage of labeled and unlabeled cells in different phases of the cell cycle at different time intervals after labeling can reveal the progression of cells in and out of the different phases.

Both BrdUrd pulse-chase and continuous labeling studies can be performed using differential fluorescence analysis with only minor modifications of our original procedure. Essentially, the technique represents a quantitative approach for measuring a small decrease in HO fluorescent intensity when the dye is bound to BrdUrd incorporated into DNA either during short pulse labeling or continuous labeling periods. Quantitative FCM measurement of the HO quenching is accomplished by electronic cell-by-cell subtraction of the HO fluroescent signal from the signal of a second dye, such as the G-C binding stain, MI, that is not affected by BrdUrd. Also HO and MI have different spectral and DNA binding properties and neither dye–dye interference nor energy transfer between the MI and HO significantly affects FCM BrdUrd analysis. Both HO and MI bind stoichiometrically to DNA content in non-BrdUrd-labeled cells; therefore, the integrated areas of the fluorescent signals of both HO and MI will be equal, if instrument gain settings are adjusted so that the G_1 peak of the HO-DNA content and the G_1 peak of the MI-DNA content histograms are at the same channel number. When the two fluorescent signals are subtracted electronically, on a cell-by-cell basis, bivariate profiles plotted for the MI minus HO fluorescent signal differences vs MI-DNA content will show zero values for the BrdUrd content of all cells in the population. However, in a cell population that has been pulse labeled with BrdUrd (i.e., only S phase cells are labeled), zero values for the MI minus HO fluorescent signal differences are obtained only for the G_1 and $G_2 + M$ cells in which the fluorescent intensity of both dyes remains proportional to DNA content. Cells in S phase containing BrdUrd produce HO fluorescent signals that are reduced in proportion to the quantity of BrdUrd-substituted DNA, while the MI fluorescent intensity in these same cells still remains proportional to the DNA content. Differential fluorescence analysis of S phase cells will then yield MI minus HO differences that are proportional to BrdUrd content.

III. Materials

Stock solutions of BrdUrd (Sigma Chemical Co.) are freshly prepared in distilled water (1.0 mg/ml) and sterilized by filtration. The BrdUrd solution should be refrigerated in vials wrapped in foil to protect it from light. Saline GM solution and 95% ethanol are prepared and used for ethanol fixation (70% final concentration) of cells as we describe in Chapter 13. Hoechst 33342 (Calbiochem, San Diego, CA), dissolved in distilled water at 1.0 mg/ml, can be stored in the refrigerator for at least 1 month in a foil-wrapped container. MI (Pfizer Co., Groton, CT), 2.5 mg/vial, is dissolved in PBS at 1.0 mg/ml and can be stored in the refrigerator for at least 1 month. Solutions of $MgCl_2$ (250 mM) are prepared in distilled water and may be stored at room temperature for at least 1 month.

IV. Cell Preparation and Staining Procedure

For ethanol fixation, cells are harvested from culture by centrifugation and then *thoroughly* resuspended, to prevent cell aggregation, in one part cold "saline GM" (g/liter: glucose, 1.1; NaCl, 8.0; KCl, 0.4; $Na_2HPO_4 \cdot 12\ H_2O$, 0.39; KH_2PO_4, 0.15) containing 0.5 mM EDTA for chelating free calcium and magnesium ions. Then three parts cold, 95% nondenatured ethanol are added to the cell suspension with mixing to produce a final ethanol concentration of about 70%. Prior to staining, fixed cell samples are centrifuged and the ethanol fixative is removed by aspiration. The cells are resuspended in a dye solution containing HO and MI at concentrations of 0.5 and 5.0 μg/ml, respectively, in PBS supplemented with 5.0 mM $MgCl_2$. The cell density should be adjusted to about 7.5×10^5 cells/ml stain solution. Cell samples can be analyzed in the stain solution from 30 min to 5 hr following staining at room temperature.

V. Critical Aspects

For pulse-chase studies, CHO cells are routinely labeled in culture with 30 μM BrdUrd for 1 hr at 37°C. Cultures are not exposed to strong light over this period since the BrdUrd is light sensitive and can deteriorate. Also, cells incorporating BrdUrd into their DNA can suffer chromatin damage and lose viability when exposed to strong light and thus fail to incorporate detectable quantities of BrdUrd for FCM detection.

The concentration of BrdUrd and the length of the labeling period can vary depending on the cell-cycle time and the rate of DNA synthesis. Rapidly growing L1210 cells (i.e., $T_G = 12$ hr), for example, may require only a 30-min pulse, but some slow growing cells may require pulses of 1–2 hr before they have incorporated detectable quantities of BrdUrd. The pulse duration as well as the BrdUrd concentration must be determined empirically. Excessively high concentrations of BrdUrd can be toxic to cells, and comparative growth studies and FCM analyses of BrdUrd-treated and untreated cell populations are necessary to determine optimal conditions for BrdUrd labeling. These same criteria apply for continuous BrdUrd labeling and, in particular, the BrdUrd concentration must be reduced compared to pusle-chase studies, since the cellular BrdUrd content is continuously increasing during the course of these studies. For CHO and human skin fibroblast cells, a concentration of 5 μM is used for no longer than a 24 hr labeling period and 2–3 μM BrdUrd for longer labeling time periods. The sampling intervals after pulse chase or continuous labeling can be gauged from the cell-cycle doubling time and the particular cycle intervals of interest.

VI. Controls and Standards

Staining experiments performed initially on untreated cells are necessary to determine the lowest concentrations of HO and MI that can be combined to obtain DNA content histograms of good quality with low coefficient of variation (CV) values. Ideally, both the HO- and the MI-DNA content histograms of nonlabeled cells will yield the same percentage of cells in the various phases of the cell cycle. Under these conditions the dye–dye interactions will be minimal and the subtraction of the HO fluorescent signal from the MI signal in unlabeled cells, on a cell-by-cell basis, will be as close to the ideal value of zero as possible. Using the optimized HO and MI dye concentrations for staining and analysis of BrdUrd-labeled cell populations will ensure the most sensitive and accurate FCM analysis.

VII. Instruments

A multilaser, FCM system (Steinkamp *et al.*, 1982) was used and two lasers were operated, one in the UV (333.6–363.8 nm) and one tuned to 457.9 nm. The laser beams were separated by 250 μm to provide sequential excitation and analysis of each fluorochrome (i.e., HO and MI, respectively). The HO fluorescence was measured over the 400- to 500-nm range, while the MI fluorescence was measured above 500 nm. A 500-nm long-pass dichroic filter was used for these studies. The electronic signal gains were adjusted so that the G_1 peaks of the FCM-generated HO- and MI-DNA content histograms were initially at the same channel number. The fluorescence signals are then subtracted electronically (i.e., MI minus HO) on a cell-by-cell basis using a differential fluorescence analysis method described by Steinkamp and Stewart (1986) for subtracting background cellular autofluorescence during analysis of stained cells. These difference measurements reflect quantitatively the quenching of

HO fluorescence, which is directly proportional to cellular BrdUrd content. The arbitrary adjustment of the G_1 peak positions for both dyes in the same channel sets the zero value for the subtraction process for all cells across the cell cycle. The assignment of the zero value is valid since the G_1 cells in pulse-labeled cells do not contain BrdUrd, thus do not exhibit HO quenching, and, therefore, the MI minus HO value should be zero.

In pulse-chase and continuous labeling studies, postmitotic, BrdUrd-labeled cells will eventually appear at some given time after the initial labeling period. When a storage oscilloscope (*x-y* dot plot) is used to visualize the HO vs MI displays during data acquisition, the cells that exhibit HO quenching can easily be discriminated so that alignment of the HO- and MI-G_1 peaks is based only on the unlabeled G_1 cells. The analytical procedure for continuous labeling studies varies from this procedure only when all cells become labeled; then a non-BrdUrd-labeled sample is used for G_1 peak alignments, prior to running the labeled sample.

VIII. Results

Data from a pulse-chase experiment performed on BrdUrd-labeled Chinese hamster (line CHO) cells are shown in Fig. 1. The DNA content, based on fluorescence from MI, is on the *x*-axis, and the BrdUrd content, as derived from the MI minus the HO difference measurements, is represented on the *y*-axis, while the time intervals shown are the harvest periods after a 1-hr pulse with 30 μM BrdUrd at 37°C and rinsing and growth of cells in BrdUrd-free medium. For these displays the zero value for BrdUrd content is offset slightly so that unlabeled G_1 and G_2/M cells lie on a plane slightly above and parallel to the *x*-axis for better visualization of these subpopulations. The 0.5-hr sample is similar to the 0-hr sample (1 hr pulse-labeled sample, data not shown) while the 1-hr sample shows the progression of the most highly labeled S phase cells toward G_2/M. At 2 hr (arrow) there is the first appearance of labeled postmitotic G_1 cells and the relative proportion of this subpopulation increases during the 4- (arrow) and 6-hr interval. The appearance of labeled G_1 cells at 2 hr, after the 1-hr pulse (3 hr total time), shows the sensitivity of the technique, since the G_2 period is about 2 hr in CHO cells and mitosis about 45 min. By 6 and 8 hr, a significant number of unlabeled S phase cells are noted and by 12 hr there appears a mixture of unlabeled S phase cells and labeled S phase cells, which originate from the postmitotic, labeled G_1 cells. The 24-hr sample, harvested about 1.6 generations after pulse labeling, shows the rapid evolution of heterogeneity in the population as well as the dilution of the BrdUrd content in the individual cells. Data shown at the 0.5- and the 1-hr interval can be used to determine the rate of DNA synthesis using the calculations of Begg *et al.* (1985).

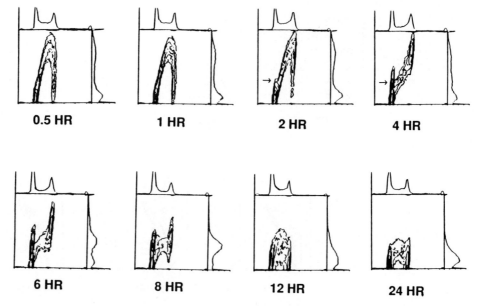

Fig. 1 Bivariate contour diagrams for CHO cells treated in culture with 30 μM BrdUrd for 1 hr, rinsed, cultured in BrdUrd-free medium, and harvested at the time intervals indicated. The *x*-axis, MI fluorescence, proportional to DNA content and the *y*-axis, BrdUrd content, derived by differential fluorescence analysis, are linear relative units. The difference amplifier gain for the BrdUrd content (*y*-axis) was initially adjusted and fixed to give a maximum expansion of the fluoroescence difference signal range for the cell population at the 0.5-hr chase period. All the samples were then analyzed at that same gain setting.

CHO cells were continuously labeled for 24 hr with 5 μM BrdUrd and harvested at the times shown in Fig. 2. The axes are as described for Fig. 1: the *x*-axis is DNA content, and the *y*-axis, BrdUrd content. The distribution for the 1-hr sample is similar to that for the 0.5-hr sample in Fig. 1, except the gain for BrdUrd content channel is reduced so that all of distributions could be displayed at the same gain setting. As in Fig. 1, labeled postmitotic G_1 cells appear at 2 hr but they are not as apparent at these gain settings. Unlabeled G_2 cells (UL arrow) are seen at 2 hr, and at 4 hr there is an increase in the number of labeled G_1 cells. Also at 4 hr there are still a few unlabeled G_2 cells as well as labeled G_2 cells with varying amounts of BrdUrd (region designated L), depending on the cycle position of the cells at the initial time of labeling. Cells in G_2 that began to incorporate BrdUrd in early S phase have the greater BrdUrd content while cells labeled that were in late S phase have a significantly lower content and subsequently lie slightly above the unlabeled G_2 cells within the region marked L. From 4 to 6 hr more cells have divided and the cells initially labeled in late S phase, with low amounts of label, appear first within the G_1 phase, but at 6 hr more G_1 cells appear with increased BrdUrd content since these cells were at an earlier position in S phase and were labeled

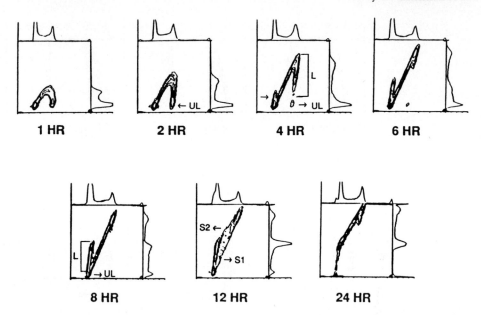

Fig. 2 Bivariate contour diagrams for CHO cells treated continuously in culture with 5 μM BrdUrd and harvested at the time intervals indicated. The x-axis, MI fluorescence, proportional to DNA content and the y-axis, BrdUrd content, derived by differential fluorescence analysis, are linear relative units. The difference amplifier gain for the BrdUrd content (y-axis) was initially adjusted and fixed to give a maximum expansion of the fluroescence difference signal range from cells that had incorporated BrdUrd during the 12- and 24-hr continuous labeling periods. All the samples were then analyzed at that same gain setting.

over a longer time period. By 8 hr, G_1 cells with the maximum amount of BrdUrd content (i.e., the very early S phase cells) finally appear and by 12 hr these cells begin to enter a second round of DNA systhesis (S2) while a few unlabeled G_1 cells are still entering S phase (S1). The very narrow range in BrdUrd content in the cell population, such as seen in the 8-hr sample, reflects the stoichiometry of the method for BrdUrd content. The data are nearly identical to computer-simulated data shown previously by Yanagisawa *et al.* (1985). By 24 hr all but a few cells are labeled and many are engaged in the synthesizing DNA. We used these analyses to detect a subpopulation of slowly traversing G_1 phase cells following synchronization of human skin fibroblast cells (Tobey *et al.*, 1989) and to monitor the progression of HSF cells following treatment with the protein kinase inhibitor, staurosporine (Crissman *et al.*, 1991).

The simplicity, sensitivity, and accuracy of BrdUrd analysis by differential fluorescence analysis makes the procedure quite applicable for basic and clinical studies. For routine studies, sample preparation should be simple and rapid, and preparation of the samples for analysis should be free of manipulations that would induce cell loss and variability from sample to sample, due to repetitive sample handling during the preparative procedure. The preparative and analytical procedure described above fulfills these criteria: one-step ethanol fixation, preventing any cell loss; one-step staining with a stain cocktail; and differential fluorescence analysis in the stain solution. Although most commercial FCM systems do not have capabilities for differential fluorescence analysis, it is possible to acquire the data in list mode and perform the subtraction of the MI and HO fluorescence signals on a cell-by-cell basis using an appropriate computer program (Habbersett *et al.*, 1990). The only significant disadvantage is that this method requires a laser with UV capabilities for exciting HO in order to analyze fluorescence quenching by BrdUrd. Ideally dyes with properties similar to those of HO, but with excitation in the visible range, will become available soon.

Acknowledgments

This work was supported by NIH Grant R24 RR06758, the Los Alamos National Flow Cytometry Resource funded by the Division of Research Resources of NIH (Grant P41-RR01315), and the Department of Energy.

References

Begg, A. C., McNally, N. J., Shrieve, D. C., and Karcher H. (1985). *Cytometry* **6**, 620–626.

Bohmer, R. M. (1979). *Cell Tissue Kinet.* **12**, 101–110.

Bohmer, R. M., and Ellwart, J. (1981a). *Cytometry* **2**, 31–34.

Bohmer, R. M., and Ellwart, J. (1981b). *Cell Tissue Kinet.* **14**, 653–658.

Crissman, H. A., and Steinkamp, J. A. (1987). *Exp. Cell Res.* **173**, 256–261.

Crissman, H. A., and Steinkamp, J. A. (1990). *In* "Flow Cytometry" (Z. Darzynkiewicz and H. A. Crissman, eds.), pp. 199–206. San Diego, CA: Academic Press.

Crissman, H. A., and Tobey, R. A. (1974). *Science* **184**, 1297–1298.

Crissman, H. A., Stevenson, A. P., Kissane, R. J., and Tobey, R. A. (1979). *In* "Flow Cytometry and Sorting" (M. R. Melamed, P. F. Mullaney, and M. L. Mendelsohn, eds.), pp. 243–261. New York: Wiley.

Crissman, H. A., Gadbois, D. M., Tobey, R. A., and Bradbury, E. M. (1991). *Proc. Natl. Acad. Sci. U.S.A.* **88**, 7580–7584.

Dolbeare, F., Gratzner, H., Pallavacini, M., and Gray, J. W. (1983). *Proc. Natl. Acad. Sci. U.S.A.* **80**, 5573–5577.

Ellwart, J., and Dohmer, P. (1985). *Cytometry* **6**, 513–520.

Gratzner, H. (1982). *Science* **218**, 474–475.

Habbersett, R., Crissman, H. A., and Jett, J. H. (1990). *Cytometry* (Suppl. 4), Abstract **528B**, 88.

Kubbies, M., and Rabinovich, P. S. (1983). *Cytometry* **3**, 276–281.

Latt, S. A. (1973). *Proc. Natl. Acad. Sci. U.S.A.* **70**, 3395–3399.

Poot, M., Hoehn, H., Kubbies, M., Grossman, A., Chen, Y., and Rabinovitch, P. S. (1990). *In* "Flow Cytometry" (Z. Darzynkiewicz and H. A. Crissman, eds.), pp. 185–206. San Diego, CA: Academic Press.

Steinkamp, J. A., and Stewart, C. C. (1986). *Cytometry* **7**, 566–574.

Steinkamp, J. A., Stewart, C. C., and Crissman, H. A. (1982). *Cytometry* **2**, 226–231.

Tobey, R. A., Oishi, N., and Crissman, H. A. (1989). *J. Cell. Physiol.* **139**, 432–440.

Yanagisawa, M., Dolbeare, F., Todoroki, T., and Gray, J. W. (1985). *Cytometry* **6**, 550–562.

CHAPTER 23

Analysis of Intracellular Proteins

Kenneth D. Bauer* and James W. Jacobberger[†]

*Department of Immunology
Genentech, Inc.
South San Francisco, California 94080

[†]School of Medicine
Cancer Center
Case Western Reserve University
Cleveland, Ohio 44106

I. Introduction

Monoclonal antibody (mAb) probes appear well suited for immunofluorescence (IF) analysis of intracellular proteins in combination with flow cytometry (FCM). Despite the fact that the large majority of cellular proteins do not include extracellular domains, FCM analysis of intracellular proteins to date has been far slower in gaining popularity than analysis of proteins associated with the cell surface. Possible explanations for this include: (1) It is technically more difficult to stain intracellular antigens because the cells need to be permeabilized to permit free diffusion of the antibody into the cell and because the level of nonspecific binding of IgG to permeabilized cells is significantly greater than that to the surface of cells with membrane integrity intact; and (2) the largest driving force in flow cytometry to date has been to identify and sort populations of immune cells which could be most easily recognized by significant differences in expression of molecules with extracellular domains. Many factors influence the specificity of IF measurements. Some of these are illustrated in Fig. 1 and include the titer of the antibody used, the extent of nonspecific interaction of the antibody, and the degree of background autofluorescence. This chapter is an overview of methodologic considerations for optimizing the IF analysis of intracellular proteins by FCM, with emphasis on the measurement of cell proliferation-associated proteins.

A. Cell Fixation and Permeabilization

To achieve access to intracellular compartments by antibody probes, current approaches require cell permeabilization. Permeabilization has been accomplished using organic solvent or detergent solubilization of the lipid component of the cell membrane. The actions of alcohols and detergents have been reviewed comprehensively (Kiernan, 1981; Helenius *et al.*, 1979) as have fixation/permeabilization strategies for IF staining (Jacobberger, 1991; Clevenger and Shankey, 1993). In this chapter, three procedures that have been employed most often for measurement of nuclear proteins and which appear complementary in their properties are overviewed. These are formaldehyde fixation followed by Triton X-100 (Clevenger *et al.*, 1985), cold methanol fixation/permeabilization

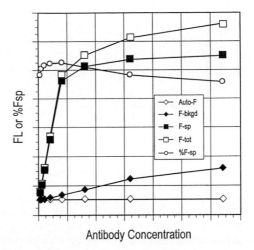

Fig. 1 Component parts of immunofluorescence. Hypothetical values representing the fluorescence components of immunofluorescently stained cells as a function of antibody titer. Autofluorescence (open diamonds) is constant and a function of the cell type, cell metabolism, fixation chemistry, and scattered excitation light. Background antibody staining (filled diamonds) results from nonspecific low-affinity binding and saturates slowly. Specific fluorescence (filled squares) is the result of high-affinity binding and saturates rapidly. Total fluorescence (open squares) is the sum of the three preceding components. These are plotted as fluorescence values (FL). The %specific fluorescence (%Fsp = specific fluorescence/total fluorescence × 100, open circles) is a measure of the sensitivity of detection. The maximum %F's occurs prior to the saturation of the high-affinity component (Srivastava *et al.*, 1992). This figure is a modified version of that published in Srivastava *et al.* (1992) and Jacobberger (1991).

(Jacobberger *et al.*, 1986), and sequential formaldehyde and methanol treatment (Pollice *et al.*, 1992; Schimenti and Jacobberger, 1992; Kurki *et al.*, 1988).

Because alcohol fixation had been shown to result in the loss or redistribution of several nuclear proteins (Clevenger *et al.*, 1985; Epstein and Clevenger, 1985), these investigators explored the use of formaldehyde fixation followed by a brief treatment with Triton X-100. The reasoning was that the formaldehyde treatment would stabilize the cell and the integrity of the cell would not be lost during the many subsequent washes associated with intracellular staining. Mann *et al.* (1987) demonstrated directly that the role of formaldehyde was to retain cell integrity and retard protein loss in studies of the soluble protein, ribonucleotide reductase. A similar finding has been reported for the relatively well-retained SV40 T antigen when a brief formaldehyde treatment is followed by MeOH (Schimenti and Jacobberger, 1992) and for soluble *Escherichia coli* β-galactosidase (Fig. 2).

Formaldehyde has been reported to form numerous and complex reactions with proteins by virtue of the fact that it can combine with a number of different functional groups (see Pearse, 1980, for a review of the chemical interactions of formaldehyde). Both reversible and irreversible chemical reactions can occur

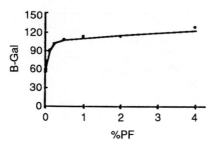 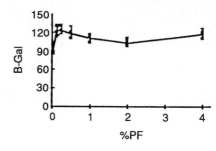

Fig. 2 Retention of nuclear and cytoplasmic antigens by formaldehyde fixation. 3T3 cells express-ing either nuclear or cytoplasmically localized *E. coli* β-galactosidase (B-Gal) were immunofluores-cently stained with a B-Gal-specific monoclonal antibody after fixation with varying levels of formaldehyde (PF) followed by 90% MeOH at −20°C. Control cells were stained without primary antibody. The samples were then stained for DNA content and analyzed by flow cytometry. The relative levels of B-Gal were calculated by subtracting mean fluorescence of the G_1 cell-cycle phase population of the control cells from the positive sample. The cytoplasmic data were generated from the normalized data (K. J. Schimenti, T. L. Sladek, and J. W. Jacobberger, unpublished).

between formaldehyde and proteins. Probably the most significant of these from the standpoint of cellular fixation previous to IF staining is the formation of crosslinks through methylene bridges. These appear to be reversible in aqueous solution. The significance of this is that when formaldehyde fixation is followed by subsequent washing and staining in aqueous solution, the resulting reversal of protein crosslinks may improve the accessibility of antibody probes to epitopes. Irreversible chemical reactions to protein groups or residual methylene bridges can impair antibody binding. While excellent for IF, formaldehyde generally yields comparatively poor-quality DNA staining when intercalating fluoro-chromes such as propidium iodide (PI) are used. This is observed as broad coefficients of variation (CV) of the G_0/G_1 peak of the DNA histogram. When formaldehyde fixation is followed by Triton X-100 treatment, the resulting CVs are surprisingly low (Clevenger *et al.,* 1985; Schimenti and Jacobberger, 1992). Poorer-quality DNA histograms have been observed for formaldehyde followed by MeOH, whereas cold MeOH fixation appears to yield histograms of inter-mediate quality (Schimenti and Jacobberger, 1992). The need to retain the protein of interest (required for accurate quantitative IF analysis) can be balanced against the requirement for high-resolution DNA content analysis: In the formaldehyde/methanol procedure antigen detection is often improved and DNA analysis can be performed (Jacobberger, 1989; Schimenti and Jacobberger, 1992).

Both alcohols and detergents can denature proteins or disrupt intermolecular interactions. Thus, these treatments can alter conformation-dependent epitopes and impair IF staining. Conversely, these treatments may uncover masked epitopes by disrupting intermolecular bonds and improve IF staining for epitopes that are not conformation dependent. Based on investigations of the SV40 T antigen (Schimenti and Jacobberger, 1992), cytokeratins, and *E. coli* β-galactosidase (J. W. Jacobberger, unpublished), experimental data suggest that detergent treatment following formaldehyde fixation may result in epitope availability that is related to the native conformation and intermolecular structure of the protein involved, whereas MeOH treatment following formaldehyde appears to result in disruption of native structure that can increase epitope availability. However, results of studies with p105 (Clevenger *et al.*, 1985, and Fig. 3) suggest that detergent treatment following formaldehyde fixation

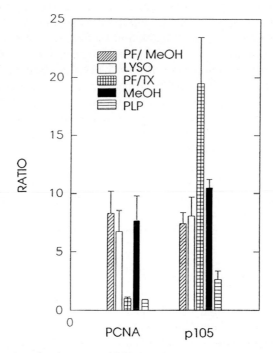

Fig. 3 Effects of various fixatives on anti-PCNA and anti-p105 IF. HL-60 cells were fixed using five different fixation methods and the ratio of mean anti-PCNA and anti-p105 IF relative to mean fluorescence of cells incubated with isotype control antibodies at the same concentration, then stained with FITC-conjugated second antibodies, are shown. PF/MeOH: 1% formaldehyde followed by absolute methanol, followed by 0.1% NP-40 fixation/permeabilization; LYSO: 1% formaldehyde plus 20 mg/ml lysolecithin, followed by absolute MeOH, followed by 0.1% MP-40; PF/TX: 0.5% formaldehyde followed by 0.1% Triton X-100; MeOH: Absolute MeOH at −70°C; PLP: 2% formaldehyde with 75 mM 1-lysine and 10 mM Na metaperiodate. Note that the strongest signal (i.e., highest ratio) for p105 IF is observed following PF/TX while PF/MeOH yields a significantly weaker signal. This finding is essentially the reverse of that observed for anti-PCNA IF following these same fixation protocols and illustrates that optimization of IF staining of different proteins may require careful screening of various fixation/permeabilization approaches (A. McNally and K. D. Bauer, unpublished results).

unmasks the epitope. Thus, at this time, the effects of a specific fixation strategy upon proteins should not be overgeneralized, and more than one fixation procedure should be examined when initiating investigation of a novel epitope.

B. Proteins Evaluated

The analysis of cell proliferation-associated proteins has been greatly facilitated by the increasing availability of mAbs from commercial sources (Table I). Among such proteins which have been analyzed by FCM are the tumor suppressor product, p53 (Kastan *et al.*, 1991a,b; Kuerbitz *et al.*, 1992), the dominant protooncogene product, c-myc (Rabbitts *et al.*, 1985; Giaretti *et al.*, 1990; Rosette *et al.*, 1990; Engelhard *et al.*, 1991); a viral oncogene product, SV40 T antigen (Sladek and Jacobberger, 1992a,b,1993); p105, an interchromatin granule-associated protein that is expressed at very high levels in M phase cells (Clevenger *et al.*, 1985,1987a,b; Bauer, 1988), p120, a nucleolar protein (Bolton *et al.*, 1992; Fonagy *et al.*, 1992); cyclin B1 (Kung *et al.*, 1993; Gong *et al.*, 1993); and cyclin E (Gong *et al.*, 1993). Proliferating cell nuclear antigen (PCNA), a cofactor for DNA polymerase δ (Kurki *et al.*, 1986,1988), and the protein or protein complex recognized by Ki-67 [(Ki-67 Ag), Gerdes *et al.*, 1984; Baisch and Gerdes, 1987] have been most commonly investigated. Given that altered cell proliferation is associated with cancer, it is not surprising that most of these studies are clinically oriented. The number of molecules that are involved in control of these cellular processes is large, and the possibilities for diagnostic and prognostic information gained through evaluation of the expression of these genes is apparent.

II. Application

FCM analysis has proven to be applicable to a wide variety of cells and tissues, including cells isolated from solid tissues. Some reports suggest that significant information can be gained by analysis of pepsin-treated nuclei from paraffin-embedded tissue (Watson, 1986; Anastasi *et al.*, 1987; Bauer *et al.*, 1986b; Morkve and Laerum, 1991). While this latter specimen type is particularly problematic (see Section V.A), the success of immunohistochemical staining of paraffin-embedded tissue following antigen retrieval techniques (Shi *et al.*, 1991) may impact significantly on the future feasibility of IF analysis using such material.

A. Multiparameter IF Analysis

Most studies of cell proliferation-associated intracellular antigens to date have coupled single-color IF with DNA content analysis. A few studies have been performed which examined more than one antigen. Kastan *et al.* (1989a,b),

Table I
Proliferation–Associated Antigen mAb Sources[a]

Antigens	Companies	Phone
c-rel, NF-$\kappa\beta$	American Bio-Technologies, Inc., Cambridge, MA	(617)547–5535
p34cdc2 kinase	Affiniti Research Products, Derbyshire, England	(602)442232
p53, raf, Ki67, p34cdc2 kinase, p13 kinase	Biodesign, Inc., Kennebunkport, ME	(207)967–4173
PCNA	Biogenix Lab., San Ramon, CA	(510)275–0550
PCNA	Boehringer-Mannheim Biochem., Indianapolis, IN	(317)849–9350
c-fos, p53	Caltag Laboratories, Inc., South San Francisco, CA	(415)873–6106
Nucleoli, c-fos, p53, p34cdc2 kinase, PCNA	Chemicon International, Inc., Temecula, CA	(714)676–8080
c-myc, p53	Cambridge Research Biochem., Chesire, England	0606-41100
PCNA, p120, p145	Coulter Cytometry, Hialeah, FL	(305)885–0131
p53	Cymbus Bioscience, Ltd., Hampshire, England	0703-766280
PCNA	Dako Corporation, Carpinteria, CA	(805)566–6655
v-fps, v-fms, v-src, v-H-ras	East-Acres Biologicals, Southbridge, MA	(508)765–9580
p34cdc2 kinase	Glentech, Inc., Lexington, KY	(606)276–2306
p53	Gibco BRL, Grand Island, NY	(716)773–0700
abl, c-fos, c-myc, c-mil, c-mos, v-mos, v-myb, v-raf, v-rel, yes	NIH BCB Respository, Quality Biotech, Inc.	(609)966–8078
bcr, c-abl, c-fos, c-H-ras, c-K-ras, c-N-ras, v-H-ras, c-myc, N-myc, p53, PAN-ras, v-fes, v-fps, v-src	Oncogene Science, Inc., Uniondale, NY	(516)222–0023
Rb, p53, Trp-E, v-abl, p34cdc2 kinase, SV40 T antigen, Adenovirus E1A	Pharmingen, San Diego, CA	(619)792–5730
myc, p53, PCNA, 780-3	Serotec LTD, Oxford, England	08675–79941
PCNA	Signet Labs, Dedham, MA	(617)329–7919
c-myb I, c-myb II, src p120 substrate, src, src p80 substrate	Upstate Biotechnology, Inc., Waltham, MA	(617)890–8845
c-fos, p53, raf-1, p34cdc2 kinase, p13 kinase, PCNA	Zymed Laboratories, South San Francisco, CA	(415)871–4494

[a] Most of these companies were located using Linscott's Directory of Immunological and Biological Reagents, 7th Ed. ISBN: 0-9604920-6-2 (Linscott's Directory, 4877 Grange Road, Santa Rosa, CA, 95404, phone 707–544–9555).

Schmid *et al.* (1991), and Drach *et al.* (1989) have performed analysis of intracellular cell proliferation-associated antigens and leukocyte surface antigens. Rosette *et al.* (1990), Houck and Loken (1985), and Begg and Hofland (1991) have stained cells for incorporated halogenated deoxyuridine, DNA, and either surface or cytoplasmic antigens. Li *et al.* (1993) have performed studies simultaneously investigating two cytoplasmic proteins (cytokeratin and c-erbB2) while Landberg and colleagues (1990) have stained for two intranuclear antigens (Ki-67 Ag and PCNA) simultaneously.

The paucity of studies to date investigating multiple intracellular proteins by FCM analysis largely reflects the lack of availability of suitable fluorophores until very recently. At present, most available antibodies to intracellular proteins are unlabeled mouse mAbs that require an indirect assay which cannot distinguish between the primary antibodies. Ideally, two directly labeled antibodies should be used. Some antibodies to cell proliferation-associated antigens are sold as directly conjugated reagents (e.g., Ki-67 Ag, PCNA, p53). In some cases, primary antibodies can be obtained labeled with biotin. In this case, labeled anti-mouse antibodies and labeled streptavidin (Srivastava *et al.,* 1992) can be used as two different secondary reagents to distinguish between the two primary antibodies.

A list of some of the properties of fluorophores in current and emerging use for IF analysis is shown in Table II. Fluorescein isothiocyanate (FITC) has a molecular weight of 389 and is by far the most popular fluorophore today for IF analysis by FCM. The most commonly used second fluorophores at present

Table II
Properties of Fluorophores Commonly Used for Immunofluorescence Staining

Fluorophore	Approximate molecular mass (daltons)	Excitation max (nm)	Emission max (nm)
Aminomethyl coumarin acetic acid (AMCA)	330	355	440
Cascade blue[a]	607	380	425
Fluorescein isothiocyanate (FITC)	389	495	520
R-Phycoerythrin (PE)	240,000	480, 545, 565	580
Peridinin-chlorophyll-a-protein (PerCP)	35,000	470	675
Indocarbocyanine[b] (CY3)	930	515, 550	570
Lissamine rhodamine[c] sulfonyl chloride (LRSC)	577	570	590
Texas red[a]	625	595	620
Indodicarbocyanine[b] (CY5)	930	600, 650	670
Allophycocyanin	104,000	650	660

[a] Registered trademark of Molecular Probes, Inc.
[b] Registered trademark of Biological Detection Systems, Inc.
[c] Registered trademark of Imperial Chemical Industries.

are phycobiliproteins (e.g., R-phycoerythrin, 240,000 kDa). While these fluorophores have the advantage of fluorescing brightly, their comparatively high molecular weight could be problematic on the basis of hindering antibody saturation of epitopes. In one case (Begg and Hofland, 1991), phycobiliprotein-labeled antibodies were successfully used to stain cytokeratins. The data presented emphasized detection rather than quantification, but suggest some utility for high-molecular-weight fluorescence labels for evaluation of cytoplasmic antigens.

A second problem with the available fluorophores is spectral overlap. While Begg and Hofland (1991) were able to distinguish propidium iodide from phycoerythrin, the antigen detected was cytokeratin, a highly expressed cytoplasmic protein which provided a bright fluorescence signal, necessitating considerable spectral compensation for dual-parameter analysis. For general use, we expect that fluorophores with less spectral overlap will be employed.

A third factor which should be considered when selecting a fluorophore for quantitative IF analysis is the fact that flavin and pyridine nucleotides, lipofuscins, and other molecules in mammalian cells exert considerable intrinsic fluorescence (autofluorescence) which is spectrally defined and varies with cell type (reviewed in Shapiro, 1988). The magnitude of autofluorescence is substantial: It has been reported that murine lymphocytes excited at 488 nm with fluorescence measured >520 nm (typical for analysis of FITC-conjugated mAb) emit autofluorescence at the equivalent of approximately 10,000 FITC molecules. A previous report (Hoffman and Houck, 1991) illustrates the limitation of this feature for the measurement of antigen expression and suggests that autofluorescence may be markedly reduced by using fluorophores which excite at >515 nm. Unfortunately, most multiparameter IF combinations using low-molecular-weight fluorophores with less spectral overlap and reduced autofluorescence will require more than one laser (e.g., Southwick *et al.*, 1990; Waggoner, 1990) and the use of DAPI or Hoechst 33258 for DNA (excitation = 360 nm). Fortunately, multilaser cytometers are commercially available that incorporate light sources compatible with these dyes.

The following procedures should provide a reasonable starting point for the analysis of intracellular proteins labeled with FITC-conjugated antibody probes in either direct or indirect procedures, including a three-layer streptavidin procedure, and DNA staining with propidium iodide.

III. Materials

A. Formaldehyde and Triton X-100 Solutions for Cell Fixation and Permeabilization

Formaldehyde solution (10%) is diluted in a mixture of phosphate-buffered saline (PBS), pH 7.2, to a final concentration of 0.5%. A highly purified solution is recommended and is available from Polysciences, Inc. (Warrington, PA; formaldehyde, 10%, ultrapure, EM grade, Cat. No. 4018). This stock solution is stable for 12 months at room temperature (RT). Diluted formaldehyde kept at 4°C is stable for months. The pH of this solution, however, should be checked occasionally as it has been documented that formaldehyde solutions acidify spontaneously through formic acid formation (Kiernan, 1981).

Triton X-100 (Sigma Chemical Co, St. Louis, MO) is diluted to a 0.1% in PBA (Triton-PBA: PBS + 0.1% Na azide + 0.1% bovine serum albumin, BSA) and stored at 4°C.

B. Alternate Fixation Using Methanol

Absolute methanol is chilled to −70°C.

C. Alternate Fixation with Formaldehyde and Methanol Used Sequentially

Prepare and store 0.1, 0.2, 0.5, and 1.0% formaldehyde solutions in PBS as above. Absolute HPLC grade methanol is chilled to −70°C. Staining and washing is done in NGS (50% PBS, pH 7.0–7.4; 50% normal serum; 0.1% azide). The serum used should be adult serum from the species of animal in which the secondary antibody was raised. The actual percentage of serum used should be determined empirically for each lot of serum by comparing the ability to detect an intracellular antigen under known staining conditions (see below) with varying levels of serum. The serum should be heat inactivated at 56°C for 1 hr, centrifuged to remove fat and large particulates, mixed with PBS, and then filtered with a 0.2 μm pore size filter. This can be stored at −20°C indefinitely.

D. Primary Antibody Solutions for Indirect IF

The primary antibody is suspended in PBA or NGS. The concentration of the antibody must be determined empirically. For quantification of epitope, it is important to stain either in saturation or just prior to saturation (for discussion, see Jacobberger, 1991; Srivastava *et al.*, 1992; and Fig. 1). To determine these end points, samples of test cells expressing the antigen of interest are stained with parallel solutions containing progressively increased dilutions (decreasing concentrations) of specific antibody and isotype control (antibody of identical isotype which does not specifically react with an epitope in the cell type under investigation), followed by secondary antibody incubation at a concentration previously determined to be saturating. Mean IF for each concentration tested of both the specific antibody and isotype control are determined by flow cytometry. The specific fluorescence at each concentration is determined by subtracting the isotype control mean from the positive mean, and then the specific fluorescence is plotted versus the quantity of antibody. Optimal staining for intracellular antigens often occurs at primary antibody concentrations of 1–4 μg per 10^6 cells for high-affinity antibodies. The use of carrier/blocking protein (see Sections III.A. and III.B above) in primary, secondary, and washing solutions will minimize nonspecific fluorescence due to nonspecific antibody binding.

E. Secondary Antibody Solutions for Indirect IF

FITC-conjugated goat anti-mouse immunoglobulin F(ab')$_2$ fragments may reduce the level of nonspecific binding and therefore provide a more specific fluorescence signal than labeled whole IgG secondary reagents. The antibody concentration should be determined empirically; the end point is a constant level of specific fluorescence for antigen expression cells as a function of secondary antibody concentration (saturation) following titering of the primary antibody to determine its saturation. Additionally, the specificity of staining may be improved by using antibody fragments which are affinity purified or adsorbed against species-specific immunoglobulin (Grinnelikhuijzen et al., 1982). If the primary/secondary reagents are being titered for the first time, the primary antibody should be retitered to ensure that the saturation curve for the primary antibody was not generated under conditions where the secondary antibody is limiting. Once the concentrations have been determined for a single primary and secondary pair, titering of additional reagents can be done with one or the other of these, thus simplifying the process.

F. DNA Staining Solutions

Ribonuclease A is required to eliminate RNA binding by PI. Since many RNase preparations are relatively crude and may contain DNases and proteases, a purified preparation is recommended. A simple method of eliminating unwanted activity is to heat a crude preparation in boiling water for 5 min. Upon cooling, the protein will renature and enzymatic activity will be retained. Otherwise, a purified RNase preparation should be purchased [e.g., Worthington Biochemical Corp., Freehold, NJ, Cat. code RASE (3600 units/ml) or Sigma Chemical Co., St. Louis, MO, Cat. No. R-5503 (0.04 Kunitz units]. RNase can be made up in PBS and stored at $-20°C$.

Propidium iodide can be purchased from several vendors (e.g., Calbiochem, San Diego, CA, Cat. No. 537059). A stock solution can be made up in distilled water (1 mg/ml) or PBS (100 μg/ml). Propidium is a suspected carcinogen and should be handled and discarded appropriately. Particular caution is recommended when weighing out pure dyes in powder form to avoid inhalation.

IV. Staining Procedures

A. Cell Fixation (Formaldehyde/Triton X-100)

1. Treat 1–2 $\times$ 10^6 cells with 1 ml 0.5% formaldehyde solution (a near isosmotic solution of 0.5% formaldehyde can be made by mixing 0.5 ml of

10% formaldehyde, 5.1 ml distilled H_2O, and 4.4 ml PBS). Duration of fixation is determined empirically (see Clevenger *et al.*, 1985).

2. Centrifuge cells 10 min at 200*g*.

3. Add 0.1% Triton-PBA solution so that the concentration of cells is 10^6/ml. Incubate for 3 min on ice. Note: It is recommended that IF be evaluated over a range of Triton-PBA concentrations (e.g., 0–0.5% Triton X-100) and durations (e.g., 1–10 min) to optimize IF staining.

B. Alternate Fixation Using Methanol

1. Centrifuge 1–2 × 10^6 cells. Pour off supernatant, resuspend cell pellet.

2. Rinse cells twice in PBS, carefully resuspending cell pellet in each case.

3. Add 100 μl PBS. Add 900 μl MeOH cooled to $-70°C$ dropwise while vortexing tube gently.

4. Incubate 30 min at $-20°C$.

5. Centrifuge, pour off supernatent, resuspend pellet.

6. Repeat step 2.

C. Cell Fixation (Formaldehyde/Methanol)

1. Wash 1–2 × 10^6 cells with PBS.

2. Resuspend in 50 μl formaldehyde solution (concentration determined empirically, see above). Incubate 10 min at 37°C.

3. Cool sample at 4°C for 10 min.

4. Stop reaction and permeabilize by adding 450 μl absolute MeOH at -20 or $-70°C$.

5. Store at $-20°C$ until ready for staining.

D. Indirect Immunofluorescence Staining

1. For Cells Fixed by Procedure A

1. Wash once in PBA (*optional*).

2. Resuspend in remaining supernatant and add 30 μl of primary mAb at appropriate concentration (determined by titration) and isotype control at the same concentration in a parallel tube. Dilute with 970 μl PBA per 10^6 cells. Incubate 20–60 min at 4°C.

3. Wash in 0.1% Triton-PBA, then in 1 ml of PBA, carefully aspirating down to cell pellet. Centrifuge.

4. Add 50 μl FITC-conjugated goat anti-mouse immunoglobulin antibody (properly titrated, e.g., 1:20, in PBA) to resuspended cell pellet. Incubate 30 min 4°C.

5. Centrifuge. Wash in 0.1% Triton-PBA.
6. Repeat step 5.

2. For Cells Fixed by Procedure B or C

1. Wash 1–2 × 10^6 cells twice with 500 μl cold PBS (*mandatory*).
2. Resuspend cells in remaining supernatant and add 50 μl primary mAb to one sample and the same amount of isotype control mAb to a second sample. Both antibodies are at the same concentration. Incubate at 37°C for 30–90 min.
3. Cool to 4°C (10 min). Wash twice at 4°C with NGS (500 μl per wash) for 15 min per wash.
4. Resuspend pellet in 50 μl secondary antibody (FITC goat anti-mouse F(ab')$_2$ fragments). Concentration is determined empirically (see above). Incubate at 37°C for 30–90 min.
5. Cool to 4°C (10 min). Wash three times at 37°C with NGS (500 μl per wash) for 15 min per wash.

Note: This procedure can be shortened with good results by washing once in step 1, eliminating cooling to 4°C in steps 3 and 5, reducing to one wash in step 3, and reducing to two washes in step 5. This will result in higher background staining; but if the antigen is expressed at sufficiently high levels, the results will be quite acceptable. This has been successful on a human prostatic carcinoma cell line, DU-145, that has been transformed with SV40 T antigen. Antigens detected were SV40 T antigen expressed from a Moloney Murine Leukemia Virus LTR, cytokeratins, PCNA, p34^{cdc2}, retinoblastoma protein (Rb), and stabilized p53.

E. Propidium Iodide Staining

1. Treat cells with 500 μl RNase (dilute RNase stock in PBA or NGS to achieve 180 units or 0.04 Kunitz units) 20 min at 37°C.
2. Centrifuge, then resuspend in PI (50 μg/ml) and incubate for 1 hr at 4°C. This concentration is a 1:20 dilution of the 1 mg/ml stock. An alternative procedure is to add 500 μl PI at 100 μg/ml in an equal volume of PBS directly to the cells in the RNase solution. Incubation time remains the same.

F. Synthetic Peptide Inhibition

This method is outlined by Kastan *et al.* (1989b).

1. Prior to addition of primary antibody, add 40 μl fetal bovine serum (final concentration 4%) to each tube.

2. Add microgram amounts of appropriate peptide to each antibody prior to adding the antibody to the sample tube (a nonreacting peptide of equal length and amino acid composition should be used as a control). The amount of peptide should be determined empirically; the end point is maximum inhibition.

V. Critical Aspects

A. Antigen Sensitivity to Proteolysis

Trypsin is a protease that hydrolyzes peptide and ester bonds at lysine or arginine residues. Pepsin has a broad range of nonspecific proteolytic activities. Among the many cleavage sites, pepsin preferentially cleaves bonds between hydrophobic residues and the amino acids leucine, phenylalanine, methionine, and tryptophan. It should be recognized that a proteolytic enzyme that is generally considered to be specific for a particular protein (e.g., collagenase) often can be contaminated with other proteases. This problem can be overcome by using highly purified enzyme preparations.

Intracellular proteins are less susceptible to the effects of proteolysis than those localized to the cell surface. IF analysis of nuclei from paraffin-embedded tissue is more problematic however, since proteases used to assist in freeing the nuclei come into direct contact with the protein of interest. When working with solid tissues or attached cells *in vitro,* it is important to understand whether routine proteolysis affects the antigen of interest. Cell lines that grow in suspension or monolayer cultures which are easily dissociated without enzymatic treatment and that express the antigen of interest can be used to address this possibility. The cells are divided into two groups (control vs enzyme-treated) and then analyzed quantitatively for IF by FCM and/or by immunoblot analysis. An example of this approach is provided by Lincoln and Bauer (1989). If sufficient fresh tissue is available, tissue can be directly (Morkve and Hostmark, 1991).

B. Choice of Fixation/Permeabilization Procedure

In this report we emphasize three strategies for cell fixation/permeabilization. It should be stressed, however, that fixation can be highly antigen specific. An example of this is shown in Fig. 3. Generalizations are: (1) Unbound or loosely bound proteins are lost from the cell by permeabilization techniques (detergent $\geqslant$ alcohol). (2) Formaldehyde fixation minimizes this problem. (3) When pursuing multiparameter (IF + DNA content) analysis, this protective effect of formaldehyde crosslinking must be counterbalanced with the fact that such crosslinking impairs high-resolution DNA staining. Aside from these, there are effects on epitope distribution and availability that are dependent on the type of detergent or organic solvent used. Therefore, the procedures discussed here represent good starting procedures. If expected results are not obtained, then other methods (overviewed in Jacobberger, 1991, and Clevenger and Shankey, 1993) should be considered. Alternate fixation procedures include the combination of formaldehyde and saponin, lysolecithin, digitonin, or *n*-octyl-*b*-D-glucopyranoside (Rabin *et al.*, 1989; Dent *et al.*, 1989; Anderson *et al.*, 1989; Hallden *et al.*, 1989); acetone (Gerdes *et al.*, 1984); and ethanol (Landberg *et al.*, 1990).

From a practical standpoint, MeOH and other alcohol-based fixatives are hypotonic and can result in cell lysis and clumping of cell suspensions submitted for FCM. Such damage can be minimized by adding the fixative slowly (e.g., dropwise) or by adding a very cold MeOH solution (e.g., $-70°C$). Cells fixed in MeOH can be stored for long periods of time. When long-term holding of cells is required for IF studies utilizing formaldehyde/Triton X-100 treatments, it is generally recommended that the cells be stored at $-20°C$ in conventional freezing medium utilized for maintaining clonogenic cell lines following formaldehyde fixation. Following thawing, Triton X-100 permeabilization can be performed. The reader will note that the procedures provided illustrate variations in the time and temperature of fixation, immunofluorescence staining, and washing. This again emphasizes the importance of empirical studies to optimize the measurement of a specific protein (epitope).

Two other strategies are noteworthy from the standpoint of recent insights into cell fixation/permeabilization considerations. First, Bruno *et al.* (1992) have demonstrated that altering the ionic strength of the fixative affects IF staining of PCNA or Ki-67. This may be due to disruption of molecular interactions during the fixing procedure itself. This finding provides an additional variable which should be considered in the optimization of methodology for intracellular protein assessment. Second, Landberg and Roos (1991) and Sasaki *et al.* (1993) have shown that residual PCNA following partial extraction with 0.1% Triton X-100 (presumably tightly bound to DNA) is highly associated with the S phase of the cell cycle relative to total PCNA. These studies provide a basis for future investigations aimed toward discriminating subcomponents of a protein of interest which may vary functionally and in terms of association with the cell cycle.

C. Sodium Dodecyl Sulfate (SDS)–Gel Electrophoresis and Immunoblotting (Verifying mAb Specificity)

It is important to emphasize that an anti-peptide mAb recognizes an epitope that consists of an oligopeptide on the order of approximately 6 to 10 amino acid residues. This same epitope may be present in more than one protein, or similar critical molecular structure may be achieved with sequences different from the immunizing peptide that will react well with the mAb. In each of these cases, the spurious epitope is crossreactive with the mAb against the antigen of interest. It is acceptable to utilize commercially available antibodies that are in wide use on cells previously tested by others to be free of crossreactive molecules. However, if the cells under study have not been previously tested, or if previously tested cells are used but the experimental treatment is unusual, or if poorly characterized mAbs are used, then there should be vertification that the antigen of interest has been exclusively analyzed. This can be accomplished with SDS–polyacrylamide gel electrophoresis (PAGE), followed by transfer of the separated proteins to nitrocellulose paper (or similar product), and immunostaining using the same mAb in combination with a secondary antibody labeled with biotin or enzymes that generate a colored precipitate (alkaline phosphatase or peroxidase). If biotin-labeled antibodies are used, an enzyme-conjugated streptavidin would be used. The correct band corresponding to the correct molecular weight should be detected without detection of other bands. The equipment and procedures to perform this analysis are available from many sources (e.g., Bio-Rad, Mellville, NY). General references are Gallagher and Smith (1992) and Winston et al. (1992).

It is also important to recognize that epitopes can be masked. This could occur through molecules that specifically react with the epitope of interest (e.g., protein: protein binding), modification of the epitope to an unreactive form (e.g., phosphorylation, farnesylation, acetylation), or restriction of antibody access (e.g., cell proliferation- or anaplasia-related changes in chromatin structure). To evaluate the possibility of these effects, parallel analyses are recommended in which IF levels by FCM are compared with immunochemical detection by SDS–PAGE followed by immunoblot analysis on cell lysates (Clevenger et al., 1987a, b). This analysis requires a standard cell and antigen that are unaffected for comparison or two (or more) samples of cells expressing different levels of the antigen of interest (Engelhard et al., 1991).

This type of verification has one other important application. For some studies, cells are treated with protein synthesis inhibitors to estimate the half-life of an antigen (Rabbitts et al., 1985; Engelhard et al., 1991). The epitope is not necessarily destroyed at the same rate as the protein itself. Thus, parallel FCM and immunoblot procedures are recommended. These results should be compared to more conventional pulse-chase experiments.

D. Cell Proliferation-Associated Expression of Intracellular Antigens

Cell proliferation-associated antigens by definition are expressed at high levels in proliferating cells and at low or undetectable levels in noncycling cells. When the difference in expression between the two cell proliferation states (G_0 and G_1) is not readily apparent, and when the difference in expression levels is not high, cell proliferation-associated expression can be determined by both FCM and immunoblot procedures. For immunoblotting, a constant amount of protein should be loaded on the gel for two or more samples representing different states of proliferation. For FCM, specific IF can be combined with total protein staining (Engelhard et al., 1991). If a difference in expression of the antigen of interest between cycling and noncycling cells exceeds the increase in total cell protein, expression of the antigen can be classified as cell proliferation associated.

It is incorrect to assume that proteins having important roles in regulating cell growth or cell-cycle traverse are necessarily expressed in a cell proliferation-associated fashion. For example, tree mammalian nuclear proteins to date have been shown to be rate limiting for G_1 traverse: c-myc (Karn et al., 1989), SV40 T antigen (Sladek and Jacobberger, 1992a, b), and cyclin E (Ohtsubo and Roberts, 1993). Cyclin E, a G_1 cyclin, is expressed in a cell-cycle-associated fashion with preferential expression in late G_1 and early S phase relative to G_2M (Gong et al., 1993). Alternatively, the levels of c-myc and SV40 T antigen accumulate approximately twofold when average levels in G_2M are compared to average levels in G_1: the average accumulation of total cell mass.

Other developments have extended approaches for assessing the cell proliferation-associated expression of intracellular proteins. First, cells stained for IF have been differentially sorted on the basis of level of expression of the protein of interest and subsequently subjected to the acidic acridine orange staining method of Darzynkiewicz and colleagues (1975) which differentiates G_0 vs cycling cell populations. The feasibility of this method was first demonstrated for p34, a cell proliferation-associated nuclear protein (Bauer et al., 1986a). This approach allows one to directly relate protein expression to the G_0 vs cycling cell-cycle compartments. A second advance comes from a study by Bruno et al., (1991) who examined IF associated with PCNA, Ki-67 Ag, p105, and p34 temporally during S phase and documented substantial variation in the expression of these cell proliferation-associated proteins during this cell-cycle phase. Finally, another, but as yet untested, approach would involve staining for two intracellular antigens, one expressed in a cell proliferation-associated fashion and the second known to be associated with cellular quiescence (e.g., statin) (Wang, 1987). This would require dual-parameter staining which may now be more feasible (see Section II).

E. Synthetic Peptide Blocking

If the epitope that a mAb recognizes is known, synthetic peptides can be used to determine the level of specific staining by inhibition of the specific IF signal. In practice, nonspecific IF is quantified and compared to total IF in the positive sample. This is not different than using an isotype control antibody. However, since the use of an isotype control is predicated on using equal amounts of specific and nonspecific antibody, synthetic peptides are useful when the level of specific antibody is not known, i.e., unevaluated hybridoma supernatant, serum, unpurified ascites containing contaminating antibody. However, this procedure does not establish specificity, since it does not control for crossreactivity (see above). Immunoblotting or immunoprecipitation is required to confirm specificity. Kastan et al. (1989a,b, 1991a,b) and Kuerbitz et *al.* (1992) have utilized this procedure to establish the level of specific staining for the protein products of several oncogenes.

F. Tissue Sampling and Handling Considerations

When analyzing tissue specimens including solid tumors, considerable cellular heterogeneity is commonly present which can impact on the interpretation of IF results. For example, a report by Crissman and colleagues (1989) suggests that colonic cancer specimens on average contain less than 27% neoplastic cells, the majority consisting of host inflammatory and stromal cells. Thus, one-parameter IF measurements performed on such specimens primarily reflect these nonneoplastic elements. Many studies (see Herman *et al.,* 1987) have appeared in the literature which advocate the use of IF analysis of intracellular proteins such as cytokeratin (a family of intermediate filament proteins specific to epithelial cells) to enrich for neoplastic cells from carcinomas in such circumstances. Li *et al.* (1993) have shown the utility for three-parameter analysis of cytokeratin (using anti-cytokeratin mAb conjugated with FITC), c-erbB2 (using anti-c-erbB2 mAb and indirect IF with AMCA), and DNA content using PI in human breast cancer specimens. This approach appears highly attractive for more precise analysis of relevant intracellular proteins, including oncoproteins when one wishes to assess expression in the neoplastic component of the specimen. An example of results obtained using this same approach applied to analysis of pan-ras expression in a colonic cancer specimen is shown on Fig. 4.

Tissue handling considerations are of fundamental importance when analysis of intracellular proteins is pursued. It is well known, for example, that prolonged holding of tissue at room temperature can result in autolysis and endogenous protein degradation, impairing subsequent measurement. In addition, many oncoproteins have short half-lives (10–30 min), and presumably each step of tissue processing can affect the detected level of protein expression. For basic investigation, compulsiveness in tissue handling is recommended as the usual rationale for performing the analysis is to extrapolate the result to the mechanics of cell function at the time the sample was obtained. For clinical studies the end point may be evaluation of differential expression allowing for the discrimination of normal and neoplastic cells, irrespective of the actual relationship to the true biology. One hypothetical example involves p53. Normal p53 has a half-life of ~15 min whereas many mutant p53 molecules are overexpressed and have significantly increased half-lives relative to wild-type p53. If the balance between synthesis and degradation is shifted toward degradation at the time of tissue removal, the time between removal, dissociation, and fixation may span many half-lives of the protein. Thus, the long interval of processing and staining of the sample could fortuitously enhance the identification of abnormal and normal cells and provide a potential diagnostic advantage.

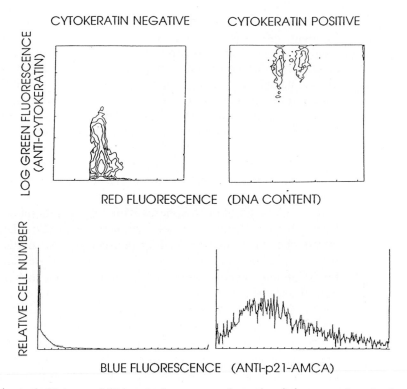

Fig. 4 Anti-p21 (pan ras) IF in colonic cancer specimens in relation to cytokeratin. A colonic adenocarcinoma specimen was mechanically dissociated, stained, and analyzed for cytokeratin (using a FITC-conjugated anti-cytokeratin mAb), p21 (pan ras, using anti p21 mAb and indirect IF using an AMCA-conjugated secondary antibody), and DNA content (propidium iodide). Following electronic gating to resolve cytokeratin-positive and cytokeratin-negative cells (upper panels), the anti-p21 fluorescence of cytokeratin-positive cells (lower right panel; neoplastic and normal mucosa cells) is approximately seven-fold greater than for cytokeratin-negative cells (lower left panel; constituting approximately 85% of total cells consisting primarily of host inflammatory and stromal cells) from the specimen (S. P. Harlow and K. D. Bauer, unpublished results).

VI. Controls and Standards

Current standards for instrument alignment most commonly utilize nonbiological particles with relatively stable and homogenous fluorescence intensity and light scattering characteristics. A second type of standard that may be crucial for many applications is a staining standard. Cells known to contain the antigen of interest can serve as a useful staining standard for quantitative IF measurements. By parallel IF staining of test cells, this type of biological standard can be used to measure relative levels of the antigen of interest. While cells continuously cultured *in vitro* can be useful for this purpose, it is important to recognize that the level of the antigen of interest can modulate with respect to culture growth phase and/or passage (Engelhard and Bauer, 1991; Kuhar and Lehman, 1991). Thus, a large number of cells from a *single* culture placed in aliquots into vials and frozen in a cryogenic freezer are preferable. These cell stocks provide a consistent biological standard to compare run-to-run variation in staining by comparison to standard beads and to normalize test sample results across time. In one study (Schimenti and Jacobberger, 1992), this staining variation resulted in a CV of 17%. This value could be reduced by careful attention to staining times etc., because the CV for several samples stained on the same day is less than 5%. A useful variation would be to have two or more standard cells with varying amounts of the antigen of interest. Such cells can be generated with expression vectors created by molecular genetics (e.g., Sladek and Jacobberger, 1992a).

Recently, a set of microspheres with varying amounts of mouse IgG binding sites (''Simply Cellular,'' Flow Cytometry Standards, Durham, NC) has been introduced. These can be used for indirect assays, to determine the amount of *primary* antibody bound to a test cell sample. A standard curve is generated for IF intensity versus number of antibody binding sites, allowing for enumeration of antibody content from the test cell sample by extrapolation from the standard curve. This capability will become increasingly important as the field advances from relative to absolute IF measurements. Knowing the number of antibody binding sites can then be used to estimate the number of epitopic sites. The usual caveats apply here with regard to interpretation of IF results, i.e., it is important to remember that available epitope is the actual measurement (see Section V,C.).

Finally, it is widely recognized that cell viability can affect IF measurements by FCM. To circumvent this problem propidium iodide has widely been used as a vital dye, with dead cells (having high levels of propidium iodide fluorescence) being excluded by electronic gating. An alternate dye, ethdium monoazide has been similarly used to assess cell viability following cell fixation (Riedy *et al.*, 1991). A recent report documents that ethidium monoazide staining can be coupled with IF analysis to evaluate differential intracellular IF in viable vs dead cells (Reddy *et al.*, 1992).

VII. Instruments

Simultaneous analysis of DNA content and cell proliferation-associated antigens is most often performed with the combination of propidium iodide and fluoresceinated antibodies. Both fluorochromes are excited by the 488-nm line of argon ion lasers. Any commercial cytometer equipped with an argon laser, three detectors, and a filter set composed of a narrow band pass at 525–530 nm for fluorescein, a 640-nm long-pass for propidium iodide, and a dichroic at 550 nm or greater to split the emission is sufficient for this analysis. In the future, as the field advances toward simultaneous multiantigen analysis, multiple lasers most likely will be employed. Instruments that have the optical capabilities to excite at 326–361, 488, 530–560, and 633–647 nm and detect the resulting emission of fluorophores of interest are well suited for these studies. At present, commercial instruments do not have the capability to separate more than two interrogation points in space. However, three or four lasers can be employed by coincident alignment of pairs of lasers. The implementation of the mixed-gas argon and krypton laser and the frequency-doubled diode-pumped neodymium:YAG laser further extends such possibilities.

VIII. Multiparameter Analysis of Intracellular Antigens

To date, the analysis of cell proliferation-associated nuclear proteins has focused primarily on two-parameter analysis of IF and DNA content, where a

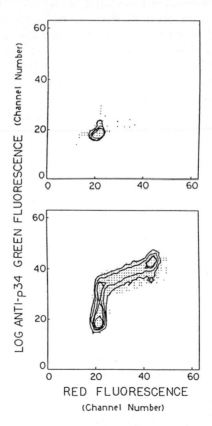

Fig. 5 Dual-parameter contour plots of peripheral blood lymphocytes (PBL) fixed with 0.5% formaldehyde–0.1% Triton X-100 and stained with anti-p34 mAb (indirect IF) and PI. (A) 0 hr PBL. (B) Cultures stimulated with pokeweed mitogen (5 μg/ml) in culture medium for 72 hr. Note expression of p34 in G_1, S, and G_2M cell cycle compartments following mitogen stimulation (panel B) relative to quiescent PBL (panel A).

"G_0"-like population is identified on the basis of reduced or indistinguishable expression of the nuclear protein of interest, relative to the cycling cells. An example of this type of analysis comparing unstimulated and pokeweed mitogen-stimulated peripheral blood leucocytes is shown in Fig. 5. The development of low-molecular-weight fluorophores for antibody conjugation that absorb and emit in regions other than 520–640 nm now permits multiparameter IF analysis of such proteins. It is our intention in this section to present some early (and hopefully, interesting) results that may inspire the reader toward this goal.

Figure 6 displays results from an experiment in which pbls were stained with a primary conjugate (FITC) anti-PCNA and primary conjugate (AMCA) anti-Ki-67 Ag. The left panel shows the results for control (unstimulated) cells and the right panel shows the results after 24 hr of stimulation which is prior to the onset of DNA synthesis. The results show that there is a preferential increase in the expression of PCNA relative to Ki-67 antigen. At 24 hr, 35% of the cells are positive for PCNA expression, whereas 2% are expressing Ki-67 at levels above the threshold determined by background IF. This indicates that PCNA is expressed prior to Ki-67 antigen during cell-cycle traverse. While much work is still required to take this type of analysis to the current level of understanding of single IF analysis, it should be clear that analysis of the ratio of expression levels of two rate controlling gene products (e.g. for the rate of cell proliferation or rate of differentiation) provides significant potential for discriminating normal from aberrant cell growth. Further, when combined with modern molecular genetic methodology, such multiparameter FCM should prove to be valuable

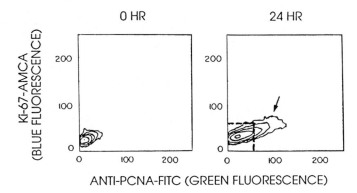

Fig. 6 Simultaneous analysis of PCNA and Ki-67 Ag. Dual-parameter contour plots of human peripheral blood leukocytes without mitogenic stimulation (left panel) or following 24 hr of stimulation with phytohemagglutinin (right panel). PCNA staining was accomplished using a FITC-conjugated mAb, whereas Ki-67 staining was performed using indirect immunofluorescence, in combination with a goat anti-mouse AMCA-conjugated secondary antibody. Staining was performed following cold methanol fixation. The dashed line illustrates the region encompassed by 0-hr cultures within the 24-hr cultures. The arrow illustrates the subpopulation of cells with markedly enhanced expression of PCNA with little or no change in Ki-67 Ag (A. McNally and K. D. Bauer, unpublished results).

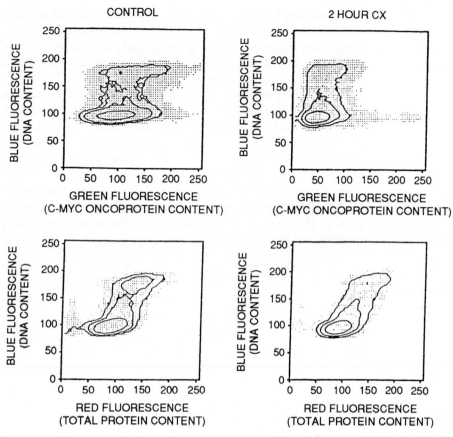

Fig. 7 Simultaneous analysis of c-myc oncoprotein, total cellular protein, and DNA content. Dual-parameter contour plots of HL-60 cells in exponentially grown (left panels) or following 2 hr of treatment with 2.5 μg/ml cycloheximide (right panels). Green (FITC) fluorescence (anti-c-myc IF) vs blue (DAPI) fluorescence (DNA content) is displayed on the top row; whereas red (sulforhodamine 101) fluorescence (total protein) vs DNA content is displayed on the bottom row. Note that whereas the cycloheximide has little effect on total cellular protein after 2 hr, the level of c-myc IF is reduced by a factor of approximately 0.5 after this period. (Adapted from Engelhard, Krupka, and Bauer, 1991)

tools for investigations of altered growth control mechanisms that result in neoplastic transformation at the molecular level. In this regard, the effects on the cell cycle of introducing a gene can be measured in 24 hr if retroviral vectors are used (Sladek and Jacobberger, 1990). If one is studying drug or cytokine treatments in G_0 cells, and the end point is increased expression of one or more of the immediate early response genes, the effects of a treatment can be quantitatively determined within one to several hours. The ratio between the expression of two rate controlling genes has the potential advantage of increasing sensitivity however, so that the treatment should not be limited to quiescent cells and thus the effects may be observed much more readily. While this is conjecture at this point, the reagents and instrumentation are now available to verify whether this notion is correct. One study that suggests that this contention may be correct is presented in Fig. 7.

As discussed (Section V,D) when evaluating the expression of cell proliferation-associated proteins, parallel measurement of cell mass (i.e., total cellular protein) is useful in interpreting changes of IF. Examples of changes in the c-myc oncoprotein in relation to both DNA content and total cellular protein following treatment with (2.5 μg/ml) cycloheximide are shown on Fig. 7. Note that whereas no discernible change in total cellular protein is observed following 2 hr of treatment with this protein synthesis inhibitor, an approximately 50% reduction in c-myc, a protein with a known short half-life is seen. Other examples of the analysis of intracellular proteins are referenced throughout this chapter.

References

Anastasi, J. A., Bauer, K. D., and Variakojis, D. V. (1987). *Am. J. Pathol.* **128**, 573–581.

Anderson, P., Blue, M., O'Brien, C., and Schlossman, S. F. (1989). *J. Immunol.* **143**, 1899–1904.

Baisch, H., and Gerdes, J. (1987). *Cell Tissue Kinet.* **20**, 387–391.

Bauer, K. D. (1988). *Pathol. Immunopathol. Res.* **7**, 371–380.

Bauer, K. D., Clevenger, C. V., Williams, T. J., and Epstein, A. L. (1986a). *J. Histochem. Cytochem.* **34**, 245–251.

Bauer, K. D., Clevenger, C. V., Endow, R. K., Murad, T. M., Epstein, A. L., and Scarpelli, D. G. (1986b). *Cancer Res.* **46**, 2428–2434.

Begg, A. C., and Hofland, I. (1991). *Cytometry* **12**, 445–454.

Bolton, W. E., Mikulka, W. R., Healy, C. G., Schmittling, R., and Kenyon, N. (1992). *Cytometry* **12**, 117–126.

Bruno, S., Crissman, H. A., Bauer, K. D., and Darzynkiewicz, Z. (1991). *Exp. Cell Res.* **196**, 99–106.

Bruno, S., Gorczyca, W., and Darzynkiewicz, Z. (1992). *Cytometry* **13**, 496–501.

Clevenger, C. V., and Shankey, T. V. (1993). *In* "Clinical Flow Cytometry: Principles and Application" (K. D. Bauer, R. E. Duque, and T. V. Shankey, eds.), pp. 157–175. Williams & Wilkins, Baltimore, MD.

Clevenger, C. V., Bauer, K. D., and Epstein, A. L. (1985). *Cytometry* **6**, 208–214.

Clevenger, C. V., Epstein, A. L., and Bauer, K. D. (1987a). *J. Cell. Physiol.* **130**, 336–343.

Clevenger, C. V., Epstein, A. L., and Bauer, K. D. (1987b). *Cytometry* **8**, 280–286.

Crissman, J., Zarbo, R., Ma, C., and Visscher, D. (1989). *Pathol. Annu.* **24**(Part II), 103–147.

Darzynkiewicz, Z., Traganos, F., Sharpless, T., and Melamed, M. R. (1975). *Exp. Cell Res.* **90**, 411–417.

Dent, G. A., Leglise, M. C., Pryzwansky, K. B., and Ross, D. W. (1989). *Cytometry* **10**, 192–198.

Drach, J., Gattringer, C., Glassl, H., Schwarting, R., Stein, H., and Huber, H. (1989). *Cytometry* **10**, 743–749.

Engelhard, H. H., and Bauer, K. D. (1991). *Cytometry Supplement* **5**, 74.

Engelhard, H. H., Krupka, J. L., and Bauer, K. D. (1991). *Cytometry* **12**, 68–76.

Epstein, A. L., and Clevenger, C. V. (1985). *In* "Recent Advances in Non-histone Protein Research" (I. Bekhor, ed.), Vol. 2, pp. 117–137. CRC Press, Boca Raton, FL.

Fonagy, A., Swiderski, C., Wilson, A., Bolton, W. E., Kenyon, N., and Freeman, J. W. (1992). *J. Cell. Physiol.* **154,** 16–27.

Gallagher, S. R., and Smith, J. A. (1992). *In* "Current Protocols in Immunology" (J. E. Coligen, A. M. Kruisbeck, D. H., Margulies, E. M. Shevach, and W. Strober, eds.), pp. 8.4.1–8.4.21. Wiley, New York.

Gerdes, J., Lemke, H., Baisch, H., Wacker, H. H., Schwab, U., and Stein, H. (1984). *J. Immunol.* **133,** 1710–1715.

Giaretti, W., Di Vinci, A., Geido, E., Marsano, B., Minks, M., and Bruno, S. (1990). *Cell Tissue Kinet.* **23,** 473.

Gong, J., Traganos, F., and Darzynkiewicz, Z. (1993). *Int. J. Oncol.* **3,** 1037–1042.

Grinnelikhuijzen, C. J. P., Dierickx, K., and Boer, G. J. (1982). *Neuroscience* **7,** 3191–3199.

Hallden, G., Andersson, U., Hed, J., and Johansson, S. G. O. (1989). *J. Immunol. Methods* **124,** 103–109.

Helenius, A., McCaslin, D. R., Fries, E., and Tanford, C. (1979). *In* "Methods in Enzymology" (S. Fleischer and L. Packer, eds.), Vol. 56, pp. 734–749.

Herman, C. J., McGraw, T. P., Marder, R. J., and Bauer, K. D. (1987). *Arch. Pathol. Lab. Med.* **111,** 505–512.

Hoffman, R. A., and Houck, D. W. (1991). *Cytometry Supplement* **5,** 116.

Houck, D. W., and Loken, M. R. (1985). *Cytometry* **6,** 531–538.

Jacobberger, J. W. (1989). *In* "Flow Cytometry: Advanced Research and Clinical Applications" (A. Yen, ed.), pp. 305–326. CRC Press, Boca Raton, FL.

Jacobberger, J. W. (1991). *Methods* **2,** 207–218.

Jacobberger, J. W., Fogleman, D., and Lehman, J. M. (1986). *Cytometry* **7,** 356–364.

Karn, J., Watson, J. V., Lowe, A. D., Green, S. M., and Vedeckis, W. (1989). *Oncogene* **4,** 773–787.

Kastan, M. B., Slamon, D. J., and Civin, C. I. (1989a). *Blood* **73,** 1444–1451.

Kastan, M. B., Stone, K. D., and Civin, C. I. (1989b). *Blood* **74,** 1517–1524.

Kastan, M. B., Radin, A. I., Kuerbitz, S., Onyekwere, O., Wolkow, C. A., Civin, C. I., Stone, K. D., Woo, T., Ravindranath, Y., and Craig, R. W. (1991a). *Cancer Res.* **51,** 4279–4286.

Kastan, M. B., Onyekwere, O., Sidransky, D., Vogelstein, B., and Craig, R. W. (1991b). *Cancer Res.* **51,** 6304–6311.

Kiernan, J. A. (1981). "Histological and Histochemical Methods: Theory and Practice," pp. 8–20. Pergamon, Oxford.

Kuerbitz, S. J., Plunkett, B. S., Walsh, W. V., and Kastan, M. B. (1992). *Proc. Natl. Acad. Sci. U.S.A.* **89,** 7491–7495.

Kuhar, S. G., and Lehman, J. M. (1991). *Oncogene* **6,** 1499–1506.

Kung, A. L., Sherwood, A. L., and Schimke, R. T. (1993). *Proc. Natl. Acad. Sci. U.S.A.* **268,** 23072–23080.

Kurki, P., Vanderlaan, M., Dolbeare, F., Gray, J., and Tan, E. M. (1986). *Exp. Cell Res.* **166,** 209–219.

Kurki, P., Ogata, K., and Tan, E. M. (1988). *J. Immunol. Methods* **109,** 49–58.

Landberg, G., and Roos, G. (1991). *Cancer Res.* **51,** 4570–4574.

Landberg, G., Tan, E. M., and Roos, G. (1990). *Exp. Cell Res.* **187,** 111–118.

Li, B. D. L., Bauer, K. D., Carney, W. P., and Duda, R. B. (1993). *J. Surg. Res.* **54,** 179–188.

Lincoln, S. T., and Bauer, K. D. (1989). *Cytometry* **10,** 456–462.

Mann, G. J., Dyne, M., and Musgrave, E. A. (1987). *Cytometry* **8,** 509–517.

Morkve, O., and Hostmark, J. (1991). *Cytometry* **12,** 622–627.

Morkve, O., and Laerum, O. D. (1991). *Cytometry* **12,** 438–444.

Ohtsubo, M., and Roberts, J. M. (1993). *Science* **259,** 1908–1912.

Pearse, A. G. E. (1980). *In* "Histochemistry: Theoretical and Applied" (A. G. E. Pearse, ed.), Vol. 1, 4th ed., pp. 97–158. Churchill Livingstone Press, New York.

Pollice, A. A., McCoy, J. P., Shackney, S. E., Smith, C. A., Agarwal, J., Burholt, D. R., Janocko, L. E. Hornicek, F. J., Singh, S. G., and Hartsock, R. J. (1992). *Cytometry* **13,** 432–444.

Rabbitts, P. H., Watson, J. V., Lamond, A., Forster, A., Stinson, M. A., Evan, G., Fischer, W., Atherton, E., Sheppard, R., and Rabbitts, T. H. (1985). *EMBO J.* **4,** 2009–2015.

Rabin, H., Trimpe, K. L., and Hamer, P. J. (1989). *Cancer Cells* **7,** 157–160.

Reddy, S., Bauer, K. D., and Miller, W. M. (1992). *Biotechnol. Bioeng.* **40,** 947–964.

Riedy, M. C., Muirhead, K. A., Jensen, C. P., and Stewart, C. C. (1991). *Cytometry* **12,** 133–139.

Rosette, C. D., DeTeresa, P. S., and Pallavicini, M. G. (1990). *Cytometry* **11,** 547–551.

Sasaki, K., Kurose, A., and Ishida, Y. (1993). *Cytometry* **14,** 876–882.

Schimenti, K. J., and Jacobberger, J. W. (1992). *Cytometry* **13,** 48–59.

Schmid, I., Uittenbogaart, C. H., and Giorgi, J. V. (1991). *Cytometry* **12,** 279–285.

Shapiro, H. M. (1988). *In* "Practical Flow Cytometry," (H. Shapiro, ed.), 2nd ed., pp. 156–157. Liss, New York.

Shi, S., Key, M. E., and Kalra, K. L. (1991). *J. Histochem. Cytochem.* **39,** 741–748.

Sladek, T. L., and Jacobberger, J. W. (1990). *J. Virol.* **64,** 3135–3138.

Sladek, T. L., and Jacobberger, J. W. (1992a). *Oncogene* **7,** 1305–1313.

Sladek, T. L., and Jacobberger, J. W. (1992b). *J. Virol.* **66,** 1059–1065.

Sladek, T. L., and Jacobberger, J. W. (1993). *Cytometry* **14,** 23–31.

Southwick, P. L., Ernst, L. A., Tauriello, E. W., Parker, S. R., Mujumdar, R. B., Mujumdar, S. R., Clever, H. A., and Waggoner, A. S. (1990). *Cytometry* **11,** 418–430.

Srivastava, P., Sladek, T. L., Goodman, M. N., and Jacobberger, J. W. (1992). *Cytometry* **13,** 711–721.

Waggoner, A. S. (1990). *In* "Flow Cytometry and Sorting" (M. R. Melamed, T. Lindmo, and M. L. Mendelsohn, eds.), 2nd ed., pp 209–225. Wiley-Liss, New York.

Wang, E. (1987). *J. Cell. Physiol.* **133,** 151–157.

Watson, J. V. (1986). *Cytometry* **7,** 400–410.

Winston, S. E., Fuller, S. A., and Hurrell, J. G. R. (1992). *In* "Current Protocols in Immunology" (J. E. Coligen, A. M. Kruisbeck, D. H. Margulies, E. M. Shevach, and W. Strober, eds.), pp. 8.10.1–8.10.17. Wiley, New York.

CHAPTER 24

"Washless" Procedures for Nuclear Antigen Detection

Jørgen K. Larsen

Finsen Laboratory
Rigshospitalet
DK-2100 Copenhagen, Denmark

I. Introduction

Simple and rapid washless staining procedures have for many years been applied for flow cytometric (FCM) quantitation of cell-surface antigens, membrane functions, cytoplasmic enzymes, and DNA and RNA content. This paper deals with washless staining procedures for FCM dual-parameter analysis of the expression of a nuclear antigen versus DNA content, or the expression of two different nuclear antigens, in particular with regard to proliferation-associated antigens.

In the first step of the staining procedures, the cells are lysed with detergent into a suspension of pure nuclei. This is in order to optimize the exposure of DNA and the specific nuclear antigen to the staining reagents without denaturating the antigen and to minimize the amount of extranuclear material that might contribute nonspecifically to the measured nuclear fluorescence. If fluoro-chrome-conjugated specific antibodies are available, the staining may theoretically be accomplished by adding a mixture of all reagents to the cell suspension in one single step. In practice, it may be organized in two steps: the first step for lysing the cells and the second step for application of the specific antibody and an irrelevant control antibody, respectively. If only nonconjugated specific antibodies are available, an antibody sandwich is built up in the single nucleus by addition of a specific monoclonal nonconjugated antibody and a secondary FITC-conjugated antibody, in a second and a third step, sequentially. In this case, the success of the method is based at least on the following assumptions:

1. The specific nuclear antigen does not dissolve from the nuclear matrix as a result of the lysis procedure.
2. The specific antibody distributes freely in and out of the nucleus, and to the antigenic site in the nucleus.
3. Localization of the FITC-conjugated secondary antibody to nuclear-bound primary antibody is not prevented by a limited surplus of primary antibody in solution.
4. The measurement of increased green fluorescence, corresponding to increased expression of specific antigen in the individual nucleus, is not prevented by a limited surplus of FITC-conjugated antibody in the solution.
5. The antibody sandwich is stable during the necessary period of measurement.

For the special purpose of staining bromodeoxyuridine (BrdUrd) with an antibody recognizing BrdUrd in single-stranded DNA, two additional steps have to be inserted in the procedure before the application of specific antibody, in order to denature the DNA by HCl and subsequently to restore the pH.

II. Applications

So far, applications have been limited to normal and pathological cells of the hemo- and lymphopoietic system and to corresponding cell lines. Washless procedures for Ki-67–DNA (Section III,A and B) and BrdUrd–DNA analysis (Section III,C) have been applied to phytohemagglutinin (PHA)-stimulated normal human peripheral blood lymphocytes (PBL), isolated as the mononuclear fraction by density centrifugation of peripheral blood, and to leukemia cell lines (HL-60 and K-562) (Larsen et al., 1991). The Ki-67–DNA method has been used clinically by Drach et al. (1992) in analysis of multiple myelomas. Giaretti et al. (1990) and Minks et al. (1992) have analyzed the p62-c-myc oncoprotein–DNA distribution in normal and transformed murine mast cells, using the monoclonal Myc 1-6E10 antibody (Cambridge Research Biochemicals) as primary antibody, with the procedure described in Section III,B. The BrdUrd–DNA analysis has furthermore been applied to monolayer cultures of butyrate–treated bladder cancer cell lines (Larsen, 1992).

Landberg and Roos (1991, 1992, 1993) used a slightly modified procedure (Section III,D) for detection of the nuclear antigens, PCNA (with a human autoantiserum), Ki-67, p105, MPM-2, and fibrillarin, applied to various cell types: PBL, human lymphoma cells, and human hemo- and lymphopoietic cell lines (Jurkat, Daudi, MN-60, HL-60, CCRF-CEM, MOLT-4, and lymphoblastoid B cell lines 158-B4 and UM-42).

Furthermore, Landberg and Roos demonstrated that combined detection of proliferating cell nuclear antigen (PCNA) and Ki-67 reveals the distribution of G_0, G_1, S, and G_2M phase cells, thus enabling a cell-cycle analysis not based on measurement of DNA content (Section III,E).

The washless staining of unfixed nuclei makes it possible in a simple way to extend the FCM DNA analysis (cell-cycle distribution, DNA ploidy of stemlines) with a supplementary cell kinetic parameter, enabling the distinction of DNA synthesizing cells, the distribution of cycling–noncycling, S–non-S phase, or mitotic–interphase cells, or by expression of cell-cycle-associated oncoprotein. As a probe for a supplementary cell kinetic parameter, either antibody mentioned above (except anti-fibrillarin) recognizes a nuclear antigen naturally expressed in cycling cells or in specific cell-cycle phases and therefore might be more practical for clinical investigations than the anti-BrdUrd antibody. This is because their use is not dependent on complicated vital or supravital labeling and subsequent denaturation of the labeled molecule. However, for the use of Ki-67-DNA analysis on biopsy specimens from heterogeneous solid tumors, the present methods need modification, because complete lysis of the cytoplasm and permeabilization of the nucleus are hard to accomplish for all present cell populations simultaneously.

Using the washless staining procedures, cycling cells are discriminated by their expression of Ki-67, p105, and MPM-2, and S phase cells are discriminated by expression of PCNA. It should, however, be noted that discrimination of S phase cells according to nuclear expression of PCNA has only been possible using a specific human autoantiserum and has unfortunately not been possible with the monoclonal anti-PCNA antibodies tested (19F4, TOB7, and PC10). In addition to discrimination of cycling cells by their moderate expression of p105 as well as MPM-2, mitotic cells were discriminated by significantly higher expression of these antigens, and the estimated mitotic fractions were highly correlated with data from countings of mitotic figures in the microscope (Landberg and Roos, 1992).

In comparison with established methods using fixed cells, the presented methods for washless staining of unfixed nuclei have the following characteristics (Landberg and Roos, 1991, 1992; Larsen et al., 1991):

1. Cell suspensions can be frozen and stored.
2. The staining procedure is simple and might be suited for automatic processing.
3. Staining time is generally <2 hr for BrdUrd–DNA and <1 hr for other nuclear antigen–DNA.
4. Small samples (consisting of $\leq 10^5$ cells) can be analyzed.
5. The loss of nuclei is negligible.
6. DNA is measured with high precision (CV = 2–4%).
7. There are few false signals of DNA hyperploidy (minimum aggregation).
8. The fraction of the antigen-positive subpopulation is adequately and reproducibly measured and is similar to that measured in fixed samples (nuclei of cycling cells are Ki-67, p105, and MPM-2 positive; nuclei of S phase cells are PCNA positive; mitoses are highly p105 and MPM-2 positive; nuclei of continuously labeled cells are BrdUrd positive).

III. Materials, Cell Preparation, and Staining

Monodisperse cell suspensions (harvested from cultures of lymphocytes, leukemia cell lines, or trypsinized monolayer cell cultures) are centrifuged, resuspended in ice-cold freezing buffer [sucrose 250 mM, dimethyl sulfoxide (DMSO) 5% v/v, Na citrate 40 mM, pH 7.6 (Vindeløv *et al.*, 1983a)] to a minimum concentration of 10^5 cells in 50 μl, and stored at −80°C. Cell suspensions from trypsinized monolayer cultures are washed once in calcium- and magnesium-free Dulbecco's phosphate-buffered saline (PBS) with trypsin inhibitor or serum before storage.

A. Ki-67/DNA Staining (Direct)

Samples of frozen, unfixed cells are thawed, and aliquots of ~10^5 cells are stained in a series of steps, during which the sample tubes, protected against intensive light, are slowly agitated in an ice bath mounted on a mixer (~100 rpm). No washings are applied. The reagents are added stepwise on top of each other to a final sample volume of 250–300 μl. The samples are not filtered before FCM.

1. For 15 min add 200 μl lysis–DNA staining solution [PBS with Nonidet P40 (NP-40) 0.5% v/v, propidium iodide (PI; Sigma Chemical Co., St. Louis, MO) 20 μg/ml, RNase (Sigma, Cat. No R-5503) 0.2 mg/ml, EDTA 0.5 mM, pH 7.2)].
2. For at least 30 min, add 25 μl FITC-conjugated monoclonal Ki-67 antibody [DAKO Corp., Carpinteria, CA., Cat. No F-788, diluted 10× in PBS with 1% bovine serum albumin (BSA) to ~10 μg/ml mouse IgG] or equivalent

amount of FITC-conjugated isotype control antibody (DAKO Corp., Cat. No. X-927).

B. Ki-67/DNA Staining (Indirect)

1. For the first 15 min, add 200 μl lysis–DNA staining solution.
2. For the subsequent 15 min, add 25 μl monoclonal Ki-67 antibody (DAKO Corp., Cat. No M-722, diluted 10× in PBS with 1% BSA to ~10 μg/ml mouse IgG) or equivalent amount of isotype control antibody (DAKO Corp., Cat. No X-931).
3. For at least 15 min, add 25 μl FITC-conjugated rabbit anti-mouse antibody [DAKO Corp., Cat. No. F-313, F(ab')$_2$ fragment, diluted 10× in PBS with 5% normal rabbit serum (DAKO Corp., Cat. No. X-902)].

C. BrdUrd/DNA Staining

For the BrdUrd–DNA analysis, cell cultures are incubated with and without BrdUrd (5-bromo-2'-deoxyuridine) at a final concentration of 10–50 μM for a period of 4–96 hr. Cells should be protected from intensive light exposure during the incubation with BrdUrd as well as during the staining procedure. For elimination of nonincorporated BrdUrd, the thawed cells are washed once in PBS with 1% BSA before staining. Cells are resupended to a concentration of ~10^5 cells in 50 μl PBS with 1% BSA. During all steps, except step 2 below (HCl, room temperature, RT), sample tubes are slowly agitated in ice bath.

1. For 15 min, add 100 μl lysis–DNA staining solution.
2. Add 25 μl 1 N HCl, and agitate for 30 min at RT.
3. Add 75 μl 1 M Tris (TRIZMA base, Sigma, Cat. No T-1503), pass the sample carefully through a pipette (Gilson Pipetman, 200-μl tip), and agitate for 5 min (return to ice bath).
4. For 15 min, add 25 μl anti-BrdUrd antibody [DAKO Corp., Cat. No M-744, clone Bu20a, diluted 10× in PBS with 1% BSA; alternatively anti-bromodeoxyuridine antibody, Becton–Dickinson, San Jose, CA., Cat. No. 347580, clone B44, diluted 10×] or equivalent amount of isotype control antibody (e.g., DAKO Corp., Cat. No. X-931).
5. For at least 15 min, add 25 μl FITC-conjugated rabbit anti-mouse antibody, diluted 10× in PBS with 5% normal rabbit serum.

D. Alternative Nuclear Antigen–DNA Staining

This modification of the indirect Ki-67–DNA staining procedure is described by Landberg and Roos (1991, 1992), who used it for detection of the following nuclear antigens in fresh cell suspensions of PBL, human hematopoietic cell lines, and human lymphoma cells:

1. PCNA, using a human autoantiserum obtained from a patient (AK) with systemic lupus erythematosus (SLE), as the specific primary antibody, and FITC-conjugated anti-human IgG as secondary antibody.
2. Ki-67, using monoclonal DAKO M-722 antibody.
3. Fibrillarin, using monoclonal MPM-2 antibody from culture supernatant (prepared by Dr. P. N. Rao).
4. p105, using monoclonal 780-39 antibody from culture supernatant (prepared by Dr. A. L. Epstein).

The staining procedure is as follows:

1. Samples of 2×10^5 fresh cells suspended in PBS are mixed with 100 μl lysing solution (0.5% Triton X-100, 1% BSA, and EDTA 0.2 μg/ml in PBS) supplemented with 50–100 μl of an appropriate dilution of the respective primary antibody and incubated for 30 min at RT.

2. Subsequently, 100 μl of an appropriate dilution of the secondary FITC-conjugated antibody is added, together with PI to final concentration 10 μg/ml and RNase to final concentration 20 μg/ml, and the samples are incubated. During the staining procedure the samples are occasionally gently agitated.

E. PCNA–Ki-67 Staining

This procedure is described by Landberg and Roos (1991, 1992).

1. Samples of 2×10^5 fresh cells suspended in PBS are mixed with 100 μl lysing solution (0.5% Triton X-100, 1% BSA, and EDTA 0.2 μg/ml in PBS), supplemented with 50 μl of human PCNA autoantibody (AK) diluted 1 : 100 and 5 μl Ki-67 antibody (DAKO M-722), and incubated for 30 min.

2. Subsequently, 50 μl FITC-conjugated anti-human IgG (diluted 1 : 10) and 5 μl phycoerythrin-conjugated anti-mouse IgG (TAGO, undiluted) are added, and the samples are incubated in the dark at RT for 30 min.

IV. Critical Aspects of the Procedure: Controls, Standards, and Instruments

In any series of measurements, the functioning of the reagents, the staining technique, and the flow cytometer should be controlled by analysis of simultaneously stained samples with known proportions of antigen-positive and -negative cells. Repeated measurement of the control samples makes it possible to observe changes in the level of green fluorescence detection that might occur during measurement of a long series of samples. Chicken (CRBC) and trout erythrocytes (TRBC), having ~35 and 80% of the human diploid G_0/G_1 DNA content and being Ki-67 negative and BrdUrd negative, may be useful as internal standards when added to the sample before staining (Vindeløv *et al.*, 1983c). The antigen-positive fractions estimated from stained samples kept in the refrigerator and measured the following day are slightly changed in comparison to those measured at the same day (Larsen *et al.*, 1991).

With respect to the tolerance in composition of the lysis–DNA staining solution, the Ki-67 staining of nuclei of PHA-stimulated lymphocytes and HL-60 leukemia cells is adequate within the limits of 0.1–1% NP-40, 0.5–10 mM EDTA, 0–1% BSA, 10–100% PBS, and pH 7.2–8.0. However, the staining of Ki-67 is impaired by treatment with either 1% formaldehyde, 0.15 *N* HCl (the dose used for BrdUrd staining; Section III,C), 0.1% citric acid, 0.01% dithiothreitol, or trypsin (as in solution A of Vindeløv *et al.*, 1983b). The staining can be speeded up by increasing the lysis efficiency (more detergent, less BSA, lower ionic strength, higher pH) and/or the force of sample agitation. Centrifugation of the unfixed nuclear suspension results in more-or-less selective aggregation and loss of nuclei (Larsen *et al.*, 1991).

To ensure operation within the range of stoichiometric detection of antigen expression, it is important to maintain an optimal proportion between the amount of antibody and the number of nuclei, for approaching saturation of the antigen also in the samples with highest fractions of antigen-positive nuclei. At the same time, the dose of FITC-conjugated antibody should be kept below the critical level, at which background fluorescence from free FITC-conjugated antibody submerges the green fluorescence pulses from antigen-negative nuclei.

With respect to instrumentation, measurement of the nuclear FITC fluorescence should be optimized as for measurement of DNA content according to PI fluorescence, by using a relatively narrow core stream and a critically limited observation area for the fluorescence collecting optics.

With the direct Ki-67–DNA staining technique, a lack of balance between dose of antibody and number of nuclei, as might be indicated by an unexpectedly high count rate on the flow cytometer, can be corrected by addition of supplementary volumes of lysis—DNA staining solution and FITC-conjugated Ki-67 antibody.

With the indirect method, it is important that the samples to be compared contain an equal number of cells, because supplementary dosage of primary antibody is not successful, after the secondary antibody has been added. For each particular antigen, one must find the optimal dosage by running samples with various mixtures of antigen-positive and -negative cells, at various combinations of doses of primary and secondary antibodies (Larsen *et al.*, 1991).

In terms of sensitivity, the washless staining procedure for BrdUrd detection is only suited for experiments based on continuous labeling with BrdUrd for several hours. Therefore, alternative methods for BrdUrd detection using fixed cells (e.g., the methods of Dolbeare *et al.*, 1983, or Jensen *et al.*, 1994) are recommended for experiments based on BrdUrd pulse labeling.

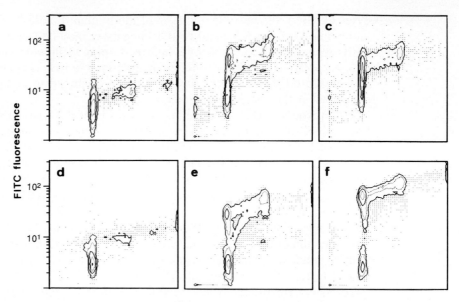

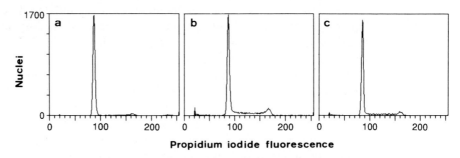

Fig. 1 Normal blood lymphocytes were stimulated for 0, 48, and 96 hr with PHA and stained for Ki-67–DNA analysis, by the procedure of Section III,B (a–c), and for BrdUrd–DNA analysis by the procedure of Section III,C (d–f). FITC fluorescence (515–540 nm, log scale) and PI fluorescence (>620 nm, linear scale) were measured in a Becton–Dickinson FACS IV (488 nm, 400-mW argon laser excitation; trigger threshold in red fluorescence; 70-μm flow nozzle; distilled water as sheath fluid; sample flow rate 0.5 μl/sec; 100–200 counts/sec). The contour levels in the bivariate plots (Consort 30; 10,000 counts) are 1 (dot), 4, 16, 64, 256, and 1024 (lines).

Fig. 2 Univariate PI fluorescence distributions (a–c) from the Ki-67–DNA stained samples shown in Figs. 1a–1c, respectively (Section III,B).

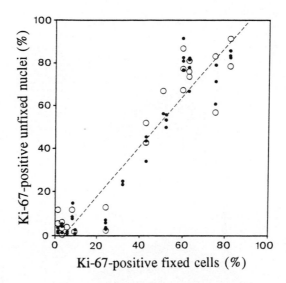

Fig. 3 Correlation between Ki-67-positive fraction in preparations of unfixed nuclei (○, direct staining, Section III,A; ●, indirect staining, Section III,B) and that of fixed cells (according to Baisch and Gerdes, 1987) from the same cultures of PHA-stimulated PBL ($r = 0.95$; $P < 0.001$).

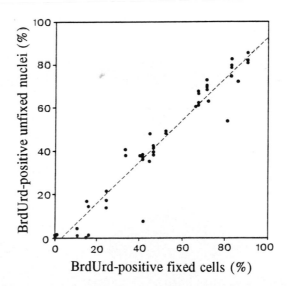

Fig. 4 Correlation between BrdUrd-positive fraction in preparations of unfixed nuclei (Section III,C) and that of fixed cells (according to Dolbeare *et al.*, 1983) from the same cultures of PHA-stimulated PBL ($r = 0.97$; $P < 0.001$).

FIXED UNFIXED

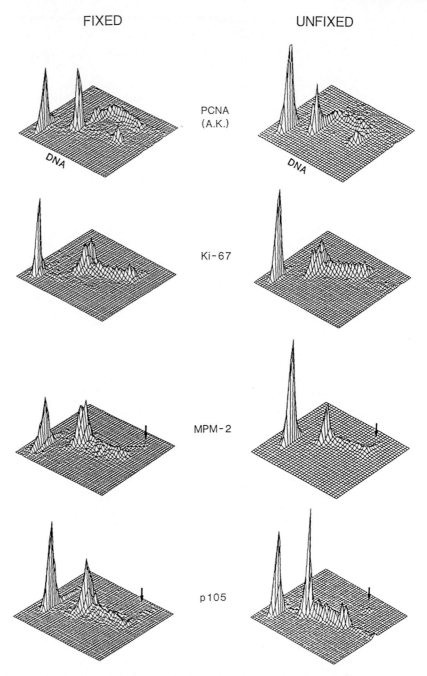

Fig. 5 Dual-parameter analysis of unfixed and fixed MOLT-4 cells mixed with normal lympho-cytes. DNA content in relation to nuclear antigen expression was studied using the antibodies specified in the figure. Arrows indicate mitotic cells. Reprinted after Landberg and Roos (1992) with permission from the authors.

In case of doubt, and always when new cell lines or biopsy materials are to be studied, it is recommended that the quality of the nuclear suspension under the fluorescence microscope be controlled, and it should be taken into account that even nuclei looking very pure are not necessarily permeable enough to give optimal access for the antibody to the antigenic sites.

Moreover, safety precautions should be taken to avoid contamination of personnel with hazardous biological or chemical material, e.g., PI.

V. Results

Figure 1 shows examples of the estimates of the Ki-67–DNA distribution (Figs. 1a–1c), and the BrdUrd–DNA distribution (Figs. 1d–1f) obtained after washless staining of PHA-stimulated PBL (Section III,A–C). Examples of the corresponding DNA distributions are shown in Fig. 2. The nuclei of unstimulated PBL (Figs. 1a and 1d) are confined to a single peak with low FITC fluorescence and low PI fluorescence (G_0 phase). After stimulation with PHA for 48 and 96 hr, an increasing fraction of the PBL nuclei occurs with significantly increased FITC fluorescence and is distributed throughout the G_1, S, and G_2M phases according to their PI fluorescence. The nuclei of all cells recruited into the cell cycle are thus recognized as Ki-67 positive, in concordance with the results obtained with alternative methods (Baisch and Gerdes, 1987; Palutke et al., 1987; Sasaki et al., 1987; Drach et al., 1989; Jacob et al., 1991). The BrdUrd–DNA distribution of PBL incubated with 50 μM BrdUrd from 24 hr after addition of PHA, at which time the cells start entering the cell cycle, shows at 48 hr (Fig. 1e) two different lanes of S phase nuclei with differing FITC fluorescence, indicating cells in the first and second S phase transit, whereas at 96 hr (Fig. 1f) no more cells are entering S from G_0.

As shown in Figs. 3 and 4, the Ki-67- and BrdUrd-positive fractions obtained using the washless staining procedures (Section III,A–C) on PHA-stimulated PBL, are highly correlated with respective estimates obtained using established staining procedures for fixed cells (Baisch and Gerdes, 1987; Dolbeare et al., 1983).

Likewise, as shown in Fig. 5, it is demonstrated by Landberg and Roos (1992), analyzing a mixture of PBL (noncycling) and MOLT-4 cells (cycling) with regard to expression of PCNA, Ki-67, MPM-2, and p105, that similar distributions were obtained using washless staining of unfixed nuclei and traditional staining of fixed cells.

Acknowledgments

This work was supported by the Danish Medical Research Council, the Danish Cancer Society, the Simon Spies Foundation, and the Mogens and Jenny Vissing Foundation.

References

Baisch, H., and Gerdes, J. (1987). *Cell Tissue Kinet.* **20,** 387–391.

Dolbeare, F., Gratzner, H., Pallavicini, M. G., and Gray, J. W. (1983). *Proc. Natl. Acad. Sci. U.S.A.* **80,** 5573–5577.

Drach, J., Gattringer, C., Glassl, H., Schwarting, R., Stein, H., and Huber, H. (1989). *Cytometry* **10,** 743–749.

Drach, J., Gattringer, C., Glassl, H., Drach, D., and Huber, H. (1992). *Hematol. Oncol.* **10,** 125–134.

Giaretti, W., di Vinci, A., Geido, E., Marsano, B., Minks, M., and Bruno, S. (1990). *Cell Tissue Kinet.* **23,** 473–485.

Jacob, M. C., Favre, M., and Bensa, J.-C. (1991). *Cytometry* **12,** 550–558.

Jensen, P. Ø., Larsen, J. K., Christensen, I. J., and van Erp, P. E. J. (1994). *Cytometry* **15,** 154–161.

Landberg, G., and Roos, G. (1991). *Acta Oncol.* **30,** 917–921.

Landberg, G., and Roos, G. (1992). *Cytometry* **13,** 230–240.

Landberg, G., and Roos, G. (1993). *Cell Prolif.* **26,** 427–437.

Larsen, J. K. (1992). *Nouv. Rev. Fr. Hematol.* **34,** 317–335.

Larsen, J. K., Christensen, I. J., Christiansen, J., and Mortensen, B. T. (1991). *Cytometry* **12,** 429–437.

Minks, M., Di Vinci, A., Bruno, S., Geido, E., Avignolo, C., and Giaretti, W. (1992). *Cancer Letters* **62,** 243–249.

Palutke, M., KuKuruga, D., and Tabaczka, P. (1987). *J. Immunol. Methods* **105,** 97–105.

Sasaki, K., Murakami, T., Kawasaki, M., and Takahashi, M. (1987). *J. Cell. Physiol.* **133,** 579–584.

Vindeløv, L. L., Christensen, I. J., Keiding, N., Spang-Thomsen, M., and Nissen, N. I. (1983a). *Cytometry* **3,** 317–322.

Vindeløv, L. L., Christensen, I. J., and Nissen, N. I. (1983b). *Cytometry* **3,** 323–327.

Vindeløv, L. L., Christensen, I. J., and Nissen, N. I. (1983c). *Cytometry* **3,** 328–331.

CHAPTER 25

Light Scatter of Isolated Cell Nuclei as a Parameter Discriminating the Cell–Cycle Subcompartments

Walter Giaretti* and Michael Nüsse[†]

*Istituto Nazionale per la Ricerca sul Cancro
Laboratorio di Biofisica e Citometria
16132 Genova
Italy
[†]GSF-Forschungszentrum für Umwelt und Gesundheit
Institut für Biophysikalische Strahlenforschung
D-85758 Oberschleissheim
Germany

I. Introduction

Cell structural properties measured by flow cytometry have been used to discriminate subpopulations of white blood cells (Hoffman *et al.*, 1980), to resolve subpopulations in tumors (Dolbeare *et al.*, 1983), to separate mitotic cells from interphase cells (Darzynkiewicz *et al.*, 1977, 1987; Benson *et al.*, 1984; Larsen *et al.*, 1986; Zucker *et al.*, 1988), and to evaluate chromatin conformational changes in isolated nuclei (Papa *et al.*, 1987). The molecular basis for the observed alterations in cytoplasm and chromatin conformation/ structure is unknown but could involve the nuclear matrix and the cytomatrix as recently reviewed (Pienta *et al.*, 1989).

A new flow cytometric method (Giaretti *et al.*, 1989; Nüsse *et al.*, 1989) combining light scattering measurements and detection of bromodeoxyuridine (BrdUrd) incorporation via fluorescent antibodies and quantitation of cellular DNA content by propidium iodide (PI) allows identification of additional compartments in the cell cycle. Thus, while cell staining with anti-BrdUrd antibodies and PI reveals the G_1, S, and G_2 + M phases of the cell cycle, differences in light scattering allow separation of G_2 phase cells from M phase cells and subdivision of G_1 phase into two compartments, i.e., G_1A representing post-mitotic cells which mature to G_1B cells ready to initiate DNA synthesis (see Appendix for the comparison of mitotic index values obtained by this method and by an independent one) (Pfeffer and Vidali, 1991; Pfeffer *et al.*, 1991;Di Vinci *et al.*, 1993).

For the analysis of the G_1, S, G_2, and M phases a second method is presented additionally using a preparation of isolated nuclei (Nüsse *et al.*, 1990,1992). Flow cytometric DNA content in combination with side scatter measurements are performed to discriminate interphase cells (G_1, S, and G_2 phases) from cells in mitosis due to differences in chromatin structure of both types of cells. Postmitotic G_1 phase cells (G_1A) can additionally be discriminated from late G_1 cells (G_1B). The same technique can also be used for a measurement of micronuclei induced in cells by ionizing radiation or chemicals (see Nüsse *et al.*, Chapter 9 in volume 42).

The first method involves fixation of cells in 70% ethanol, extraction of histones with 0.1 *N* HCl at 0°C, and thermal denaturation of DNA at 80–95°C for variable durations in relation to cell type. Mitotic cells show lower 90° scatter than G_2 phase cells, and G_1 postmitotic cells (G_1A) show lower 90° scatter than G_1 cells about to enter the S phase (G_1B). In addition, depending on the conditions of thermal denaturation, M phase cells can show lower propidium iodide fluorescence emission. The thermal denaturation appears to enhance the differences in chromatin structure of cells in the various phases of the cell cycle to the extent that cells could be separated on the basis of the 90° scatter which is mainly dependent on reflective and refractive components in the nucleus. Light scattering is correlated with chromatin condensation, as judged by micro-scopic evaluation of cells sorted on the basis of light scatter. The method has the advantage over the parental BrdUrd/DNA bivariate analysis (Dolbeare *et al.*, 1983, 1985; Beisker *et al.*, 1987) in allowing the G_2 and M phases of the cell cycle to be separated and the G_1 phase to be analyzed in more detail.

The second method is a simple two-step method to obtain a suspension of nuclei. Cells are treated with a salt solution containing a detergent, and then a second solution is added containing citric acid and sucrose. Both solutions contain ethidium bromide as DNA-specific dye. By this two-step treatment cellular membranes and the cytoplasm are destroyed and nuclei are released in suspension. The nuclear membrane is maintained; mitotic cells, however, are not destroyed (Nüsse *et al.*, 1990).

II. Application

In the first method, fixation is with 70% ethanol (Dolbeare *et al.*, 1983, 1985; Beisker *et al.*, 1987). Different cell types grown either as monolayers or as suspension culture gave similar results (see Sections III and VII). Thermal denaturation conditions, as used in our protocol, were analyzed by a systematic investigation (see Section IV). Only two different instruments have been used so far (see Section V).

In the second method the cells are not fixed but have to be prepared directly after culture. The samples can, however, be stored at 4°C for several weeks. Nuclei prepared according to this method could be analyzed in all types of flow cytometers offering possibilities of light scattering measurements. The preparation method is fast and especially useful in those cases where a division delay due to a mitotic block has to be studied (Nüsse *et al.*, 1992).

III. Materials

A. Cell Lines

1. A strain of hyperdiploid Ehrlich ascites tumor cells (EAT, clone F5, DNA index, DI = 1.22) were grown in suspension. The cells were maintained at 37°C under 6% CO_2 in a special medium (A2 medium) supplemented with 20% horse serum (Nüsse, 1981). A quasi-continuously growing culture was obtained by daily dilution of the cells in fresh medium. Cell-cycle data of these cells were published earlier (Nüsse, 1981).

2. V79 and Chinese hamster embryo cells (obtained from Dr. Scott Cram, Los Alamos National Laboratory, Los Alamos, NM) were maintained at 37°C under 6% CO_2 in minimum essential medium (Gibco, Grand Island, NY) supplemented with 15% fetal calf serum (FCS). The cells were subcultured three times a week.

3. MCF-7 cells (courtesy of Dr. Angelo Nicolin) were maintained in Dulbecco's minimum essential medium (DMEM) (Flow Laboratories, Ayrshire, UK) supplemented with 10% FCS, L-glutamine, and nonessential amino acids (Gibco). The cells were subcultured weekly in Falcon plastic T flasks at 2.7×10^3 cells—cm^2 seeding density. Cell-cycle data of these cells were published earlier (Bruno *et al.*, 1988).

4. PB-3 and PB-1 cells, purchased from J. F. Conscience (Ball *et al.*, 1983; Conscience and Fischer, 1985), were maintained in RPM1 1640 medium (Gibco)

supplemented with 10% glutamine, and nonessential amino-acids (Gibco). In the case of PB-3 cells, 20% of serum-free medium conditioned by the WEHI-3 myelomonocytic leukemia cell line as a source of IL-3 was added.

5. K562 cells were grown in RPMI 1640 medium (Gibco) and supplemented with 10% FCS.

B. Enrichment of Mitotic Cells and BrdUrd Pulse Labeling

Colcemid was added at a concentration of 0.2 μg/ml for several time intervals. Control cells with no colcemid were run in parallel. Mitotic indices (MI) were measured in a microscope using conventional staining techniques. The cells were pulse labeled with BrdUrd [in all experiments 10 μM BrdUrd was added to the cultures for 15 min at 37°C; no additional substances reported to increase incorporation of BrdUrd (Beisker et al., 1987) were used]. The cells were trypsinized after the BrdUrd pulse, if necessary, washed with PBS, and fixed in cold 70% ethanol.

For the second method no BrdUrd has to be added to the medium.

C. Monoclonal Anti-BrdUrd Antibodies

We have used monoclonal anti-BrdUrd antibodies obtained from:

1. Dr. Frank Dolbeare, Lawrence Livermore National Laboratory, Livermore, California (IU-1 and IU-4; Vanderlaan and Thomas, 1985; Vanderlaan et al., 1986).
2. Becton–Dickinson (Mountain View, California).
3. Partec (Arlesheim, Switzerland).
4. Eurodiagnostics (Apeldoorn, The Netherlands).

D. Staining Procedures

1. Method 1

For staining, 2-3 $\times$ 10^6 fixed cells were sedimented and incubated in 2 ml phosphate-buffered saline (PBS) containing RNase (1 mg/ml) at 37°C for 20 min. The cells were then sedimented and incubated in ice-cold 0.1 N HCl for 10 min, washed once in cold distilled water, resuspended in 2 ml distilled water, and heated for different time intervals (usually between 20 and 40 min.) at 95°C. The cells were rapidly cooled in ice water; next, 5 ml of PBS containing 0.5% Tween 20 (PBST) was added and the cells were sedimented. The cells were resuspended for 30 min at room temperature in 0.4 ml of anti-BrdUrd antibody diluted between 1 : 100 and 1 : 1000 in PBST containing 0.5% bovine serum albumin (BSA). After being washed with PBST the cells were stained with 0.4 ml FITC-conjugated goat anti-mouse IgG antibody (Sigma) diluted

1 : 50 in PBST containing 0.5% BSA. After 20 min, the cells were washed again and resuspended in PBS containing 20 μg/ml PI.

2. Method 2

Two solutions are used for the preparation of a suspension of nuclei. Solution I (prepare 500 ml stock):

584 mg/liter NaCl

1000 mg/liter Na citrate

10 mg/liter RNase from bovine pancreas (Serva)

0.3 ml/liter Nonidet P40.

Solution II (prepare 500 ml stock):

15 mg/liter citric acid

0.25M sucrose.

These solutions are made up using distilled water and should be filter sterilized. They can be stored at 4°C for at least several months. Ethidium bromide (EB) is added before use from a stock solution of 1 mg/ml in distilled water. Both solutions are used at room temperature.

A total of $0.5-1 \times 10^6$ cells are centrifuged at around 800–1000 rpm for 5 min. The supernatant medium should be removed completely by aspirating or carefully pouring off. The small cell pellet should be shaken slightly, without using a mixer, to resuspend the cells in the remaining medium (less than 50μl).

Solution I (1 ml) containing 25 μg/ml EB (prepared before use and filtered again) is added to the cell pellet; the cell suspension can be vortexed for a short time (2 sec).

After about 1 hr at room temperature, 1 ml of solution II containing 40 μg/ml EB (prepared before use and filtered again) is added to the cell suspension; the suspension is vortexed again (2 sec) and stored at 4°C before flow cytometric measurement. This suspension of nuclei can be measured directly after preparation or can be stored at 4°C for at least 2–3 days before measurement.

The CV of the G_1 peak in several different cell cultures was usually found to be between 2 and 3%.

IV. Critical Aspects of the Procedure

A. Method 1

Sometimes cell loss by clumping of cells during heat treatment and a rather large coefficient of variation of the G_1 peak were observed. These two effects depended both on the cell type under study and the temperature and duration of the thermal treatment. The thermal denaturation, which is necessary to produce single-stranded DNA in order to make BrdUrd accessible to the anti-BrdUrd antibodies, appears to cause both structural changes and altered fluorescence emission of the cells, in the various phases of the cell cycle. Usually 95°C and 20 min treatment resulted in a good separation of M versus G_2 and G_1A versus G_1B cells. No separation was observed after a heat treatment with a temperature below 80°C. Prolongation of thermal treatment (up to 50 min) induced much lower propidium iodide fluorescence emission of M cells without affecting extensively their 90° scattering signals. In conclusion, the preparation conditions for different cell types should be optimized by carefully varying temperature and duration of the heat treatment. The other steps in our staining procedure, such as 0.1 HCl treatment, or the use of different anti-BrdUrd antibodies, did not influence the results.

B. Method 2

A couple of aspects have been found to influence the preparation of nuclei:

1. If more than $1.5-2 \times 10^6$ cells were treated with solution I and II (1 ml each), a fraction of the nuclei may not be completely isolated and may be surrounded by cytoplasm and the cellular membrane. Such a suspension could show a higher side scatter and a nonoptimal separation of mitotic cells from interphase cells.

2. Before measurement, the suspension of nuclei should be checked in a fluorescence microscope using additional phase-contrast optics. All nuclei should be free of cytoplasm and membranes. If this is not the case, three reasons should be considered:

A. The cell number is too high (see 1).
B. Solution I needs to be prepared again. Because of the high viscosity of Nonidet P40 it could be that solution I did not contain the proper concentration of the detergent.
C. The cell pellet contains too much medium. In this case try to remove the medium completely or use 2 or 3 ml of solution I and II instead of 1 ml each.

V. Instruments

A. Method 1

This procedure was tested using three different instruments.
A FACS 440 dual-laser flow sorter (Becton–Dickinson) was located in the Laboratory of Biophysics (IST) (Genoa, Italy). With this instrument we have performed list-mode analysis and cell sorting. The four parameters, red (PI)

and green (FITC) fluorescence as well as forward (FSC) and 90° scatter were measured simultaneously, the data being stored in real time as four parameters per cell in list mode. Any combination of two of the four parameters was studied after the measurements. Bivariate BrdUrd/DNA or scatter/DNA distributions were displayed as "contour plots" (lowest contour line between 2 and 5% of maximum). Red fluorescence and scatter were always displayed on a linear scale, and green fluorescence on a log or linear scale. At least 20,000 cells were measured for each histogram. Mitotic and interphase nuclei were sorted on slides according to 90° scatter signals and PI fluorescence. The sorted cells were than analyzed in a fluorescence microscope to separate mitotic and interphase cells by means of the nuclear morphology.

A Cytofluorograf 30 L (Ortho Diagnostic) equipped with a data acquisition system to store and analyze two parametric histograms (no list mode) was also used.

For both instruments, excitation of PI and FITC was provided by the 488-nm line (500 mW) of an argon ion laser (Model 2025, Spectra Physics). The emitted fluorescence was collected at two wavelength ranges, 510–560 nm (band pass filter) for FITC antibody staining and wavelengths longer than 610 nm for PI fluorescence.

B. Method 2

EB fluorescence (pulse height and pulse area) and side scatter as well as forward scatter of micronuclei and nuclei were measured simultaneously in list mode using a FACSstar cell sorter (Becton–Dickinson). Excitation of EB was provided by the 488-nm line (500 mW) of an argon laser; EB fluorescence was collected with a long-pass filter (combination of KV 550 and OG 590). Similar results were obtained using a Cytofluorograf (Ortho Instruments).

VI. Results

Figure 1A shows a typical BrdUrd (FITC)/DNA (PI) content bivariate plot for exponentially growing EAT cells (Giaretti *et al.*, 1989). Three separated cell populations are shown: cells which have incorporated BrdUrd and show high FITC fluorescence (greater than channel 20 on the y-axis) which represent the S phase cells, and two groups of unlabeled cells which show different content of double-stranded DNA (equivalent to 18 and 36 channels on the x or DNA axis), which represent G_1 and $G_2 + M$ phase cells, respectively.

By electronic gating on the BrdUrd/DNA parameters, it was possible to analyze further any group of cells for the forward and 90° light scatter parameters (B-D FACS 440 instrument). With the gating set as shown in Figs. 1A and 1C, only unlabeled cells are selected. The 90° light scattering by these cells is shown in Figs. 1B and 1D. The G_2 and M cells exhibit markedly different light scattering

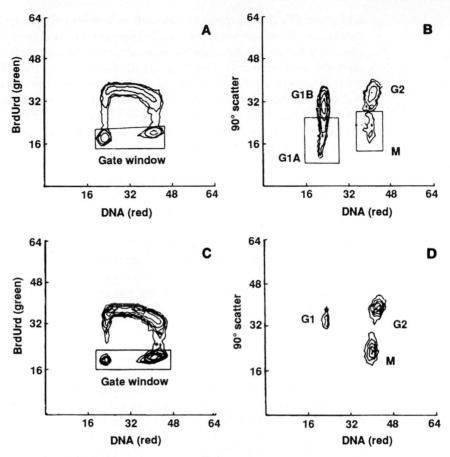

Fig. 1 Bivariate contour plots obtained from list-mode FCM analysis of EAT cells subject to BrdUrd pulse labeling (refer to text for detailed explanation). In (B) the presence of early G_1 cells (G_1A) separated from late G_1 cells (G_1B) and G_2 cells separated from M cells can be closely observed. In (D) the accumulation of M cells by colcemid and the disappearance of G_1A cells can be clearly observed.

properties. The scatter from G_2 phase cells is almost double that of M phase cells. The identity of these populations was confirmed by determination of mitotic index and by cell sorting (B-D FACS 440 instrument). More than 95% of cells sorted from the group labeled M exhibited mitotic figures whereas none of the G_2 phase cells showed a mitotic figure.

Additionally and depending mainly on the duration of heat treatment, G_2 and M phase cells can display different PI fluorescence intensities although they have the same DNA content. M phase cells show less PI fluorescence compared to G_2 phase cells. Under optimal conditions, M phase cells are found between G_1 and G_2 phase cells (for details, see Nüsse *et al.*, 1989). Interestingly, the

90° light scatter of unlabeled G_1 phase cells also showed two populations although not as well resolved as the G_2 and M populations. As seen in Fig. 1C, colcemid treatment, which blocks cells in mitosis, resulted in the disappearance of G_1 phase cells with low 90° light scatter (G_1A). The reappearance of this population after release of the block was also observed (not shown). Discussions on the potential applications of this technique in cell biology are presented in other reports (Nüsse *et al.*, 1989; Giaretti *et al.*, 1989).

Figure 2 shows typical DNA/side scatter (SSC) histograms of nuclei prepared from untreated Chinese hamster cells (a) and cells treated for 2 hr with colcemid (b) using the second method. The DNA distribution of all nuclei and the side scatter distribution of G_1 phase nuclei (indicated by the window) are presented additionally. The DNA distribution shows the typical shape of a growing cell population with G_1, S, and G_2/M phase cells. Several subpopulations can, however, additionally be identified in the bivariate DNA/SSC histogram according to differences in 90° light scatter intensity: Two populations with G_2/M-DNA content and two populations with G_1-DNA content, one with high

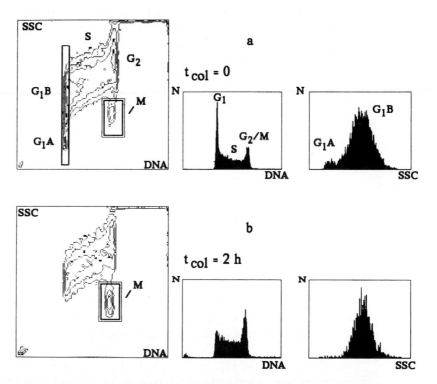

Fig. 2 DNA-SSC distributions (left) of asynchronously growing Ehrlich ascites tumor cells before (a) and after (b) colcemid treatment (2 hr, 0.2 μg/ml). DNA distributions of all cells and side scatter distributions of G_1 phase cells only (indicated by the window in the bivariate DNA-SSC distribution) are shown additionally (right). The position of M phase cells is indicated by the window.

and one with low 90° scatter. Sorting of the two populations with G_2/M-DNA content revealed that nuclei with low side scatter were mitotic cell nuclei with clearly visible chromosomes, while the particles with high 90° light scatter signal showed the typical shape of interphase nuclei with decondensed chromatin (G_2 phase cells). During growth of the cells in the presence of colcemid the M phase subpopulation increased as a function of time (see Nüsse *et al.*, 1990). The mitotic index could be measured by calculating the fraction of cells indicated by the window in Fig. 1. These values agreed with those obtained by microscopic scoring.

The 90° light side scatter distributions presented in Figs. 2a and 2b show that the subpopulation with G_1-DNA content and low values of 90° scatter ("G_1A") disappeared during colcemid treatment. Since colcemid arrests cells in metaphase and inhibits cell division it is assumed that these G_1A nuclei are obtained from early G_1 phase cells which progress during colcemid treatment into the normal G_1 phase ("G_1B") increasing thereby their side scatter intensities. Kinetic experiments showing the durations of the different subpopulations are discussed in detail elsewhere (Nüsse *et al.*, 1990).

In conclusion, the low side scatter intensity of M phase cell nuclei and G_1A phase nuclei compared to nuclei from interphase cells as demonstrated by both techniques was probably caused by the different chromatin structure in these cells as mentioned earlier. The first technique using anti-BrdUrd antibodies has the advantage that S phase cells are detected unambiguously in addition to the detection of M phase and early G_1A phase cells. The second technique has, however, the advantage of a fast and easy preparation leading to DNA distribution of high quality. The first technique is rather time consuming, and it cannot always be excluded that cells are lost due to clumping.

VII. Appendix

We have recently described a novel protein (AF-2), conserved between fission yeast and man, and we have shown by flow cytometry (FCM) that AF-2 is highly accessible to specific monoclonal antibodies (MAbs) in mitotic and post-mitotic early G_1 phase cells (Pfeffer and Vidali, 1991; Pfeffer *et al.*, 1991).

For this reason, the M indexes obtained by the first of the two methods described here were compared with those from the AF-2–DNA method as described in detail elsehwere (Pfeffer *et al.*, 1991; Di Vinci *et al.*, 1993).

Detection of mitotic (M) and postmitotic G_1 phase (early G_1) cells after specific staining of the AF-2 antigen by monoclonal antibodies and of nuclear DNA by PI is shown in Fig. 3. The method was based on ethanol fixation. Green fluorescence values measured with a logarithmic amplifier are from FITC-conjugated monoclonal antibodies to IgM isotype control and to the primary monoclonals against the AF-2 antigen (Di Vinci *et al.*, 1993). Red fluorescence values in linear scale are from PI. Human erythroleukemic K562 cells were

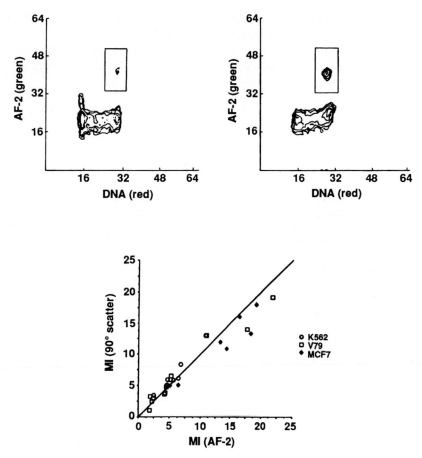

Fig. 3 Bivariate contour plots (top) showing a third method to the quantitative evaluation of the mitotic index based on the use of a new AF-2 monoclonal antibody (Pfeffer and Vidali, 1991; Pfeffer *et al.*, 1991; Di Vinci *et al.*, 1993). The mitotic index values shown at bottom were obtained using the first of the two methods described here and the method based on the AF-2 monoclonal antibody. K562, V79, and MCF-7 cells in log phase were treated for various time intervals (1–8 hr) before harvest (Di Vinci *et al.*, 1993). The straight line was drawn for visual reference at 45°.

used either in exponential growth (middle) or after treatment with colcemid which blocked the cells in mitosis (right). A total of 15,000 cells was measured in every plot. Control cells incubated with an IgM isotype control antibody prior to FITC staining remained lower than 16 a.u. (not shown). Contour plots are with lowest contour lines at 2% of maximum. The percentage of M phase cells (MI) was as calculated from the rectangular windows indicated as M.

Figure 3 (bottom) shows the MI as evaluated according to the methods based on 90° scatter (first of the two methods described here) and AF-2 antigen content. The data points refer to K562, V79, and MCF-7 cells. The cells in log phase were treated with colcemid for various time intervals (1–8 hr) before harvest. The straight line was drawn for visual reference at 45°. The coefficient of correlation was $r = 0.93$. The mitotic indices were also evaluated by visual microscopic analysis ($r = 0.94$, not shown).

References

Ball, P. E., Conroy, M. C., Heusser, C. H., Davis, J. M., and Conscience, J. F. (1983). *Differentiation (Berlin)* **24,** 74.

Beisker, W., Dolbeare, F., and Gray, J. W. (1987). *Cytometry* **8,** 235–239.

Benson, M. C., McDougal, D. C., and Coffey, D. S. (1984). *Cytometry* **5,** 515–522.

Bruno, S., Di Vinci, A., Geido, E., and Giaretti, W. (1988). *Breast Cancer Res. Treat.* **11,** 221–229.

Conscience, J. F., and Fischer, F. (1985). *Differentiation (Berlin)* **28,** 291.

Darzynkiewicz, Z., Traganos, F., Sharpless, T., and Melamed, M. R. (1987). *J. Histochem. Cytochem.* **25,** 875–880.

Darzynkiewicz, Z., Traganos, F., Carter, S. P., and Higgins, P. J. (1987). *Exp. Cell Res.* **172,** 168–179.

Di Vinci, A., Geido, E., Pfeffer, U., Vidali, G., and Giaretti, W. (1993). *Cytometry* **14,** 421–427.

Dolbeare, F., Gratzner, H., Pallavicini, M. G., and Gray, J. W. (1983). *Proc. Natl. Acad. Sci. U.S.A.* **80,** 5573–5577.

Dolbeare, F., Beisker, W., Pallavicini, M. G., Vanderlaan, M., and Gray, J. W. (1985). *Cytometry* **6,** 521–530.

Giaretti, W., Nüsse, M., Bruno, S., Di Vinci, A., and Geido, E. (1989). *Exp. Cell Res.* **182,** 290–295.

Giaretti, W., Di Vinci, A., Geido, E., Marsano, B., Minks, M., and Bruno, S. (1990). *Cell Tissue Kinet.* **23,** 473–485.

Hoffman, R. A., Kung, P. C., and Hansen, P. (1980). *Proc. Natl. Acad. Sci. U.S.A.* **77,** 4914.

Larsen, J. K., Munch-Peterson, B., Christansen, J., and Jorgensen, K. (1986). *Cytometry* **7,** 54–63.

Nüsse, M. (1981). *Cytometry* **2,** 70–79.

Nüsse, M., Juelch, M., Geido, E., Bruno, S., Di Vinci, A., Giaretti, W., and Ruoss, R. (1989). *Cytometry* **10,** 312–319.

Nüsse, M., Beisker, W., Hoffmann, C., and Tarnok, A. (1990). *Cytometry* **11,** 813–821.

Nüsse, M., Recknagel, S., and Beisker, W. (1992). *Mutagenesis* **7,** 57–67.

Papa, S., Capitani, S., Matteucci, A., Vitale, M., Santi, P., Martelli, A. M., Maraldi, N. M., and Manzoli, F. A. (1987). *Cytometry* **8,** 595–601.

Pfeffer, U., and Vidali, G. (1991). *Exp. Cell Res.* **193,** 411–419.

Pfeffer, U., Di Vinci, A., Geido, E., Vidali, G., and Giaretti, W. (1991). *J. Cell. Physiol.* **149,** 567–574.

Pienta, K. J., Partin, A. W., and Coffey, D. S. (1989). *Cancer Res.* **49,** 2525–2532.

Vanderlaan, M., and Thomas, C. B. (1985). *Cytometry* **6,** 501–515.

Vanderlaan, M., Watkins, B., Thomas, C., Dolbeare, F., and Stanker, L. (1986). *Cytometry* **7,** 499–507.

Zucker, R. M., Elstein, K. H., Easterling, R. E., and Massaro, E. J. (1988). *Cytometry* **9,** 226–231.

CHAPTER 26

Simultaneous Analysis of Cellular RNA and DNA Content

Zbigniew Darzynkiewicz

Cancer Research Institute
New York Medical College
Valhalla, New York 10595.

I. Introduction

Acridine orange (AO) and pyronine Y (PY) are the two most common fluorochromes of RNA. Used either alone (AO) or in combination with the DNA-specific fluorochrome Hoechst 33342 (PY), these dyes have found application primarily for simultaneous, correlated (bivariate) analysis of cellular RNA and DNA content. The mechanisms of interaction with nucleic acids, and in particular the spectral changes upon binding to DNA or RNA, are very much different for each of the dyes. The binding characteristics and the specific features of the respective methods, one based on the use of AO and another utilizing PY, therefore, are described separately in this chapter.

A. Acridine Orange

AO is a unique fluorochrome in many respects. It is one of the most versatile dyes and can be used to stain a variety of different constituents of the cell (Darzynkiewicz and Kapuscinski, 1990). Because of the multitude and complexity of its interactions with different substrates, basic knowledge of the mechanisms of these interactions is required to successfully apply this dye in any staining reaction. There are three major types of application of AO in flow cytometry:

1. Supravital Cell Staining

During exposure of live cells to a low concentration of AO (<5 μM), due to a low pH (high proton concentration) within lysosomes, this dye is specifically entrapped in these organelles. This manifests by their red luminescence. Therefore, when applied to live cells, AO can be used to estimate the efficiency of the proton pump (pH) of lysosomes (see Chapter 12 of this volume);

2. Differential Staining of Double-Stranded (ds) versus Single-Stranded (ss; Denatured) DNA

This reaction is performed after cell fixation, extraction of RNA by RNase, and induction of partial DNA denaturation by acid, heat, or other means (see Chapter 33 of this volume).

3. Differential Staining of RNA and DNA

This application is described in the present chapter. Applications 2 and 3 are based on the quite unique propensity of AO to differentially stain ds versus ss nucleic acids (Kapuscinski *et al.*, 1982; Kapuscinski and Darzynkiewicz, 1984; Darzynkiewicz and Kapuscinski, 1990). Namely, AO shows a considerable change in its absorption and emission spectra when bound to ds, compared to ss, nucleic acids. The shift in the absorption and/or emission spectrum when the dye binds to different substrates received the name metachromasia, and, thus, AO is a metachromatic fluorochrome. The metachromatic behavior of AO is the cause for its wide applicability as a probe of the content and conformation of nucleic acids in cytochemistry and biochemistry. AO binds to ds nucleic acids by intercalation, and fluoresces green in the intercalated form when excited in blue light. The maximum absorption of AO, when bound by intercalation to DNA, is at 500–506 nm, and emission is at 520–524 nm; this is classical fluorescence emission (S_1–S_0 transitions), with a short (5 nsec) lifetime (Table I).

Interaction of AO with ss nucleic acids is a complex, multistep process, which is initiated by AO intercalation between the neighboring bases, neutralization of the polymer charge by this cationic dye, and subsequent condensation and agglomeration (precipitation; solute-solid state transition) of the product (Kapuscinski *et al.*, 1982). The condensation reaction is highly cooperative. In the final product, AO molecules are interspaced with bases, forming stacks of alternating dye-base composition, which by virtue of their solid-state form are protected, to a limited degree, from interaction with oxygen or water molecules. The absorption spectrum of AO in these precipitated products is blue shifted, compared to the intercalated AO, with the maximum range between 426 and 458 nm, depending on the base composition of the nucleic acid. The emission of AO in these complexes also varies between 630 and 644 nm, depending on the base composition. The lifetime of the red emission, at room temperature, is >20 nsec. These spectral properties of AO in complexes with ss nucleic acids are suggestive of the intersystem crossing (phosphorescence; T_1–S_0 transition) rather than fluorescence. The solid-state nature of the complexes, as when freezing AO in solution (Zanker, 1952), may facilitate the intersystem crossing by partially eliminating the collision quenching, which otherwise occurs due to the long lifetime of the T_1-excited state in the presence of oxygen and solvents.

In the cell, large sections of rRNA and tRNA have ds conformation. Therefore, to obtain differential staining of DNA versus RNA with AO, these sections have to be selectively denatured, under conditions in which DNA still remains double-stranded. Cell treatment with AO in the presence of the chelating agent EDTA results in such selective denaturation of RNA (Darzynkiewicz *et al.*, 1975). By breaking RNA–protein interactions in ribosomes which stabilize dsRNA, EDTA promotes denaturation of dsRNA, which occurs as a result of interaction with AO. The RNA-selective denaturing properties of AO result from the fact that this ligand has a higher affinity to ssRNA than to ssDNA

Table I
Properties of AO and PY and Their Complexes with Natural and Synthetic Nucleic Acids

		Absorption		Emission		Binding	
Dye	Nucleic acid	λ_{max} (nm)	$E_{max} \times 10^{-4}$ (M^{-1} cm^{-1})	λ_{max} (nm)	Q_R	n	$K_1 \times 10^4$ (M^{-1})
AO (m)	None	492[a]	6.85[a]	525[a]	1.00[a]	—	—
AO (d)	None	466[a]	4.57[a]	—	<0.01[a]	—	—
AO	Calf thymus DNA	502[a]	5.85[a]	522[a]	2.22[a]	4.0[a]	5.0[a]
AO	Poly(dA-dT)·poly(dA-dT)	504[a]	6.23[a]	522[a]	2.13[a]	4.0[a]	9.2[a]
AO	Poly(dG-dC)·poly(dG-dC)	503[a]	6.64[a]	522[a]	2.06[a]	4.0[a]	7.9[a]
AO	Poly(dA-dC)·poly(dG-dT)	500[a]	5.17[a]	520[a]	1.83[a]	4.0[a]	7.4[a]
AO	rRNA	—	—	638[b, d]	[e]	1.0[c]	—
AO	Poly(rA)	457[c]	4.76[c]	630[b, d]	[e]	1.0[c]	20.6[c, f]
AO	Poly(rC)	426[c]	1.62[c]	644[b, d]	[e]	1.0[c]	11.5[c, f]
AO	Poly(rU)	438[c]	2.28[c]	643[b, d]	[e]	1.0[c]	3.6[c, f]
PY	None	547[g]	10.13[g]	565[g]	1.00[g]	—	—
PY	Calf thymus DNA	559[g]	7.55[g]	569[g]	0.30[g]	4–6[g]	1.74[g]
PY	Poly(dA-dT)·poly(dA-dT)	559[g]	7.23[g]	573[g]	1.26[g]	—	—
PY	Poly(dG-dC)·poly(dG-dC)	560[g]	7.68[g]	569[g]	0.22[g]	—	—
PY	Poly(dA-dC)·poly(dG-dT)	559[g]	8.18[g]	574[g]	0.36[g]	—	—
PY	rRNA (16S + 23S)	560[g]	7.04[g]	573[g]	0.29[g]	4–5	6.96[g]
PY	Poly(rA)·poly(rU)	562[g]	8.95[g]	573[g]	1.02[g]	—	—
PY	Poly(rI)·poly(rC)	563[g]	7.05[g]	574[g]	0.95[g]	—	—
PY	Poly(rG-rC·poly(rG-rC)	560[g]	7.90[g]	565[g]	0.24[g]	—	—

Note. λ_{max}, maximum wavelength; E_{max}, molar extinction coefficient at maximum; Q_R, relative quantum yield, as compared to AO monomer (Kapuscinski and Darzynkiewicz, 1987b), or in the case of PY, relative to PY free in solution (Kapuscinski and Darzynkiewicz, 1987a) n, binding site size (nucleotides), for ds nucleic acids per mole of base pairs; K_1, association constant; m, AO monomer; d, AO dimer (Kapuscinski and Darzynkiewicz, 1987b).

[a] In 0.15 N NaCl, pH 7.0 (Kapuscinski and Darzynkiewicz, 1987b).
[b] Suspensions in 0.1 N NaCl, pH 7.0 (Kapuscinski et al., 1983).
[c] In 10 mM phosphate buffer, pH 7.0; 25% (v/v) ethanol as a cosolvent (Kapuscinski et al., 1983).
[d] Noncorrected for emission monochromator and photomultiplier response.
[e] Phosphorescence.
[f] Cooperative association constant (M^{-1} nucleotides).
[g] In 10 mM NaCl, pH 7.0; PY concentration 5–6 μM; $D:P \sim 0.05$ (Kapuscinski and Darzynkiewicz, 1987a).

(Table I). The denaturation itself results from the fact that at an increased dye:phosphate ratio ($D:P$), binding of AO to ss nucleic acids becomes thermodynamically preferable; the weaker but more numerous 1:1 ($D:P$) interactions of AO with ss sections dominate thermodynamically over the stronger but fewer (1:4) sites of AO intercalation to ds regions (Kapuscinski and Darzynkiewicz, 1989). Thus, increasing $D:P$ promotes denaturation of ds RNA sections.

Cell staining with AO is generally done in salt solutions of relatively high ionic strength (0.1–0.2 M NaCl). Because of the significant electrostatic component in binding of AO to nucleic acids, resulting in competitive interactions between the AO$^+$ and Na$^+$, it is the concentration of free AO in solution (rather than the absolute *D:P* calculated based on molar ratios of the dye and nucleic acid in the sample) which is of importance for selective RNA denaturation (Darzynkiewicz and Kapuscinski, 1990).

To recapitulate, AO has two distinct functions in the mechanism of metachromatic staining of RNA. Namely, it: (i) denatures ds RNA, and (ii) differentially stains RNA (after its conversion to ss form) versus DNA. Selective RNA denaturation can be achieved only within a narrow concentration range of free dye, at a given ionic strength. Too low an AO concentration produces incomplete RNA denaturation (both DNA and portions of RNA then stain green), whereas too high a concentration also leads to denaturation of DNA (both RNA and DNA stain red). Acridine orange thus, in contrast to most other dyes used in cytometry, requires very stringent conditions, especially in regard to its concentration and the ionic strength of the staining solutions. The protocol for cell staining described in this chapter has been established after a variety of ionic conditions, pH, and dye concentrations was tested. (Darzynkiewicz *et al.*, 1976; Traganos *et al.*, 1977).

B. Pyronine Y

PY has been widely used in absorption microscopy as a dye which, in combination with methyl green, specifically stains cellular RNA (Brachet, 1940). More recently, it was also found to have an application in flow cytometry as a fluorochrome of RNA (Tanke *et al.*, 1980; Shapiro, 1981; Pollack *et al.*, 1982). The interactions of PY with nucleic acids, which are responsible for its specificity to RNA, as in the case of AO, also are complex (Kapuscinski and Darzynkiewicz, 1987). The following binding and spectral characteristics of PY are of importance for its role as RNA fluorochrome:

1. Intercalation of PY

PY binds by intercalation to ds nucleic acids. Its binding affinity to dsRNA is severalfold higher than to dsDNA (Table I). In the intercalated form, regardless of whether bound to RNA or DNA, the dye has maximal absorption between 547 and 563 nm and fluoresces with maximum emission between 565 and 574 nm; the variation is due to differences in the base composition of the nucleic acids (Kapuscinski and Darzynkiewicz, 1987a). Its quantum yield also varies, very widely, with changes in base composition. PY thus, when used as a fluorochrome, in contrast to AO which stains total cellular RNA, can detect only ds sections of RNA and is sensitive to the AU:GC base ratio.

2. Condensation of Nucleic Acids

Binding of PY to ss nucleic acids, as in the binding of AO, results in condensation (precipitation) of the product. The mechanism of the condensation, structure of the complexes, etc., are all similar to those generated by AO. In contrast to AO, however, fluorescence of PY is nearly totally quenched in these complexes (Kapuscinski and Darzynkiewicz, 1987). In absorption microscopy these products are characterized by lavender color.

3. Stoichiometry of RNA Detection

The stoichiometry and thermodynamics of binding of PY to ds and ss nucleic acids is similar to that of AO. PY, thus, can denature the ds sections of nucleic acids, rendering them ss. Upon binding PY, these ss sections undergo condensation and agglomeration (Kapuscinski and Darzynkiewicz, 1987a). At the increasing PY concentration, therefore, the fluorescence of PY bound to ds RNA is suppressed because of the progressive denaturation of the ds sections ("self-extinguishing" effect of PY). As in the case of AO, therefore, selective stainability of RNA with PY can be obtained at a relatively narrow range of dye concentration.

4. DNA Stainability with PY

PY, having affinity to dsDNA, can also counterstain DNA. Its binding to DNA, however, can be suppressed by DNA-specific ligands such as methyl green or Hoechst 33342 (Tanke *et al.*, 1980; Shapiro, 1981). In the presence of these dyes, therefore, PY can be used as a specific RNA fluorochrome. Dual-cell staining with Hoechst 33342 and PY provides the basis for simultaneous detection of DNA and RNA in flow cytometry (Shapiro, 1981; Darzynkiewicz *et al.*, 1987).

II. Applications

In the majority of cell types, approximately 80% of total cellular RNA is rRNA. Most of the remaining RNA is tRNA, and only a minor fraction of the total RNA is mRNA. The content of total cellular RNA, therefore, is primarily an indicator of the number of ribosomes per cell, and it reflects the translational potential of the cell. Nuclear RNA, being predominantly pre-rRNA, is also associated with cell capacity to synthesize proteins, and its increase often precedes the buildup of the ribosomal machinery in the cytoplasm. RNA content measurement, thus, either of the whole cell or of the isolated nucleus, is a marker of the overall translational capacity of the cell.

Differences in RNA content between individual cells have two different origins. One is associated with the tissue type- or cell differentiation-related constitutive level of the translational activity. Thus, cells known to secrete, or produce for internal use, large quantities of the tissue-specific protein (e.g., plasma cells or neurons) are generally characterized by a high RNA content. In these cases, the cellular RNA content is a reflection of a phenotype of the differentiated cell. Its measurement, therefore, can discriminate between cells of different tissues in the sample or be a marker of cell differentiation. The most common application of RNA measurement, in this context, is to identify reticulocytes, the immature red blood cells which retain RNA (in contrast to mature red blood cells that have no measurable RNA) (Tanke *et al.*, 1980).

The second cause of variability in RNA content is related to cell reproduction. Dividing cells double their constituents, including the number of ribosomes, during the cell cycle. Progression through the cell cycle is thus associated with an increase in cellular RNA content, and the increase occurs throughout interphase, at a relatively constant rate, proportional to the rate of cell proliferation (Darzynkiewicz *et al.*, 1979). Cellular RNA content, therefore, is a reflection of cell maturity in the cycle, and, e.g., it allows discrimination of early and late G_1 cells (Darzynkiewicz *et al.*, 1980a,b). Because cell growth in size (number of ribosomes) and the rate of proliferation are generally coupled, the RNA parameter, therefore, is also an indirect marker of cell proliferation (Darzynkiewicz, 1988).

The cells withdrawn from the cell cycle (quiescent cells, G_0, G_{1Q}) have on the average a 5- to 10-fold fewer ribosomes than their cycling counterparts (Johnson *et al.*, 1974). Thus, the difference in RNA content allows one to identify noncycling cells and can be used as a marker of their mitogenic stimulation (Darzynkiewicz *et al.*, 1976). This is the most common application of the RNA methodologies so far (Darzynkiewicz, 1990).

Measurements of cellular RNA have several potential applications in clinical oncology. RNA content of tumor cells has been shown to be a prognostic marker in several malignancies (Darzynkiewicz, 1988). This is not surprising, given that tumor progression is a consequence of uncontrolled cell proliferation and/or defective differentiation; both, as mentioned, correlate with cellular RNA content. It is also common knowledge that all markers of nucleolar activity, i.e., activity related to the synthesis of preribosomal RNA, have prognostic value in tumors. Because many antitumor drugs are cell-cycle specific, cellular RNA content may also be predictive of the tumor cell sensitivity to such drugs. Furthermore, unbalanced growth of cells treated with antitumor drugs can be estimated from RNA content measurements to evaluate cell response to the drug and as a factor predictive of cell death or recovery (Traganos *et al.*, 1982). Thus, the RNA assays may be useful in customizing therapy to achieve the maximal therapeutic effect with minimal toxicity to the patient and in monitoring the treatment.

III. Materials

A. Stock Solution of Acridine Orange

Dissolve AO in distilled water (dH_2O) to obtain 1 mg/ml concentration. It is essential to have AO of high purity. Molecular Probes (Eugene, OR) offers AO of high purity (Cat. No. A 1301). The stock solution may be kept in the dark (in dark or foil-wrapped bottles) at 4°C for several months without deterioration.

B. First Step: Solution A

Triton X-100, 0.1% (v/v)
HCl, 0.08 M (final concentration)
NaCl, 0.15 M (final concentration).

Solution A may be prepared by adding 0.1 ml of Triton X-100 (Sigma Chemical Co., St. Louis, MO), 8 ml of 1.0 M HCl, 0.877 g NaCl, and dH$_2$O to a final volume of 100 ml. This solution may be stored at 4°C for several months.

C. Second Step: Solution B

Acridine orange, 6 μg/ml ($\sim$20 μM)
EDTA-Na, 1 mM
NaCl, 0.15 M
Phosphate-citric acid buffer, pH 6.0.

Solution B may be made as follows:

1. Prepare 100 ml of buffer by mixing 37 ml of 0.1 M citric acid with 63 ml of 0.2 M Na$_2$ HPO$_4$.
2. Add 0.877 g NaCl, stir until dissolved.
3. Add 34 mg of EDTA disodium salt. Equivalent amounts of tetrasodium EDTA may also be used. EDTA in acid form may be used, but it requires a longer time to dissolve. Stir until dissolved.
4. Add 0.6 ml of the stock solution of AO (1 mg/ml).

Solution B is also stable and can be stored for several months at 4°C in dark. Solutions A and B may be kept in automatic dispensing pipette bottles set at 0.4 ml for solution A and 1.2 ml for solution B; the latter in a dark bottle.

D. Nuclear Isolation Solution

Prepare a solution containing 10 mM Tris buffer (pH 7.6), 1 mM sodium citrate, 2 mM MgCl$_2$, and 0.1% (v/v) non-ionic detergent Nonidet (NP-40).

E. PY and Hoechst 33342 Staining Solution

In 100 ml of HBSS (containing Ca^{2+} and Mg^{2+}), dissolve 0.2 mg of Hoechst 33342 and 0.4 mg of high-purity PY. Hoechst 33342 is available from Molecular

Probes. Relatively pure PY is offered by Polysciences, Inc. (Warrington, PA; Cat No. 18614). Most batches of PY available from other sources have 30–40% impurities and should be purified by extractions with chloroform and recrystallization from methanol (Kapuscinski and Darzynkiewicz, 1987a).

IV. Staining Procedures Employing AO

A. Staining of Unfixed Cells

1. Transfer a 0.2-ml aliquot of the original cell suspension (not more than 2×10^5 cells) to a small tube (e.g., 2- or 5-ml volume). Chill on ice.
2. Add gently 0.4 ml of ice-cold solution A. Wait 15 sec, keeping on ice.
3. Add gently 1.2 ml of ice-cold solution B. Measure cell luminescence during the next 2-10 min (equilibrium time).

The sample should be kept on ice prior to and during the measurement. Vortexing or syringing cells when immersed in solution A, especially in the absence of any serum or proteins in the original cell suspension, results in disintegration of plasma membrane and isolation of cell nuclei. RNA content of isolated nuclei, therefore, can be measured in this way. Visual inspection of the nuclei under phase contrast or UV light microscopy is essential to estimate the efficiency of the isolation, which can be controlled by selecting optimal time and speed of vortexing, or number of syringings.

B. Staining of Fixed Cells

1. Fix cells in suspension in 70% ethanol, on ice.
2. Centrifuge cells, remove all ethanol, rinse once, and resuspend in Hanks' buffered saline (HBSS), at a cell density $<2 \times 10^6$ per 1 ml.
3. Withdraw 0.2 ml of cell suspension and stain, using solutions A and B, as described already for fresh, unfixed cells. In the case of fixed cells, the presence of Triton X-100 in solution A is not necessary, although it does not interfere with their staining.
4. To assess the contribution of RNA to the detected luminescence, after removal of ethanol and suspension in HBSS, the cells can be incubated with RNase A (5 Kunitz units per 1 ml) for 30 min at 37°C, prior to being stained with AO.

The advantage of staining fixed cells is the stability of the fluorescence intensity during the measurement, due perhaps to inactivation of the endogenous nucleases by ethanol. Also, due to the possibility of control treatments with exogenous RNase or DNase, the specificity of the staining reaction can be estimated more reliably. The disadvantage is, however, lower resolution of

DNA measurements reflected by higher coefficient of variation (CV) of the mean green fluorescence of $G_{0/1}$ cells and increased cell aggregation.

C. Isolation and Staining of Unfixed Nuclei from Solid Tumors

1. The tissues may be kept frozen (below $-40°C$) prior to nuclear isolation. Place the freshly resected (or frozen and then thawed) tissue in the nuclear isolation solution and trim to remove the necrotic, fatty, and other undesirable portions.

2. Transfer the trimmed tissue to a new aliquot of the isolation solution and mince finely with scalpel or scissors. Mix small tissue fragments by vortexing and/or pipetting with a Pasteur pipette or syringe with a large-gauge needle. Observe the release of nuclei by sampling the suspension and viewing it under the phase-contrast microscope. If the release of nuclei is inadequate, transfer the minced tissue suspended in the isolation solution into homogenizer with a glass or Teflon pestle. Homogenize by pressing the pestle several times; check the efficiency of nuclear isolation. The nuclei should be clean, unbroken, and lacking cytoplasmic tags.

Collect the nuclear suspension from the above remaining tissue fragments (allow to sediment for ~1 min) with a Pasteur pipette and filter through 40- to 60-μm-pore nylon mesh. Dilute the suspension with the isolation buffer, if necessary, so that no more than 2×10^6 nuclei per 1 ml of the isolation solution remains in the final suspension. All isolation steps should be done on ice.

3. Withdraw a 0.2-ml aliquot of nuclei suspension and stain identically as described for whole cells; that is, mix with 0.4 ml of solution A and then with 1.2 ml of solution B.

4. Control samples can be incubated with RNase A. To this end, 1 ml of final nuclear suspension is treated with 50 units of RNase A for 20 min at $24°C$, and 0.2-ml aliquots of this suspension are then processed in exactly the same way as nuclei that were not subjected to treatment with RNase.

RNA leaks from the isolated, unfixed nuclei into the buffer. It is advisable, therefore, to stain the nuclei as soon as possible after isolating them. After being stained RNA while complexed with AO is less soluble.

V. Staining RNA and DNA with PY and Hoechst 33342

1. Fix cells in suspension in 70% ethanol on ice.
2. Centrifuge cells, remove all ethanol, rinse once with and resuspend in HBSS containing Ca^{2+} and Mg^{2+}, at a cell density $<2 \times 10^6$ per 1 ml.
3. Admix 0.5 ml of this suspension with 0.5 ml of a solution of HBSS containing 2 μg/ml of Hoechst 33342 and 4 μg/ml of PY, prepared as described in Section III,E.
4. Measure cell fluorescence after 20 min.

══ VI. Critical Aspects of the Procedures

A. Acridine Orange Procedure

1. Preservation of Intact Cells

To ensure that unfixed cells are permeabilized but do not disintegrate, the presence of serum (proteins) or serum albumin in the first step is required. To this end, prior to adding solution A, all cells must be suspended in a salt solution or tissue culture medium that contains 10–20% (v/v) serum or albumin. Thus, the cells can be taken directly from tissue culture without prior centrifugation, washing, etc., and stained, as described in the procedure. Vigorous shaking, pipetting, or vortexing of cell suspensions after addition of the detergent breaks the cells and results in a release of the nuclei.

2. Critical Concentration of AO

Differential staining of DNA versus RNA requires a proper concentration of free (unbound) AO in the final staining solution and at the time of the actual act of measurement under equilibrium ($\sim$ 20 mM). The following problem associated with this requirement may occur:

When the cell number (density) in the original suspension exceeds 2×10^6 cells per 1 ml (or even less when cells are highly hyperdiploid and/or have excessive RNA content), the amount of bound AO is high and therefore the free dye concentration may be significantly reduced (the "mass action" law). The RNA denaturation is then incomplete and part of the RNA can stain green.

Solution: Dilute the original cell suspension to have fewer cells in the sample.

3. Diffusion of AO from Sample during Flow

With some instruments (most cell sorters) in which cell measurements take place outside the nozzle (i.e., in air), a significant diffusion of dye from the sample to the sheath fluid takes place after the stream leaves the nozzle, prior to its intersection with the laser beam. This breaks the equilibrium and lowers the actual AO concentration in the sample at the time of cell measurement. Dye diffusion is also a problem in some instruments that have a narrow sample stream and long flow channel (e.g., Cytofluorograf 50 made by Ortho Diagnostics).

Solution: Increase the AO concentration (up to 20 μg/ml) in solution B and increase sample flow rate to compensate for the diffusion. Wherever possible, use channels with favorable geometry (wider sample stream and/or shorter distance between the nozzle and intersection with the laser beam). The optimal dye concentration for a particular instrument can be established by preparing a series of solution B with different AO concentrations (e.g., from 5 to 20 μg/

ml) and testing at which concentration cells in the $G_{0/1}$ cell cluster have the same green fluorescence [the lowest CV of the green fluorescence mean value, corresponding to a lack of correlation between green and red luminescence; the $G_{0/1}$ cell cluster ought to be horizontal (or vertical if axes are reversed), but never skewed (diagonal)].

4. Overlap of Emission Spectra of AO: Sensitivity of RNA Detection

One of the limitations of the AO technique is the relatively low sensitivity of RNA detection. This is primarily due to the emission spectrum overlap: the green fluorescence of AO intercalated to DNA has a long "tail" toward a higher wavelength, which is significant to even as far as 620–670 nm. Therefore, RNA measurements in cells (or cell nuclei) characterized by a high DNA:RNA ratio lack sensitivity, being obscured by the high component due to AO bound to DNA.

Solution: The following points should be considered to improve the measurement:

Use long-pass filters transmitting above 640 or 650 nm, rather than 610 or 620 nm, to measure red luminescence. This significantly reduces the DNA-associated spectral component.

Due to significant differences in absorption spectra (Table I), it is impossible to achieve maximum excitation of AO bound to both DNA and RNA simultaneously, using a single laser. The strategy for optimal excitation is described in Section VIII.

5. Contamination of the Tubing by AO

As a cationic, strongly fluorescing dye, AO is absorbed in the surface of the sample flow tubing. Binding of a similar nature is also exhibited by rhodamine 123 and some other dyes. Release of such dyes interferes with measurement of the subsequent samples, especially if the cells have low fluorescence.

Solution: Rinse the sample flow line with bleach (e.g., 10% Clorox), then 50% ethanol, and then PBS, each 10 min. Alternatively, if possible, replace the sample flow tubing after using AO, and keep this tubing for use only with AO.

B. Pyronine Y – Hoechst 33342 Procedure

In many respects the critical points of staining with PY are similar to those of the AO methodology. The most critical aspect of the Hoechst 33342-PY procedure is the stringent requirement for the appropriate concentration of PY. Too low a concentration of the dye (or too dense a cell suspension) cannot ensure stoichiometry of the staining because of the paucity of PY in the solution (dye:binding site ratio <1.0). Too high a concentration of PY triggers denaturation and condensation of RNA, which quenches its fluorescence. Paradoxically, thus, by increasing the PY concentration one can completely suppress RNA fluorescence and (in the absence of Hoechst 33342) induce PY intercalation to DNA; under these conditions PY can be used as a DNA-specific fluorochrome (Portela and Stockert, 1979).

Because of these denaturing properties of PY, the RNA stainability is very sensitive to its native conformation. Under appropriate conditions, the procedure allows one to discriminate between polyribosomal RNA (which is more resistant to denaturation and shows more extensive double strandedness) and rRNA in dispersed ribosomes (Traganos *et al.*, 1988). The staining procedure, thus, can be used to measure disaggregation of polyribosomes which occurs during mitosis or hyperthermia (Fan and Penman, 1970).

It should be stressed that PY taken up by live cells is partially localized in the mitochondria and lysosomes (Darzynkiewicz *et al.*, 1987). The specificity of staining of RNA with PY in live cells, therefore, is uncertain.

VII. Controls and Standards

A. Specificity of DNA and RNA Detection

Specificity of cell staining can be assayed by measuring parallel samples, prefixed in ethanol and then suspended in PBS containing Mg^{2+} and either RNase or DNase, as follows:

1. Dissolve 5 Kunitz units of RNase A (DNase-free RNase or RNase boiled for 5 min) in 1 ml of PBS, and incubate cells in this solution for 30 min at 37°C.
2. Dissolve 0.5 mg of DNase I in 1 ml of PBS and incubate cells in this solution for 30 min at 37°C.

Following these incubations, the cells should be stained with AO, passing through solutions A and B, as described earlier in the chapter, or with PY-Hoechst 33342. The percentage loss of red and green luminescence as a result of treatment with RNase or DNase (AO procedure) or red and blue (PY-Hoechst 33342) is an indication of the specificity of staining of RNA or DNA, respectively.

In the case of unfixed cells, cell suspensions already treated with solutions A and B (AO procedure) may be subsequently treated with 5 Kunitz units of RNase A and incubated for 30 min at 24°C prior to fluorescence measurements.

B. RNA Content Standards

RNA and DNA content of the measured cells can be expressed quantitatively, by comparing them with standard cells. Nonstimulated, peripheral blood lymphocytes appear to offer the best standard. The RNase-treated and untreated lymphocytes should be measured to establish the extent of RNase-specific red luminescence. The cells to be compared have to be measured under conditions identical to those used for lymphocytes. Their RNA index should be expressed as a multiplicity (or fraction) of RNA content of lymphocytes.

Lymphocytes are not uniform, and there is a minor difference in RNA content between B and T cells. A more accurate standard would, therefore, be purified populations of B or T lymphocytes. If the measured cells have a severalfold higher RNA content than lymphocytes (which is often the case), or if they are aneuploid, it is convenient to use lymphocytes as an internal standard.

Note that because the extent of the spectrum overlap (higher in the case of the AO methodology) depends on the excitation wavelength and the emission filters used, it can vary from instrument to instrument, depending on minor differences in the filter specifications. Therefore, the proportion of the RNase-specific luminescence to total red luminescence varies as well. This variation does not allow one to express RNA content as a ratio of the mean intensity of red luminescence of the measured cell population to that of the lymphocytes (without an adjustment for the RNase specificity), because such an index cannot be compared between different laboratories or flow cytometers.

The lymphocytes may also serve as a standard of DNA content, to estimate the DNA index from the mean (modal) intensity of the green fluorescence of the $G_{0/1}$ population.

C. Cell Staining on Slides

Observation of cells under UV light microscopy is often required to reveal localization and confirm specificity of cell staining. The cells may be stained with AO on microscopic slides (cytospin preparations, smears, etc.) by rinsing fixed specimens with solution A and mounting under coverslips in a drop of solution B, in equilibrium with AO. Because of AO absorption by the glass surface, the dye concentration is lowered. This can be compensated for by higher initial concentrations of the dye in solution B ($\sim$20 μg/ml). Alternatively, the slides with cells should be rinsed several times with solution B, to saturate the binding sites with AO, prior to mounting.

VIII. Instruments

A. Staining with AO

Maximum absorption of AO is at $\sim$455–490 nm. The 488-nm line of the argon ion laser is the most commonly used excitation wavelength. In the instruments illuminated by a mercury or xenon lamp, blue excitation filters can be used (BG 12, band pass combination filters transmitting light between 460 and

500 nm). Optimal excitation can be achieved using two lasers, one tuned to 488 nm (DNA detection) and another to 457 nm (RNA detection).

The emission filters and dichroic mirror setup should discriminate green fluorescence (measured at 530 + 15 nm) and red luminescence (measured preferably above 640 or 650 nm). Electronic compensation of the spectrum overlap, while it improves the "cosmetics" of the results, should rather be avoided. If the degree of the compensation varies between the samples, the method cannot be quantitative or properly standardized.

As discussed before, the geometry of the flow channel affects the staining reaction (the "diffusion problem"). The author has experience with the following instruments, which can be categorized into three groups:

1. FC-4800, FC-4801 (Biophysics); FC-200, FC201, and ICP-22 (all Ortho Instruments Co.); FACS analyzer, FACScan (Becton–Dickinson); Profile and Elite (Coulter Electronics); and PAS III (PARTEC). Very good differential stainability of RNA–DNA with AO was observed using all of the listed instruments. They have no diffusion problem and do not require an increased AO concentration compared to this protocol. Mercury lamp illumination in the ICP-22, FACS analyzer, and PARTEC instruments makes them especially well designed for measurement of cells with low RNA content.

2. FACS II, FACS III, FACStar (Becton–Dickinson); EPICS IV and C (Coulter Electronics); and 30H Cytofluorograph (Ortho) require higher (8–20 μg/ml) concentrations of AO in solution B to compensate for the dye diffusion.

3. Due to extensive dye diffusion, it is difficult to obtain good stainability using the Ortho 50H cytofluorograph with the original sorting channel. Substitution of this channel with the nonsorting one (as in the 30H cytofluorograph) improves the staining.

B. Staining with PY–Hoechst 33342

Absorption of Hoechst 33342 is in the UV light spectrum range and optimal excitation can be achieved with a 350/356-nm wavelength line of the krypton ion laser, 351/363 argon laser, or UV filters (UG1) in the mercury lamp illumination systems (Shapiro, 1981, 1988). PY can be excited with any wavelength between 488 and 530 nm, by several lines available in either krypton or argon ion lasers (Crissman et al., 1985). Differential staining of DNA and RNA by Hoechst and PY, thus, requires dual excitation, provided either by two lasers or a laser and a mercury lamp.

IX. Results

Stainability of DNA and RNA with AO of HL-60 leukemic cells undergoing myeloid differentiation in the presence of dimethyl sulfoxide (DMSO) is shown in Fig. 1. A significant decrease in RNA content accompanies cell differentiation and cell arrest in the G_1 compartment. The horizontal, nonslanted position of the contour representing G_1 cells is evidence of proper staining.

Changes in cellular DNA and RNA content during mitogenic stimulation of lymphocytes are shown in Fig. 2. Based on differences in RNA content, it is possible to distinguish nonstimulated, quiescent cells from cells entering the cell cycle. Their progression through the cell cycle is paralleled by a further increase in RNA content. Correlated measurements of RNA and DNA offer a sensitive assay of lymphocyte stimulation, providing information regarding both the initial steps of stimulation (exit from G_0) and cell-cycle progression. Since stimulation of lymphocytes is a multistep process that does not always result in cell proliferation, the traditional assays based on radioactive thymidine incorporation, in contrast to the present technique, cannot detect the early steps of the stimulation process and thus are useless in such situations.

Stainability of RNA and DNA with PY and Hoechst 33342 is shown in Fig. 3. The staining reaction is very sensitive to the conformation of RNA in the cell. Therefore, as is evident, cells in mitosis (M) have lower stainability with PY than G_2 cells, despite the fact that the RNA content of M cells is somewhat higher than in most G_2 cells. The hypochromicity of RNA in M cells with PY is a consequence of the denaturing properties of this dye: RNA of the polyribosomes (which are more numerous in interphase than in metaphase) is more resistant to denaturation and thus it stains more intensely compared to RNA in mitotic cells. During mitosis, polyribosomes

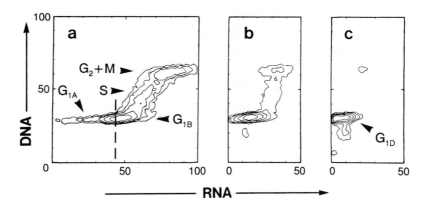

Fig. 1 The isometric contour maps representing the distribution of HL-60 cells with respect to their RNA and DNA contents after staining with AO. (a) Exponentially growing cells, (b) cells growing in the presence of 1.5% (v/v) DMSO for 3 days, and (c) cells cultured in the presence of DMSO for 5 days. In the exponentially growing population, two G_1 compartments, G_{1A} and G_{1B}, can be distinguished: Cells enter S phase from G_{1B}, whereas cells in G_{1A} are early G_1 cells. A decrease in cell proliferation observed during cell differentiation is accompanied by a decrease in RNA content. Cells in G_1 that exhibit the differentiated phenotype were denoted as G_{1D} (Darzynkiewicz *et al.*, 1980a).

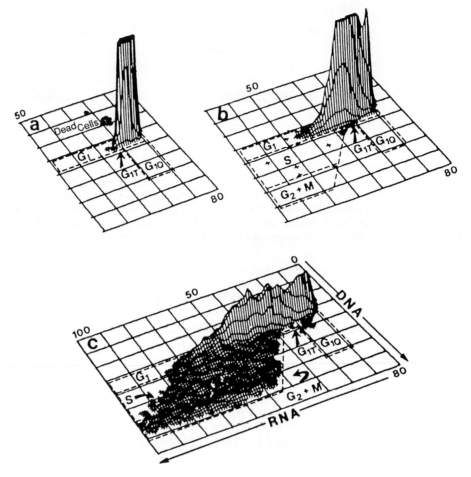

Fig. 2 Differential AO stainability of DNA and RNA in lymphocytes, unstimulated and stimulated to proliferation. Bivariate distribution (RNA *versus* DNA) of human peripheral blood lymphocytes, unstimulated (a) and incubated with the mitogenic agent, phytohemagglutinin for 24 hr (b) and 72 hr (c). Stimulated lymphocytes are distinguished from quiescent cells (G_{1Q}) by the increased RNA content, early (24 hr) as cells in transition to the cycle (G_{1T}), and later (72 hr) as cells progressing through the cycle, in G_1, S, and G_2 + M.

disaggregate and RNA of individual ribosomes is more extensively denatured by PY, which leads to quenching of this dye's fluorescence (Traganos *et al.*, 1988).

Other examples of the results and descriptions of other applications of the RNA–DNA staining techniques are presented in several reviews (Darzynkiewicz, 1988; Darzynkiewicz and Kapuscinski, 1990; Traganos, 1990).

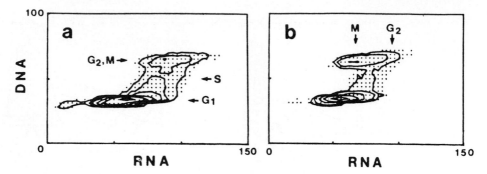

Fig. 3 Stainability of DNA and RNA with HO342-PY in CHO cells at different phases of the cell cycle. (a) Exponentially growing CHO cells and (b) cells from the colcemid-treated culture, containing approximately 20% cells in mitosis. Decreased PY stainability of mitotic cells (M) compared to G_2 is due to the fact that PY stains preferentially polyribosomal RNA and during mitosis polyribosomes dissociate (Traganos *et al.*, 1988).

X. Comparison of the Methods

Different advantages and limitations characterize each of the methods and each should be considered when making the choice:

A. RNA Quantitation

Stoichiometry of RNA measurement is better assured by the AO methodology (Bauer and Dethlefsen, 1980). This is due to the fact that the total RNA content is stained by AO, and there is less variation in quantum yield resulting from differences in base composition or conformation of RNAs than in the case of PY.

B. Sensitivity

Because there is significant spectrum overlap between AO bound to DNA and that to RNA, the ability to detect small amounts of RNA under the standard measuring conditions is better provided by the HO342-PY technique. Excitation of the AO-stained cells at two different wavelengths, however, raises the sensitivity of the AO method to or above the level of PY. Quantum yield and fluorescence intensity of cells stained are higher with AO than with PY.

C. Specificity

HO342 is a more specific DNA stain than AO; the latter, in addition to DNA and RNA, also stains glycosaminoglycans (Darzynkiewicz and Kapuscinski,

1990). Resolution of the DNA measurements by AO, thus, in cells containing excessive amounts of glycosaminoglycans (primary fibroblasts, mast cells, keratinocytes) is low.

D. Analysis of RNA Conformation

Both AO and PY can be used to reveal changes in conformation of RNA. The HO342-PY technique detects changes associated with association of polyribosomes (Traganos *et al.*, 1988), whereas the AO methodology can be applied to measure the degree of double-strandedness of RNA and its sensitivity to heat denaturation (melting profile) (Darzynkiewicz *et al.*, 1975).

E. Instrument Requirements

The possibility of excitation with a single wavelength, or by use of a mercury lamp as the only source of illumination, gives the AO methodology the advantage that it can be used with simpler and less costly instruments.

Acknowledgments

Supported by NCI Grant RO1 CA28704, the "This Close" Foundation, the Carl Inserra Fund, and the Chemotherapy Foundation. I thank Drs. Frank Traganos and Jan Kapuscinski for their comments and suggestions.

References

Bauer, K. D., and Dethlefsen, L. A. (1980). *J. Histochem. Cytochem.* **28**, 493–498.

Brachet, J. (1940). *C. R. Seances Soc. Biol. Ses Fil.* **133**, 88–90.

Crissman, H. A., Darzynkiewicz, Z., Tobey, R. A., and Steinkamp, J. (1985). *Science* **228**, 1321–1324.

Darzynkiewicz, Z. (1988). *Leukemia* **2**, 777–787.

Darzynkiewicz, Z. (1990). *In* "Flow Cytometry and Sorting" (M. R. Melamed, T. Lindmo, and M. L. Mendelsohn, eds.), pp. 469–501. Wiley-Liss, New York.

Darzynkiewicz, Z., and Kapuscinski, J. (1990). *In* "Flow Cytometry and Sorting" (M. R. Melamed, T. Lindmo, and M. L. Mendelsohn, eds.), pp. 293–314. Wiley-Liss, New York.

Darzynkiewicz, Z., Traganos, F., Sharpless, T., and Melamed, M. R. (1975). *Exp. Cell Res.* **95**, 143–153.

Darzynkiewicz, Z., Traganos, F., Sharpless, T., and Melamed, M. R. (1976). *Proc. Natl. Acad. Sci. U.S.A.* **73**, 2881–2884.

Darzynkiewicz, Z., Evenson, D. P., Staiano-Coico, L., Sharpless, T., and Melamed, M. R. (1979). *J. Cell. Physiol.* **100**, 425–438.

Darzynkiewicz, Z., Sharpless, T., Staiano-Coico, L., and Melamed, M. R. (1980a). *Proc. Natl. Acad. Sci. U.S.A.* **77**, 6696–6700.

Darzynkiewicz, Z., Traganos, F., and Melamed, M. R. (1980b). *Cytometry* **1**, 98–108.

Darzynkiewicz, Z., Kapuscinski, J., Traganos, F., and Crissman, H. A. (1987). *Cytometry* **8**, 138–145.

Fan, H., and Penman, S. (1970). *J. Mol. Biol.* **50**, 655–670.

Johnson, L. F., Abelson, H. T., Green, H., and Penman, S.(1974). *Cell* (*Cambridge, Mass.*) **1,** 95–100.

Kapuscinski, J., and Darzynkiewicz, Z. (1984). *J. Biomol. Struct. Dyn.* **1,** 1485–1499.

Kapuscinski, J., and Darzynkiewicz, Z. (1987a). *Cytometry* **8,** 129–137.

Kapuscinski, J., and Darzynkiewicz, Z. (1987b). *J. Biomol. Struct. Dyn.* **5,** 127–143.

Kapuscinski, J., and Darzynkiewicz, Z. (1989). *In* ''Biological Structure, Dynamics, Interactions and Stereodynamics'' (R. H. Sarma and M. H. Sarma, eds.), pp. 267–281. Adenine Press, Schenectady, NY.

Kapuscinski, J., Darzynkiewicz, Z., and Melamed, M. R. (1982). *Cytometry* **2,** 201–212.

Kapuscinski, J., Darzynkiewicz, Z., and Melamed, M. R. (1983). *Biochem. Pharmacol.* **32,** 3679–3694.

Pollack, A., Prudhomme, D. L., Greenstein, D. B., Irvin, G. L., III, Claflin, A. J., and Block, N. L. (1982). *Cytometry* **3,** 28–35.

Portela, R. A., and Stockert, J. C. (1979). *Experientia* **35,** 1663–1665.

Shapiro, H. M. (1981). *Cytometry* **2,** 143–150.

Shapiro, H. M. (1988). ''Practical Flow Cytometry,'' 2nd ed. Liss, New York.

Tanke, H. J., Niewenhuis, I. A. B., Koper, G. J. M., Slats, J. C. M., and Ploem, J. S. (1980). *Cytometry* **1,** 313–320.

Traganos, F. (1990). *In* ''Flow Cytometry and Sorting'' (M. R. Melamed, T. Lindmo, and M. L. Mendelsohn, eds), pp. 773–801. Wiley–Liss, New York.

Traganos, F., Darzynkiewicz, Z., Sharpless, T., and Melamed, M. R. (1977). *J. Histochem. Cytochem.* **25,** 46–56.

Traganos, F., Darzynkiewicz, Z., and Melamed, M. R. (1982). *Cytometry* **2,** 212–218.

Traganos, F., Crissman, H. A., and Darzynkiewicz, Z. (1988). *Exp. Cell Res.* **179,** 535–544.

Zanker, V. (1952). *Z. Phys. Chem.* **200,** 250–292.

CHAPTER 27

Analysis of DNA Content and Cyclin Protein Expression in Studies of DNA Ploidy, Growth Fraction, Lymphocyte Stimulation, and the Cell Cycle

Zbigniew Darzynkiewicz, Jianping Gong, and Frank Traganos

Cancer Research Institute
New York Medical College
Valhalla, New York 19595

I. Introduction

Several proteins that are major components of the complex machinery regulating and propelling cells through the cycle progression have recently been identified (e.g., Hartwell and Weinert, 1989; Nurse, 1990; Pines and Hunter, 1991; Norbury and Nurse, 1992; Sherr, 1993). Some of these proteins, namely cyclins, are expressed selectively in particular phases of the cycle. Thus, cyclin B1, being a member of the G_2 family of cyclins, is the key molecule that controls the cell entrance into mitosis; it is synthesized in G_2 and is abruptly degraded, through a ubiquitin-mediated pathway, during anaphase. During G_2, cyclin B1 associates with the serine/threonine specific protein kinase p34^{cdc2}. The kinase activity of this complex is regulated by sequential phosphorylations and dephosphorylations of threonine-14 and tyrosine-15 of p34^{cdc2}, at the location of the ATP binding site. Cyclin E belongs to the class of G_1 cyclins, and its expression peaks at the G_1/S transition. It associates with p33^{cdk2}, and the complex phosphorylates histone H1 and other substrates late in G_1 and early in S phase (Norbury and Nurse, 1992; Sherr, 1993).

The development of antibodies to cyclin proteins made it possible to investigate their expression in individual cells by flow cytometry (Gong *et al.*, 1993a,b, 1994; Kung *et al.*, 1993). Because of their cell-cycle phase specificity, analysis of the expression of cyclin proteins can be used, in addition to DNA content, as an independent marker of the cells' position in the cycle. The methods of detection of intracellular antigens combined with DNA content measurements, developed by Bauer and his colleagues (1986) and by Jacobberger *et al.* (1986) (see Chapter 23 of this volume), can be applied to the analysis of cyclin proteins. In this chapter we present the optimal conditions for immunocytochemical detection of cyclin B1 and E and describe certain applications of this methodology, some of them published previously (Gong *et al.*, 1993a,b, 1994a).

II. Bivariate Analysis of DNA Content versus Cyclin B1 or Cyclin E Expression

A. Reagents

1. *Cyclin antibodies.* Antibodies to cyclin proteins are offered by different vendors. We have tested antibodies from a variety of sources and have found that only a few were satisfactory for flow cytometric detection of cyclin proteins. Namely, the mouse monoclonal antibodies to cyclin B1 (Cat. No. 14541A and 14551A) and to cyclin E (Cat. No. 14591A; both provided by PharMingen, San Diego, CA) were tested by us on a variety of cell types of human origin, including clinical tumor and leukemic samples, and found to be cell-cycle phase specific in all these cell types, except

when cyclins were expressed in unscheduled (ectopic) fashion (Gong *et al.*, 1994b).

2. *FITC-conjugated goat anti-mouse IgG* (available, e.g., from Sigma Chemical Co., St. Louis, MO).

3. *Mouse IgG1* (isotypic control; available, e.g., from Sigma).

4. *Cell fixative.* Prepare 80% ethanol. Keep in the freezer (approx. $-20°C$) prior to use.

5. *Phosphate-buffered salt solution* (PBS).

6. *Bovine serum albumin* (BSA).

7. *Triton X-100.*

8. *RNase A* (DNase-free RNase, available, e.g., from Sigma).

9. *Propidium iodide* (PI; available from Molecular Probes, Eugene OR).

B. Procedure

1. Fix cells in suspension by pipetting 1 ml of cells suspended in PBS (approximately 10^6–10^7 cells) into tubes containing 10 ml of cold (approximately $-20°C$) 80% ethanol. The cells may be stored in this fixative in the cold (-20 to $-40°C$) for 2–24 hr.

2. Centrifuge cells, rinse once with PBS containing 1% BSA, centrifuge again, and suspend the cell pellet ($<10^6$ cells) in 1 ml of 0.25% solution of Triton X-100 in PBS. Keep on ice for 5 min. Add 5 ml of PBS. Centrifuge.

3. Suspend cell pellet in 100 μl of 1% BSA in PBS, containing cyclin B antibody diluted 1:400, or cyclin E antibody, diluted 1:100. Incubate at 4°C overnight.

4. Rinse the cells with 1% BSA in PBS, centrifuge, and suspend the cell pellet in 100 μl of 1% BSA in PBS containing FITC-conjugated goat antimouse IgG antibody (diluted 1:40). Incubate for 30 min at room temperature.

5. Rinse the cells with 1% BSA in PBS and suspend the cell pellet in a solution containing 10 μg PI and 1 mg/ml of RNase A in PBS. Incubate 20 min at room temperature before measurement.

6. Control cells should be treated identically, except instead of using cyclin antibody, the cells should be incubated with the isotypic antibody, at the same titer.

7. Measure cell green (530 + 20 nm; FITC) and red (>590 nm; PI) fluorescence.

C. Results

Figure 1 illustrates expression of cyclin B1 and cyclin E in exponentially growing cells of the human leukemic line MOLT-4. Expression of cyclin proteins is related to cell position in the cell cycle, identified by DNA content. Note the high specificity of cyclin B1 with respect to G_2 + M cells: a high level of

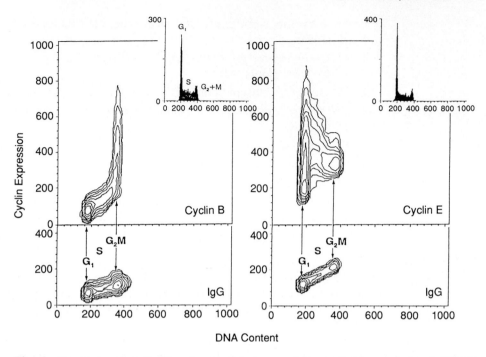

Fig. 1 Expression of cyclin B1 and cyclin E in exponentially growing human leukemic MOLT-4 cells. The lower panels represent the respective isotypic controls. Note that cyclin B1 is expressed exclusively by G_2 + M cells. Expression of cyclin E, on the other hand, peaks in G_1 and is decreasing during S. During stathmokinesis induced by vinblastine, it becomes apparent that the G_1 cyclin E-negative cells are the postmitotic cells, while the cells that express cyclin E maximally are late G_1 cells (reprinted with permission from Gong *et al.*, 1993a).

expression of cyclin B1 is observed exclusively in G_2 + M cells. Cells in G_1, in early and mid S are cyclin B negative, while late S cells show minimal expression of cyclin B.

The cell-cycle phase specificity is less pronounced in the case of cyclin E (Fig. 1). This protein is maximally expressed during G_1; the population of G_1 cells, however, is very heterogenous, ranging from cyclin E-negative cells to cells that have the highest expression of cyclin E.

III. Application of Cyclin B1 Analysis in DNA Ploidy and Chromatin Cycle Estimates

A. Discrimination between G_1 and G_2 + M Cells with the Same DNA Content: Elimination of Doublets of G_1 Cells

The most common analysis of the cell cycle is based on cellular DNA content measurement followed by deconvolution of the DNA content frequency histograms, using one of many available algorithms (see Chapter 18 of this volume). Bivariate analysis, which takes into account, in addition to DNA, the presence or content of other constituents in the cell, often allows one to discriminate between cells having the same DNA content but in different compartments of the cycle, e.g., between G_2 and M, G_1 and G_0, or early G_1 (G_{1A}) and late G_1 (G_{1B}) phase (e.g., Darzynkiewicz et al., 1976). All these approaches, however, fail to discriminate between cells in G_2 phase of a lower DNA ploidy and G_1 phase at an increased DNA ploidy level, e.g., between diploid G_2 (G_{2D}; cells with 4C DNA) versus tetraploid G_1 (G_{1T}; cells also with 4C DNA content). The presence of such mixed cell populations is often apparent in cultures of tumor cells that grow at two different DNA ploidy levels, when treated with drugs that impair cytokinesis and induce DNA re-replication and polyploidization, or in multiploid tumors.

Figure 2 illustrates an application of cyclin B antibodies to discriminate between G_2 cells of lower DNA ploidy and G_1 cells of the higher ploidy. In this experiment, MOLT-4 cells were grown in the presence of the protein kinase, inhibitor staurosporine (SSP), which impairs cytokinesis and induces the transition of cells of this cell line to enter the cell cycle at higher DNA ploidy levels (Bruno et al., 1992). As is evident, such treatment leads to the appearance of cells with an increased DNA content, above that of G_2 + M phase cells. Among the cells with a DNA content equivalent to that of G_2 + M, a subpopulation of cells that do not express cyclin B became very prominent (Fig. 2C). This population, as judged by its continuity on the bivariate DNA content/cyclin B distributions with the S_T population, represents cells which, following mitosis, entered G_1 phase at a tetraploid DNA level (G_{1T} cells). It is also clear that entrance of cells to the G_2 + M phase at the tetraploid DNA level is also accompanied by an increase in cyclin B expression (Fig. 2E).

Such an approach can also be used to discriminate between G_2 + M cells (which express this protein) and doublets of G_1 cells (which are cyclin B1 negative) (Gong et al., 1993a).

B. Assay of the Chromatin Cycle When Cytokinesis Is Impaired

Bivariate analysis of cyclin B1 expression and DNA content makes it possible to estimate the kinetics of cell progression through various compartments of the cell cycle at two DNA ploidy levels, simultaneously, in situations when cell division is prevented (endoreplication), as it is shown in Fig. 2. This is done by measuring the percentage of cells in each subpopulation identified on the bivariate DNA content/cyclin B distributions by gating analysis at different times of treatment with SSP. The plot of such data reveals the rates of cell entrance to G_2, G_{1T}, S_T, G_{2T}, G_{10}, etc., as shown in Fig. 3. Other details of this assay are presented elsewhere (Gong et al., 1993a).

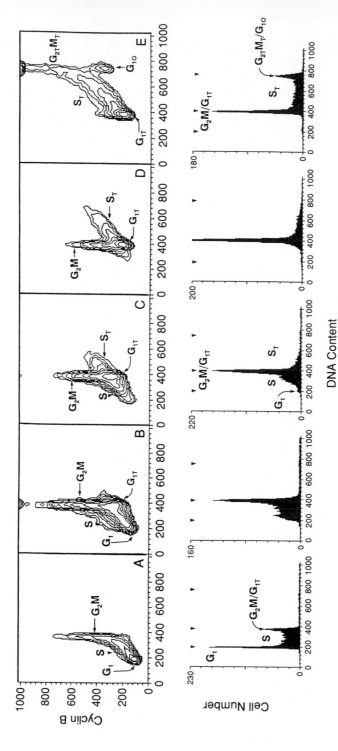

Fig. 2 Changes in DNA content and cyclin B1 expression of MOLT-4 cells untreated (A) or treated with 0.1 μM staurosporine for 4 (B), 6 (C), 8 (D), and 12 hr (E) to induce DNA endoreduplication. Top: Bivariate contour maps representing DNA content/cyclin B1 distributions. Note the appearance of the cyclin B1-negative cell population with G_2 + M DNA content (tetraploid; G_{1T}), which is already evident after 4 hr, and cyclin B1-negative cells with octaploid DNA content (G_{1O}), after 12 hr. Bottom: DNA content frequency histograms of the same cultures. The short arrows indicate the position of cells with a 2C, 4C, and 8C DNA content, respectively (reprinted with permission from Gong *et al.*, 1993b).

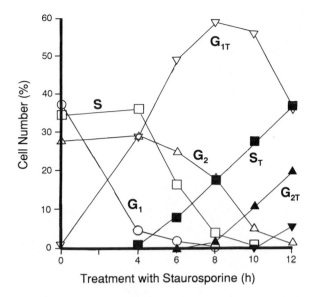

Fig. 3 The rates of MOLT-4 cell progression through the chromosome cycle during DNA endore-duplication induced by 0.1 μM staurosporine. As shown in Fig. 2, the bivariate analysis of DNA content/cyclin B1 expression makes it possible to discriminate between G_2 cells of lower DNA ploidy and G_1 cells of higher ploidy, i.e., between cells having the same DNA content but differing in progression through the cycle (G_2 versus G_{1T}, tetraploid; G_{2T} versus G_{1o}, octaploid). This approach can be used to reveal the proportions of cells in the respective phases of the cell cycle under conditions of DNA endoreduplication, e.g., when the cytokinesis is impaired, and to estimate the rates of cell progression through the chromosome cycle (reprinted with permission from Gong *et al.*, 1993b).

IV. Lymphocyte Stimulation

Peripheral blood lymphocytes are quiescent G_0 cells, which can be stimulated *in vitro* by a variety of mitogens to enter and progress through the cell cycle. The assays of lymphocyte response to mitogenic stimuli have wide application in experimental and clinical immunology, and one of the methods is presented in Chapter 26 of this volume.

Nonstimulated lymphocytes are cyclin E negative. Expression of cyclin E, which is associated with their progression through G_1 and entrance to S phase, appears to be a specific and relatively early event of lymphocyte stimulation (Fig. 4). Maximum expression is observed 24 and 40 hr after addition of the polyvalent mitogen phytohemagglutinin (Gong *et al.*, 1993a). The expression peaks just prior to entrance to S phase and decreases with cell progression through S.

Stimulation of lymphocytes *in vitro*, thus, can be estimated based on the percentage of cells expressing cyclin E between 24 and 48 hr after the addition

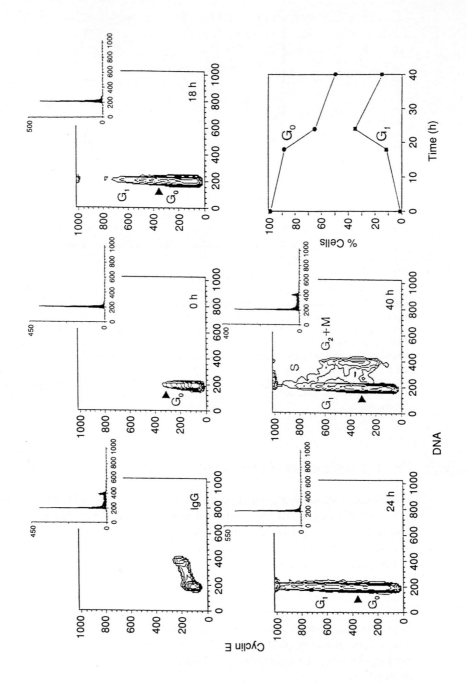

of mitogen. Because stimulated lymphocytes start to enter S phase 24 hr after addition of the mitogen (Darzynkiewicz *et al.*, 1976), no early G_1, postmitotic cells (i.e., stimulated but cyclin E negative) are present 24 hr after addition of the mitogen. Furthermore, administration of the S phase blocker (e.g., aphidicolin) 24 hr following the mitogen prevents cell progression through S and therefore precludes the appearance of cyclin E-negative, G_1 postmitotic cells, even 48 hr after stimulation. The percentage of lymphocytes expressing cyclin E, thus, directly corresponds to the percentage of lymphocytes activated by a given mitogen during 48 hr of continuous exposure to the mitogen.

The assay of lymphocyte stimulation based on the percentage of cells expressing cyclin E offers the following advantages over the standard analysis of stimulation, which employs cell labeling with radioactive thymidine and radioactivity measurements in bulk, in acid-insoluble material:

1. no radioactive materials are used;

2. the absolute number of cells entering the cell cycle can be quantified with high accuracy;

3. cell entrance to G_1 can be discriminated from the progression through S phase; the rate of cell transition from G_0 to G_1, as well as the rate of cell entrance to S phase, thus, can be estimated;

4. the phenotype of the responding lymphocytes can be revealed by simultaneous counterstaining with antibody to surface antigens and multivariate analysis;

5. the percentage of cells that undergo apoptosis in the same cultures can be evaluated; apoptotic cells have fractional DNA content, under the present staining conditions (See Chapter 2 of this volume).

V. Tumor Growth Fraction Analysis

The fraction of proliferating cells in tumors (Mendelsohn, 1962) is currently estimated primarily by Ki-67 antibody (Gerdes *et al.*, 1984). The role of the antigen recognized by this antibody in the cell cycle, however, is poorly defined. Furthermore, the late G_1 cells, just prior to entrance to S phase, are character-

Fig. 4 Expression of cyclin E during stimulation of lymphocytes. Human peripheral blood lymphocytes were stimulated with phytohemagglutinin for 0, 18, 24 or 40 hr fixed, and stained for cyclin E and DNA. The presence of cyclin E-positive cells was already apparent by 18 hr and their proportion was increased after 24 hr. By 40 hr, cells had already entered S and G_2 + M. The increase in percentage of cyclin E-positive G_1 cells and the concomitant decrease in cyclin E-negative (G_0) cells is plotted as a function of time after administration of PHA (bottom right panel); the discrimination between cyclin E-negative and -positive cells is indicated by the arrows to the left of the distributions in each panel.

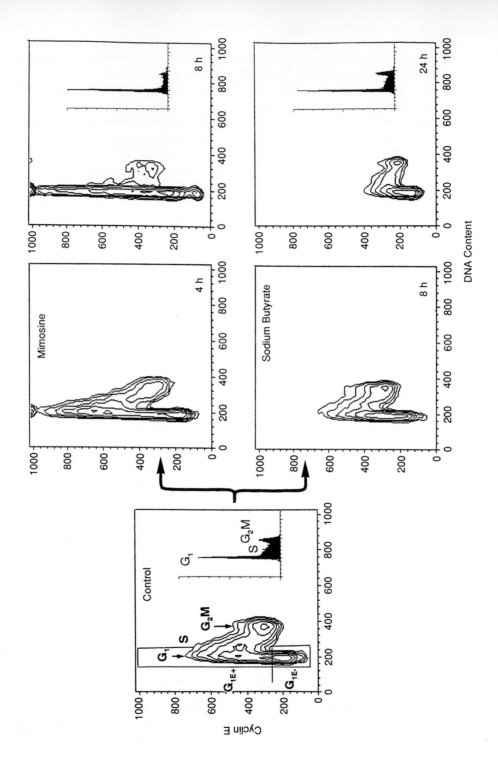

ized by very low expression of Ki-67 (Lopez *et al.*, 1991) and therefore can mistakenly be identified as noncycling cells.

Simultaneous labeling of cells with cyclin E antibody (which marks G_1 cells committed to enter S phase and early and mid S phase cells) and cyclin B1 antibody (marker of G_2 and M cells) can be used to identify proliferating cells. Such labeling, therefore, can be an alternative marker of the tumor growth fraction. The advantage of these antibodies is that they interact with very well-defined constituents of the cell-cycle machinery and, therefore, are very specific to cycling cells. The ambiguities, therefore, resulting from application of an antibody detecting an antigen of unknown function, are eliminated with the use of antibodies to cyclin proteins (Gong *et al.*, 1994b).

VI. Mapping the Cell Cycle: Mechanism of Action of Cytostatic Drugs

Numerous antitumor agents arrest cells in particular phases of the cycle. The point of action of these drugs in the cycle is generally characterized in terms of time interval, for example, between administration of the drug and cell entrance to S phase (e.g., Gadbois *et al.*, 1992) or mitosis (Tobey, 1975). Such characterization is imprecise for several reasons: (1) the temporal mapping may vary, depending on the inherent variation in the duration of G_1 of the particular cell type, e.g., when cells which have a G_1 duration of 4 hr are compared with cells having a G_1 duration several times longer; (2) the cell entrance rate to S is asynchronous (stochastic), which makes the estimate of a defined time interval difficult; (3) cell arrest, by a particular agent, often results in unbalanced growth. The degree of unbalanced growth varies, depending on the cell type and the duration of the arrest (Traganos *et al.*, 1982). Upon release from the arrest, the rate of cell progression through the cell cycle is affected by the degree of growth imbalance (Darzynkiewicz *et al.*, 1979). In summary, temporal mapping has drawbacks when used as a yardstick to characterize the point of action of a particular drug when different cell types are compared or different degrees of growth unbalance are induced.

The landmarks in cell progression through G_1 phase are the onsets of synthesis of G_1 cyclins such cyclin E or D. Cyclin E is synthesized relatively late in G_1

Fig. 5 Changes in cyclin E expression in MOLT-4 cells treated for 4 and 8 hr with 0.3 mM mimosine (upper panels) and for 8 and 24 hr with 1 mM sodium butyrate (lower panels). The decreased proportion of G_1 cyclin E-negative cells (G_{1cyE-}; marked as G_{1E}) and an increase in expression of cyclin E accompany arrest by mimosine. In contrast, an increased proportion of G_{1cyE-} cells and a decrease in cyclin E expression are evident in the case of sodium butyrate-treated cells.

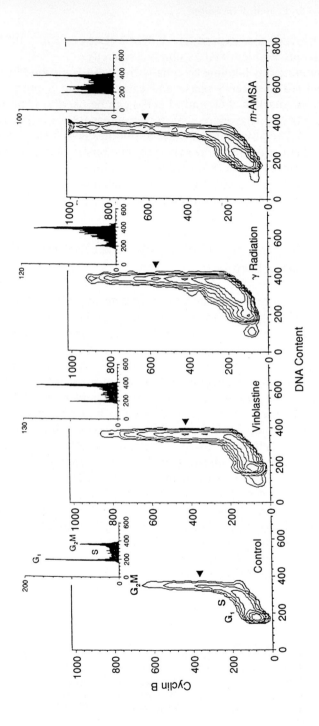

while synthesis of cyclin D precedes that of cyclin E (Sheer, 1993). Likewise, the onset of cyclin B synthesis is a landmark in G_2. It is possible, therefore, to use these landmarks as a reference to the points of cell arrest in the cycle by a particular drug. Figure 5 illustrates such an application of the cyclin E antibody. Namely, the onset of cyclin E synthesis (cyclin E restriction point) was chosen as a reference to the arrest of cells in G_1 by mimosine and sodium butyrate. Mimosine, a plant amino acid, has been shown to arrest several cell types late in G_1, approximately 15 min before the G_1/S border (Watson *et al.*, 1992). The mechanism by which mimosine perturbs progression through G_1 is unknown, but it may involve its interaction with a protein which is essential for initiation of DNA replication (Mosca *et al.*, 1992). As is evident from the data shown in Fig. 5, cell arrest by mimosine is beyond the cyclin E restriction point (G_{1cyE+}) inasmuch as cells arrested by this drug accumulate in G_1 with an increasing cyclin E content.

The mechanism by which sodium butyrate arrests cells in G_1 is not well understood. This agent is an inhibitor of histone deacetylase, and incubation of cells in its presence leads to hyperacetylation of nucleosomal core histones while histone H1 is hypophosphorylated (D'Anna *et al.*, 1980). It is not known whether the changes in chemical modification of histone proteins are the cause of cell arrest in G_1. The data in Fig. 5, however, clearly indicate that cell arrest by *n*-butyrate is prior to the cyclin E restriction point (G_{1cyE-}).

Figure 6 illustrates the expression of cyclin B1 in cells arrested in G_2, by γ radiation or by *m*-AMSA, and in mitosis, by vinblastine. Here again, the data, which demonstrate accumulation of cyclin B1 in cells arrested by these agents, indicate that they affect the cell cycle at a point beyond the onset of cyclin B1 synthesis (G_{2cyB1+}).

This type of mapping of the cell cycle has been done for other agents, such as quercetin, lovastatin, staurosporine, and cycloheximide; a full description of the data and further discussion of this application are the subject of a separate publication (Gong *et al.*, 1994a).

VII. Critical Points of the Procedure

Immunocytochemical detection of intracellular antigens and their quantitative assessment are possible under the following conditions:

Fig. 6 Bivariate distributions of DNA content and expression of cyclin B1 of MOLT-4 cells, growing exponentially (control), treated with 0.2 μM vinblastine for 6 hr, 5 Gy of radiation followed by 6 hr growth in culture, or 0.1 μM *m*-AMSA for 3 hr followed by 6 hr growth in drug-free medium. The cells arrested in M by vinblastine or in G_2 by radiation and *m*-AMSA are characterized by the elevated expression of cyclin B1; the arrows to the right on each distribution indicate the position of the peak value of $G_2 + M$ cyclin B1 expression (reprinted with permission from Gong *et al.*, 1993a).

1. cell fixation does not destroy the native structure of the epitope;
2. cell fixation stabilizes the antigen in the cell, preventing its extraction during sequential centrifugations and washings;
3. the cell is permeable to the antibody; and
4. the epitope is accessible to the antibody and not sequestered, e.g., by interactions with neighboring macromolecules.

All these conditions, especially the choice of fixative, which can be critical for the detection of intracellular macromolecules, are discussed in detail in Chapter 23 of this volume. For the detection of cyclins E and B1, fixation in cold 80% ethanol appears to be optimal (Gong *et al.*, 1993a,b). Equally good is fixation in a 1:1 mixture of cold ethanol and acetone; although it induces more cell aggregation compared with 80% ethanol. It is quite possible, however, that for the detection of other cyclins, fixatives other than 80% ethanol may be required.

Another critical point, which should be stressed, is that some antibodies available from particular vendors may not be suitable for immunocytochemical detection of the antigen. Often, the antibody which is very specific and useful for Western blotting is ineffective for the detection of the antigen *in situ*. This is due to the fact that the same epitope of the protein while denatured may react with the antibody, but in the native state, or somewhat altered by fixation, may no longer bind the antibody.

VIII. Conclusions

The availability of antibodies to cell-cycle phase-specific cyclin proteins and their application in multivariate analysis, to reveal expression of these proteins in relation to cell-cycle position or in relation to other cell constituents, open new possibilities for the study of the cell cycle. A few applications, outlined in this chapter, appear to be the most obvious and they certainly will be further explored for different cell types, different drugs, etc. The most valuable, however, may be the application of these antibodies in the clinic, to identify tumors or stages of some proliferative diseases, for classification and prognosis. It is likely that the unscheduled expression or overexpression of these proteins, the key regulatory molecules of the cell cycle, will be shown to be patognomic to the stage or the disease and be valuable prognostic markers. Such unscheduled (ectopic) expression of cyclins, namely expression of a G_1 cyclin during G_2, and vice versa, was recently observed in several tumor cell lines (Gong *et al.*, 1994b). It should be stressed that analysis of cyclin B1 expression for DNA ploidy and chromatin cycle estimates (Section III) is limited to the cells which express these cyclins in a scheduled fashion.

Acknowledgments

This work was supported by U.S.P.H.S. Grant CA 28704. Dr. Gong, on leave from the Department of Surgery, Tongji Hospital, Wuhan, China, is supported by the fellowship from the "This Close" for Cancer Research Foundation.

References

Bauer, K. D., Clevenger, C. V., Williams, T. J., and Epstein, A. L. (1986). *J. Histochem. Cytochem.* **34,** 245–250.

Bruno, S., Ardelt, B., Skierski, J., Traganos, F., and Darzynkiewicz, Z. (1992). *Cancer Res.* **52,** 470–473.

D'Anna, J. A., Tobey, R. A., and Gurley, L. R. (1980). *Biochemistry* **19,** 2656–2671.

Darzynkiewicz, Z., Traganos, F., Sharpless, T., and Melamed, M. R. (1976). *Proc. Natl. Acad. Sci. U.S.A.* **73,** 2881–2884.

Darzynkiewicz, Z., Evenson, D. P., Staiano-Coico, L., Sharpless, T. K., and Melamed, M. R. (1979). *J. Cell. Physiol.* **100,** 425–438.

Gadbois, D. M., Crissman, H. A., Tobey, R. A., and Bradbury, M. E. (1992). *Proc. Natl. Acad. Sci. U.S.A.* **89,** 8626–8630.

Gerdes, J., Lemke, H., Baisch, H., Wacker, H., Schwab, U., and Stein, H. (1984). *J. Immunol.* **133,** 1710–1715.

Gong, J., Traganos, F., and Darzynkiewicz, Z. (1993a). *Int. J. Oncol.* **3,** 1037–1042.

Gong, J., Traganos, F., and Darzynkiewicz, Z. (1993b). *Cancer Res.* **53,** 5096–5099.

Gong, J., Traganos, F., and Darzynkiewicz, Z. (1994a). *Int. J. Oncol.* **4,** 803–808.

Gong J., Li, X., Traganos, F., and Darzynkiewicz, Z. (1994b). *Cell Prolif.* (in press).

Hartwell, L. H., and Weinert, T. A. (1989). *Science* **246,** 629–634.

Jacobberger, J. W., Fogleman, D., and Lehman, J. M. (1986). *Cytometry* **7,** 356–364.

Kung, A. L., Sherwood, S. W., and Schimke, R. T. (1993). *J. Biol. Chem.* **268,** 23072–23080.

Lopez, F., Belloc, F., Lacombe, F., Dumain, P., Reiffers, J., Bernard, P., and Boisseau, M. R. (1991). *Cytometry* **12,** 42–49.

Mendelsohn, M. L. (1962). *J. Natl. Cancer Inst. (U.S.)* **28,** 1015–1029.

Mosca, P. J., Dijwell, P. A., and Hamlin, J. L. (1992). *Mol. Cell. Biol.* **12,** 4375–4383.

Norbury, C., and Nurse, P. (1992). *Annu. Rev. Biochem.* **61,** 441–470.

Nurse, P. (1990). *Nature (London)* **344,** 503–508.

Pines, J., and Hunter, T. (1991). *J. Cell Biol.* **115,** 1–17.

Sherr, C. J. (1993). *Cell (Cambridge, Mass.)* **73,** 1059–1065.

Tobey, R. A. (1975). *Nature (London)* **254,** 245–247.

Traganos, F., Darzynkiewicz, Z., and Melamed, M. R. (1982). *Cytometry* **21,** 212–218.

Watson, P. A., Hanauske-Abel, H. H., Flint, A., and Lalande, M. (1992). *Cytometry* **12,** 242–246.

CHAPTER 28

Oxidative Product Formation Analysis by Flow Cytometry

J. Paul Robinson,★ Wayne O. Carter, and Padma Kumar Narayanan

★ Department of Physiology and Pharmacology
School of Veterinary Medicine
Purdue University
West Lafayette, Indiana 47907

I. Introduction

Prior to flow cytometric methods, measurement of oxidative function was labor intensive and often very difficult to quantitate. The *Staphylococcus aureus* killing methods (Alexander *et al.,* 1968) required a large volume of blood and at least 48 hr to complete. More recent methods such as chemiluminescence, while quantitative and rapid, have been difficult to interpret (DeChatelet *et al.,*

1982; Cheung *et al., 1983*). The advantages of the flow-based methods are that a significantly reduced cell number is required, the procedure is relatively easy, and, if a flow cytometer is already available, the procedure is inexpensive to perform. A significant feature of the flow-based methods not available using any other technique is the capability to determine heterogeneity of response (Taga *et al., 1985*; Neill *et al., 1985*). Subpopulations defined by either light scatter or fluorescence intensity can be identified and quantified. Thus unresponsive or poorly responsive cell populations are easily discerned.

II. Oxidative Burst

A. Application

A variety of reactive oxygen species is produced by several cells. Our principal application for measuring oxidative systems is for functional evaluation of ''oxidative or respiratory bursting'' in neutrophils. The respiratory burst results from activation of the membrane-bound NADPH oxidases via an electron transfer reaction. Two electrons are transferred from NADPH through an FAD-flavoprotein utilizing cytochrome b$_{-245}$ to oxygen. Superoxide anion is produced and then dismutates to hydrogen peroxide (H_2O_2) either spontaneously or by superoxide dismutase (SOD). The reactive oxygen species (ROS) and the hydrogen peroxide produced are necessary for normal bactericidal mechanisms in the neutrophil. These oxidative mechanisms exist in many cells and we have used these techniques to evaluate a variety of cell types including neutrophils, macrophages, HL-60 cells (human leukemia-60 cells, see Vol. 42, chapter 25), and endothelial cells.

A calibration curve can be generated based upon spectrophotometric data and flow cytometric measurements. This allows for conversion of flow cytometry fluorescence channels into quantitative estimations of H_2O_2, if necessary (Bass *et al., 1983*).

The assays described below utilize two dyes. In the first assay, 2′,7′-dichlorofluorescin diacetate (DCFH-DA), which is freely permeable, is incorporated into hydrophobic lipid regions of the cell (Bass *et al., 1983*). The acetate moieties are cleaved off leaving the nonfluorescent 2′,7′-dichlorofluorescin (DCFH). Hydrogen peroxide and peroxidases produced by the cell oxidize DCFH to 2′,7′-dichlorofluorescein (DCF) which is fluorescent (530 nm). The green fluorescence produced is thus proportional to the H_2O_2 produced.

The second assay utilizes hydroethidine (HE) which can be directly oxidized to ethidium bromide by superoxide anions produced by the cell. The third assay incorporates both dyes and has been used to detect selective defects in phagosomal oxidation following lysosomal degranulation as has been reported to occur in sepsis (Rothe and Valet, 1990). There are several advantages of HE over the DCF assay.

B. Materials

1. Hanks' balanced salt solution (HBSS)

Stock HBSS $10\times$ concentration: NaCl, 40 g; KCl, 2.0 g; Na_2HPO_4, 0.5 g; $NaHCO_3$, 0.5 g; q.s. to 500 ml.

Stock Tris 1.0 M: Tris base, 8.0 g; Tris–HCl, 68.5 g; q.s. to 500 ml, pH to 7.3.

Preparation of 100 ml HBSS:

Stock HBSS, $10\times$	10 ml
Distilled water	80 ml
Tris, 1.0 M	2.75 ml
$CaCl_2$, 1.1 M	170 μl
$MgSO_4$, 0.4 M	200 μl
Dextrose	220 mg

adjust pH to 7.4 and q.s. to 100 ml.

2. PBS gel

- Stock PBS gel

EDTA (disodium salt), 0.2 M	7.604 g
Dextrose, 0.5 M	9.0 g
10% Gelatin (Difco)	10 g
Distilled water	100 ml

- Heat water to 45–50°C and slowly add gelatin while mixing with a magnetic stirrer. Continue stirring and add EDTA and dextrose. Do not exceed 55°C because gelatin and glucose will "caramelize." Store in 1.2-ml aliquots at -20°C.
- Working solution PBS gel (make daily as needed). Warm 1 ml gel to 45°C. Add 95 ml warm PBS (phosphate buffered saline) and mix. Adjust pH to 7.4 and q.s. to 100 ml.

3. Erythrocyte lysing solution (preparation of 100 ml):

NH_4Cl, 0.15 M	0.8 g
Na HCO_3, 10 mM	0.084 g
EDTA (disodium), 10 mM	0.037 g
Distilled water	95 ml

adjust pH to 7.4 and q.s. to 100 ml.

4. DCFH-DA (MW 487.2) [Molecular Probes, Inc., Eugene, OR], 20 mM solution:

- Weigh 2–9 mg of DCFH-DA and place in a foil-covered 12 $\times$ 75-mm tube.
- Add absolute ethanol in a volume equivalent to the weight in milligrams of the DCFH-DA divided by 9.74.
- Cap the tube, mix, cover in foil, and store at 4°C until use.

5. HE (MW 315) [Molecular Probes Inc., Eugene, OR], 10 m*M* solution:
   • Stock solution, 10 m*M* in dimethylformamide (3.15 mg/ml).

6. PMA (phorbol 12-myristate 13-acetate) [Sigma Chemical Co., St. Louis, MO]: PMA is toxic and carcinogenic; additionally dimethyl sulfoxide (DMSO) is readily absorbed through the skin. Wear gloves while handling solutions, prepare solutions in a hood, and be extremely cautious!

   Stock PMA (2 mg/ml in DMSO): Mix well and aliquot 15–20 μl of stock PMA in small capped polypropylene bullets. Store at -20°C.

   Working PMA solution (make daily as needed): 5 μl PMA stock in 10 ml PBS gel (1000 ng/ml PMA solution). A final PMA concentration of 100 ng/ml will predictably result in maximal cell stimulation (e.g., 900 μl of cells in solution and 100 μl of working PMA solution).

C. Instrumentation

Excitation is at 488 nm for both the above probes. Emission filters should be 525 nm for DCF and 590 nm for HE. Collection of forward light scatter and 90° scatter, as well as both fluorescence wavelengths, is necessary. Where possible collect list-mode data for further analysis. If performing a kinetic assay, time may be required.

D. Methods

DCFH-DA Assay (using whole blood):

1. Place 2 ml of preservative-free heparinized whole blood in a 50-ml conical tube.
2. Add 48 ml of erythrocyte lysing solution.
3. Gently mix solution for 10 min at 25°C on a hematology rotator.
4. Centrifuge for 10 min at 350g and 4°C.
5. Decant supernatant and resuspend in 5 ml of PBS gel (working solution).
6. Centrifuge for 10 min at 350g and 4°C. Decant supernatant and resuspend in 2.5 ml HBSS.
7. Count leukocytes and adjust cell suspension to 2.0 × 10^6 cells/ml.
8. Add 1 μl 20 m*M* DCFH-DA per ml of cell suspension to be loaded.
9. Incubate loaded cells at 37°C for 15 min.
10. Stimulate cells with PMA: add 100 μl PMA (working solution) to 900 μl of cell suspension (final PMA concentration 100 ng/ml). Reserve some loaded, unstimulated cell suspension for a control.
11. Maintain cell sample at 37°C and run stimulated and unstimulated samples every 10 min on the cytometer for a total of 40 min.

Hydroethidine Assay: Procedures 1–7 are identical (as above).

8. Add 1 μl HE per ml of cell suspension to be loaded.
9. Incubate loaded cells at 37°C for 15 min.

Procedures 10 and 11 are identical (as above)

Combined DCFH-DA and Hydroethidine Assay: Procedures 1–7 are identical (as above)

8. Add 1 μl 20 mM DCFH-DA per ml of cell suspension to be loaded.
9. Incubate loaded cells at 37°C for 5 min.
10. Add 1 μl HE per ml of cell suspension to be loaded.
11. Incubate loaded cells at 37°C for an additional 15 min.
12. Stimulate cells with PMA: add 100 μl PMA (working solution) to 900 μl of cell suspension (final PMA concentration 100 ng/ml). Reserve some loaded, unstimulated cell suspension for a control.
13. Maintain cell sample at 37°C and run stimulated and unstimulated samples every 10 min on the cytometer for a total of 40 min.

E. Critical Aspect

Clumped cells are obviously detrimental to the assay and may potentially plug the flow cell. Additionally, clumped cells cannot be measured as "functionally normal" and must be eliminated. Clumping may be related to highly activated cells and may therefore provide important information regarding the context of the assay. Keeping the cell sample at 4°C (on ice) and using PBS gel will help prevent clumping of these reactive cells. Additionally, the dextrose, calcium, and magnesium in the HBSS are necessary for optimal cell function. Preservative-free heparin is also important to prevent any alteration in normal cell function. Dimethylformamide, used for preparation of the hydroethidine solution, will dissolve plastics and should be stored in glass. Alternatively DMSO can be used to prepare the hydroethidine solution. Another crucial aspect of these assays is that an unstimulated cell sample must be used for comparison at all time points.

There is a steady increase in oxidation of the DCFH-DA and hydroethidine even in unstimulated cells. This spontaneous oxidation varies with the cell type and with the dye and is primarily mitochondrial in origin. Mitochondrial oxidation can be blocked with azide or cyanide. In addition to measuring an increase in red fluorescent ethidium bromide, a decrease in blue fluorescent hydroethidine (excitation 350–380 nm, emission 418–500 nm) can be simultaneously measured in cells. One of the advantages of hydroethidine is the measurement of a ROS, superoxide anion, which occurs earlier in the cascade of oxidation events than hydrogen peroxide. An advantage of the combined assay is the identification of different subpopulations of neutrophils with disparate oxidative function.

F. Results

Cells must be preloaded for the correct time before activation. The time period varies with the cell type. Neutrophils are usually fully "loaded" within 15 min at 37°C.

Figures 1–3 show a typical histogram of an unstimulated and a stimulated population of human neutrophils. A 4- to 20-fold increase in green fluorescence is expected and evident above.

III. Phagocytosis

A. Application

Phagocytosis involves a series of stages by a cell progressing from particle attachment to ingestion. Any step in the phagocytic process can potentially fail. By using both opsonized and unopsonized particles in the assay we can determine whether abnormal phagocytosis is due to defective ingestion or opsonization. The technique described below employs the use of opsonized and

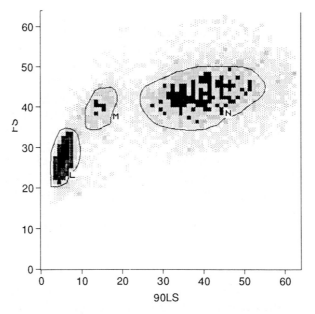

Fig. 1 Histogram with forward angle light scatter (FS) on the *y*-axis and 90° light scatter (90LS) on the *x*-axis. The cell sample was whole blood (human) prepared by erythrocyte lysis with ammonium chloride. Three different cell populations are readily identified and labeled: L, lymphocytes; M, monocytes; N, neutrophils. The gate (N) drawn around the neutrophil population was applied to additional histograms to identify the fluorescent changes in only the neutrophils.

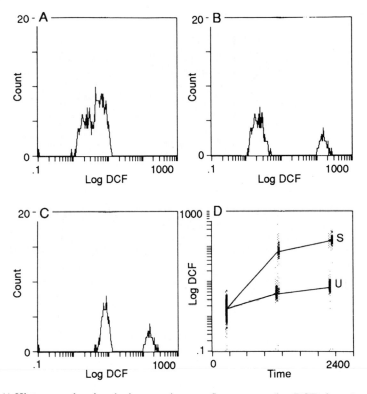

Fig. 2 (A) Histogram showing the increase in green fluorescence (log DCF) from 0 to 40 min in unstimulated neutrophils. (B) Similar histogram to A except the neutrophils are stimulated with PMA. (C) Histogram indicating the difference in fluorescence (log DCF) at 40 min between unstimulated and PMA-stimulated neutrophils (a 4- to 20-fold increase in fluorescence is expected in normal neutrophils). (D) Overview histogram of three time points (time in seconds) versus green fluorescence (log DCF on *y*-axis) for both unstimulated (U) and PMA-stimulated (S) neutrophils. The lines are drawn through the mean channel fluorescence of each time point. As indicated in the text, unstimulated neutrophils will oxidize (primarily mitochondrial oxidation) some DCFH-DA to DCF and increase the intensity of green fluorescence. Nevertheless, the change in fluorescence is much greater in PMA-stimulated neutrophils.

unopsonized bacteria (Bjerknes and Bassoe, 1984). Alternatively, fluorescent beads may be used or FITC (fluorescein isothiocyanate)-labeled yeast. After phagocytosis of the FITC-labeled yeast, ethidium bromide can be added to stain the extracellular yeast red so they can be differentiated on the flow cytometer.

A related version of this assay can also be used for evaluating phagocytosis in macrophage populations. This has particular importance in evaluating lavage fluids such as peritoneal lavage or bronchoalveolar lavage.

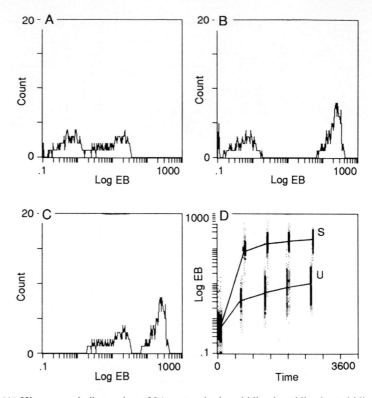

Fig. 3 (A) Histogram similar to that of 2A except hydroethidine is oxidized to ethidium bromide (EB) in unstimulated neutrophils. (B) Similar histogram to A except the neutrophils are stimulated with PMA. The change in red fluorescence (log EB) is evident from 0 to 40 min in PMA-stimulated neutrophils. (C) Difference in red fluorescence (log EB) between unstimulated and PMA-stimulated neutrophils at 40 min. (D) Overview histogram showing five time points (time in seconds) versus EB fluorescence for unstimulated and PMA-stimulated neutrophils. The lines are drawn through the mean channel fluorescence of each time point. Similar to the changes with DCF, an increase in EB fluorescence is evident in unstimulated neutrophils. However, the fluorescence change is much greater for PMA-stimulated neutrophils. More time points were collected for the HE assay than the DCF assay (Fig. 2) but the result is the same.

B. Materials

Bacterial culture

Blood agar plate

Brain–heart infusion broth

Brain–heart infusion agar slant

Neutrophil isolation medium (NIM) [Cardinal Associates, Santa Fe, NM]

HBSS (as above)

Pooled human serum

Glycerin

Carbonate/bicarbonate buffer (pH 9.5):

Na$_2$CO$_3$, 0.5 M (5.3 g/100 ml)	1 volume
NaHCO$_3$, 0.5 M (4.2 g/100 ml)	3 volumes

FITC: 0.02 mg/ml in carbonate/bicarbonate buffer (make a concentrated solution and dilute).

0.9% NaCl–0.02% EDTA:

NaCl	900 mg
EDTA disodium	20 mg
Distilled water, q.s. to	100 ml

Trypan blue (Gibco Laboratories, Grand Island, NY):

Stock, 4 mg/ml in saline

Working solution, 3 mg/ml (dilute stock 3:4 with saline).

C. Methods

1. Bacterial Culture

Bacteria are cultured in a blood agar plate and then subcultured to obtain discrete colonies. A discrete colony is then cultured on a brain–heart infusion agar slant and again subcultured. Using a sterile loop, some of the colonies are transferred into broth and cultured overnight in a 37°C incubator.

2. Labeling Bacteria with FITC

The bacteria are washed with HBSS and centrifuged at 10,000g for 10 min and the supernatant is decanted. The bacterial slurry is heat killed for 1 hr at 60°C. Enumeration of the bacteria is achieved by serial dilutions and subsequent plating and correlation using spectrophotometry so that the final concentration is approximately 10^9/ml. The bacteria are resuspended in carbonate/bicarbonate buffer such that the absorbance is, for example, 0.35 at 620 nm (this may differ depending upon the bacterial species and the media composition). The volume of the bacterial suspension is doubled with 0.02 mg/ml FITC in carbonate/bicarbonate buffer and incubated at 37°C with end-over-end rotation for 30 min. The bacteria are washed three times with HBSS, counted, and resuspended at a final concentration of 1 × 10^8/ml. The bacteria are aliquoted into sterile Eppendorf microcentrifuge tubes at a volume of 1 ml/vial; a drop of sterile glycerin is added to each vial and frozen at −70°C. The labeled bacteria will last for several years at −70°C.

3. Phagocytosis Assay (Using a Purified Population of Neutrophils)

1. Overlay 5 ml of EDTA or preservative-free heparinized whole blood on NIM.

2. Centrifuge for 30 min at 400g and 25°C.

3. Remove neutrophil layer and wash in PBS gel; centrifuge 250g for 10 min at 4°C.

4. Remove supernatant and resuspend neutrophil pellet in HBSS.

5. Count neutrophils and adjust cell suspension to 1×10^6 cells/ml.

6. Thaw one vial of bacteria; sonicate five times for 10-sec intervals at approximately 75% power. Cool bacteria on ice between cycles.

7. For opsonization, dilute 1 ml of pooled human serum with 3 ml HBSS (1:4 dilution).

8. Mix 1 ml of bacterial solution wth 4 ml of diluted human serum and incubate at 37°C for 15 min with end-over-end rotation.

9. Mix 5 ml of neutrophil solution with opsonized bacteria (adjust volumes to approximate a 20:1 bacteria:neutrophil ratio) and immediately remove 1 ml of the mixture and place into a 12 × 75-mm tube containing 1 ml 0.9% NaCl–0.02% EDTA solution at 4°C to stop phagocytosis.

10. Maintain the remaining neutrophil/bacteria mixture at 37°C and remove 1-ml aliquots at 10, 15, 30, and 60 min each time mixing the aliquot with 1 ml cold 0.9% NaCl–0.02% EDTA solution at 4°C in a 12 × 75-mm tube to stop phagocytosis.

11. For a control with unopsonized bacteria, mix 100 μl of bacterial solution with 400 μl of HBSS. Incubate for 15 min at 37°C, then add 500 μl of neutrophil solution and incubate the tube for 30 min at 37°C. Add 1 ml cold 0.9% NaCl–0.02% EDTA solution at 4°C in a 12 × 75-mm tube to stop phagocytosis.

12. Run on cytometer measuring green fluorescence at 525 nm emission.

13. Immediately after measuring the green fluorescence of each tube, add 1 ml of trypan blue (3 mg/ml) and repeat measurements on the cytometer. The trypan blue quenches the fluorescence of the extracellular bacteria and thus allows for measurement of only the intracellular bacteria.

D. Critical Aspects

As with the oxidative burst, proper cell handling and care are necessary to prevent clumping. Since phagocytosis is an active functional assay, dextrose, calcium, and magnesium in the HBSS are necessary for normal cell function. This assay can be performed with a leukocyte mixture instead of a pure neutrophil suspension. The neutrophil population is relatively easy to separate on the cytometer using forward angle and 90° light scatter. However, it may be difficult to separate the neutrophil phagocytosis from the monocyte phagocytosis. The optimal ratio of bacteria to neutrophils is 20:1. Small particles such as bacteria may be difficult to see using light scatter. It is necessary to trigger using green fluorescence to detect these small particles (1 μm or less).

IV. Controls and Standards

It is important to establish a standard procedure for running oxidative burst assays. This can be achieved by finding a fluorescent bead which falls generally within the range of fluorescence of activated cells. This bead is then used to set up the flow cytometer each time setting the HV of the PMTs based upon the bead fluorescence. If a full calibration is performed, the mean channel fluorescence can then be equated with the quantity of H_2O_2 formed per cell.

Acknowledgment

Funding for these studies was provided by Grants P42 ES04911 and GM38827 from the National Institutes for Health.

References

Alexander, J. W., Windhorst, D. B., and Good, R. A. (1968). *J. Lab. Clin. Med.* **72,** 136–148.

Bass, D. A., Parce, J. W., DeChatelet, L. R., Szejda, P., Seeds, M. C., and Thomas, M. (1983). *J. Immunol.* **130,** 1910–1917.

Bjerknes, R., and Bassoe, C.-F. (1984). *Blut* **49,** 315–323.

Cheung, K., Archibald, A., and Robinson, M. (1983). *J. Immunol.* **130,** 2324–2329.

DeChatelet, L. R., Long, G. D., Shirley, P. S., Bass, D. A., and Thomas, M. J., Henderson, F. W., and Cohen M. S. (1982). *J. Immunol.* **129,** 1589–1593.

Neill, M. A., Henderson, W. R., and Klebanoff, S. J. (1985). *J. Exp. Med.* **162,** 1634–1644.

Rothe, G., and Valet, G. (1990). *J. Leuk. Biol.* **47,** 440–448.

Taga, K., Seki, H., Miyawaki, T., Sato, T., Taniguchi, N., Shomiya, K., Hirao, T., and Usui, T. (1985). *Hiroshima. J. Med. Sci.* **34,** 53–60.

CHAPTER 29

Flow Cytometric Determination of Cysteine and Serine Proteinase Activities in Living Cells with Rhodamine 110 Substrates

**Sven Klingel,★ Gregor Rothe,★ Wolfgang Kellermann,†
and Günter Valet★**

★ Arbeitsgruppe Zellbiochemie
Max-Planck-Institut für Biochemie
D-82152 Martinsried, Germany
† Institut für Anästhesiologie
Klinikum Grosshadern der Universität
D-81377 Munich, Germany

I. Introduction

Proteases are of essential importance for the homeostasis of intra- and extracellular protein metabolism. They are distinguishable by their cutting mode as endo- and exopeptidases or by their reactive sites as serine, cysteine, aspartic, and metalloproteases (Powers *et al.*, 1993).

Naturally occurring proteins like hemoglobin and casein or synthetic chromogenic or fluorogenic molecules are in use as experimental protease substrates (Bergmeyer, 1984). Chromogenic substrates are frequently used for cuvette assays. Fluorogenic substrates are one or two orders of magnitude more sensitive and can be used in cuvette or cellular assays.

Cellular assays are of particular interest for flow or image cytometrical measurements on single cells in heterogeneous suspensions of peripheral blood cells as well as of cells from hemato- and immunopoietic tissues, benign or malignant tumors, or normal body tissues.

Incubation of viable or fixed cells with 4-methoxy-2-napthylamine (MNA) amino acid or peptide derivatives as substrate, followed by *in situ* precipitation of the MNA reaction product by 5-nitrosalicylaldehyde, results in a greenish reaction product upon UV excitation at 365 nm as expression of intracellular protease activities (Dolbeare and Smith, 1977; Dolbeare and Vanderlaan, 1979). Protease activities in human gynecological malignancies have been determined in this way (Haskill and Becker, 1982; Haskill *et al.*, 1983a,b; Becker *et al.*, 1983).

7-Amino-4 methylcoumarin (Zimmerman *et al.*, 1976; Kanaoka *et al.*, 1977) or 7-amino-4-trifluoromethylcoumarin (Smith *et al.*, 1980; Cox and Eley, 1987) as alternative fluorophores are useful for cuvette assays. Their use for cytometric assays is, however, problematic due to overlapping substrate and product fluorescence spectra and diffusion of uncharged reaction product through cell membranes at physiological pH values (Waggoner, 1990).

A new class of fluorogenic rhodamine 110 (R110) substrates for cuvette assays was introduced by Leytus (Leytus *et al.*, 1983a,b). These substrates show a high potential for cytometric measurements due to practically complete fluorescence quenching in the substrates; high quantum yield of free R110, similar to fluorescein; excitation at 488 nm, e.g., by argon ion laser light; and practical independence of fluorescence intensity between pH 3 and 10. The slightly positive charge of the amphiphilic free R110 reaction product at physiological pH values causes autoaccumulation in cytoplasm and mitochondria due to the internally negative transmembrane and mitochondrial electrical potentials (Rothe *et al.*, 1992; Assfalg-Machleidt *et al.*, 1992).

II. Materials and Methods

A. Synthesis of R110 Substrates

1. Chemicals

Amino acid derivatives were purchased from Bachem (Heidelberg, Germany), rhodamine 110 from Exciton Chemical Company (Dayton, OH), and 33% HBr in acetic acid and 1-ethyl-3-(3-dimethylaminopropyl)-carbodiimide (EDC) from E.Merck (Darmstadt, Germany), and all other reagents and anhydrous solvents were from Aldrich (Steinheim, Germany).

2. Synthesis of (Z-Arg)$_2$-R110

(Z-Arg)$_2$-R110 [compound (COMP) No. 1, MW, 983.9] was synthesized according to Leytus *et al.* (1983b), using a 10-fold instead of a 25-fold molar excess of protected amino acid and EDC in relation to R110. The reaction was complete within 10 hr as judged by RP-HPLC. Reaction progress and purity of all products was checked by RP-HPLC, ^{1}H-NMR, and FAB mass spectroscopy. The HPLC analysis was performed on an two-pump LKB Bromma system (Pharmacia, Freiburg, Germany) with UV detector (254 nm) and a Shimadzu fluorescence detector RF 535 (Latek GmbH, Germany) with excitation monochromator at 480 nm and emission monochromator at 525 nm. The HPLC column (4 × 250 mm) was filled with Nucleosil 100 C_{18} (M.Grom, Herrenberg, Germany). Arginine containing substrates were analyzed in acetonitril (MeCN)/ $0.25M$ phosphate buffer, pH 3.5 (gradient: 0–3 min 15% MeCN, 3–18 min to 60% MeCN). A MeCN/0.2 M triethylammonium acetate, pH 7.0 (gradient: 0–3 min 50% MeCN, 3–18 min to 80% MeCN) buffer system was used for all other substrates.

3. Synthesis of N,N'-Bis-L-arginyl-rhodamine 110 tetrahydrobromide, COMP No. 2 [(Arg)$_2$-R110 ★ 4 HBr]

Chilled 33% HBr in acetic acid (30 ml) is dropped into a stirred solution of COMP No. 1 (1.0 g; 1 mmol) in methanol (10 ml) at − 15°C. The resulting brown suspension is diluted after 2 hr at room temperature with acetone (300 ml) and centrifuged at 1000*g* for 5 min. The pellet is redissolved in methanol (30 ml) and reprecipitated with acetone (300 ml). Following centrifugation the pellet is redissolved three times in methanol (30 ml), reprecipitated with diethyl ether (400 ml), recentrifuged, and finally dried *in vacuo* to yield 0.93 g of a yellow powder (90%, MW, 966.4) which appears homogeneous upon HPLC analysis.

4. Synthesis of N,N'-Bis-(N$_\alpha$-benzyloxycarbonyl-L-arginyl)-rhodamine 110 tetrahydrochloride, COMP No. 3 [(Z-Arg-Arg)$_2$-R110 ★ 4 HCl;]

N_α-Benzyloxycarbonyl-L-arginine hydrochloride (2 g, 5.8 mmol) is dissolved in an anhydrous 1:1 mixture of dimethylformamide and pyridine (50 ml) at 0°C. Five minutes after addition of EDC (1.0 g, 5.2 mmol) a solution of COMP No. 2 (500 mg; 0.5 mmol) in the same solvent (5 ml) is added. After 15 hr the solvent is reduced *in vacuo* to about 20 ml and the product is precipitated by addition of ethyl acetate (200 ml). The substrate is further purified by precipitation with 1 *N* HCl (150 ml) from dimethylformamide (10 ml; two times) and diethyl ether (100 ml) from methanol (10 ml). The product is collected by centrifugation and dried *in vacuo* to give 0.52 g of a light orange powder (80%) with a purity of over 98% (MW, 1368.22) as controlled by HPLC.

5. Synthesis of *N,N'*–Bis–(benzyloxycarbonyl-L-alanyl)–rhodamine 110, COMP No. 4 [(Z-Ala)$_2$-R110]

Benzyloxycarbonyl-L-alanine (1.5 g, 6.7 mmol) is dissolved at 0°C in an anhydrous 1 : 1 mixture of dimethylformamide and pyridine (25 ml). Five minutes after the addition of EDC (1.2 g, 6.3 mmol) a solution of R110 hydrochloride (0.25 g, 0.68 mmol) in the same solvent mixture (2 ml) is added. After 10 hr at room temperature the solvent is removed *in vacuo* and the resulting orange oil is redissolved in a mixture of ethyl acetate (100 ml) and water (50 ml). The organic phase is washed (30-ml portions each) two times with 10% sodium carbonate solution, one time with water, three times with 1 *N* HCl, and again three times with water. The organic phase is dried over anhydrous sodium sulfate. After evaporation of the solvent *in vacuo* the crude orange product is dissolved in ethyl acetate (5 ml) and precipitated with hexane (80 ml). After 1 hr at 4°C the suspension is centrifuged at 2000*g* for 10 min and the product is dried *in vacuo* to yield 0.47 g of a pale pink powder (93%; MW, 740.78).

6. Synthesis of *N,N'*–Bis–L–alanyl–rhodamine 110 dihydrobromide, COMP No. 5 [(Ala)$_2$-R110 ★ 2HBr]

To a solution of COMP No. 4 (0.45 g, 6.1 mmol) in methanol (10 ml) is added 33% HBr in acetic acid (10 ml) with stirring at − 15°C. As controlled by HPLC the reaction is complete after 2 hr at room temperature and the suspension is diluted with ethyl acetate (200 ml). The precipitate is collected by centrifugation at 1000*g* for 5 min. The pellet is redissolved in methanol (5 ml) and precipitated with diethyl ether (100 ml). After centrifugation (1000*g*, 5 min) the precipitate is washed three times with diethyl ether (50 ml) and dried in an evacuated desiccator over sodium hydroxide yielding 0.32 g analytical pure orange product (84%; MW, 634.34).

7. Synthesis of *N,N'*–Bis–(benzyloxycarbonyl-L-alanyl-L-alanyl)–rhodamine 110, COMP No. 6 [(Z-Ala-Ala)$_2$-R110]

Benzyloxycarbonyl-L-alanine (1.5 g, 6.7 mmol) is dissolved in an ice bath in an anhydrous 1 : 1 mixture of dimethylformamide and pyridine (25 ml). Five minutes after the addition of EDC (1.2 g, 6.3 mmol) a solution of COMP No. 5 (0.25 g, 0.28 mmol) in the same solvent mixture (2 ml) is added. After 12 hr the solvent is removed *in vacuo* and the resulting orange oil is redissolved in a mixture of ethyl acetate (100 ml) and water (50 ml). The organic phase is washed (30-ml portions each) two times with 10% sodium carbonate solution, one time with water, three times with 1 *N* HCl, and again three times with water. The ester phase is dried over anhydrous sodium sulfate. After evaporation of the solvent *in vacuo* the red crude product is dissolved in ethyl acetate (5 ml) and precipitated with hexane (80 ml). After 1 hr at 4°C the precipitate is collected by centrifugation and dried *in vacuo* to yield 0.17 g (70%; MW, 882.94) of final product.

B. Cell Preparation

1. Cell Culture Cells

Cell culture cells are washed twice in 50 ml saline, buffered to pH 7.35 with 10 mM Hepes (N-[2-hydroxyethyl]-piperazine-N'-[2-ethanesulfonic acid]) and resuspended in the same buffer (HBS) to a concentration between 1 and 10 × 10^6 cells/ml.

2. Organ Tissue Cells

A total of 50–100 mg organ tissue is covered by 0.1 ml HBS buffer to avoid drying and minced several times with an electrical tissue chopper (McIlwain, Gomshall, UK) equipped with five parallel razor blades at 0.7 mm distance. The sample support is turned each time by 90° to achieve maximum cutting. The chopped tissue is transferred from the Teflon chopping table with 3–5 ml HBS buffer into a 50-ml Falcon tube with conical buttom. The suspension is sucked back and forth with moderate speed using an Eppendorf-type pipette with a plastic tip previously cut to obtain an opening between 1.5 and 2 mm. Bubble formation is carefully avoided. The cell suspension is filled up to 50 ml with HBS, filtered through a steel sieve with 60-μm wire distance, followed by two centrifugal washes for 10 min at 200g and 0–4°C. The cells are finally resuspended in HBS buffer to a concentration between 1 and 10 × 10^6 cells/ml and stored at 0–4°C until staining.

3. Blood Cells

A total of 2–3 ml of heparinized whole blood (10–30 IU/ml) is carefully layered from the syringe on top of 5 ml Ficoll-Hypaque solution (Sigma Chemicals, St. Louis, MO) in a 10-ml glass or polycarbonate test tube, followed by 30-45 min cell sedimentation at room temperature. Erythrocytes aggregate at the blood/Ficoll-Hypaque interface and sediment rapidly leaving thrombo-, lympho-, mono-, and granulocytes behind. The upper 1–1.5 ml blood plasma is removed, washed in 15 ml HBS buffer by centrifugation immediately or prior to staining, and resuspended in HBS buffer to a concentration between 1 and 10 × 10^6 cell/ml. Besides relative ease, this procedure most importantly avoids any contact of nonerythrocytic cells with the separation medium which otherwise, e.g., significantly stimulates mono- or granulocytes and thereby potentially influences subsequent enzyme activity measurements.

C. Protease Assays

A total of 250 μl cells (1–10 $\times$ 10^6/ml) is incubated with 5 μl dye cocktail containing between 0.02 and 0.2 mM (Z-Arg-Arg)$_2$R110, (Z-Phe-Arg)$_2$R110 or (Z-Ala-Ala)$_2$R110 together with 3 mM propidium iodide (PI, Sigma) in dimethylformamide (DMF) for 15 to 45 min at 20 or 37°C.

The incubation temperature and substrate concentration depends of the cell type. Blood cells require 0.2 mM substrate concentration, while, e.g., 0.02 mM is sufficient for guinea pig kidney cells. The plateau phase of substrate cleavage is usually reached after 15 to 30 min at 37°C depending again on the cell type.

The specificity of the intracellular R110 substrate cleavage is controlled by preincubation of the cells for 15 min at 20 or 37°C with 5 μl of a 0.5 mM solution of (Z-Phe-Ala)-diazomethylketone (DMK) cysteine proteinase inhibitor (Bachem, Heidelberg, Germany) in dimethyl sulfoxide (DMSO) while serine proteinases are inhibited by preincubation with 5 μl of a 50 mM diisopropylfluorophosphate (DFP, Aldrich) solution in DMSO. Extreme care has to be taken when working with DFP because of volatility and high neurotoxicity. (Proceed according to manufacturer's recommendations: hood, gloves, inactivation in 4–6 N NaOH, antidote: atropine).

D. Flow Cytometry

The assays can be directly measured in the flow cytometer. Blood cells require 488 nm argon ion laser light (5–50 mW) for sufficient fluorescence excitation while high-pressure mercury arc lamp (HBO-100) light using a 460- to 500-nm band pass excitation filter is sufficient in the case of larger cells especially for measurements with the serine proteinase substrate (Z-Ala-Ala)$_2$R110. R110 fluorescence is collected in the F1 (FITC) channel between 512–542 nm while the signals of the PI-stained DNA of dead cells are registered in the F3 (PI,PERCP,CY5) channel between 600 and 700 nm. Simultaneous forward (FSC) and sideward (SSC) scatter signals or alternatively electric volume signals (ECV) from hydrodynamically focused measuring orifices should be additionally collected. Three or four decade logarithmically amplified fluorescence, FSC or ECV signals are stored together with linearly amplified SSC signals in list-mode data files. A long-term standardization of the flow cytometer by monosized yellow/green (YG) fluorescent particles (Polysciences, Warrington, PA) permits later automated classification of the data based results (Valet *et al.*, 1993).

III. Results

The peptide residues of the R110 substrates (Fig. 1) are sequentially cleaved by proteases. Single peptide cleavage on either side of the R110 substrate results

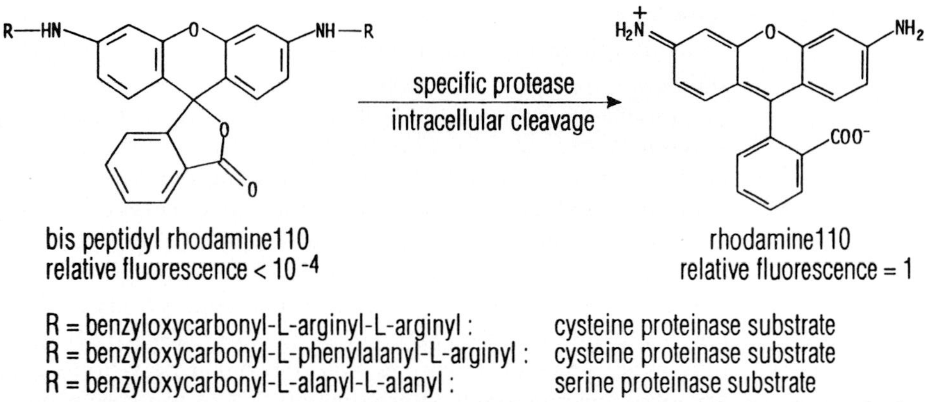

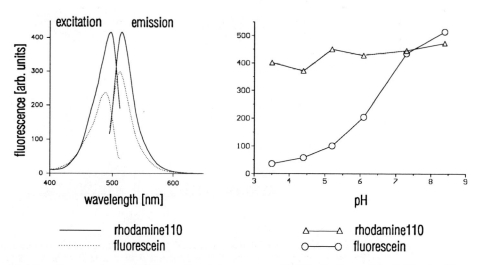

bis peptidyl rhodamine110
relative fluorescence < 10^{-4}

rhodamine110
relative fluorescence = 1

R = benzyloxycarbonyl-L-arginyl-L-arginyl :   cysteine proteinase substrate
R = benzyloxycarbonyl-L-phenylalanyl-L-arginyl :   cysteine proteinase substrate
R = benzyloxycarbonyl-L-alanyl-L-alanyl :   serine proteinase substrate

Fig. 1   Schematic representation of R110 protease substrate cleavage.

in 1/10 of the fluorescence as compared to cleavage of both peptide residues. While complete substrate cleavage in cuvette assays, where many substrate molecules have to be cleaved by comparatively few enzyme molecules, sometimes requires several hours incubation at 37°C, HPLC analysis of blood cell lysates following 15 to 45 min incubation at 37°C with R110 protease substrates shows that intracellular proteinases *in situ* completely cleave R110 substrate molecules resulting in full R110 fluorescence.

The fluorescence excitation and emission spectra of R110 virtually superimpose the fluorescein spectra (Fig. 2, left) i.e., 488-nm fluorescence excitation

Fig. 2   Fluorescence excitation and emission spectra of R110 and fluorescein (left) and of their pH-dependent fluorescence (right).

is optimal. As a further advantage R110 fluorescence unlike fluorescein fluorescence is practically independent of pH values in the range pH 3.5–9 (Fig. 2, right).

Cysteine proteinase activity in human blood cells is practically only observed in monocytes (Fig. 3A). This activity increases during inflammatory processes as they occur, e.g., 24 hr following liver tranplantation (Fig. 3B).

Serine proteinase activity, in contrast, is observed in lympho-, mono-, and granulocytes (Fig. 3C). Lymphocytes frequently consist of two clusters of high- and low-activity cells (Fig. 3D,) in diseased as well as in apparently healthy persons (Fig. 4C). Monocyte serine proteinase activity is also increased following transplantation (Fig. 3D). The activity of high-activity granulocytes remains practically constant (Figs. 3C and 3D) but a new population of low-activity granulocytes containing approximately 1/3 of the serine proteinase activity of high-proteinase granulocytes is observed 24 hr after liver transplantation (Fig. 3D) with the low-activity cells carrying an increased amount of CD63 antigen on the cells surface (not shown).

The R110 cysteine and serine proteinase are specifically cleaved since preincubation of the cells with DMK or DFP protease inhibitors practically com-

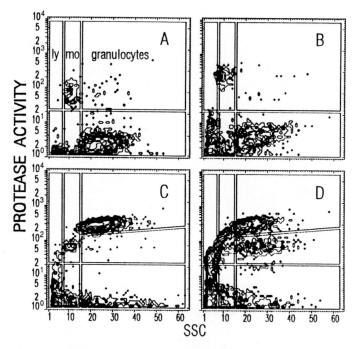

Fig. 3 SSC against R110 fluorescence of peripheral blood lympho-, mono-, and granulocytes in HBS buffer following 30 min 37°C incubation with 4 μM (Z-Arg-Arg)$_2$R110 (A,B) or (Z-Ala-Ala)$_2$R110 (C,D) substrate before (A,C) and 24 hr after liver transplantation (B,D) as determined in a FACScan (Becton–Dickinson) flow cytometer.

pletely inhibits substrate cleavage (Figs. 4B and 4D) in comparison to the uninhibited assays (Figs. 4A and 4C).

IV. Discussion

R110 protease substrates are advantageous for cytometric single-cell protease activity measurements because of specific cleavage, low toxicity, absent fluorescence background, autoaccumulation due to positive electrical charge at physiological pH, and pH-independent fluorescence.

R110 protease substrates will permit, e.g., detailed investigations of protease activity regulation during cell cycle and differentiation, in lymphocyte subpopulations (Fig. 3d), during monocyte/macrophage activation (Rothe *et al.*, 1992) and apoptosis (Elsherif *et al.*, 1994), in premalignant and malignant tumors or in aquatic microplankton cells (Sieracki *et al.*, 1993). Monocyte activation can be followed from an increase of the ratio of (Z-Phe-Arg)$_2$R110 over (Z-Arg-Arg)$_2$R110 cleavage due to preferential cleavage of (Z-Phe-Arg)$_2$R110 by cathepsin L and (Z-Arg-Arg)$_2$R110 by cathepsin B and H (Rothe *et al.*, 1992).

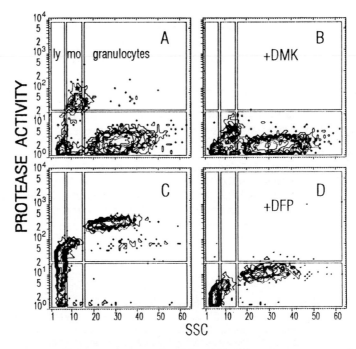

Fig. 4 SSC against R110 fluorescence of peripheral blood lympho-, mono-, and granulocytes preincubated for 15 min at 37°C with HBS buffer (A,C), 10 μM DMK cysteine proteinase inhibitor (B), or 1 mM DFP serine proteinase (D) inhibitor, followed by 30 min 37°C incubation with 4 μM (Z-Arg-Arg)$_2$R110 (A,B) or (Z-Ala-Ala)$_2$R110 (C,D) protease substrate.

The appearance of a distinct low protease granulocyte population raises the question whether circulating granulocytes upon stimulation secrete in a discrete step approximately 2/3 of their serine proteinase activity, whether a newly produced granulocyte population appears in circulation due to a general stimulus, or whether these granulocytes represent transfused blood granulocytes. Although the mechanism of low-activity granulocyte formation is presently unclear, the increased expression of CD63 antigen on the cell surface points toward involvement of a secretory mechanism.

Besides cysteine and serine proteinase substrates, additional R110 substrates for the determination of leucine and phenylalanine aminopeptidases (metalloproteinase) and cathepsin D (aspartic proteinase) have been recently synthesized (Klingel *et al.*, 1994). The intracellular cathepsin D measurement is performed as the coupled enzyme reaction with cathepsin D endopeptidase cuts followed by aminopeptidase digestion of remaining residues until liberation of free R110. Coupled protease reaction, at least in principle, will allow the specific determination of many other intracellular endopeptidases.

Altogether, the cytometric use of R110 proteinase substrates will open up a new area of studies on the mechanism and regulation of cellular protein and peptide catabolism.

Note: The authors will be glad to provide substrate samples for research purposes. Very recently (Z-Ala-Ala)$_2$R110 (R-6504) has become commercially available through Molecular Probes (Eugene, OR).

References

Assfalg-Machleidt, I., Rothe, G., Klingel, S., Banati, R., Mangel, W. F., Valet, G., and Machleidt, W. (1992). *Biol. Chem. Hoppe-Seyler* **373**, 433–440.

Becker, S., Halme, J., and Haskill, S. (1983). *RES* **33**, 127–138.

Bergmeyer, H. U. (1984). "Methods of Enzymatic Analysis." Vol. V. "Enzymes 3: Peptidases, Proteinases and Their Inhibitors" (J. Bergmeyer and M. Graßl, eds., H. Fritz, ed. consult.). Weinheim: Verlag Chemie.

Cox, S. W., and Eley, B. M. (1987). *Archs Oral. Biol.* **32**, 599–605.

Dolbeare, F. A., and Smith, R. E. (1977). *Clin. Chem.* **23**, 1485–1491.

Dolbeare, F., and Vanderlaan, M. (1970). *J. Histochem. Cytochem.* **27**, 1493–1495.

Elsherif, T., Kahle, H., Klingel, S., Ganesh, S., and Valet, G. (1994). *J. Anal. Cell. Pathol.*, **6**, 256.

Haskill, S., and Becker, S. (1982). *RES* **32**, 273–285.

Haskill, S., Becker, S., Johnson, T., Marro, D., Nelson, K., and Propst, R. H. (1983a). *Cytometry* **3**, 359–366.

Haskill, S., Kivinen, S., Nelson, K., and Fowler, Jr., W. (1983b). *Cancer Res.* **43**, 1003–1009.

Klingel, S., Ganesh, S., Kahle, H., and Valet, G. (1994). *J. Anal. Cell. Pathol.*, **6**, 257.

Kanaoka, Y., Takahashi, T., and Nakayama, H. (1977). *Chem. Pharm. Bull.* **25**, 362–363.

Leytus, S. P., Patterson, W. L., and Mangel, W. F. (1983a). *Biochem. J.* **215**, 253–260.

Leytus, S. P., Melhado, L. L., and Mangel, W. F. (1983b). *Biochem. J.* **209**, 299–307.

Powers, J. C., Odake, S., Oleksyszyn, J., Hori, H., Ueda, T., Boduszek, B., and Kam, Ch. M. (1993). *In* "Proteases, Protease Inhibitors and Protease-Derived Peptides" (J. C. Cheronis and J. E. Repine, eds.), AAS 42, pp. 3–18. Basel: Birkhäuser.

Rothe, G., Klingel, S., Assfalg-Machleidt, I., Machleidt, W., Zirkelbach, Ch., Banati, R., Mangel, W. F., and Valet, G. (1992). *Biol. Chem. Hoppe-Seyler* **373,** 547–554.

Sieracki, M., Valet, G., and Cucci, T. (1993). *Signal Noise* **6,** 1–2.

Smith, R. E., Bissel, E. R., Mitchell, A. R., and Pearson, K. W. (1980). *Thromb. Res.* **17,** 393–401.

Valet, G., Valet, M., Tschöpe, D., Gabriel, H., Rothe, G., Kellermann, W., and Kahle, H. (1993). *Ann. N.Y. Acad. Sci.* **677,** 233–251.

Waggoner, A. S. (1990). *In* "Flow Cytometry and Sorting" (M. R. Melamed, T. Lindmo, and M. L. Mendelsohn, eds.), 2nd Ed., pp. 209–225. New York: Wiley–Liss.

Zimmerman, M., Yurewicz, E., and Patel, G. (1976). *Anal. Biochem.* **70,** 258–262.

CHAPTER 30

Leucine Aminopeptidase Activity by Flow Cytometry

John J. Turek* and J. Paul Robinson[†]

*Department of Veterinary Anatomy
[†]Department of Physiology and Pharmacology
School of Veterinary Medicine
Purdue University
West Lafayette, Indiana 47907

I. Introduction

Macrophage activation or maturation is an important parameter in the study of cell-mediated immunity (Assreuy and Moncada, 1992; Buchmüller-Rouiller et al., 1992; Fortier et al., 1992; Hamilton et al., 1991; Higginbotham et al., 1992; Lim and Stewart, 1991; Michelacci and Petricevich, 1991; Murray, 1992; Nelson et al., 1992; Oswald et al., 1992; Singh and Sodhi, 1991; Valdez et al., 1990). Some of the methods to assess macrophage activation include measurement of various enzyme levels (Morahan et al., 1980), phagocytic ability (Abel et al., 1989, 1991; Cannon and Swanson, 1992), chemotaxis (Litwin et al., 1992), microbicidal activity (Fortier et al., 1992; Kiderlen and Kaye, 1990; Murray, 1992), production of reactive oxygen intermediates (Decker, 1990; Johnson et al., 1986), and nitric oxide production (Marotta et al., 1992). One of the enzymes associated with macrophage activation is leucine aminopeptidase (LAP). LAP is an ectoenzyme located in the outer cell membrane that is found in a variety of cells and tissues and appears to have higher levels of activity in phagocytic cells (Nagaoka and Yamashita, 1984). Measurement of LAP by methods other than flow cytometry has been used to assess macrophage or leukocyte activation/differentiation in a variety of studies (Dempsey *et al.*, 1988; Johnson *et al.*, 1986; Morahan *et al.*, 1980; Nagaoka and Yamashita, 1984; Volkman *et al.*, 1983).

A flow cytometric assay to stain for LAP activity was originally described by Dolbeare and Smith (1977) and the methods included here are essentially those of Haskill and Becker (1982). The flow cytometric assay has been used to study macrophage heterogeneity (Haskill and Becker, 1982; Becker *et al.*, 1983), the effects of interferon on monocyte and macrophage development (Becker, 1984), activation states of tumor-associated macrophages (Mahoney *et al.*, 1983), and the effects of dietary fatty acids on peritoneal macrophage activation (Turek *et al.*, 1991). The advantage of the assay is that it requires a minimum of preparation time and may be performed very rapidly. Further, the assay provides accurate information related to the activation status of macrophages, despite the heterogeneous nature of these cells, particularly when isolated from the lung or peritoneal cavity. An additional advantage is that the use of UV light excitation reduces the very real problem of autofluorescence levels associated with macrophages.

II. Application

This assay is primarily used to assess monocyte or macrophage activation, but it may also be used for other cell types. It is a simple analytical assay that is very reproducible if the conditions under Section V are followed.

III. Materials

Material	Supplier	Location
Leucine 4-methoxy-2-napthylamine	Enzyme systems products	Livermore, CA
5-Nitrosalicylaldehyde	Kodak Chemical	Rochester, NY
2-[N-morpholino]ethanesulfonic acid	ICN Biochemicals	Cleveland, OH
Ficoll (Histopaque-1077)	Sigma Chemical	St. Louis, MO

IV. Methods

1. Cell preparation: Alveolar and peritoneal macrophages and peripheral blood monocytes may all be assayed. Heparinized blood will require separation

of the monocytes using a Ficoll gradient. Lung and peritoneal macrophages should be collected by lavage with calcium and magnesium-free Hanks' balanced salt solution. The cells from peritoneal and lung lavage solutions often do not require separation on a Ficoll gradient, but gradient separation may be performed if the samples contain excessive debris. All cells should be washed twice with cold 0.15 M 2-[N-morpholino]ethane sulfonic acid (MES) buffer (pH 6.5) and adjusted to 2×10^6 cells/ml prior to the procedures listed below.

2. Prepare a 0.15 M solution of leucine 4-methoxy-2-napthylamine by dissolving 1 mg of this compound in 20 μl of N,N-dimethylformamide.

3. Prepare a 0.5 M solution of 5-nitrosalicylaldehyde by dissolving 1.67 mg of this compound in 20 μl of N,N-dimethylformamide.

4. Add 5 ml of 0.15 M MES buffer (pH 6.5) to 20 μl of each of the above (2 and 3).

5. Mix the two reagents 1 : 1 and add 1 ml of cells ($\sim$2 $\times$ 10^6) to 1 ml of the reagent mixture.

6. Incubate for 10–25 min. at 37°C.

7. Wash the cells one time in cold sodium acetate buffer (0.1 M, pH 5.2) containing 0.02% Triton X-100 and resuspend the cells in cold sodium acetate buffer without Triton X-100. The cells should be kept on ice and assayed within 2 hr. The cells may be assayed cold or warmed to room temperature just prior to analysis. Collect data on 20,000 to 50,000 cells.

V. Critical Aspects of the Procedure

For reproducibility of data, the instrument operating parameters, reagent concentrations, and cell incubation time need to be rigorously controlled. Since such small amounts of leucine 4-methoxy-2-napthylamine and 5-nitrosalicylaldehyde are required, it is best to weigh out more than is needed, add an amount of N,N-dimethylformamide that would be equivalent to the above concentrations, and then use 20 μl of that solution to prepare the staining reagents. Incubation time is also crucial. Longer incubation times (e.g., 25 min) will produce an increase in fluorescence and may decrease the sensitivity of the assay to detect small differences between cell populations. If freshly isolated cells are frozen and stored in liquid nitrogen using methods for preserving tissue culture cells, LAP activity may also be assayed after thawing of cells, but there is a slight decrease in overall LAP staining.

VI. Controls

Unstained cells that have been washed and incubated in the different buffers should be used to determine endogenous fluorescence levels.

VII. Instruments

A flow cytometer with a UV laser is required. The assay should be run with 352–363 nm excitation and an emission band pass filter at 525 nm. When the assay is first set up, the laser wattage and photomultiplier tube settings should be adjusted so that the endogenous fluorescence of control cells falls within the first decade of the log signal. Cells stained for LAP will then typically fall within the second and third decade of the log signal. These settings should remain the same for all samples to be compared.

VIII. Results

Figures 1 and 2 demonstrate LAP staining from 50,000 cells contained in lavage fluid from porcine lung. The ellipse in Fig. 1 denotes the gate chosen for analysis of a macrophage population. It is important that identical forward angle light scatter and 90° light scatter (90LS) gating be used to compare different samples as larger macrophages will be brighter due to a greater cell surface area. Figure 3 represents the log fluorescence signal due to LAP staining from

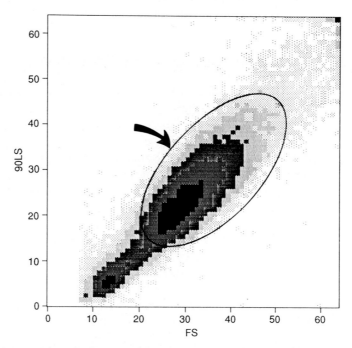

Fig. 1 Fluorescence resulting from staining for LAP in porcine alveolar cells. The ellipse (curved arrow) indicates the gating used for further analysis.

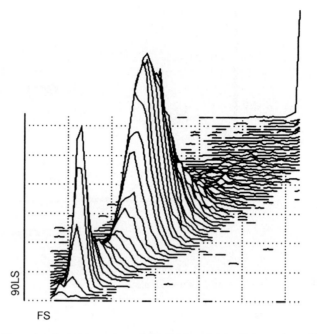

Fig. 2 Three-dimensional representation of the LAP staining of the entire cell population in Fig. 1.

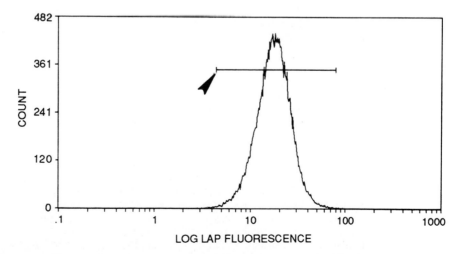

Fig. 3 Histogram of the log fluorescence due to LAP staining from the gated population in Fig. 1. The gate (arrowhead) is used to determine the linear equivalent of the mean log fluorescence channel.

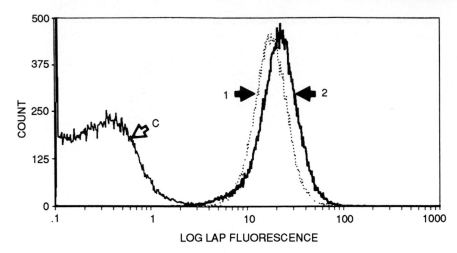

Fig. 4 Composite histogram showing endogenous fluorescence for a control sample (C) and from samples stained for LAP (1, 2). The linear equivalent of the mean log fluorescence channel is 17.8 for sample 1 and 22.9 for sample 2.

the gated population in Fig. 1. To determine the degree of fluorescence, a gate extending from base to base of the peak is used to determine the linear equivalent of the mean channel for the log fluorescence signal (Fig. 3). This gate will need to be centered over the peak for each sample. Figure 4 represents an overlay of the log fluorescence signals from a control sample (endogenous fluorescence) and samples from two different animals. The mean linear equivalent of the log channel difference of these two LAP-stained samples in Fig. 4 is typical of the results with this assay.

References

Abel, G., Szöllösi, J., Chihara, G., and Fachet, J. (1989). *Int. J. Immunopharmacol.* **11,** 615–621.
Abel, G., Szöllösi, J., and Fachet, J. (1991). *Eur. J. Immunogenet.* **18,** 239–245.
Assreuy, J., and Moncada, S. (1992). *Br. J. Pharmacol.* **107,** 317–321.
Becker, S. (1984). *Cell. Immunol.* **84,** 145–153.
Becker, S., Halme, J., and Haskill, S. (1983). *J. Reticuloendothel. Soc.* **33,** 127–138.
Buchmüller-Rouiller, Y., Corradin, S. B., and Mauël, J. (1992). *Biochem. J.* **284,** 387–392.
Cannon, G. J., and Swanson, J. A. (1992). *J. Cell Sci.* **101,** 907–913.
Decker, K. (1990). *Adv. Exp. Med. Biol.* **283,** 507–520.
Dempsey, W. L., Hwu, P., Russell, D. H., and Morahan, P. S. (1988). *Life Sci.* **42,** 2019–2027.
Dolbeare, F. A., and Smith, R. E. (1977). *Clin. Chem. (Winston-Salem, N.C.)* **23,** 1485–1491.
Fortier, A. H., Polsinelli, T., Green, S. J., and Nacy, C. A. (1992). *Infect. Immun.* **60,** 817–825.
Hamilton, J. A., Vairo, G., Knight, K. R., and Cocks, B. G. (1991). *Blood* **77,** 616–627.
Haskill, S., and Becker, S. (1982). *J. Reticuloendothel. Soc.* **32,** 273–285.
Higginbotham, J. N., Lin, T. L., and Pruett, S. B. (1992). *Clin. Exp. Immunol.* **88,** 492–498.
Johnson, W. J., DiMartino, M. J., and Hanna, N. (1986). *Cell. Immunol.* **103,** 54–64.

Kiderlen, A. F., and Kaye, P. M. (1990). *J. Immunol. Methods* **127,** 11–18.

Lim, W. H., and Stewart, A. G. (1991). *Int. Arch. Allergy Appl. Immunol.* **95,** 77–85.

Litwin, D. K., Wilson, A. K., and Said, S. I. (1992). *Regul. Pept.* **40,** 63–74.

Mahoney, K. H., Fulton, A. M., and Heppner, G. H. (1983). *J. Immunol.* **131,** 2079–2085.

Marotta, P., Sautebin, L., and Di Rosa, M. (1992). *Br. J. Pharmacol.* **107,** 640–641.

Michelacci, Y. M., and Petricevich, V. L. (1991). *Comp. Biochem. Physiol. B* **100B,** 617–625.

Morahan, P. S., Edelson, P. J., and Gass, K. (1980). *J. Immunol.* **125,** 1312–1317.

Murray, H. W. (1992). *J. Interferon Res.* **12,** 319–322.

Nagaoka, I., and Yamashita, T. (1984). *Comp. Biochem. Physiol. B* **79,** 147–151.

Nelson, B. J., Belosevic, M., Green, S. J., Turpin, and Nacy, C. A. (1992). *Adv. Exp. Med. Biol.* **319,** 77–88.

Oswald, I. P., Afroun, S., Bray, D., Petit, J.-F., and Lemaire, G. (1992). *J. Leukocyte Biol.* **52,** 315–322.

Singh, R. K., and Sodhi, A. (1991). *Immunol. Lett.* **28,** 127–134.

Turek, J. J., Schoenlein, I. A., and Bottoms, G. D. (1991). *Prostaglandins, Leukotrienes Essent. Fatty Acids* **43,** 141–149.

Valdez, J. C., De Alderete, N., Meson, O. E., Sirena, A., and Perdignon, G. (1990). *Immunobiology* **181,** 276–287.

Volkman, A., Chang, N. C., Strausbauch, P. H., and Morahan, P. S. (1983). *Lab. Invest.* **49,** 291–298.

CHAPTER 31

Enzyme Kinetics

James V. Watson* and Caroline Dive†

*MRC Clinical Oncology Unit
The Medical School
Cambridge CB2 2QH, United Kingdom
†Department of Molecular Pharmacology
Department of Physiological Science
University of Manchester
Manchester M13 9PT, United Kingdom

I. Introduction

Enzymes, in one form or another, perform the majority of dynamic molecular interactions responsible for function in the intact living cell. Measurement of such dynamic cellular events in intact cells under near-physiological conditions, irrespective of the assay system, is arguably one of the most powerful techniques in cell biology as physiological and pathological processes are not static but are continuously variable. Clearly, an understanding of the dynamics of enzyme activity will be of crucial importance in attempts to modulate pathological states. This chapter addresses flow cytometric analysis procedures to measure the dynamics of enzyme activity.

II. Classical Enzyme Kinetics

A general schematic for the conversion of substrate to product by enzymatic action is shown in Fig. 1. Substrate combines with enzyme to form an enzyme–substrate complex with rate constant k_1. This may revert to free enzyme and substrate with rate constant k_{-1} or dissociate into product and free enzyme with rate constant k_2. The differential equations describing the reaction are as follows.

$$\frac{\delta[ES]}{\delta t} = k_1[E][S] - k_{-1}[ES] - k_2[ES] \tag{1}$$

$$\frac{\delta[P]}{\delta t} = k_2[ES] \tag{2}$$

$$\frac{\delta[E]}{\delta t} = -k_1[E][S] + k_2[ES] \tag{3}$$

$$\frac{\delta[S]}{\delta t} = -k_1[S] \tag{4}$$

$$[E_0] = [E] + [ES]. \tag{5}$$

If you think you do not understand these equations it is only because you are not familiar with the symbols, but if you know all about these things then skip the rest of this section. In this section we go through the first equation, (1), in detail starting on the left. $\delta[ES]/\delta t$ means the rate of change of ES, the enzyme–substrate complex. In this context it is the "speed" at which ES is being formed. Rates of change are less daunting if they are put into the context of everyday experience. The road sign shown in Fig. 2 means that there is a gradient of 1 in 10. Thus, for every 10 meters you travel horizontally you go up, or down depending on your direction, by 1 meter. This sign could have

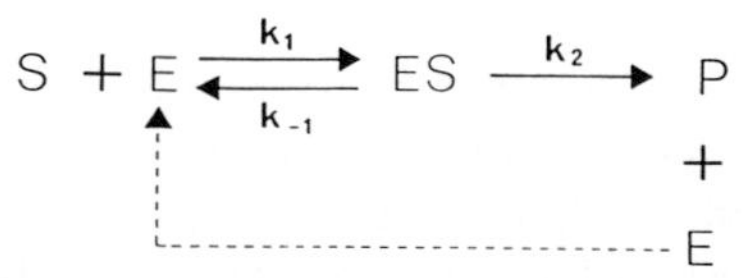

Fig. 1 Schematic for conversion of substrate to product via an intermediate enzyme–substrate complex.

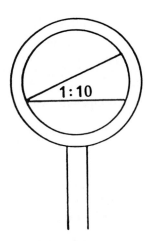

Fig. 2   Road sign for 1 in 10 gradient.

been written $\delta\text{Road}/\delta D = 0.1$ where D is distance. Clearly, if you are moving then distance and time are related so the D in the expression δD could be replaced by t for time to give $\delta\text{Road}/\delta t = ?$. I have replaced the 0.1 of $\delta\text{Road}/\delta D = 0.1$ by ? because the time it will take you to get up or down the incline in the road is dependent on your speed. $\delta\text{Road}/\delta D$ is a constant of the road at that particular point but, $\delta\text{Road}/\delta t$ depends on you. If you are a jogger and are running up the incline your speed, $\delta\text{Road}/\delta t$, depends on the incline, including both the constant of the incline, $\delta\text{Road}/\delta D$, and its length, your fitness, and your desire to get up the incline.

Returning to equation (1) we can see that the speed at which ES is being formed is given by the first term on the right-hand side, $k_1[E][S]$, where $[E]$ and $[S]$, respectively, are the concentrations of free enzyme and substrate which are available to react and where k_1 is the rate constant for that reaction which is analogous with $\delta\text{Road}/\delta D$, the incline constant of the road. The speed at which ES is dissociating is in two parts. The second term $k_{-1}[ES]$ is the speed at which the ES complex is reverting to free enzyme E and substrate S with rate constant k_{-1}, and the term has a minus sign in front of it to indicate that ES is dissociating. The last term $k_2[ES]$ is the speed at which ES is dissociating into product and free enzyme with rate constant k_2. This term is also preceded by a minus sign to represent dissociation of the complex ES. The second rate equation $\delta[P]/\delta t = k_2[ES]$ (Eq. 2) represents the speed at which the product P is being formed from ES; this occurs at the same rate at which ES is dissociating into product and free enzyme, the last term in the first rate equation, but the sign is now positive. The concentration of free enzyme at any given time is obtained by rearranging Eq. (5) to give $[E] = [E_0] - [ES]$, where $[E_0]$ is the initial enzyme concentration and where *ES* represents enzyme bound to substrate. We can now substitute for [*E*] in Eq. (4) to give

$$\frac{\delta[ES]}{\delta t} = k_1([E_0] - [ES])[S] - k_{-1}[ES] - k_2[ES],$$

which upon rearrangement yields

$$\frac{\delta[ES]}{\delta t} = k_1[E_0][S] - k_1[ES][S] - [ES](k_1 + k_2)$$
$$= k_1[E_0][S] - [ES](k_1[S] + k_{-1} + k_2). \qquad (6)$$

When a steady state exists, the speed at which ES is being formed is equal to the speed at which it is dissociating; thus, the rate of change of [*ES*], $\delta[ES]/\delta t$, will be zero and so Eq. (6) reduces to

$$[ES](k_1[S] + k_1 + k_2) = k_1[E_0][S],$$

from which we obtain the following expression for [*ES*] in terms of the remaining parameters:

$$[ES] = \frac{k_1[E_0][S]}{k_1[S] + k_{-1} + k_2}.$$

We now divide through the right-hand side by k_1, which gives

$$[ES] = \frac{[E_0][S]}{[S] + \left[\dfrac{k_{-1} + k_2}{k_1}\right]}.$$

The expression $(k_{-1} + k_2)/k_1$ is the Michaelis constant K_m so the above equation reduces to

$$[ES] = \frac{[E_0][S]}{[S] + K_m}. \qquad (7)$$

We now return to Eq. (2) which when rearranged gives

$$[ES] = \frac{\delta[P]}{\delta t}\frac{1}{k_2}. \qquad (8)$$

Equations (7) and (8) can now be equated which eliminates [*ES*] to give the rate of change of [*P*]:

$$\frac{\delta[P]}{\delta t} = \frac{k_2[E_0][S]}{[S] + K_m}. \qquad (9)$$

$\delta[P]/\delta t$ is the rate at which product is formed and represents the reaction velocity, v. At infinite substrate concentration at the very start of the reaction (this is theoretical), when all the enzyme is in its free form [E_0], the reaction velocity will be at its maximum, V, which is given by $k_2[E_0]$ and the equation may be written as

$$v = V \frac{[S]}{[S] + K_m}.$$ (10)

This gives the initial reaction velocity v for a given substrate concentration, $[S]$, with the Michaelis constant, K_m, for that particular reaction.

Plots of initial reaction velocity, v, versus increasing substrate concentration are shown in Fig. 3 which is the so-called Michaelis–Menten rectangular hyperbola although Michaelis and Menten (1913) did not describe this mathematical analysis. As the substrate concentration increases so the reaction velocity becomes closer and closer to the theoretical maximum, V, and K_m is the substrate concentration where $v = V/2$. The initial velocities are obtained by regression analysis of the initial slopes of enzyme progress curves obtained with increasing substrate concentrations and a representation is shown in Fig. 4. It is difficult to estimate the Michaelis constant directly from $V/2$ in the type of plot shown in Fig. 3 as V cannot be defined with any degree of precision. Moreover, there will always be some uncertainty in the results which makes an estimate from the type of plot in Fig. 3 even more difficult. However, help is at hand with a variety of linear transforms of the Michaelis–Menten rectangular hyperbola.

The first of these is the Lineweaver–Burke double-reciprocal $1/v$ versus $1/S$ plot (Lineweaver and Burke, 1934). If we take the reciprocal of the Michaelis–Menten hyperbola (Eq. 10) we get

$$\frac{1}{v} = \frac{[S] + K_m}{V[S]}.$$

We now rearrange both components of the right-hand side to give

$$\frac{1}{v} = \frac{[S]}{V[S]} + \frac{K_m}{V[S]} = \left[\frac{K_m}{V}\right] \times \frac{1}{S} + \frac{1}{V}.$$

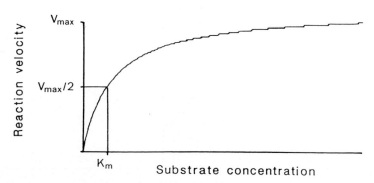

Fig. 3 The Michaelis–Menten rectangular hyperbola of reaction velocity versus substrate concentration.

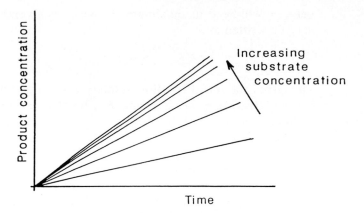

Fig. 4 Enzyme progress curves where product concentration is plotted against time. As the substrate concentration increases, the slopes approach the limit of V_{max}.

This reciprocal equation is now of the form $Y = mX + C$, where $1/v$ is equivalent to Y, K_m/V is equivalent to the slope, m, and $1/V$ is the intercept on the y-axis, C. Thus, by plotting $1/v$ versus $1/S$ as in Fig. 5 and carrying out a regresion analysis we obtain a straight line which cuts the y-axis at $1/V$; hence, $V = 1/C$, the reciprocal of the intercept. As the slope, m, equals K_m/V we get $K_m = m/C$ and the kinetic parameters of the reaction are fully defined.

The second linear transform is due to Eadie (1942) and Hofstee (1952). The hyperbola (Eq. 10) is multiplied through on both sides by $S + K_m$ to give

$$v([S] + K_m) = V[S]$$

$$v[S] + vK_m = V[S].$$

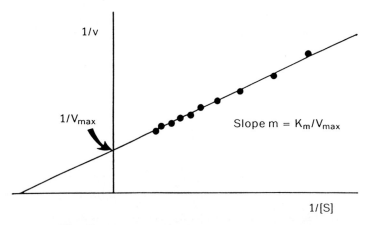

Fig. 5 Lineweaver–Burke plot of $1/v$ versus $1/[S]$.

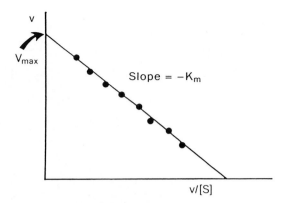

Fig. 6 Eadie-Hofstee plot of v versus $v/[S]$.

We now divide through by $[S]$ and rearrange, which gives

$$v = -K_m \frac{v}{[S]} + V.$$

This equation is now also in the form $Y = mX + C$, where v is equivalent to Y, the slope, m, is $-K_m$, X is equivalent to $v/[S]$, and the intercept, C, is equivalent to V. Thus by plotting v versus $v/[S]$ and performing regression analysis, as shown in Fig. 6, we obtain K_m and V.

The third transform is due to Eisenthal and Cornish-Bowden (1974) who rearranged the hyperbola to give the maximum velocity, V, in terms of the remaining parameters. This is effected by rearranging the Eadie–Hofstee transform to give

$$V = v + (v \times K_m)/S.$$

For each value of v which is associated with its substrate concentration, S, a line can be plotted which crosses the abscissa at $-S$ and the ordinate at v. With a perfect data set the various lines for each v and S combination will all intersect at a point with (x,y) coordinates of (K_m, V). This is the "direct linear plot" of Eisenthal and Cornish-Bowden which is illustrated in Fig. 7.

III. Incorporation of Time

Incorporation of time in the data base is the most important single feature in the measurement of dynamic cellular processes using flow cytometry and a number of different techniques have been implemented for this. Watson *et al.* (1977) used a continuous interrupted sampling technique where the cells were continuously flowing through the instrument. The data acquisition computer

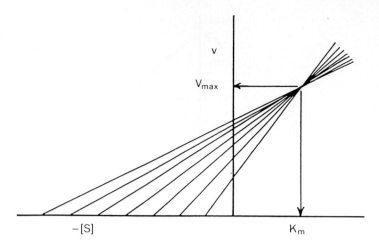

Fig. 7   Direct linear plot of Eisenthal and Cornish-Bowden.

was instructed to record for 5 sec and then wait for 10 sec before recommencing acquisition. During the interval between recordings the median of the population was calculated and printed out. This method had resolution problems, particularly for "fast" reactions, as the fluorescence from the population was increasing during the data acquisition interval hence the medians tended to be overestimated. Martin and Swartzendruber (1980) overcame this very elegantly by incorporating time directly into the data base. Their initial method used a voltage ramp generator where the potential increased linearly with time and a recording of the voltage was made from the ramp generator each time a cell was analyzed in the flow chamber. This method had one disadvantage, namely, that time could only be recorded over an interval of about 15 min. However, immediately following the introduction of continuous time recording by Martin and Swartzendruber (1980) we incorporated the time stamp from the data acquisition computer directly into the data base. Hence, events were timed automatically with 50-msec resolution by the computer clock and this time, in the form of the number of clock "ticks" (one tick per 50 ms), was recorded for each event analyzed.

Keij *et al.* (1989) have incorporated "pseudo" time into list-mode data by inference from the computer-recorded duration of the run. The latter was divided sequentially into 255 equal channels which produced a time-related factor proportional to time. This is not the ideal method for incorporating time into the data base as extreme care must be taken to keep the flow rate constant. However, it is the only compromise possible for commercial instruments which do not include the facility to incorporate time in the data base from the computer clock.

IV. Flow Cytoenzymology

Flow cytoenzymology is now an accepted term for the study of enzyme reaction kinetics in intact cells using flow systems. This term should be used when describing results obtained with these instruments as these can differ considerably from those obtained with classical techniques. The first nonkinetic description of enzyme measurements by flow cytometry (Hulett *et al.*, 1969) exploited the phenomenon of fluorochromasia (Rotman and Papermaster, 1966) in which the lipophilic, membrane-permeable fluorogenic substrate fluorescein diacetate (FDA) was converted to the comparatively polar fluorescent product fluorescein by intracellular esterases. One of the first kinetic assays where the development of fluorescence with time was recorded also used FDA (Watson *et al.*, 1977). Since then there have been considerable advances which will continue with the further developments of probes.

A. Cytoplasmic Enzymes

A schematic for the conversion of substrate to product by enzyme action within whole cells is shown in Fig. 8. This differs from the schematic shown in Fig. 1 in that substrate external to the cell, S_e, must first cross the external membrane to enter an intracellular pool, S_i, then interact with the enzyme generating the reaction product which can then be measured if it remains inside

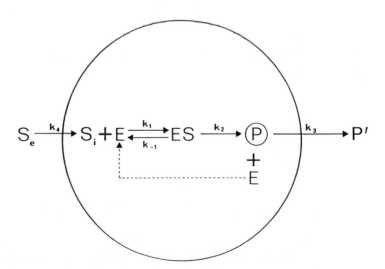

Fig. 8 Schematic for intracellular conversion of substrate to product. Only product which is cell associated is ''seen'' by flow cytometers.

or associated with the cell. Unlike classical biochemical assays only product which is cell associated, *P,* is "seen" by the instrument. Any product which diffuses away from the cell, *P',* is not scored. The ideal fluorogenic substrate for flow enzymology is lipophilic, nontoxic, and nonfluorescent. During the reaction this is converted to a highly polar fluorescent product. Substrate lipophilicity generally is required for rapid transport across the external cell membrane and high polarity of the product is desirable in order to "trap" the product within the cell but few substrates meet all these criteria. However, there is now a considerable variety of such substrates based on a number of different fluorophores including fluorescein, rhodamine, naphthol, naphthylamine, quinolines, methylumbelliferone, and monochlorobimane. These have been used for studying esterases, lipases, sulfatases, phosphatases, glucuronidases, transferases, peroxidases, peptidases, transpeptidases, galactosidases, arylamidases, glucosidases, and glutathione S-transferases and the list is growing all the time.

In general, the fluorescein- and rhodamine-based substrates and monochlorobimane are either weakly or nonfluorescent which is ideal. In contrast, the 4-methylumbelliferone conjugates are generally fluorescent but both their excitation and their emission wavelengths are much shorter than those of the released product 4-methylumbelliferone and with correct optical design and/or electronic compensation any "overlap" can be minimized or eliminated. Techniques using the naphthol derivatives involve "trapping" the released product within the cell by coupling with 5-nitrosalicylaldehyde which forms an insoluble fluorescent complex with an emission spectrum that is shifted into the red (Dolbeare and Smith, 1971). A comprehensive list of substrates and the reactions to release the associated fluorophores is not in order here and the reader is referred to the Molecular Probes Catalogue (Molecular Probes, Inc. Eugene, OR) and an excellent volume, "Applications of Fluorescence in the Biomedical Sciences" (Lansing-Taylor *et al.,* 1986).

1. Esterases

Esterases are a ubiquitous set of enzymes present in all cells in varying quantities but there are no specific substrates for specific esterases. The most widely used substrate for this class of enzyme is FDA and a protocol for quantitating nonspecific esterase activity using flow cytometry in viable intact cells is as follows.

1. Prepare a single cell suspension adjusted to 10^6 cells ml^{-1} in medium.

2. Dissolve AR grade FDA in 1.0 ml spectrograde acetone to give a stock solution with a concentration of 12 mM which must be kept in the dark at $-10°C$.

3. Take 20 μl of the stock solution and add to 50 ml phosphate-buffered saline (PBS). This gives a concentration of 4.8 μM which is the maximum that can be attained without precipitating flocculation.

4. Now prepare 1.0-ml aliquots of working substrate solution at various concentrations by dilution with PBS. Generally, six different substrate concentrations are used.

5. Mix 1.0 ml of the highest substrate concentration with 1.0 ml of the cell suspension which halves the concentration of the substrate solution. The reaction mixture is now introduced into the instrument which has been tuned to the 488-nm argon excitation line. The system must be triggered on forward or right angle light scatter.

6. Adjust the voltage of the "green" photomultiplier (520- to 560-nm band pass filter) and electronic gain of the instrument so that half the population is still "on-scale" after the longest time interval to be investigated. Do not subsequently change the settings.

7. FDA is adsorbed to plastic tubing so flush the instrument through with dilute bleach. This must be followed with extensive flushing with water or PBS as all the bleach must be removed before reintroducing cells.

8. Start the experiment with the lowest substrate concentration, mixing 1.0 ml of this with 1.0 ml cells. Introduce the reaction mixture into the instrument as fast as possible. In our instrument (Watson, 1980, 1991) "as fast as possible" means a "dead" time of about 15 sec to reestablish stable flow after surge pumping the reaction mixture into the flow chamber.

9. The kinetic measurements that can be obtained depend on the capability of your flow cytometer. With primitive instruments, by which we mean all those which do not incorporate the time stamp from the computer clock for each cell in the data base, you will have to record a green fluorescence histogram containing between 5000 and 10000 cells at known intervals.

10. Plot the medians of the fluorescence distributions versus time for data obtained by the continuous interrupted sampling or discrete sampling (see step 9 above) techniques to generate an enzyme progress curve at each substrate concentration. An example is shown in Fig. 9.

11. In instruments which incorporate the computer time stamp in the data base the progress curve is obtained by plotting the fluorescence from each cell versus the time at which it was recorded either as a dot plot or as a contour plot. An example from bone marrow is shown in Fig. 10.

12. The initial reaction velocities are obtained by regression analysis of data such as those shown in Figs. 9 and 10.

13. The initial reaction velocities are now plotted versus substrate concentration as a Michaelis–Menten plot (see Fig. 11) from which K_m and V_{max} may be obtainable.

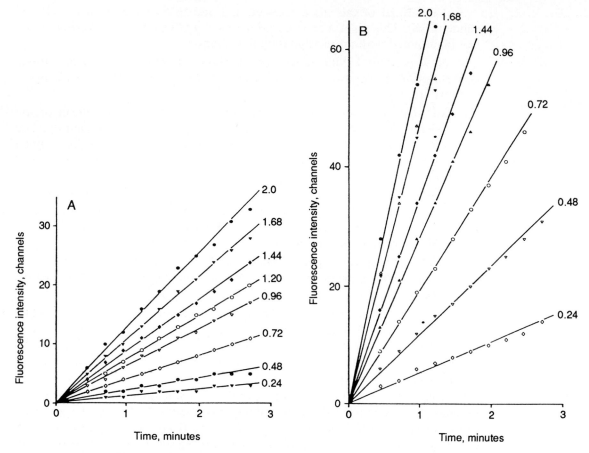

Fig. 9 Enzyme progress curves of esterase activity in log and plateau phase EMT6 cultures using FDA as substrate (A and B respectively). The micromolar concentrations of substrate are shown against each progress curve.

The continuous interrupted sampling technique (Watson *et al.*, 1977) with FDA as substrate for esterases was used to show that cells in late plateau phase EMT cultures, in which more than 95% of the population was arrested in a G_1/G_0 state, exhibited considerably higher esterase activity than exponentially growing cells (Watson *et al.*, 1978). These progress curves are shown in Fig. 9, with the data from exponentially growing and plateau phase populations in panels A and B, respectively. The micromolar concentrations of substrate are shown against each progress curve. Figure 11 shows the substrate-dependent velocity plots associated with these data, which theoretically should follow the Michaelis–Menten rectangular hyperbola (Michaelis and Menten, 1913). There was a highly abnormal "double-sigmoid" pattern for the plateau phase cells

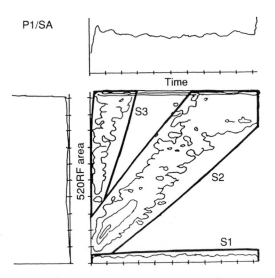

Fig. 10 Contour plots of fluorescence versus time for a subset in mouse bone marrow in which three major subsets are apparent. These are labeled S1, S2, and S3, which correspond to increasing enzyme activity.

but the exponentially growing cells exhibit less abnormal kinetic behavior. Krisch (1971) has reported abnormal kinetic behavior in certain esterase preparations and it has been suggested that the reaction mechanism may need two or more interacting catalytic sites. Substrate activation, where the reaction proceeds most rapidly when more than one substrate molecule is bound to a single enzyme molecule, may also be involved (Adler and Kistiakowsky, 1961). Furthermore, it is highly probable that more than one esterase is involved with the hydrolysis of FDA. Guibault and Kramer (1966) have shown that FDA hydrolysis is catalyzed by a number of enzymes including α- and γ-chymotrypsin and lipase. However, further factors must also be considered for the abnormal kinetic behavior shown in Fig. 11. First, cellular and subcellular permeability barriers are likely to exist in the intact cell which could limit the availability of substrate at the enzyme site. Second, there could be active transport mechanisms for the substrate with intracellular accumulation. Each of these factors would result in a difference between the substrate concentration external to the cell and at the site of enzyme reaction and all of these various factors could contribute to the upward concavity ("lag kinetics") seen for plateau phase cells in Fig. 9.

Complex tissues such as mouse bone marrow contain many different subsets which exhibit considerable diversity in esterase activity. Figure 12 shows forward (488FS WDTH) versus right angle scatter (488RS WDTH) where one population labeled P1 has been gated. The esterase activity of this particular

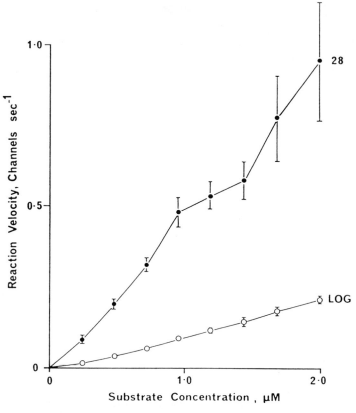

Fig. 11 Substrate-dependent velocity plots associated with the data in Fig. 9 which should theoretically follow the Michaelis–Menten rectangular hyperbola.

subset was used as the illustration in Fig. 10 and the data reveal three major subsets, labeled S1, S2, and S3 with increasing enzyme activity. Furthermore each subset responds differently to esterase inhibitory agents (see Section IV,B) which has implications in cancer chemotherapy.

2. Glutathione S Transferase(s)

Glutathione (GSH) is a critical determinant in control of cellular response to anticancer drugs and radiation. Depletion of protective GSH results in cell sensitization, especially to free radicals. Kosower *et al.* (1979) were the first to use fluorescent bimanes to study intracellular thiols and Rice *et al.* (1986) using flow cytometry have shown that monochlorobimane (MClB) can act as a reporter molecule in GSH metabolism. As an extension to this we developed a sensitive multiparametric flow cytoenzymological assay using continuous time

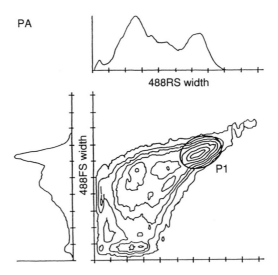

Fig. 12 Contour plot of forward light scatter on the *y*-axis (488FS WDTH) versus right angle scatter (488RS WDTH) on the *x*-axis obtained from mouse bone marrow in which one gated region, labeled P1, has been set. The associated monodimensional histograms are displayed against the respective axes.

measurements to determine reaction kinetics for conjugation of mClB with GSH in populations of intact viable cells catalyzed by the enzyme family, the glutathione-S-transferases (Workman and Watson, 1987; Workman *et al.*, 1988). Monochlorobimane is a nonfluorescent probe which after conjugation with GSH results in a relatively insoluble fluorescent complex excited by UV light with emission in the blue (460–510 nm). The reaction is illustrated in Fig. 13. As with esterases there is no specific substrate for a specific glutathione-S-transferase and monobromobimane (mBrB) can be substituted in the following protocol.

1. Prepare a single cell suspension adjusted to 10^6 cells ml^{-1} in medium.

2. Dissolve mClB in PBS at a concentration of 240 μM to give the stock solution.

3. Dilute the stock solution of mClB to give a number of working substrate concentrations.

4. The GSH–mClB conjugate must be excited with UV light and an argon, krypton, or He-Cd laser must be used. Tune the former two to the UV lines.

5. Add 50 μl of propidium iodide (PI) made up at a concentration of 2 mg ml^{-1} to 1.0 ml of the cell suspension. PI enters cells with nonfunctional external membranes (dead) and the PI–DNA complex is excited by UV light which is a means of excluding dead cells.

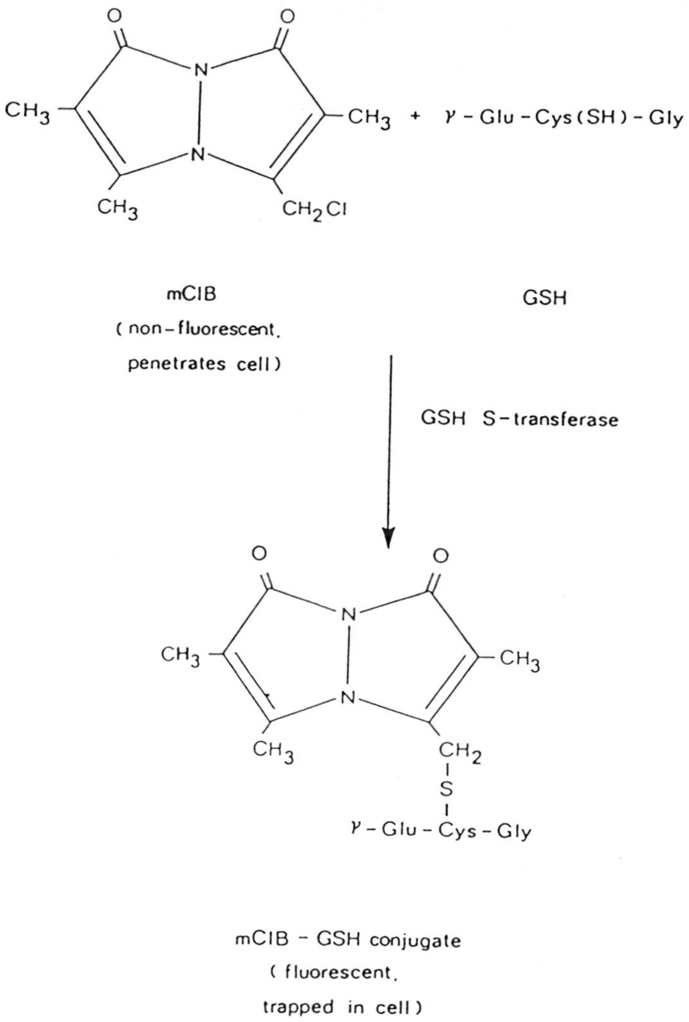

Fig. 13 Reaction of monochlorobimane with the tripeptide glutathione mediated by transferases to produce the insoluble fluorescent conjugate.

6. Add 1.0 ml of the highest working substrate concentration solution to 1.0 ml of the cell suspension and introduce the mixture into the instrument. Again, the system must be triggered on light scatter. However, solid-state detectors are relatively insensitive to UV light; thus, instruments which have solid-state forward scatter detectors should be triggered on right angle scatter signals.

7. Adjust the voltage of the "blue-green" photomultiplier (460- to 510-nm band pass filter) and electronic gain of the instrument so that half the population is still "on-scale" after the longest time interval to be investigated which may be as long as 60 min. Do not subsequently change the settings.

8. Start the experiment with the lowest substrate concentration, mixing 1.0 ml of this with 1.0 ml cells, as in the protocol for esterases.

9. Record red fluorescence (PI-DNA, dead cells), blue-green fluorescence (mClB-GSH conjugate), and forward and right angle scatter together with time.

10. Different cell types exhibit considerable variation in their GSH and GSH-S-transferase content. In cells with "low" GSH-S-transferase content it is sufficient to use the discrete sampling technique where individual samples are run through the instrument at known times after cells are mixed with substrate. In such cases the medians of the population, after gating on PI-positive cells to exclude the dead component, are plotted against time to obtain the enzyme progress curve. In contrast, populations with "high" GSH-S-transferase activity would be better analyzed using continuous time recording as the fluorescence would be changing during the recording interval.

11. Theoretically, this reaction should produce a progress curve which reaches a well-defined substrate concentration-dependent asymptote that is directly proportional to thiol content. However, this may not always be the case as the reaction is complex and the kinetic analyses have not yet been fully elucidated. Some of the analytical problems are discussed below.

The rate of this reaction is dependent on substrate concentration (mClB or mBrB), intracellular GSH content, and the activity of the transferases(s). An illustration of the reaction, which includes the rate constants and the possibility that GSH may be produced during the reaction, is shown in Fig. 14. The differential rate equations describing the system are given below.

$$\frac{\delta[EGS]}{\delta t} = k_1[E][G][S] - k_1[EGS] - k_2[EGS]$$

$$\frac{\delta[P]}{\delta t} = + k_2[EGS]$$

$$\frac{\delta[S]}{\delta t} = - k_1[E][G][S] + k_{-1}[EGS]$$

$$\frac{\delta[E]}{\delta t} = - k_1[E][G][S] + k_2[EGS]$$

$$\frac{\delta[G]}{\delta t} = - k_1[E][G][S] + k_1[EGS] + k_3[GP]$$

$$[E] = [E_0] - [EGS].$$

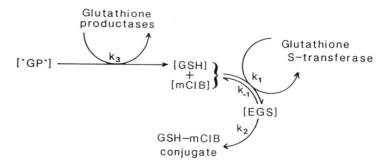

Fig. 14   Schematic for the conversion of mClB and GSH to the conjugate by transferases. An additional GSH production pathway (GP) has been included.

The short-hand notation in this equation matrix is interpreted as follows. $[E]$, $[G]$, and $[S]$ represent the concentrations of free enzyme, glutathione, and substrate, respectively. $[EGS]$ represents the concentration of the complex of enzyme plus glutathione plus substrate. $[E]$ is the concentration of free enzyme at any given time and $[E_0]$ is the initial concentration of enzyme. $[GP]$ represents the "glutathione production system" which is considerably more complex than is represented but this very simplified version has been included for partial completeness. However, the possibility of metabolic breakdown of the glutathione conjugate with reutilization of the three peptides which make up glutathione within the "glutathione production system" has not been included. Moreover, it is possible that the conjugate may be secreted by the cell and this too has not yet been included.

The equation complex above is the simplest model that can be postulated and, clearly, it is considerably more complicated than that shown in Section III as $\delta[EGS]/\delta t$ is dependent on three concentrations, namely, $[E]$, $[S]$, and $[G]$. This equation complex is not readily amenable to the type of solution presented in Section III. However, this is no problem in this day and age as there are numerical integrating algorithms available to tackle exactly this type of differential equation matrix.

The progress curves generated from the above matrix reach an asymptote when all intracellular GSH is converted. Figure 15 shows the results from a Ficoll-Paque lymphocyte-enriched preparation obtained from heparinized whole blood of a normal subject which was mixed with 120 μM mClB plus propidium iodide at a concentration of 50 μg ml^{-1} (Workman *et al.*, 1988). Forward and 90° scatter signals were collected together with blue and red fluorescence. The latter is also excited by UV from the DNA of dead cells which do not exclude propidium iodide. Figure 15A shows forward versus 90° scatter in which three regions can be defined. Regions 1 and 2 correspond to lymphocytes and monocytes, respectively. Region 3 represents debris and propidium iodide-positive dead cells. Figures 15B and 15C show the accumula-

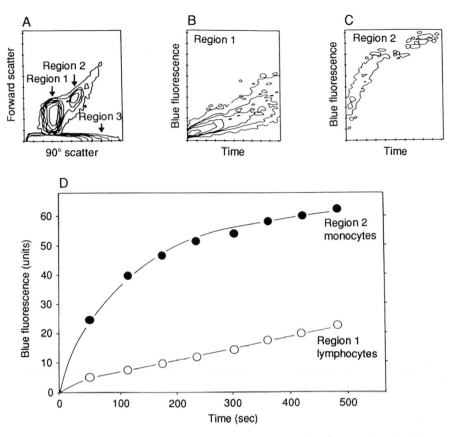

Fig. 15 Ficoll-Paque lymphocyte-enriched preparation obtained from heparinized whole blood of a normal subject which was mixed with 120 μM mClB plus propidium iodide at a concentration of 50 μg ml^{-1}. Forward and 90° scatter signals were collected together with blue and red fluorescence. The latter is also excited by UV from the DNA of dead cells which do not exclude propidium iodide. (A) forward versus 90° scatter in which three regions can be defined. Regions 1 and 2 correspond to lymphocytes and monocytes, respectively. Region 3 represents debris and propidium iodide-positive dead cells. (B and C) The accumulation of blue fluorescence with time as frequency contour plots where it can be seen that monocytes (C) exhibited greater activity than lymphocytes (B) consistent with larger quantities of GSH and/or GSH S-transferase. A more conventional display, derived from the data in B and C, shows medians of the distributions at discrete time intervals plotted against (D). The dead cells and debris in region 3 exhibited no activity.

tion of blue fluorescence with time as frequency contour plots where it can be seen that monocytes (C) exhibited greater activity than lymphocytes (B) consistent with larger quantities of GSH and/or GSH-S-transferase. A more conventional display, derived from the data in panels B and C, shows medians of the distributions at discrete time intervals plotted against time in panel D. The dead cells and debris in region 3 exhibited no activity. The data in Fig. 15

represent a five-dimensional set as two scatter and two fluorescence measurements were recorded with the fifth parameter, time.

We have consistently found that monocytes reach a fairly well-defined asymptope but lymphocytes and some tumor cells frequently do not. One explanation is that monocytes may not have the capacity to replenish intracellular GSH levels and, therefore, the progress curves reach a well-defined asymptope. However, any cells which have the capacity to generate glutathione in response to depletion during the reaction will not reach a well-defined asymptope. If this is true the initial phase of the reaction may reflect GSH transferase activity more accurately and in the later phases the continuing increase in blue fluorescence levels would reflect GSH production. A second explanation is substrate nonspecificity. Monochlorobimane plus any thiol can act as substrates for the appropriate transferases. Hence, we might expect departures from the predicted progress curve in cells which are thiol rich, other than in GSH, including antibody producing cells.

B. Membrane Enzymes

An automated fluorimetric method for assaying alkaline phosphatase using 3-O-methylfluorescein phosphate (MFP) was developed by Hill *et al.* (1968). An adaption of this technique has been applied to intact cells using flow cytometry by Watson *et al.* (1979) using the following protocol.

1. Prepare a single cell suspension adjusted to 10^6 cells ml^{-1} in medium.

2. Dissolve 10 mg MFP in 25 ml PBS. Use chromic acid-washed glassware as this substrate can undergo "spontaneous" hydrolysis in a dirty apparatus.

3. Make up a number of aliquots of substrate at varying concentrations by dilution with PBS.

4. Mix equal volumes of the cell suspension and substrate and introduce into the instrument tuned to the 488-nm laser line.

5. Record the data as described previously and plot the medians of the distributions versus time.

6. Discard all unused substrate solution.

Results for EMT6 mouse mammary tumor cells are shown in Fig. 16. These are obviously very different from those for intracellular esterases shown in Fig. 9 and they presented a considerable interpretative problem which was not resolved until a sample was viewed under the fluorescence microscope. It was then seen that the fluorescence was concentrated as "halos" at the external cell membrane with no fluorescence from the interior of the cells. Subsequent incubation of cells with product, 3-O-methylfluorescein, failed to demonstrate entry into intact cells. The viability of EMT6 cells decreases considerably after 3 hr of incubation with protein-free phosphate PBS. In a sample of cells so treated it was found that the fluorescence from 3-O-methylfluorescein was no

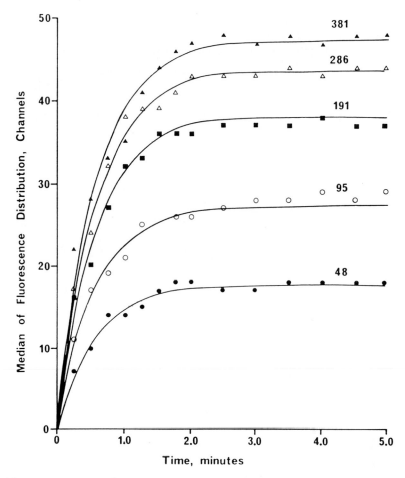

Fig. 16 Progress curves for hydrolysis of 3-*O*-methylfluorescein phosphate. The micromolar concentrations of substrate are shown adjacent to each curve.

longer located at the cell surface but was being emitted from granular structures surrounding the nucleus. This fluorescence was considerably more intense than that emitted from the cell surface, so much so that the latter could no longer be seen. These various data suggested that substrate hydrolysis in intact viable cells takes place at or within the cell by phosphatases located in the plasma membrane and that the product was lost from the immediate vicinity of the cells by diffusion and consequently was not detected by the instrument. Thus a steady state would be reached in which the rate of production of fluorescent product would equal the rate of loss, giving rise to asymptotic fluorescence responses from the population at each substrate concentration.

Assuming that this hypothesis is correct, it is possible to modify Eq. (9) describing the rate of change of product (see Section III) by adding to the term $-k_3 [P]$ to the right-hand side which gives

$$\frac{\delta[P]}{\delta t} = \frac{k_2[E_0][S]}{[S] + K_m} - k_3[P],$$

where k_3 is the rate constant for product loss assuming first-order kinetic processes. Rearrangement of this equation gives

$$\frac{\delta[P]}{\delta t} + k_3[P] = \frac{k_2[E_0][S]}{[S] + K_m}$$

and on integration we obtain the following inverted exponential (the full derivation of which is given in Appendix 1):

$$P(t) = P_\infty \times \frac{[S]}{[S] + K_m} \times (1.0 - \exp(-k_3 \times t)). \qquad (11)$$

$P(t)$ is the fluorescence response at time t, P_∞ is the theoretical maximum asymptotic response at infinite substrate concentration, K_m is the Michaelis constant, and k_3 is the rate constant for product loss assuming a first-order kinetic process. When a steady-state exists (i.e., beyond 2.5–3 min in Fig. 16), the term $\exp(-k_3 \times t)$ in Eq. (11) will tend to zero and the asymptotic fluorescence response will be given by the expression $P_\infty \times [S]/([S] + K_m)$ and will vary with substrate concentration. P_∞ can be eliminated from this expression by taking ratios; thus,

$$\frac{P_2(t_\infty)}{P_1(t_\infty)} = \frac{[S_2]}{[S_1]} \times \frac{([S_1] + K_m)}{([S_2] + K_m)} = R_{12}, \qquad (12)$$

where $P_1 (t_\infty)$ and $P_2 (t_\infty)$ are the asymptotic fluorescence values associated with substrate concentrations $[S_1]$ and $[S_2]$ respectively, and where R_{12} is the ratio of $P_2 (t_\infty)$ to $P_1 (t_\infty)$. Equation (12) can be rearranged to give

$$R_{ij} = K_m \times \left[\frac{1}{[S_i]} - \frac{R_{ij}}{[S_j]} \right] + 1,$$

where for N substrate concentrations I varies from 1 to $(n - 1)$ and j varies from $(i + 1)$ to N, to give a triangular matrix for the ratios containing $N(N - 1)/2$ values. Thus, by plotting R_{ij} against $(1/[S_i]) - (R_{ij}/[S_j])$, a line with slope K_m is obtained which intersects the ordinate at unity (see Fig. 17). In three separate analyses, values of 110.0 ± 31, 122.3 ± 31, and $120.5 \pm 24 \, \mu M$ were obtained, where the limits were calculated at two standard errors and in all cases the ordinate intercept did not differ from unity ($p > 0.1$). It can be seen from Eq. (12) that if P_∞ is the asymptotic fluorescence response at infinite substrate

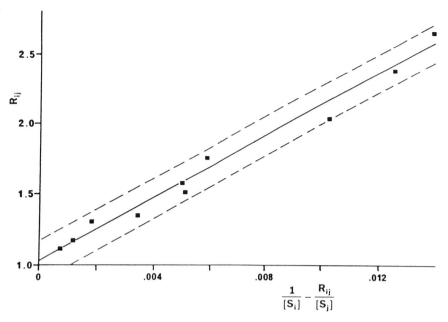

Fig. 17 Derivative plots of R_{ij} versus $(1/[S_i]) - (R_{ij}/[S_j])$, see text. The slope of the regression lines gives the value of K_m and the 95% confidence limits are shown by the dashed lines. The ordinate intercept does not differ from unity, $p > 0.1$.

concentration $[S_\infty]$ then

$$\frac{P_\infty}{P_n} = -\frac{[S_\infty]}{[S_n]} \times \frac{([S_n] + K_m)}{([S_\infty] + K_m)},$$

where P_n is the asymptotic response at $[S_n]$. The term $[S_\infty]/([S_\infty] + K_m)$ is unity, so $P_n \times ([S_n] + K_m) = P_\infty \times [S_n]$. By plotting $P_n \times ([S_n] + K_m)$ against $[S_n]$ a line of slope P_∞ is obtained which intersects the origin (see Fig. 18). Thus P_∞ can be obtained after a value for K_m is found. The maximum reaction velocity V is equal to the product of $P_\infty \times k_3$, which can be calculated after k_3 is defined. Equation (11) is an inverted exponential, and since K_m and P_∞ have been obtained we can use a linear transform of the inverted exponential to obtain values for k_3 at each substrate concentration, S. The transform for this particular problem is

$$\log_e\left[1 - \left(\frac{P(t)}{P_\infty} = \frac{[S] + K_m}{[S]}\right)\right] = -k_3 \times t.$$

These regressions are shown in Fig. 19. As we can see, the slopes k_3 are very similar and none of the intercepts differed from zero, $p > 0.999$ in all cases.

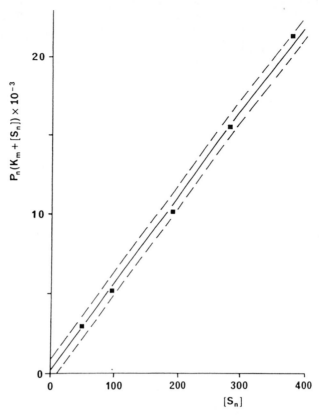

Fig. 18 Regression of $P_n (k_m + [S_n])$ on $[S_n]$, see text. The slope gives P_x, defined as the asymptotic fluorescence intensity at infinite substrate concentration. The 95% confidence limits are shown as the dashed lines and the ordinate intercept does not differ from zero, $p > 0.1$.

As the values obtained for k_3 were all so similar and independent of $[S]$ (which they should be), we now perform a simultaneous regression analysis to obtain a common value for k_3. This, and a subsequent common curve fitting analysis, gave a rate constant of 1.98 min^{-1}; thus V can be defined in the arbitrary units of channels min^{-1}. The three analyses gave maximum reaction velocities of 107.0, 110.0, and 109.2 channels per minute. The curves shown in Fig. 16 were calculated with the various parameters obtained here.

The progress curves in Figs. 9 and 16 are very different, and the major differences between the fluorogenic substrates and their products should be considered. First, 3-O-methylfluorescein phosphate is highly polar, ionized at physiological pH, and hydrophilic. In contrast fluorescein diacetate is essentially nonpolar, is not ionized at physiplogical pH, and is lipophilic. Thus fluo-

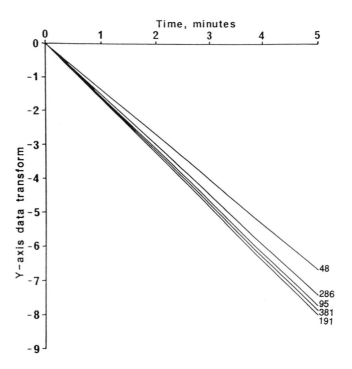

Fig. 19 Regression lines for the individual inverted exponential linear transform analyses of the data points shown in Fig. 16.

rescein diacetate will penetrate the cell without prior hydrolysis by any membrane esterases whereas 3-*O*-methylfluorescein phosphate would not be expected to traverse the cell membrane unless there was either an active transport mechanism for it or it was dephosphorylated at the membrane. Second, 3-*O*-methylfluorescein, the reaction product of 3-*O*-methylfluorescein phosphate, is considerably less polar than fluorescein, the reaction product of FDA. Thus, if 3-*O*-methylfluorescein enters the cell we would expect the rate constant for leakage from the cell to be greater than that for fluorescein. Studies with EMT6 cells have shown that fluorescein leaks out of these cells with a half-time of 7–8 min (Watson *et al.* 1979; Watson, 1980), thus the leakage rate constant for 3-*O*-methylfluorescein is about 20 times greater than that for fluorescein. The direct observations made with the fluorescence microscope suggest that 3-*O*-methylfluorescein either does not enter the cell or diffuses out of the cell very rapidly, although some remains associated with the external membrane. Both possibilities are compatible with the magnitude of the loss rate constant whatever the mechanism.

C. Inhibition Kinetics

Chloroethylnitrosoureas (Cnus) are an important class of anti-tumor agent. Passive diffusion is the means by which CCNU, BCNU (Begleiter *et al.*, 1977), and chlorozotocin (Lam *et al.*, 1980) gain access to the cell interior as the uptake of each is unsaturable, independent of temperature, and unaltered by metabolic inhibitors. Their exact mode of action is not known but chloroethylation of DNA and subsequent crosslinking has been implicated (Gibson *et al.*, 1985). However, protein carbamoylation may also be involved as organic isocyanates as well as alkylating species are formed when Cnus decompose in aqueous environments. Carbamoylating agents have been shown clearly to inactivate a number of enzymes, and inactivation of glutathione reductase has been used to determine the protein carbamoylation potential of Cnus (Babson and Reed, 1978). Because of the susceptibility of serine hydrolases to isocyanate inactivation (Brown and Wold, 1973), Paul Workman then of our laboratories proposed that flow cytometric measurement of esterase activity might form a basis for the determination of intracellular carbamoylation by Cnu-derived isocyanates (Dive, 1988; Dive *et al.*, 1987, 1988). This proved to be the case and an example of esterase inhibition induced by BCNU-derived isocyanate is shown in Fig. 20 (see figure legend for full explanation). By using a number of different drug doses as in Fig. 21 it is possible to obtain an estimate of the drug concentration which causes a 50% reduction in FDA hydrolysis activity, defined as the I_{50} value. This makes it possible to compare directly the inhibitory effects of a number of Cnus and related cytotoxic agents. Such comparisons are shown in Table I which includes both flow cytometric and conventional spectrofluorimetric results from sonicated preparations together with the ratios of the results.

The results in this table are interesting. The I_{50} ratios for CHI, CCNU, BCNU, and CEI indicate that only about half the concentration of drug is required to

Table I
I_{50} **Values for a Variety of Nitrosoureas and Related Isocyanates**

	I_{50} values (M)		
Drug	Intact cells	Sonicates	Ratio
CHI	1.1×10^{-5}	2.6×10^{-5}	0.423
CCNU	3.8×10^{-5}	8.3×10^{-5}	0.458
BCNU	5.0×10^{-5}	9.2×10^{-5}	0.543
CEI	1.0×10^{-5}	1.6×10^{-5}	0.625
TCNU	1.8×10^{-4}	9.9×10^{-5}	1.818
ACNU	7.3×10^{-3}	3.2×10^{-3}	2.281
CHLOZ	$>1.5 \times 10^{-2}$	2.4×10^{-3}	>6.250
GANU	$>1.5 \times 10^{-2}$	2.4×10^{-3}	>6.250

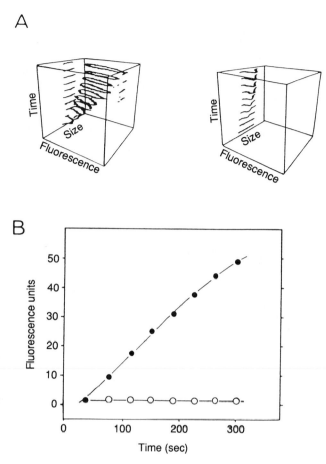

Fig. 20 Inhibition of esterase activity by BCNU-derived isocyanate. The top two diagrams (A) show size versus fluorescence on the x and z horizontal axes with time (which increases upward on these displays) on the y-axis. Contour plots of size versus fluorescence for 10 equal "time slices" are shown within these three-dimensional data spaces. Top left is the control and top right shows the inhibition due to BCNU. (B) Medians of the fluorescence distributions of each "time slice" plotted against time.

obtain 50% esterase activity inhibition in whole cells compared with sonicates. With TCNU and ACNU this is reversed with double the drug concentration required in whole cells compared with sonicates; however, with chlorozotozin and GANU this has increased to >six-fold. These results almost exactly parallel the hydrophilicity of the agents. CHI, CCNU, BCNU, and CEI are lipophilic, but chlorozotozin and GANU are hydrophilic and readily soluble in water at physiological pH and temperature and would not be expected to cross the

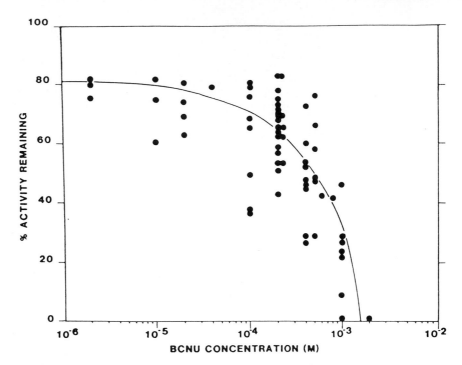

Fig. 21 Enzyme activity for multiple concentrations of BCNU. The dose which produces 50% inhibition of enzyme activity, I_{50}, can be estimated from data such as these.

external cell membrane without an active transport mechanism. We have seen earlier that nitrosoureas enter cells by passive diffusion; thus, we would not expect chlorozotozin and GANU to enter cells. This conclusion would seem to be substantiated by the data in Table I and the method seems capable of assessing not only the ability of such agents to cross the external cell membrane, but also their capability of inducing damage having entered the cell.

In view of the implication that intracellular carbamoylation might be responsible for the myelosupressive qualities of CNUs we applied this assay to investigate the inhibitory effects of BCNU on the cellular esterases of mouse bone marrow. The data in Fig. 10 demonstrate that three subsets, each derived from a single homogenous cluster in the light scatter data space of Fig. 12, can be identified on their esterase activities. After treating the marrow with 10^{-4} M BCNU, subsets S2 and S3 (see Fig. 10) exhibited reductions in esterase activity of 79 and 49%, respectively. Subset S1 exhibited minimal initial esterase activity; hence, inhibition was difficult to assess reliably. Further subsets defined on both esterase activity and light scatter properties exhibited reductions in enzyme activity of 75, 9, and 38%, respectively. These results are very interesting as they demonstrate considerable heterogeneity, from only 9% inhibition to 79% inhibition, with the same dose of BCNU. These differences may be due to disparate accessibilities of the inhibitor to the intracellular target enzymes and/or may reflect intrinsic variation in BCNU inhibitory sensitivity.

D. Short Time Scale Kinetics

Each of the various methods described in Section II for incorporating time into the data base suffers from a considerable disadvantage, which is the finite time taken between mixing cells with substrate or ligand and recording the first event. This "dead time" is due to mixing, location of the containing vessel in the instrument, surge pumping the reaction mixture through the sample feed tube into the flow chamber, and restoration of stable flow. In our instrument the absolute minimum dead time is 15 sec if everything proceeds smoothly, but more often this was about 20 sec. There was a clear need to reduce this temporal delay, so as to facilitate the analysis of biochemical events taking place over the time period of a few seconds. Such a development would be valuable not only in flow cytoenzymology (e.g., for substrate diffusion kinetics, very rapid reaction, and membrane enzyme analysis), but also for a variety of other biological applications involving very short time scales (e.g., drug uptake and ion flux analysis).

Kachel *et al.* (1982) described a flow chamber in which time-resolved measurements could be made within 1 sec but this was not variable. Recently, we have developed a system which markedly reduces the temporal reference frame so that this can be varied within the interval 1 to 20 sec (Watson *et al.,* 1988; Watson, 1992). The technique employs computer-controlled precision drive syringe pumps, one for substrate, the other for cells, together with variable length tubing between a mixing chamber and analysis point, selected by an array of zero dead-space pinch valves. The different tube lengths at a given pump flow rate give different times between mixing and analysis. A schematic of the system is shown in Fig. 22.

Calibration was effected using two sets of microbeads with different fluorescence intensities (Polysciences, Inc., Warrington, PA). Pumps 1 and 2 were filled with beads of the higher and lower intensity, respectively. The concentrations were adjusted to give a flow rate of 250 beads per second with both pumps running at 100 μl min^{-1}. The concentration of beads in pump 2 (lower intensity) was about 1.4 times greater than that in pump 1. Both pumps were activated before data collection in order to fill the input pipes to the mixing chamber, and they were then stopped. The chamber was then flushed through and pump 1 was restarted before data collection. Pump 2, containing the lower intensity beads, was started during the run after about 1,500 of the 10,000 requested events had been recorded. This procedure was repeated for each tube length at various flow rates.

Figure 23 shows fluorescence versus time data as frequency contour plots at the 3-, 6-, and 12-event levels with the fluorescence and flow rate histograms

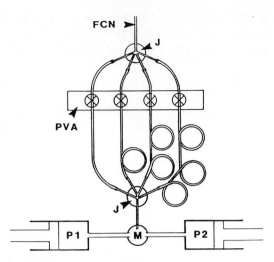

Fig. 22 Schematic of "stop-flow" device which allows kinetic measurements to be made during the first few seconds of a dynamic reaction.

adjacent to their respective axes. The sawtooth pattern of the flow rate histogram is caused by interruption of data collection due to time sharing on the PDP 11/ 40 computer which dumps the data to disk, displaying data on the screen between each buffer dump and checking the pump flow rates before recommencing data collection. These data were recorded with flow rates of 100 μl min^{-1} from each pump and Figs. 23A–23D, respectively, were obtained for tube lengths of 5, 10, 20, and 40 cm. Full scale on the abscissa is 40 sec, and each division on the ordinate represents 100 digitization steps. The long vertical lines drawn through each panel represent the time which is flagged in the data base when pump 2 was activated. The immediate flow rate increase is apparent and the sudden surge is also manifest by a widening of the distribution of the higher intensity beads. The short vertical lines show the time at which the beads from pump 2 first appeared in the data record. The interval between these lines represents the time for the lower intensity beads from pump 2 to travel from the mixing chamber to the analysis point.

Plots of time from mixing chamber to analysis point versus tube length are shown in Fig. 24 for pump flow rates of 100, 150, 200, and 250 μl min^{-1}. The regression lines for these pump flow rates extrapolated almost to a common point and all the intersections are included in the region defined by the open circle shown in Fig. 24. Therefore, the four equations defined by the regression analyses were solved simultaneously to give a common intersection point with x and y coordinates of -24.73 cm and 0.6 sec, respectively. This value of 24.73 cm represents the volume of the "nonvariable" (dead-space) section of the system which comprises an individual pair of arms of the four-way junctions,

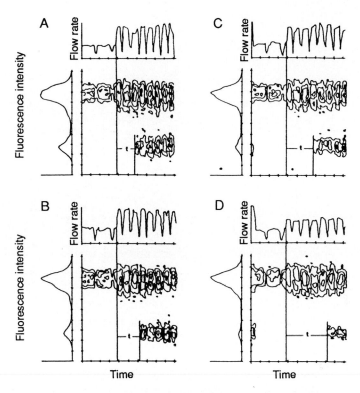

Fig. 23 Fluorescence versus time data as frequency contour plots at the 3-, 6-, and 12-event levels with the fluorescence and flow rate histograms adjacent to their respective axes.

the flow chamber injection needle, the mixing chamber, and its outflow. The volume of unit length of the tubing contained in the variable section was measured using Hamilton syringe injection of eight different lengths. The average of the eight readings showed that 1.0 cm was equivalent to 1.046 μl. The volume of the dead space, excluding the mixing chamber (18.1 μl), was measured (again by Hamilton syringe injection) to be 8.1 μl. Thus the total dead-space volume was 26.2 μl, which corresponds well with the theoretical value of 25.8 μl (24.7 cm $\times$ 1.046 μl cm^{-1}) represented by the intersection point in Fig. 24.

The vertical displacement of the intersection point, 0.6 sec, constituted a puzzle until we discovered that the assembler language routines controlling the pump flow rates, which had been developed for a different application, not only tested, but also corrected for, any "backlash" before commencing delivery. This involves a feedback hysteresis loop to the computer and the time taken for backlash correction is inversely proportional to the requested flow rate. At 100 μl min^{-1} the backlash assessment and correction time is between 500 and 600 msec, which accounts for the observed vertical displacement. This

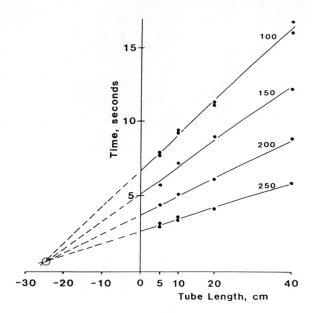

Fig. 24 Plots of time from mixing chamber to analysis point versus tube length for pump flow rates of 100, 150, 200, and 250 μl min^{-1}.

minor problem was only encountered when the second pump was started during data collection when backlash had to be tested for and corrected.

Figure 25 shows the slopes of the four regression lines plotted against both the square root and the log of the respective flow rates. Regression analyses gave correlation coefficients of 0.9999 and 0.991, respectively. From this we assumed that the square root transformation was more appropriate, which gave a slope of -0.0283 and intercept of 0.528. From these various data it was possible to express time in relation to flow rate and length of tubing between mixing and analysis points. This is given by the equation

$$T = ((0.528 - (0.0283 \times \sqrt{F})) \times (L + 24.73)) + 0.6,$$

where T is time, F is the average flow rate of the two pumps, and L is tube length. The expression $0.528 - (0.283 \times \sqrt{F})$, derived from Fig. 25, is equivalent to the slope, m, in the equation $Y = mX + C$, and $L + 24.73$ is equivalent to X, derived from Fig. 24. The constant C, 0.6 sec, can be dropped when both pumps are started before data collection.

The method was tested with EMT6 mouse mammary tumor cells in exponential growth adjusted to a concentration of 2×10^5 cells ml^{-1} and introduced into pump 1 (see Fig. 22). FDA was made up at a concentration of 2 μM and loaded into pump 2. Both pumps were started and the high tension voltage of the green detector was adjusted to record the fluorescein emission histogram with a mean in about channel 600 at pump flow rates of 100 μl min^{-1} for a tube

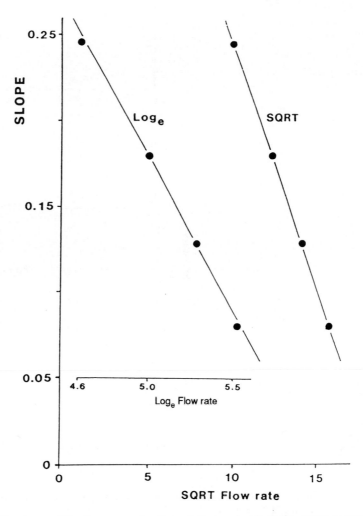

Fig. 25 The slopes of the four regression lines plotted against both the square root and the log of the respective flow rates. Regression analyses gave correlation coefficients of 0.999 and 0.991, respectively.

length of 40 cm. Recordings were then made at each of the four tube lengths at the same flow rate. From the data shown in Fig. 26 it can be seen that the fluorescence distribution is progressively shifted to higher values with increasing tube length. It should also be noted that the distribution remains constant with time as the period taken for cells to flow from the mixing chamber to analysis point is constant. This overcomes one of the potential problems associated with continuous time recording of "fast" reactions where, because of the reaction velocity, only relatively few cells can be recorded with a given fluorescence

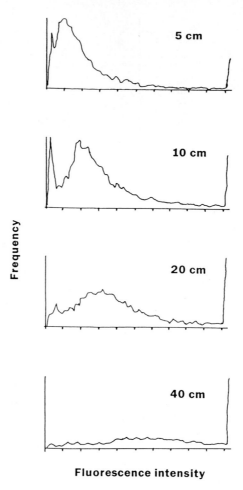

5 cm

10 cm

20 cm

40 cm

Frequency

Fluorescence intensity

Fig. 26 FDA hydrolysis fluorescence distributions for four tube lengths at the same flow rate from EMT6 cells.

intensity. A number of further records were made with various tube lengths and pump flow rates. The medians of the green fluorescence distributions so obtained are plotted against time in Fig. 27 which shows that the accumulation of fluorescence was biphasic over the first 16 sec of the reaction. A linear regression analysis (dashed line) was carried out for the data beyond 5 sec and this gave a correlation coefficient of 0.99 which is compatible with a linear increase in fluorescein with time between 5 and 16 sec. However, this is not really describing the biological observation and a nonlinear regression analysis was carried out using a linear transform of the logistic equation (see Appendix 2) and the result of this analysis is shown in Fig. 28. The uninterrupted curve

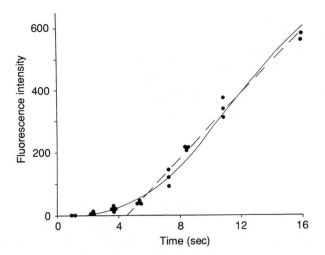

Fig. 27 Biphasic FDA hydrolysis over the first 16 sec of the reaction shown by the dashed curve obtained by nonlinear regression. The data from 5 to 16 sec can also be fitted with a straight line (noninterrupted) but this is less likely to represent the biological reality.

in Fig. 27 generated with the parameters obtained from the nonlinear logistic regression fit is more compatible with the biological reality than is the linear regression from 5 to 16 sec. The sigmoid nature of the data is probably a reflection of the time taken for substrate to diffuse across the external cell membrane and cytoplasm to the site(s) of enzyme action.

Appendix 1

The rate equation encountered in Section IV,B which requires integration is reproduced below.

$$\frac{\delta P}{\delta t} + k_3 P = \frac{k_2 [E_0][S]}{[S] + K_m}.$$

In order to simplify let the expression $([E_0][S])/([S] + K_m)$ be represented by the symbol B; hence, equations can now be written as

$$\frac{\delta P}{\delta t} + k_3 P = k_2 B.$$

Unfortunately, this equation is not in an integratable form and the trick is to transform it so that it can be integrated and to do this we have to make a slight diversion. The Law of Leibnitz states that the rate of change of the product of two variables is equal to the product of the first variable and the rate of change

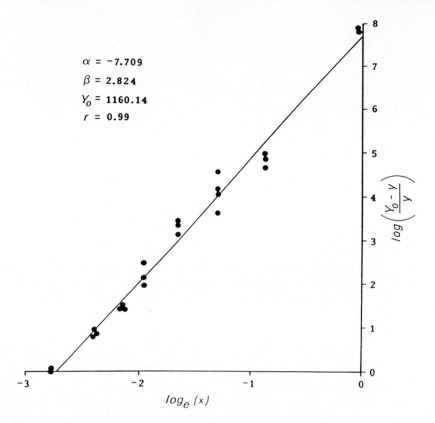

$\alpha = -7.709$

$\beta = 2.824$

$Y_0 = 1160.14$

$r = 0.99$

Fig. 28 Linear transform analysis of the logistic equation applied to the data in Fig. 27.

of the second added to the product of the second variable and the rate of change of the first. This is a bit of a brainful which is more easily appreciated in symbolic form as follows:

$$\frac{\delta(UV)}{\delta t} = Ux\frac{\delta V}{\delta t} + Vx\frac{\delta U}{\delta t}.$$

We transform the equation by multiplying each term by $e^{k_3 t}$ to give

$$(e^{k_3 t})\frac{\delta P}{\delta t} + k_3 (e^{k_3 t}) P = k_2 (e^{k_3 t}) B.$$

Thus, P in our equation is equivalent to V in Leibnitz's equation and $\delta P/\delta t$ is equivalent to $\delta V/\delta t$. The really crafty bit is that the term $e^{k_3 t}$ in our equation is equivalent to U, which, when differentiated, gives $k_3 (e^{k_3 t})$, the rate of change of U, $\delta U/\delta t$. Thus, our equation now reduces to

$$\frac{\delta[P(e^{k_3t})]}{\delta t} = k_2 B (e^{k_3t}).$$

This can now be integrated, remembering to add the constant C, to give

$$P (e^{k_3t}) = \frac{k_3}{k_2} B (e^{k_3t}) + C$$

and we now divide through by e^{k_3t} to get

$$P = \frac{k_3}{k_2} B + C e^{-k_3t}.$$

At time $t = 0.0$ the expression e^{-k_3t} is unity and there will be no substrate in the cells; therefore, $P = 0.0$. Hence,

$$C = -\frac{k_3}{k_2} B.$$

Substituting this solution for C in the previous equation gives

$$P = \left[\frac{k_3}{k_2} B\right] - \left[\frac{k_3}{k_2} B\right] e^{-k_3t}.$$

On simplification this now reduces to the familiar inverted exponential

$$P = \frac{k_3}{k_2} B(1 - e^{-k_3t}).$$

We now make the resubstitutions for B and get

$$P(t) = \frac{k_3[E_0] \times [S]}{k_2([S] \times K_m)} \times (1 - e^{-k_3t}).$$

Now, at infinite substrate concentration, $[S_\infty]$, the ratio $[S]/([S] + K_m)$ tends to unity and at infinite time the expression e^{-k_3t} tends to zero. Hence $[P_\infty]$, the asymptotic value for $P(t)$ at $t = \infty$ for $[S_\infty]$, will be given by

$$P_\infty = \frac{k_3}{k_2} E_0.$$

If we now substitute this into the previous equation we obtain

$$P(t) = P_\infty \left[\frac{[S]}{[S] + K_m}\right] \times (1 - e^{-k_3t}).$$

Appendix 2

The logistic equation is shown below.

$$y = Y_0 \times (1 + \exp(-(\alpha + (\beta \times \log_e(x)))))^{-1}.$$

This can often be used to approximate nonlinear data behavior dependency on the independent variable as the sets of curves generated by varying the constants Y_0, α, and β can assume a large variety of different forms. The linear transform of this equation is effected by dividing through by Y_0, taking the reciprocal of both sides, and then subtracting 1.0 from both sides, in that order, which gives

$$\frac{Y_0}{y} - 1 = \exp(-(\alpha + (\beta = \log_e(x)))).$$

We now take logs to the base e to get

$$\log_e\left(\frac{Y_0}{y} - 1\right) = -(\beta \times \log_e(x)) + \alpha.$$

This completes the linear transformation which is in the form of $Y = mX + C$. The whole of the expression on the left-hand side is equivalent to Y; β is the slope, m; and α is the y-axis intercept, C. Hence, by performing regression analysis of $\log_e((Y_0/y) - 1)$ versus $-\log_e x$, we obtain a straight line of slope β which intersects the y-axis at C from which we get $\alpha = -C$.

Although the nontransformed equation has three constants, namely, Y_0, α, and β, the transform is only dependent on one of these, Y_0, as the manipulation enables α and β to be obtained as a consequence of the regression. The value of Y_0 is found by successive iteration which maximizes the correlation coefficient.

References

Adler, A. J., and Kistiakowsky, G. B. (1961). *J. Biol. Chem.* **236**, 3240–3245.
Babson, J. R., and Reed, D. J. (1978). *Biochem. Biophys. Res. Commun.* **83**, 754–762.
Begleiter, A., Lam, H. Y. P., and Goldenberg, G. J. (1977). *Cancer Res.* **37**, 1022–1027.
Brown, W. E.,, and Wold, F. (1973). *Biochemistry* **12**, 835–840.
Dive, C. (1988). Ph.D. Thesis, Council for National Academic Awards.
Dive, C., Workman, P., and Watson, J. V. (1987). *Biochem. Pharmacol.* **36**, 3731–3738.
Dive, C., Workman, P., and Watson, J. V. (1988). *Biochem. Pharmacol.* **37**, 3987–3993.
Dolbeare, F. A., and Smith, R. E. (1971). *Clin. Chem.* (Winston-Salem, N.C.) **23**, 1485–1491.
Eadie, G. S. (1942). *J. Biol. Chem.* **146**, 85–93.
Eisenthal, R., and Cornish-Bowden, A. (1974). *Biochem. J.* **139**, 715–720.
Gibson, N. W., Mattes, W. B., and Hartley, J. A. (1985). *Pharmacol. Ther.* **31**, 153–163.
Guibault, G. G., and Kramer, D. N. (1966). *Anal. Biochem.* **14**, 28–32.
Hill, H. D., Sumner, G. K., and Waters, M. D. (1968). *Anal. Biochem.* **24**, 9–14.
Hofstee, B. H. S. (1952). *Science* **116**, 329–331.
Hulett, H. R., Bonner, W. A., Barrett, J., and Herzenberg, L. A. (1969). *Science* **166**, 747–749.
Kachel, V., Glossner, E., and Schneider, H. (1982). *Cytometry* **3**, 202–212.
Keij, J. F., Griffioen, A. W., The, T. H., and Rijkers, G. T. (1989). *Cytometry* **10**, 814–817.

Kosower, N. S., Kosower, E. M., Newton, G. L., and Ranney, A. M. (1979). *Proc. Natl. Acad. Sci. U.S.A.* **75**, 3382–3386.

Krisch, K. (1971). *In* "The Enzymes" (P. D. Boyer, ed.), Vol. 5, pp. 43–69. Academic Press, New York.

Lam, H. Y. P., Talgoy, M. M., and Goldenberg, G. J. (1980). *Cancer Res.* **40**, 3950–3955.

Lansing-Taylor, D., Waggoner, A. S., Murphy, R. F., Lanni, F., and Birge, R. R. (1986). *In* "Applications of Fluorescence in the Biomedical Sciences" (D. Lansing-Taylor, A. S. Waggoner, R. F. Murphy, F. Lanni, and R. R. Birge, eds.). Liss, New York.

Lineweaver, H., and Burke, D. J. (1934). *J. Am. Chem. Soc.* **56**, 658–666.

Martin, J. C., and Swartzendruber, D. E. (1980). *Science* **207**, 199–200.

Michaelis, L., and Menten, M. L. (1913). *Biochem. Z.* **49**, 333–369.

Rice, G. C., Bump, E. A., Shrieve, D. C., Lee, W., and Kovacs, M. (1986). *Cancer Res.* **46**, 6105–6110.

Rotman, B., and Papermaster, B. W. (1966). *Proc. Natl. Acad. Sci. U.S.A.* **55**, 134–141.

Watson, J. V. (1980). *Cytometry* **1**, 143–151.

Watson, J. V. (1991). "Introduction to Flow Cytometry," Chapter 15. Cambridge Univ. Press, Cambridge, UK.

Watson, J. V. (1992). "Flow Cytometry Data Analysis: Basic Concepts and Statistics." Cambridge Univ. Press, Cambridge, UK.

Watson, J. V., Chambers, S. H., Workman, P., and Horsnell, T. S. (1977). *FEBS Lett.* **81**, 179–192.

Watson, J. V., Workman, P., and Chambers, S. H. (1978). *Br. J. Cancer* **37**, 397–402.

Watson, J. V., Workman, P., and Chambers, S. H. (1979). *Biochem. Pharmacol.* **28**, 821–827.

Watson, J. V., Cox, H., Hellon, C., Workman, P., and Dive, C. (1988). *Cytometry, Suppl.* **2**, 93a.

Workman, P., and Watson, J. V. (1987). *Cytometry, Suppl.* **1**, 48a.

Workman, P., Cox, H., and Watson, J. V. (1988). *Cytometry, Suppl.* **2**, 47a.

CHAPTER 32

On-Line Flow Cytometry: A Versatile Method for Kinetic Measurements

Kees Nooter, * **Hans Herweijer,** * **Richard R. Jonker,** *
and Ger J. van den Engh †

* Department of Medical Oncology
University Hospital
3015 GD Rotterdam
The Netherlands

† Department of Molecular Biotechnology
University of Washington
Seattle, Washington 98915

I. Introduction

Since the early 1970s, flow cytometry (FCM) has become a highly developed cell analysis technique. Its use is widespread among a great variety of disciplines and new areas of application are being exploited continuously. The FCM technique is ideally suited for the quantitation of the uptake of virtually any fluorescent substance by cells or cell organelles. With classical FCM, quantitative analysis of the uptake of fluorescent dyes by cells has to be done by examining cell samples at intervals. This method becomes very difficult, if not impossible, to perform when the speed of the kinetic phenomenon of interest is very fast (seconds instead of minutes). Then the most appropriate approach is to analyze a single sample continuously. We here describe an FCM technique that enables the study of real-time, kinetic processes in biological materials in time periods of seconds up to a few hours.

The RELACS-III (three-laser Rijswijk experimental light-activated cell sorter) was modified to allow for on-line measurements. The term ''on-line'' is used here to describe conditions where the cells from an individual sample are passing the excitation light beam continuously for a predefined time, and all necessary additions of reagents are performed during this uninterrupted measurement. To perform on-line measurements, the flow cytometer must be equipped with an adequate sampling system and a dedicated combination of data acquisition electronics and software. The method of on-line FCM offers especially the great advantage of visualizing the very early and rapid phases of kinetic processes. Furthermore, compared to classical kinetic measurements, no intersample variations or irregular delays between time of addition of reagents and time of measurement are of influence.

In this article a technical description will be given of the flow cytometer used to perform real-time kinetic studies, including the software necessary to acquire and analyze data. This is followed by a description of two applications of on-line FCM. First, the real-time analysis of chromatin structure changes by enzymatic digestion of the DNA is shown. Changes in the fluorescence intensity of bound DNA stains occurred within minutes. These experiments therefore demonstrate the applicability of on-line flow cytometry to study fast kinetic changes. Second, an assay is described for measuring uptake kinetics of anthracycline drugs. Uptake profiles and changes therein after addition of specific membrane transport inhibitors can show the presence of typical multidrug-resistant cells. This assay is much more accurate and sensitive than classical FCM.

II. Instrumentation

A. On-Line Flow Cytometer

Figure 1 is a flow diagram of an on-line flow cytometer. The cells (or particles) of interest are in a reaction tube, which is surrounded by a water jacket that can be maintained at a constant temperature (mostly 37°C). The cells are kept in suspension and the added reagents are mixed by a small magnetic stirrer. The stopping of the reaction tube has two inlets and one outlet. Adjustable air pressure on one of the inlets forces the cells to pass the tubing of the outlet toward the flowthrough cuvette of the flow cytometer. At any desired time reagents can be added to the cell suspension by the second inlet, using a syringe. From the very moment of addition, effects of the reagents on the measured parameter(s) can be studied. The time for cells to travel from the sample tube

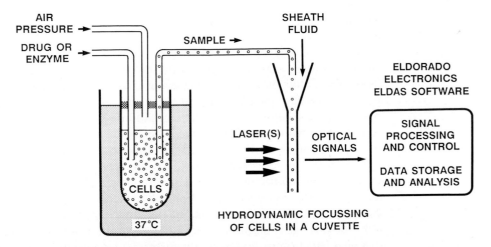

Fig. 1 A schematic representation of an on-line flow cytometer. Cells (or particles) are suspended in a thermostated tube and forced to the measuring point of the flow cytometer by means of air pressure. Additives can be mixed into the suspension at any time via an extra inlet in the reaction tube. Cells are kept in suspension and reagents are mixed by a magnetic stirrer.

to the measuring point of the flow cytometer is in the order of 5–20 sec, depending on the flow rate of the sample and the length of the sample-line tubing.

The RELACS-III flow cytometer is equipped with three lasers: two argon ion lasers (Coherent Innova 90-5, Palo Alto, CA, and Spectra Physics 2020/5, Mountain View, CA) and one helium-neon laser (Shanghai Institute of Laser Technology, Shanghai, PRC), thus enabling all standard excitation wavelengths to be used in single-, double-, or triple-wavelength excitation experiments. The flowthrough cuvette (Hellma, Müllheim, FRG) has a diameter of 250 μm. A maximum of eight parameters per cell can be measured.

B. Electronics

The electronics of the RELACS flow cytometers show two unique features. Neither is an absolute requirement for kinetic studies, but both do increase accuracy and ease of operation. The first of these features is a complete parallel processing system (van den Engh and Stokdijk, 1989). In a multiparameter flow cytometer, the incoming signals are handled most efficiently when each detector is equipped with its own electronics for pulse processing and analog-to-digital conversion (ADC). The input channels can then operate in parallel, and all signals can be processed simultaneously. The several pulses measured from one particle may reach the detectors at different times (multilaser measurements). The RELACS electronics allow for parallel processing of these in time-separated signals by storing the data temporarily in first-in/first-out (FIFO) buffers. After a particle has been seen by all detectors, the stored values are combined and transferred from the FIFO units over a bus to the acquisition computer. The data are transferred via a 16-bit parallel interface and stored under direct-memory access (DMA) in the computer memory. This allows for an effective transfer of 1 Mbyte of data per second, corresponding to a list-mode file of 60,000 four-parameter events. As a result of the scheme of parallel-pulse processing, the dead time of the system is independent of the number of parameters measured or the number and time separation of the excitation light beams. The instrument has a cycle time of 5 μsec, which corresponds to a maximal throughput rate of 200,000 events per second.

The second feature of the RELACS electronics is the measurement of general experimental conditions. During data sampling, eight DC values are read periodically (10 times per second) and placed in between the list-mode data. These eight DC channels can contain parameters like experiment time, sample temperature or pH, laser output power, and background fluorescence. The channels can also function as counters, for instance, to determine the total number of processed events. For the determination of the kinetics of fluorescence changes, the time marker is used. The time stamps within each block of data at the several time points of a measurement allow for the construction of uptake or release curves of the parameter of interest versus the actual measurement time. Experimental errors, like a temporary clogging, can be filtered out (also see Watson, 1987).

C. Software

The ELDAS (eight-parameter list-mode data analysis system) software package is composed and implemented to allow for acquisition and analysis of data from the RELACS flow cytometers. The ELDAS package consist of several modules capable of handling data acquisition (ACQUIS); analysis of list-mode data (ANNA); analysis and graphical representation of single-parameter histograms (HISTO), bivariate histograms (BIVAR), and kinetic data (KINET). The program further serves modules for standard operations as data storage and data management. The package is implemented on Hewlett–Packard (HP) 9000 microcomputers (series 200 and 300). Data acquisition is done with HP 9000-220 computers equipped with 15-20 Mbyte hard disks for immediate data storage. Analysis of data is done with a HP 9000-330 computer, equipped with a hard disk, a color monitor, a printer, and a plotter for graphical representation of data. At present, three acquisition and three analysis computers are connected in a network system with hard disk and tape backup storage. The software is written in HP BASIC, with all calculation-intensive subroutines in HP PASCAL.

1. Data Acquisition

The data acquisition module (ACQUIS) is used to sample data from the RELACS electronic bus and to write the data to disk in a list-mode file. For kinetic measurements the program asks in advance for the total duration of the experiment, the number of sample points (blocks) in this time (usually 35), the number of cells to be measured at each sample point (usually 2000), and the distribution of the sample points over the time (linear or logarithmical). Also, some experimental details like the number of parameters, a parameter identification, a description of the sample and the additives, and a description of the experimental parameters used (like time and sample temperature) are added as a text file to the measurement data.

After alignment of the flow cytometer, and adjustment of the amplification for all parameters, a kinetic measurement starts by sampling the first block of (generally) 2000 cells. This data set is used in the analysis to set windows and is a reference since no changes have occurred yet. Thereafter, a reagent is injected through the inlet. This can be the fluorescent dye of interest or some substrate which influences an already present fluorochrome. Upon addition, the kinetic acquisition program is activated. The program then immediately starts sampling data blocks according to the predefined specifications. It is also possible to set up sampling profiles for multiple addition of substrates, each time with its own distribution of the sample points over the measurement time period. The RELACS electronics put the eight experimental (DC) parameters on the bus 10 times per second. These are acquired by the computer like normal parameters and included in the list-mode file. All parameters carry an

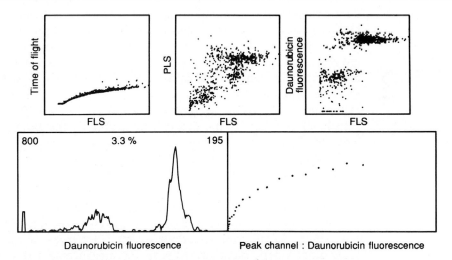

Fig. 2   Data display of the acquisition program while running a kinetic experiment. The displayed parameters of the three dot plots (top) and two histograms (bottom) can be freely chosen. Usually, histogram 1 (left) displays the data of the kinetic parameter. The peak position of this histogram is plotted in histogram 2 (right) at each time point, thus creating a provisional kinetic curve. The total number of cells in the histogram (800), the CV (3.3%), and the peak channel of the histogram (195) are plotted at the top of the histogram 1 panel.

identification code in their 16-bit data: 4-bit identification + 12-bit parameter value = 16-bit data. With these identification codes, DC parameters can be recognized apart from "normal" parameters.

Besides these basic acquisition features, the program also continuously displays the measured data. This allows for immediate interpretation of the experiment, and even more importantly, this serves as a quality control monitor. The computer displays three dot plots and two histograms (Fig. 2). The parameters of the dot plots and the histogram can be set or changed at any time (except during storage of data, which has always the highest priority). Usually, histogram 1 is used to display the parameter of interest; the peak channel of this histogram at each sample point can be plotted in histogram 2 (see Fig. 2). This shows already the kinetic plot during the measurement, although it is rather inaccurate (no windowing or averaging of data). The dot plots are refreshed after a preset number of events, and the histograms after each sample point.

2. Data Analysis

The analysis of the stored list-mode data is performed with the program module ANNA. Like all list-mode data analysis software, windows can be set on all parameters (even on the kinetic parameter if wanted). To obtain kinetic curves, the data are reduced to a set of values versus the time (sample points). That is, for each data block the cells in the window are selected and the values of the kinetic parameter are reduced to one value. The modal, mean, and median fluorescence can be computed together with the coefficient of variation (CV). During these calculations, logarithmically amplified parameter values can be recalculated to linear values or vice versa, based on the known amplifier characteristics. The data set of one value per time point thus obtained is stored in a new file, which can be used in the KINET module or transferred to a spread sheet program on a personal computer. It is also possible to store a complete histogram of the events in window per time point. The first time marker in a data block is taken as the measurement time of the time point. Excessively long sampling times for a particular time point (as will be the result of a clogging of the sample line) can be noticed, since the duration of sampling (= sample rate) is calculated using all time markers in a data block. These time points may be excluded from further analysis (Watson, 1987). The other stored experimental parameters can also be used to control the quality of the measurements.

Construction of kinetic curves is performed with the KINET program module. This allows for plotting (overlay) of data in curves of the fluorescence (or some other kinetic parameter) versus the measurement time (see Fig. 3 for an example). The program allows for scaling of both axes, plotting of several curves in one graph (see Fig. 5), and plotting of histograms of selected time points (see Fig. 6). Fitting of kinetic curves has not been implemented yet. The data structure, however, is well suited for these calculations.

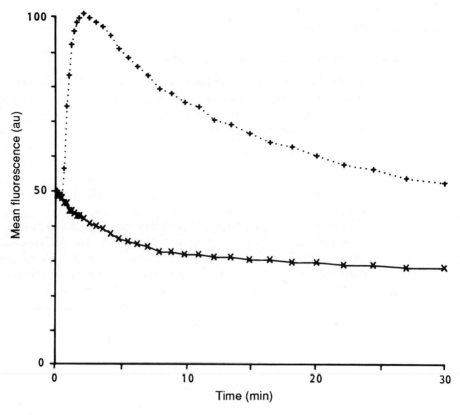

Fig. 3 DNase I digestion of mouse thymocytes: mean fluorescence of AO-stained (7.5 μg/ml) thymocytes as a function of time after addition of 10 μg/ml DNase I. The laser was set to emit 488-nm light (500 mW). Red fluorescence (+) was filtered through two RG610 long-pass filters (Schott). Green fluorescence (×) was filtered through a 530 ± 10-nm band pass filter (Melles Griot).

D. Standardization

We calibrate our flow cytometer with fluorescent microspheres (Fluoresbrite, Polysciences, Inc., Warrington, PA). Fixed laser powers are used in all experiments (set in light control modes). Within one type of experiment the same photomultipliers (PMT), filters, and (pre)amplifiers are used for each parameter each time. Using fixed amplifier settings, the PMT voltages are adjusted to measure the mean peak of the particle signals in the same channel number. This way, measurements performed on different days can be compared. The CV of the fluorescence histogram is calculated (see Fig. 2, bottom left, histogram 1) and is used as a quality control for the alignment of the flow cytometer.

III. Applications

A. Real-Time Analysis of Chromatin Structure Changes

DNA *in situ* is highly organized with nuclear proteins bound to it. The local structure of this chromatin is a function of gene activity and differentiation state of the cell. Chromatin structure can be investigated by deoxyribonuclease I (DNase I) digestion of the DNA. The sensitivity of DNA for DNase I appears to depend on the chromatin structure and is increased by damage caused by ionizing radiation. To probe for DNase I sensitivity, the DNA can be first stained to equilibrium with DNA-specific fluorochromes. Changes in DNA stainability are then measured during the enzymatic digestion (Roti Roti *et al.*, 1985).

We have attempted to find a relationship between the different binding modes of several DNA binding dyes and changes in the fluorescence of these dyes during digestion of the DNA with DNase I. Darzynkiewicz (1979) showed that digestion of DNA can be determined by monitoring the fluorescence of the DNA binding dye acridine orange (AO). The presence of AO had no effect on the digestion of the DNA by DNase I. Therefore, the measured differences in fluorescence were mainly attributed to DNase I activity and not to unwinding of the two DNA strands by the fluorochrome. Roti Roti *et al.* (1985) showed that at the time of DNase I addition to the cells there already was an increase in fluorescence. We attributed this to a delay between sampling and measuring of the nuclei. Therefore, we have chosen to perform the study of chromatin structural changes with on-line FCM, allowing for real-time measurements.

Fluorochromes with different binding modes to the DNA have been tested. AO has two binding modes. It emits green fluorescent light upon intercalation in the double helix. This requires the stacking of several dye molecules together (Kapuscinski *et al.*, 1983). AO emits red light upon external binding to single- and double-stranded nucleic acids. Propidium iodide (PI) and ethidium bromide (EB) intercalate in the DNA. Hoechst 33258 (HO258) and Hoechst 33342 (HO342) bind externally, inside the minor groove where they generally are hydrogen bonded to four adjacent AT base pairs (Teng *et al.*, 1988).

Mouse thymocytes were used in this study because these cells represent a population with uniform DNA content and chromatin structure. Thymocytes were fixed by the use of formaldehyde and ethanol. RNA was removed with RNase A and cells were stained with one of the fluorochromes. Six million cells in 3 ml DNase I buffer were measured with the RELACS-III flow cytometer. The sample rate was 500 cells per second and the temperature of the reaction mix was kept at 37°C. Besides dye fluorescence(s), forward light scattering and time-of-flight (TOF; scatter pulse width) were also measured. The TOF parameter was used in the data analysis to remove doublets of thymocytes. DNase I (10 μg/ml = 720 units/ml final concentration) was added to the reaction mix at time = 0.

In Fig. 3, the kinetics of changes in AO fluorescences are shown. At each time point the mean red and green fluorescence was computed and plotted versus the time after addition of DNase I. Within 2 min, the red fluorescence (external binding) of the thymocytes increased > twofold, whereas the green fluorescence (intercalative binding) decreased at the same time. After 2 min the red fluorescence also decreased. These steadily decreasing fluorescence signals are caused by the breakdown of the DNA by prolonged DNase I activity.

Fluorescence from PI and EB showed the same sudden increase as the red AO fluorescence, although the effect is somewhat smaller (20% increase). In Fig. 4, PI fluorescence histograms of DNase I-treated thymocytes are plotted versus the measuring time. At the maximum mean fluorescence intensity, discrete subpopulations of cells with original and with increased fluorescence were

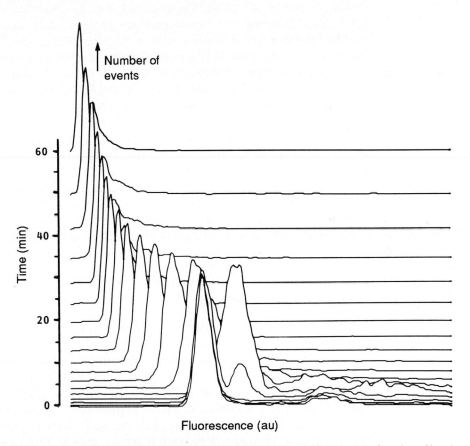

Fig. 4 DNase I digestion of mouse thymocytes: PI fluorescence histograms as function of incubation time after addition of 25 μg/ml DNase I. Thymocytes were stained with 10 μg/ml PI. The laser was set to emit 488-nm light (500 mW). PI fluorescence was filtered through a KV550 long-pass filter (Schott). The heights of the peaks indicate the number of events.

measured, indicating that the process of DNase I digestion is fast and complete within one nucleus. When HO258 or HO342 was used as the DNA stain, DNase I digestions showed a directly decreasing fluorescence intensity (data not shown).

The fluorescence maximum proved to be independent of the DNase I concentration. The time necessary to reach the maximum fluorescence level was longer with lower DNase I concentrations. Concentrations of as low as 0.1 μg/ml DNase I gave a 20% increase in PI fluorescence after 15 min. After washing away the histones with 0.1 N HCl, the initial PI fluorescence was 40% higher. When these pretreated cells were digested with DNase I, there was still an additional increase (10%), indicating that HCl only washes away proteins, leaving most of the DNA intact.

DNase I produces single-strand breaks potentially at 10-base pair intervals (Sollner-Webb *et al.*, 1978). These so-called nicks allow for unwinding of the DNA, thus enabling more fluorochromes to intercalate into the DNA. We attribute the increase in fluorescence of the intercalating dyes (PI, EB) to this effect. The denaturation of DNA decreases the number of external binding sites for the Hoechst dyes HO342 and HO258. These dyes require an intact minor groove for binding. The unwinding of the DNA also decreases the possibility of the stacking of AO molecules in the DNA helix, resulting in a decreased green fluorescence. The increase in red AO fluorescence, caused by external binding to single- and double-stranded nucleic acids, must be attributed to the unwinding of the DNA helix and the increased single-strandedness, resulting in more accessible external binding sites for AO.

Roti Roti *et al.* (1985) showed that the increase in EB fluorescence can also be obtained after large amounts of irradiation. Since chromatin structure and accessibility of DNA is very important in the distribution of radiation damage (Chiu *et al.*, 1982), this measuring system can give further insight into the relation between chromatin structure and radiation sensitivity.

B. Studies with Multidrug–Resistant Cells

Mammalian cell lines which have been selected *in vitro* for resistance to cytotoxic drugs like the *Vinca* alkaloids or the anthracycline antibiotics, doxorubicin and daunorubicin, display the so-called multidrug-resistant (MDR) phenotype (for reviews on MDR, see van der Bliek and Borst, 1989; Nooter and Herweijer, 1991). A striking feature of the classical MDR phenotype is its reduced ability to accumulate drugs, as compared to the parent cell lines. This reduced drug accumulation is most likely the main cause of MDR. It is assumed that the reduced drug accumulation is due to activity of an energy-dependent unidirectional drug efflux pump with broad substrate specificity. This drug pump is composed of a transmembrane glycoprotein (P-glycoprotein) with a molecular weight of 170 kDa, which is encoded by the so-called multidrug resistance (*mdr1*) gene. P-Glycoprotein uses energy in the form of ATP to transport drugs through a channel formed by the transmembrane segments.

There is accumulating evidence that MDR can also occur in human cancer (Goldstein *et al.*, 1989; Herweijer *et al.*, 1990). (For a review on the clinical relevance of the MDR phenotype in human cancer, see Nooter and Herweijer, 1991.) This is potentially of therapeutic importance, since a variety of substances (the so-called P-glycoprotein inhibitors) have been identified that can overcome MDR in *in vitro* model systems (Tsuruo *et al.*, 1982; Nooter *et al.*, 1989). The current hypothesis about reversing MDR for the majority of these agents is that they have an affinity for the intracellular P-glycoprotein drug efflux pump and thus can compete for outward transport and thereby restore the cellular cytotoxic drug accumulation. Since in model systems P-glycoprotein inhibitors have also been able to overcome MDR *in vivo*, agents are available now with which clinical trials can be started in typical MDR patients. Therefore, it is of great importance to determine the occurrence of MDR in human cancer. Thus, accurate, sensitive, and rapid assays for the detection of MDR cells are highly desirable. We here describe a FCM assay for the detection of MDR cells in heterogeneous cell populations that is based on the ability of MDR cells to respond to RM (Herweijer *et al.*, 1990).

1. Detection of MDR Cells

The human ovarian carcinoma A2780 anthracycline-sensitive (A2780/S) and two anthracycline-resistant A2780 cells (A2780/T10 and A2780/T100) were used as *in vitro* model systems for MDR. The A2780/T10 and A2780/T100 cell lines were made drug resistant by transfecting the drug-sensitive parent line with an expression vector containing a full length cDNA of the human *mdr1* gene and subsequent selection with the anthracycline drug doxorubicin at concentrations of 10 and 100 ng per ml tissue culture medium, respectively. The A2780/T10 and A2780/T100 cell lines have resistance factors relative to the parent line of 9 and 15 and relative *mdr1* mRNA expression levels of 50 and 150 arbitrary units, respectively (Herweijer *et al.*, 1990).

Daunorubicin uptake kinetics have been determined for both types of cells. The daunorubicin content of individual cells can be measured by FCM. The method makes use of the fluorescent properties of the anthracycline drugs (Nooter *et al.*, 1983). Upon excitation with 488-nm laser light (close to the absorption maximum), anthracyclines fluoresce, with an emission maximum around 600 nm.

For the kinetic measurements, the cells were suspended in Hepes-buffered Hanks' balanced salt solution (HHBSS) at a final concentration of 2×10^5 cells/ml. Uptake kinetics of daunorubicin (2 μM final concentration, added in 2–8 ml of cell suspension) were determined over a period of 90 min. Besides anthracycline fluorescence, forward and perpendicular light scattering were measured. Daunorubicin fluorescence was filtered through a KV550 long-pass filter (Schott, Mainz, FRG) and measured with a S20-type PMT (Thorn EMI, Fairfield, NJ). The signal was logarithmically amplified (4 decades full scale). Scattering light was filtered through 488-nm band pass filters (Melles Griot, Irvine, CA) and linearly amplified. Dead cells were identified by counterstain with the nonvital dye Hoechst 33258.

Daunorubicin net uptake curves of A2780/S and A2780/T10 and A2780/T100 cells are plotted in Fig. 5. These curves were computed from the list-mode data

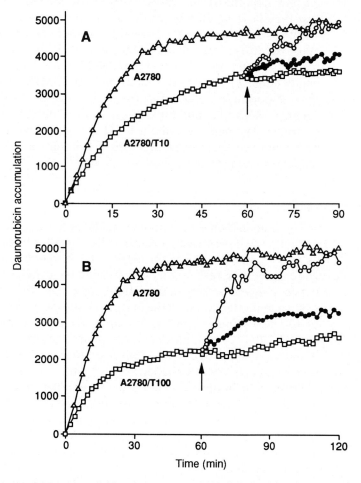

Fig. 5 Daunorubicin accumulation (expressed as fluorescence intensity in arbitrary units) by *mdr1*-transfected cell lines. At time zero, daunorubicin (final concentration, 2 μM) was added to the cell suspension. Arrows indicate the time point of addition of cyclosporin, [final concentration, 3 μM (o)], verapamil [10 μM (●)], or medium (□). Results with two *mdr1*-transfected cell lines are shown. (A) A2780/T10 and (B) A2780/T100. For comparison, the accumulation curve of the sensitive A2780 cell line (△) is included in both graphs. Appropriate amounts of cyclosporine can restore the lowered drug accumulation of both transfectants to the accumulation level of the A2780 cells.

by recalculating the logarithmically measured value to the corresponding linear value (which is possible since the amplification characteristics of the amplifiers are known) and, subsequently, averaging all values at each time point. It is clear that steady-state kinetics are reached after approximately 30–60 min. The steady-state level daunorubicin accumulation in the resistant A2780/T10 and /T100 cells is decreased by, respectively, 25 and 50%, as compared to the drug-sensitive cells. The typical MDR phenotype of the resistant A2780 cells was demonstrated by on-line addition of P-glycoprotein inhibitors. Figure 5 shows the effects of the addition of two different P-glycoprotein inhibitors, cyclosporine or verapamil, to the cells after 60 min of daunorubicin accumulation. The addition of 3 μM cyclosporin to either resistant cell line resulted in an increase of the intracellular accumulation to the level reached by the parent cell line. The addition of 10 μM verapamil only resulted in an increase to approximately 30% of this level; higher concentrations resulted in a larger increase. However, these higher concentrations were highly toxic to the cells, as detected by alterations in scattering signals and rapid accumulation of Hoechst 33258 dye. Addition of the P-glycoprotein inhibitors to the sensitive A2780/S cells did not result in changes in the daunorubicin accumulation.

To obtain an indication of the sensitivity of the on-line FCM assay for the detection of MDR cells, mixing experiments were performed (Herweijer *et al.*, 1989). Sensitive A2780/S and resistant A2780/T100 cells were mixed in different ratios, and subsequently the daunorubicin uptake kinetics and modulation by cyclosporine were measured. Plotting of the fluorescence histograms measured at several time points after addition of daunorubicin allowed for the detection of small subpopulations of MDR cells. As few as 2.5% MDR cells could readily be detected in the mixture of 2780/S and 2780/T100 cells (Fig. 6).

2. Drug Accumulation by Leukemic Cells

It is a frequent observation that leukemic cells obtained from patients with acute nonlymphocytic leukemia (ANLL) overexpress the *mdr1* gene (Nooter and Herweijer, 1991). However, from a therapeutic point of view it is important to demonstrate that *mdr1* overexpression in leukemic cells is associated with decreased drug accumulation that can be restored by P-glycoprotein inhibitors. To establish the MDR phenotype of ANLL cells expressing *mdr1*, estimated by mRNA levels (Nooter *et al.*, 1990), we monitored samples of a large group of ANLL patients for drug accumulation *in vitro* and the effects of P-glycoprotein inhibitors. In the data analysis, the leukemic cells were selected from the total cell population on the basis of their light scattering characteristics (Nooter *et al.*, 1983; Visser *et al.*, 1980). In Fig. 7 a typical example is shown of ANLL patients with high mRNA *mdr1* expression (in this case 105 arbitrary units in an RNase protection assay). Addition of saline had no effect on the accumulation of daunorubicin by leukemic blast cells, whereas that of cyclosporine was clearly increased. A much smaller increase resulted with verapamil. The

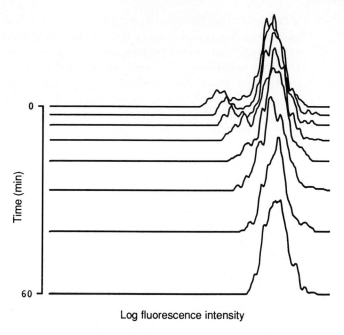

Fig. 6 Histograms of daunorubicin fluorescence intensity (logarithmic scale) of a mixture of 2780/S and 2780/R cells (24:1) measured at several drug exposure times, plotted in front of each other. Cells were incubated with daunorubicin (2 μM) at 37°C for 60 min, prior to addition of 10 μM cyclosporin A at time zero. Daunorubicin accumulation was measured for another 60 min.

upper curve in the figure represents the drug accumulation by ANLL cells from the same patient, at the time of diagnosis when the leukemic blasts had no detectable *mdr1* expression. No effects were measured after addition of cyclosporine or verapamil. These results indicate that expression of *mdr1* mRNA in ANLL cells can result in an active outward drug pump that can be inhibited effectively by clinically achievable concentrations of cyclosporine.

3. Energy Dependence of MDR Phenotype

It is generally assumed that the membrane transport of anthracycline drugs occurs as a "leak and pump" system, the leak being inward diffusion of non-ionized drug molecules and the pump, an active efflux. In 1973 Dano presented evidence for an active outward drug transport as the basis for cytotoxic drug resistance. With the technique of on-line FCM the ATP dependence of the MDR phenotype can be demonstrated very elegantly. The living cell can make ATP by glycolysis and by oxidative phosphorylation. In the next experiments glucose-free tissue culture medium in combination with the metabolic inhibitor sodium azide was used for depleting MDR cells of ATP.

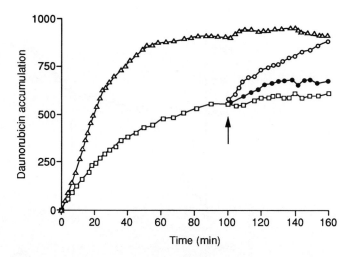

Fig. 7 Daunorubicin accumulation (expressed as fluorescence intensity in arbitrary units) by leukemia cells *ex vivo* from an ANLL patient. The upper curve was made at the time of diagnosis, when the leukemic blasts had no detectable *mdr1* expression. The other drug accumulation curves were made with leukemic blasts, expressing relatively high levels of *mdr1*, obtained from the same ANLL patient after intensive chemotherapeutic treatment, at the time that the patient was refractory to further chemotherapeutic treatment. For further details, see legend to Fig. 5. Cyclosporine can restore the lowered drug accumulation of *mdr1* expressing ANLL cells.

A2780/T100 cells were preincubated for 30 min in glucose-free medium and at time zero daunorubicin (2 μM final concentration) was added and the intracellular drug accumulation was monitored (Fig. 8A). After about 45 min, steady-state drug accumulation was reached. In order to inhibit the oxidative phosphorylation, at time 60 min, sodium azide (10 mM final concentration) was added to the incubation medium, resulting in a dramatic increase of the intracellular daunorubicin accumulation. The obvious explanation is that by incubation in glucose-free medium followed by the addition of sodium azide the cells are strongly depleted of ATP, resulting in a blockage of the efflux pump which finally leads to enhanced net drug uptake. Subsequently, the corollary experiment was performed. To the ATP-deprived cells, energy was added in the form of glucose (10 mM final concentration) (at time 90 min) and a dramatic decrease in drug accumulation was seen, with a steady-state level that is even lower than the one reached in glucose-free medium. Apparently, sufficient ATP can be generated by glycolysis in the presence of sodium azide to provide energy for the reactivation of the active efflux process.

Comparable experiments were performed with *mdr1* expressing leukemia cells from a patient with refractory ANLL (Fig. 8B). The cells were preincubated for 30 min in glucose-free medium and at time zero daunorubicin was added and the intracellular accumulation was monitored. The addition of azide,

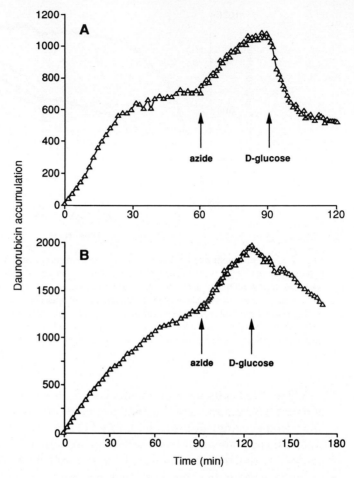

Fig. 8 Daunorubicin accumulation (expressed as fluorescence intensity in arbitrary units) by *mdr1*-transfected A2780 cells (A) and *mdr1* expressing leukemia cells *ex vivo* from a patient with refractory ANLL (B) in the absence and presence of energy. The cells had been preincubated for 30 min. in glucose-free medium. At time zero, daunorubicin (final concentration, 2 μM) was added to the cell suspension. Arrows indicate time points of addition of sodium azide (final concentration, 10 mM) and glucose (final concentration, 10 mM).

at time 90 min, resulted in an increased net uptake of daunorubicin while the addition of glucose abolished this effect.

Acknowledgments

Funds for the development of the RELACS electronics and ELDAS software were provided by the Dutch research organizations TNO and ZWO. The multidrug-resistance research program is supported by the Dutch Cancer Society, Grant No. ITRI 91-05.

References

Chiu, S. M., Oleinick, N. L., Friedman, L. R., and Stambrook, P. J. (1982). *Biochim. Biophys. Acta* **699**, 15–21.

Dano, K. (1973). *Biochim. Biophys. Acta* **323**, 466–483.

Darzynkiewicz, Z. (1979). *In* "Flow Cytometry and Sorting" (M. R. Melamed, P. F. Mullaney, and M. L. Mendelsohn, eds.), pp. 285–316. Wiley, New York.

Goldstein, L. J., Galski, H., Fojo, A. T., Willingham, M., Lai, S. L., Gazdar, A., Pirker, R., Gottesman, M. M., and Pastan, I. (1989). *J. Natl. Cancer Inst.* **81**, 116–124.

Herweijer, H., van den Engh, G. J., and Nooter, K. (1989). *Cytometry* **10**, 463–468.

Herweijer, H., Sonneveld, P., Baas, F., and Nooter, K. (1990). *J. Natl. Cancer Inst.* **82**, 1133–1140.

Kapuscinski, J., Darzynkiewicz, Z., and Melamed, M. R. (1983). *Biochem. Pharmacol.* **32**, 3679–3694.

Nooter, K., and Herweijer, H. H. (1991). *Br. J. Cancer* **63**, 663–669.

Nooter, K., van den Engh, G. J., and Sonneveld, P. (1983). *Cancer Res.* **43**, 5126–5130.

Nooter, K., Oostrum, R., Jonker, R R., van Dekken, H., Stokdijk, W., and van den Engh, G. J. (1989). *Cancer Chemother. Pharmacol.* **23**, 296–300.

Nooter, K., Sonneveld, P., Oostrum, R., Herweijer, H., Hagenbeek, T., and Valerio, D. (1990). *Int. J. Cancer* **45**, 263–268.

Roti Roti, J. L., Wright, W. D., Higashikubo, R., and Dethlefsen, L. A. (1985). *Cytometry* **6**, 101–108.

Sollner-Webb, B., Melchior, W., and Felsenfeld, G. (1978). *Cell (Cambridge, Mass.)* **14**, 611–627.

Teng, M., Usman, N., Frederick, C. A., and Wang, A. H. J. (1988). *Nucleic Acids Res.* **16**, 2671–2690.

Tsuruo, T., Iida, H., Tsukagushi, S., and Sakurai, Y. (1982). *Cancer Res.* **42**, 4730–4733.

van den Engh, G. J., and Stokdijk, W. (1989). *Cytometry* **10**, 282–293.

van der Bliek, A. M., and Borst, P. (1989). *In* "Advances in Cancer Research" (G. F. vande Woude and G. Klein, eds.), pp. 165–203. Academic Press, San Diego.

Visser, J. W. M., van den Engh, G. J., and van Bekkum, D. W. (1980). *Blood Cells* **6**, 391–407.

Watson, J. V. (1987). *Cytometry* **8**, 646–649.

CHAPTER 33

Acid-Induced Denaturation of DNA *in Situ* as a Probe of Chromatin Structure

Zbigniew Darzynkiewicz

Cancer Research Institute
New York Medical College
Valhalla, New York 10523

I. Introduction

Free DNA in aqueous solution at physiological pH and ionic strength has double-stranded conformation. Its treatment with heat, acid, or alkali causes the two strands to separate. This is known as DNA denaturation, melting, or helix–coil transition and is the result of the destruction of the hydrogen bonding between the paired bases of the opposite strands. Intrinsic sensitivity of free DNA to denaturation depends exclusively on its guanine–cytosine to adenine–thymine (GC : AT) ratio, because the GC pair confers higher stability due to an additional hydrogen bond, in comparison with AT.

DNA in chromatin is stabilized by interactions with histones, other nuclear proteins, and nuclear matrix (Darzynkiewicz, 1986, 1990). Studies on the stability of DNA *in situ,* therefore, provide insight into chromatin structure, making it possible to discern the double-helix stabilizing interactions. The classical biochemical methods to study DNA denaturation in chromatin are based on measurements of ultraviolet (UV) light absorption changes (hypochromicity) during heating of the sample (Subirana, 1973). These optical methods require chromatin isolation, shearing, and solubilization. This destroys the higher orders of chromatin structure and limits application of such methods to investigations of DNA *in situ.* The newer calorimetric approach (Touchette *et al.,* 1986) yields data that are difficult to interpret, because the method does not allow discrimination between heat-induced destruction of nucleic acid–protein bonding (e.g., dissociation of the nucleosomal core histones from DNA, "melting" of nucleosomes, Darzynkiewicz and Carter, 1989) from interstrand bond breakage (DNA melting). Furthermore, the calorimetric method, which can only be applied to whole-cell populations in bulk, cannot be used to assess intercellular heterogeneity.

A flow cytometric (FCM) method for measurement of DNA *in situ* sensitivity to denaturation, developed in our laboratory (Darzynkiewicz *et al.,* 1975, 1977, 1979), is based on the metachromatic property of the dye acridine orange (AO). This dye, under certain conditions, can differentially stain double-stranded (ds) versus single-stranded (ss) nucleic acids (Darzynkiewicz and Kapuscinski, 1990). Namely, AO intercalates into dsDNA and, in this mode of binding, upon excitation in blue light emits green fluorescence. In contrast, AO complexes with ss nucleic acids are characterized by red luminescence. The mechanism of AO interaction with ds and ss sections of nucleic acids is illustrated in Fig. 1 and is described in more detail in Chapter 26 of this volume.

In the original method, stability of DNA *in situ* was assayed by subjecting permeabilized (ethanol-fixed) and subsequently RNase-treated cells to heat or acid, followed by staining with AO (Darzynkiewicz *et al.,* 1975, 1977). After partial denaturation of DNA by heat or acid, AO, as mentioned, stains nondenatured, dsDNA sections in green (maximum fluorescence at 530 nm), whereas dye interactions with denatured sections (ssDNA) result in red luminescence (maximum emission at 640 nm). Thus, the relative proportions of red and green luminescence of cells stained this way represent AO binding to the sections of the denatured and native DNA, respectively.

This method has been adapted to unfixed cell nuclei freshly prepared from the cell nuclei of solid tumors (Kunicka *et al.,* 1987, 1989). Because isolation of nuclei from solid tumors is more convenient than isolation of whole intact cells, the adaptation extends the applicability of the method to the clinical samples.

Extensive literature exists describing studies in which AO was used to measure DNA denaturation in solution (e.g., Ichimura *et al.,* 1971) and *in situ,* in

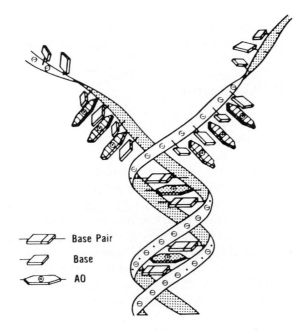

Fig. 1 Schematic representation of two major modes of AO binding to nucleic acids. At low dye concentration and low *D:P*, AO binds to ds nuclei acids by intercalation (lower part of the structure). When bound by intercalation, AO fluoresces green (dot-filled AO molecule). At increased *D:P*, AO binds also to ss sections of nucleic acids (upper part of the structure; AO molecules dash labeled). The high-density AO binding to ss sections (*D:P* = 1) results in charge neutralization of the DNA backbone which leads to condensation and agglomeration of the AO:ss nucleic acid product (not shown). In its condensed form, AO luminesces red (modified, after Darzynkiewicz and Kapuscinski, 1990).

metaphase chromosomes (e.g., Bobrow and Madan, 1973) or in nuclei of the permeabilized cells (Darzynkiewicz, 1990). Results of the *in situ* studies show great differences in DNA stability between various cell types, in cells in different phases of the cell cycle, in differentiated versus nondifferentiated cells, or even within individual chromosomes—in the latter case, reflected as chromosome banding. In general, DNA sensitivity to denaturation in cells is closely correlated with the degree of chromatin condensation: the more condensed the chromatin, the more unstable the DNA. So far, the sole exception to this rule was observed in DNA of cells undergoing spermatogenesis, where the late stages of chromatin condensation in normal spermatozoa were seen to be paralled by an increase rather than a decrease in DNA stability (Evenson *et al.*, 1980). However, during defective spermatogenesis, which leads to the development of infertile sperm cells, DNA in mature abnormal cells lacks stability, and the assay of DNA denaturation of sperm cells (Evenson *et al.*, 1980) has become a routine test of male infertility (see Chapter 10 of volume 42). DNA instability in abnormal spermatozoa, in analogy to chromatin changes during apoptosis, correlates with the presence of DNA strand breaks in these cells (Gorczyca *et al.*, 1993).

II. Application

This method can be applied in diverse studies designed to provide information about changes in nuclear chromatin. Altered sensitivity of DNA to denaturation, as mentioned, accompanies cell differentiation (e.g., Traganos *et al.*, 1979; Evenson *et al.*, 1980), parallels the cytotoxic effects of chemotherapeutic drugs (intercalators) on target tumor cells (e.g., Darzynkiewicz *et al.*, 1981), or precedes cell death induced by such different agents as tumor necrosis factor (TNF; Darzynkiewicz *et al.*, 1984) or caffeine (Kunicka *et al.*, 1990). In the latter case, the mechanism of cell death is most likely by apoptosis, and the nuclear changes typical of apoptosis (chromatin condensation) are manifested by markedly reduced stability of the DNA helix *in situ*. The method can, therefore, be used to make an early estimate of the proportion of cells undergoing apoptosis and be applicable in analysis of the cytotoxicity of a variety of antitumor drugs (Hotz *et al.*, 1992).

The variability in DNA stability to denaturation is also apparent when tumor cells are compared with normal stromal, or infiltrating, host cells, as well as between tumors of the same type but of different stages (Kunicka *et al.*, 1987, 1989). Further studies, however, are needed to evaluate the prognostic potential of this marker in the clinic.

So far the most common application of the DNA denaturability assay is in those studies that involve analysis of the cell cycle. DNA in mitotic cells is very sensitive to denaturation, as in quiescent cells characterized by condensed chromatin (Darzynkiewicz *et al.*, 1977, 1979). Conversely, the most resistant is DNA in late G_1 (G_{1B}) and early S phase cells. Thus, the method can discriminate cells in traditional phases of the cell cycle, including distinction of mitotic cells, and identify quiescent cells arrested in G_1 (G_{1Q}), S (S_Q), or G_2 (G_{2Q}) (Darzynkiewicz *et al.*, 1980). The possibility for rapid quantification of mitotic cells contributed to the application of this technique to score mitotic indices, especially in experiments in which cells are arrested in mitosis (Darzynkiewicz *et al.*, 1981). This was the basis for the development of the method which combines stathmokinesis and flow cytometry for analysis of cell-cycle kinetics and perturbations of the cycle by various drugs or physical agents. As is described in detail elsewhere (Darzynkiewicz *et al.*, 1986; Traganos and Kimmel, 1990), the stathmokinetic analysis allows one to estimate a variety of kinetic parameters, such as duration of individual phases of the cycle, the stochastic component of the cycle, and the points in the cycle at which the drug perturbs cell progression.

As in the case of differential staining of RNA versus DNA with AO (see Chapter 26 of this volume), restrictions related to specificity of DNA staining with this dye also apply to the present method. Namely, the specificity of AO to stain nucleic acids is not absolute, and the dye stains other polyanions as well. Therefore, following staining with AO, cells that contain large amounts of glycosaminoglycans or proteoglycans, such as normal fibroblasts, mast cells, chondrocytes, and differentiated keratinocytes, all have unacceptably high fluorescence unrelated to DNA. On the other hand, most cell lines proliferating in culture, especially of tumor origin, as well as lymphocytes, monocytes, leukemias, and lymphomas, or cells isolated from most solid tumors, exhibit good or at least adequate specificity of DNA stainability with AO. The degree of nonspecific luminescence can be estimated by control incubations of the cells with DNase prior to staining with AO. The assay based on isolated nuclei rather than whole cells can circumvent the problems of nonspecific stainability resulting from the presence of cytoplasmic constituents.

III. Materials

The following materials are needed for the assay of DNA denaturability in whole, fixed cells.

1. *Hanks' balanced salt solution (HBSS):* The solution should contain Mg^{2+} but no phenol red. Phosphate-buffered saline (PBS) containing 1 mM $MgCl_2$ may be used instead of HBSS.
2. *Fixative:* The cells are fixed in suspension in 70–80% ethanol. To identify apoptotic cells, preferably the cells are prefixed in 1% paraformaldehyde (methanol-free formaldehyde) dissolved in PBS, at pH 7.4 (see further).
3. *Stock solution of AO:* Dissolve AO in distilled water (dH_2O) to a final dye concentration of 1 mg/ml. Use only high-purity, chromatographically tested AO (available from Molecular Probes, Eugene, OR). The stock solution may be kept in the dark at 4°C for several months without deterioration.
4. *Solution A—0.1 M solution of HCl.*
5. *Solution B—AO staining solution:* Mix 90 ml of 0.1 M citric acid with 10 ml of 0.2 M Na_2HPO_4; the final pH of this buffer is 2.6. Add 0.6 ml of the AO stock solution (1 mg/ml) to 100 ml of this buffer; the final AO concentration is 6 μg/ml. Solution B is stable and can be stored for several months in the dark, at 4°C. Solutions A and B may be stored in automatic dispensing pipetter bottles set at 0.5 ml for solution A and 2.0 ml for solution B—the latter in a dark bottle.
6. *RNase A:* Use the DNase-free RNase. Less pure preparations require 2–3 min. heating at 100°C to inactivate DNase.

For DNA denaturability assay in isolated, unfixed cell nuclei, in addition to the materials described already, the following solution is needed.

7. *Nuclear isolation solution:* Prepare a solution containing 10 m*M* Tris buffer (pH 7.6), 1 m*M* sodium citrate, 2 m*M* MgCl$_2$, and 0.1% (v/v) nonionic detergent Nonidet P-40 (NP-40).

IV. Cell Preparation and Staining

A. Cell Fixation

1. *Cells growing in suspension, hematologic samples:* Rinse once with HBSS, and suspend in HBSS (10^6–10^7 cells per 1 ml).
2. *Cells growing attached to tissue culture dishes:* Collect cells by trypsinization and pool the trypsinized cells with cells floating in the medium (the latter consist mostly of detached mitotic and apoptotic cells). Rinse once with medium containing serum (serum is present to inactivate the trypsin; other means of trypsin inactivation such as addition of trypsin inhibitors may also be used), and suspend cells (10^6–10^7 cells per 1 ml) in HBSS.
3. *Cells dissociated from solid tumors:* Rinse free of any enzyme used for cell dissociation; suspend in HBSS.

In the final suspension in HBSS the cells should be well dispersed and not exceed a density of 10^7 cells per 1 ml. The cells are then fixed by admixture of 1 ml of this suspension into tubes containing 10 ml of fixative (80% ethanol, at 0–4°C).

In analysis of apoptotic cells, it is advisable to briefly prefix cells in paraformaldehyde. To this end, the suspension of cells in PBS (approx. 10^6 cells/ml) is admixed with 1% paraformaldehyde in PBS at pH 7.4, on ice. After 15 min the cells are rinsed with PBS and then fixed and stored in suspension in 80% ethanol, as described above. Prefixation on paraformaldehyde prevents extraction of the degraded, low MW DNA from apoptotic cells, and thus allows estimation of, at least in the early stages of apoptosis, the cell-cycle position, or DNA index, of apoptotic cells.

Cell storage in ethanol may vary from 12 hr to several weeks, at 4°C.

B. Cell Staining

1. Centrifuge the fixed cells. Suspend the cell pellet (10^6–10^7 cells) in 1 ml of HBSS. Add 5 Kunitz units of RNase A. Incubate at 37°C for 1 hr. Centrifuge and resuspend cells in 1 ml of HBSS.
2. Withdraw a 0.2-ml aliquot of cell suspension in HBSS and transfer it to a small (e.g., 5-ml volume) tube. Add 0.5 ml of 0.1 *M* HCl (solution A). After 30 sec add 2.0 ml of AO solution (B). Transfer this suspension to the flow cytometer and measure cell fluorescence. The fluorescence pattern remains stable for several hours.

Both treatment with HCl and staining with AO should be done at room temperature (RT, 20–24°C). Therefore, prior to use, solutions A and B should be adjusted to RT. DNA denaturation is incomplete when the solutions are cold.

C. Isolation and Staining of Unfixed Nuclei from Solid Tumors

This procedure has been modified after Kunicka *et al.* (1989).

1. Place the tissue in the nuclear isolation solution (Section III, step 7) and trim to remove the necrotic, fatty, and other undesirable areas. The tissues may be kept frozen (at -40 to $-60°C$) prior to nuclear isolation.

2. Transfer the trimmed tissue to a small volume of the fresh isolation solution and mince finely with scalpel or scissors. Mix small tissue fragments by vortexing and/or pipetting using a Pasteur pipette or syringe with a large-gauge needle. Observe the release of nuclei by sampling the suspension and viewing it under the phase-contrast microscope. If the release of nuclei is inadequate, transfer the minced tissue, suspended in the isolation solution, into a homogenizer with a glass or Teflon pestle. Homogenize by pressing the pestle several times; check the efficiency of nuclear isolation under a phase-contrast microscope. The nuclei should be clean, lacking cytoplasmic tags, yet unbroken.

Collect the nuclear suspension from above the remaining tissue fragments (allowed to sediment for ~ 1 min) with a Pasteur pipette, and filter through a 40- 60-μm pore nylon mesh. There should be no more than 10^7 nuclei per 1 ml of the isolation solution. All isolation steps should be done on ice.

3. Add 5 Kunitz units of RNase A per up to 10^7 nuclei in 1 ml of the solution. Incubate for 30 min at 24°C.

4. Withdraw a 0.2-ml aliquot of nuclei suspension from the RNase A incubation medium and treat with 0.5 ml of 0.1 M HCl for 30 sec at 24°C. Add 2.0 ml of AO solution (B) at 24°C. Transfer this suspension to the flow cytometer and measure the fluorescence. Since the material is unfixed, the fluorescence pattern is unstable and the nuclei should be measured shortly (up to 10 min) after addition of solution B.

V. Critical Aspects of the Procedure

Differential staining of dsDNA versus ssDNA at equilibrium with AO requires a proper concentration of free (unbound) dye in the final staining solution and during the actual measurement, that is, at the moment of cell intersection, in flow, with the laser beam. The problems of variability of AO concentration due to an excess number of cells or different geometry of flow channels ("dye

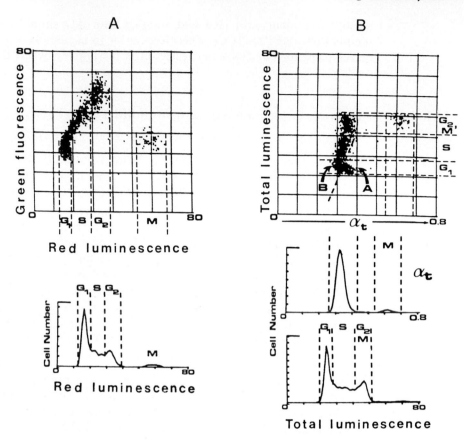

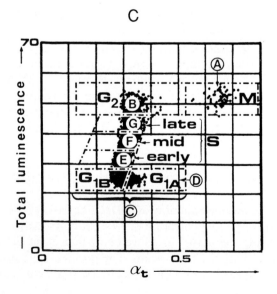

diffusion problems''), pertinent to the present method, are common to other methods utilizing AO and are discussed in detail in the description of the DNA and RNA staining method, in this volume (Darzynkiewicz, Chapter 26).

To initiate this methodology, one is advised to start with a cell population enriched in mitotic cells, for example, by staining exponentially growing cells treated for 2–3 hr, with a mitotic inhibitor such as colcemid or vinblastine. It is then easier to identify the mitotic cell population, which should differ from the interphase cells by increased sensitivity of DNA to denaturation. Under proper conditions of cell staining, luminescence excitation, and separation of the emission spectra, the results should be similar to those shown in Fig. 2. It is important, in particular, to obtain the mitotic cell cluster positioned on a diagonal axis with G_2 cells (on green vs red luminescence display, Fig. 2A) and thus having the same total luminescence (but different α_t; see legend to Fig. 2) compared to G_2 cells, as evident on the total luminescence versus α_t displays (Fig. 2B). If DNA denaturation is inadequate (e.g., in situations where the AO concentration is too low, as may be the case when the dye diffuses extensively from the cells to the flow sheath in some sorting channels), cells in mitosis are less separated from G_2 cells than shown in Fig. 2. Increased AO concentration in the staining solution compensates for the diffusion effects. On the other hand, if DNA denaturation is excessive, mitotic cells have a disproportionally high red, and an extremely low green, luminescence.

In essence, thus, the most critical points of the procedure relate to the proper choice of AO concentration and adequate separation of the green and red emission components.

VI. Standards

The essential point of the standardization is to adjust: (i) concentration of AO, and (ii) the respective green and red fluorescence detection photomultipliers' sensitivities (voltage settings) in such a way that, in the case of those cells in which DNA is extensively denatured (e.g., M cells), the increase in red luminescence, as a result of DNA denaturation, is numerically similar (increase in n number of channels) to the decrease in green fluorescence (decrease by a

Fig. 2 Luminescence of exponentially growing L1210 cells after removal of RNA, partial denaturation of DNA by acid, and subsequent staining with the metachromatic fluorochrome AO. The dsDNA stains green, while denatured ssDNA stains red. (A) Distribution of cells with respect to green versus red luminescence. (B) Total luminescence (red plus green) and α_t (red/total luminescence) of individual cells from the same culture as shown in (A). After transformation of the data the G_2 and M cell clusters are aligned along the horizontal line. Single-parameter frequency histograms of red luminescence, α_t, and total luminescence are shown below the scattergrams in (A) and (B). (C) Discrimination of various cell populations by gating analysis, based on differences in total luminescence and/or α_t. See discussion in text.

similar number of channels). Thus, for cells with the same DNA content but differing in DNA sensitivity to denaturation (e.g., G_2 vs M), the sum of red and green luminescence intensities (total luminescence; see Fig. 2) should be the same and be representative of the total DNA content.

Nonstimulated human peripheral blood lymphocytes can be used as a convenient standard that allows one to compare the measured cells, for instance, in tumor cell analysis. A batch of fixed lymphocytes can be stored at 4°C for several weeks without changing their DNA sensitivity to denaturation. These cells may then be used as an external and/or internal reference standard to be compared with the α_t value of the measured cell population. Under proper conditions of cell staining and filter selection, the photomultiplier sensitivity settings of the red- and green-channel photomultipliers should then be routinely set to obtain mean numerical values of red and green luminescence of the lymphocyte population equal to each other; hence the mean α_t value of lymphocytes is designated to be 0.5.

VII. Instruments

Specifications of the instrumentation that may be of relevance in measurements of AO luminescence are provided in Chapter 26 of this volume, describing the method for differential staining of DNA and RNA with this dye. Briefly, the optimal excitation of AO is in blue light, at 460–500 nm. Because following DNA denaturation the red luminescence of cells is generally rather strong, there is no need to try using shorter wavelengths for excitation, as was stressed in the case of weak red luminescence measurements in cells or cell nuclei having low RNA content. Thus, the 488-nm line of the argon ion laser can be used, and blue excitation filters such as BG 12 can be applied in instruments illuminated by a mercury lamp, for all samples. The exception are sperm cells which have a low overall fluorescence and may require a shorter wavelength of excitation, e.g., when using such instruments as the ICP-22 (Ortho Diagnostics, Westwood, MA) or PAS-III (PARTEC GmbG, Munich, Germany). The combination of emission filters and dichroic mirrors should separate green fluorescence at 530 ± 15 nm and red luminescence at >630 or 640 nm.

The geometry of the flow channel that affects the staining pattern due to diffusion of AO from cells to the flow sheath and methods that can compensate for the diffusion are described in Chapter 26 of this volume.

The data can be recorded in a list-mode fashion either as red and green luminescence intensities of individual cells or as total cell luminescence (red plus green) and the ratio of red to total luminescence, so-called α_t (Darzynkiewicz *et al.*, 1975). In the latter case, the analog signals from green and red photomultipliers can be added and then respectively divided by a specially designed electronic circuit board (e.g., such as used in some flow cytometers for compensation of the signals when the measured emission spectra overlap), so that the total luminescence and the α_t ratio are then digitized and recorded in the list mode.

It is also possible to develop software that transforms the originally recorded data expressed as red and green luminescence intensities of individual cells to the total luminescence and α_t values of the same cells, so the transformed file is subsequently analyzed.

VIII. Results

A. Exponentially Growing Cells

Figure 2 illustrates the typical distribution of exponentially growing cells with respect to their green and red luminescence after partial denaturation of DNA and staining with AO. The data are recorded either as bivariate distributions of green and red luminescence values (Fig. 2A) or total cell luminescence (green plus red) and α_t value (red luminescence/total luminescence) (Figs. 2B and 2C). Total cell luminescence represents total cellular DNA while α_t, which can vary from 0 to 1.0, is a measure of a portion of denatured DNA (Darzynkiewicz *et al.*, 1975). The relative intensities of red and green luminescence for each cell correlate with the extent of DNA denaturation, which, in turn, reflects the degree of chromatin condensation. Mitotic cells (M) exhibit maximal denaturation of DNA and can be easily distinguished from interphase cells on the basis of their high red and decreased (with respect to G_2 cells) green luminescence (Fig. 2A). The optimal adjustment of the staining conditions (primarily AO concentration and optical filters) and photomultiplier sensitivity settings should ensure that the imaginary line connecting M and G_2 cell clusters is diagonal (i.e., at a 45° angle with respect to red and green luminescence coordinates), and M cells have green fluorescence similar to that of G_1 cells.

The following populations can be discriminated by gating analysis, based on differences in total luminescence and/or α_t (Fig. 2C):

1. Cells in mitosis (M) have the highest α_t.
2. Cells in G_2 + M form a typical G_2 + M peak in total luminescence (DNA content) histograms: After subtracting M cells, the number of G_2 cells can be estimated.
3. G_{1A} cells are classified as having α_t significantly different from cells in early S phase. To this end, the gating window is at first located at the lowest quartile of the S population and the mean α_t and standard deviation (SD) from the mean values of these cells are established. The threshold dividing G_{1A} from the G_{1B} is then located on the α_t coordinate at the α_t value 2 SD above the mean α_t of these early S cells.
4. Gating windows can be located along the S phase cluster (total luminescence; DNA content) to identify cells in early, mid, and late S phase.

Identification of all these populations is needed for detailed analysis of stathmokinetic experiments, which yield kinetic data on rates of cell progression through different phases of the cell cycle (Darzynkiewicz *et al.*, 1986).

Based on differences in total cell luminescence it is possible to distinguish G_1, S, and G_2 + M cells, whereas differences in α_t allow one to subdivide the G_1 phase into G_{1A} and G_{1B} compartments. Cell residence times in G_{1A} have a characteristic, stochastic component (Darzynkiewicz *et al.*, 1980).

Staining of exponentially growing cells is most commonly utilized for the purpose of scoring cells in mitosis (e.g., to estimate mitotic indices in stathmokinetic experiments), to distinguish G_{1A} from G_{1B} cells, and, in the case when cells grow at two ploidy levels, to distinguish G_1 cells of higher ploidy (4C DNA content) from G_2 cells of lower ploidy. Although these cells (G_1 vs G_1) have the same DNA content, they differ in chromatin structure and, under optimal staining conditions, can be distinguished by differences in their α_t (see also Chapter 27 of this volume).

B. Quiescent versus Cycling Cells

Nonstimulated and mitogen-stimulated lymphocytes are examples of quiescent and cycling cells, respectively. Distribution of these cells with respect to their red and green luminescence after partial DNA denaturation and staining with AO is shown in Fig. 3. The changes in stainability are a reflection of

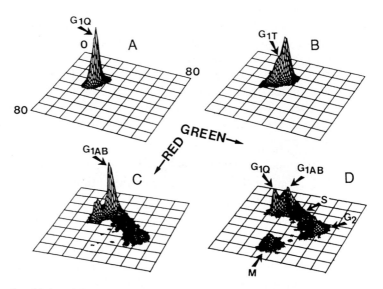

Fig. 3. Sensitivity of DNA *in situ* to acid-induced denaturation during stimulation of human lymphocytes. Nonstimulated (A) and phytohemagglutinin (PHA)-stimulated lymphocytes 18 hr (B) and 3 days (C,D) after addition of PHA. The last culture (D) was additionally treated with colcemid for 8 hr before being harvested to arrest cells in mitosis. After being stained with AO, the proportions of nondenatured and denatured DNA are represented by green and red luminescence, respectively. Subpopulations of quiescent cells (G_{1Q}) and cells during transition from quiescence to the cell cycle (G_{1T}, S, G_2, and M) can be distinguished as shown.

chromatin decondensation and cell progression through the cell cycle (DNA content increase) following mitogenic stimulation.

C. Nondifferentiated versus Differentiated Cells

Figure 4 illustrates stainability of Friend erythroleukemia cells growing exponentially (nondifferentiated) and differentiated. Their erythroid differentiation was induced by growth in the presence of dimethyl sulfoxide (DMSO) (Traganos *et al.*, 1979). Differentiated cells can be distinguished as having lower total stainability of DNA (total luminescence) and increased sensitivity to denaturation, compared to their exponentially growing counterparts. Differentiation of other cell types (e.g., HL-60 cells) may involve change in α_t but no change in total luminescence.

D. Identification of Apoptotic Cells

As mentioned, DNA in apoptotic cells is unstable, and following DNA denaturation by acid, apoptotic cells can easily be distinguished by their increased α_t (Hotz *et al.*, 1992). Figure 5 shows changes in DNA sensitivity to denaturation

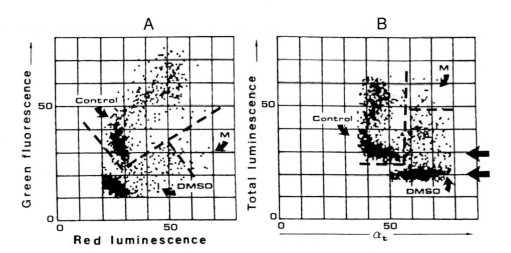

Fig. 4 Sensitivity of DNA *in situ* to acid-induced denaturation in exponentially growing (control) and differentiated (DMSO) Friend erythroleukemia cells. The cells were grown in the presence of 1.5% DMSO for 6 days, which resulted in their terminal erythroid differentiation. Cells from the control and DMSO-treated cultures were mixed in a 1 : 1 proportion, fixed, and stained as described in the text. Positions of cell clusters, as marked by arrows, were identified from the unmixed populations. M, cells in mitosis from the exponentially growing culture. (A) Red versus green luminescence; (B) total luminescence versus α_t. In (B), the positions of G_1 clusters are identified by thick arrows. Control cells have lower α_t and higher total luminescence; differentiated cells have decreased total luminescence and elevated values of α_t.

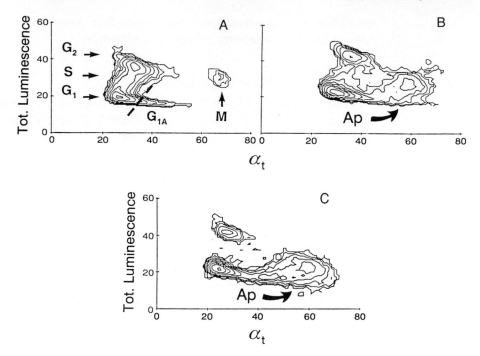

Fig. 5. Identification of apoptotic cells based on the increased DNA *in situ* sensitivity to denaturation. Exponentially growing, untreated HL-60 cells (A) and cells exposed to 0.15 μM of DNA topoisomerase I inhibitor camptothecin (an agent which preferentially triggers apoptosis of S phase cells) for 140 min (B) and 180 min (C). Apoptotic cells (Ap) are characterized by the increased α_t, which is almost as high as the α_t of M cells in the untreated culture.

in HL-60 cells undergoing apoptosis as a result of their exposure to the topoisomerase I inhibitor camptothecin: this drug preferentially triggers apoptosis of cells progressing through S phase (Del Bino *et al.*, 1992). The results shown in Fig. 5 clearly demonstrate the applicability of this method in identification of apoptotic cells. It should be stressed, however, that as with several other methods used to distinguish apoptotic cells, the specificity of cell death by apoptosis should be confirmed by additional means, such as analysis of the *in situ* DNA strand breaks, integrity of plasma membrane, or gel electrophoresis of DNA isolated from these cells (See Chapter 2 of this volume).

Acknowledgments

Supported by NCI Grant R01 CA28704, the "This Close" Foundation, the Carl Inserra Fund, the Chemotherapy Foundation, and the Dr. I Fund Foundation. I thank Dr. Frank Traganos for his comments and suggestions.

References

Bobrow, M., and Madan, K. (1973). *Cytogenet. Cell Genet.* **12**, 145–153.

Darzynkiewicz, Z. (1986). *Int. Encycl. Pharmacol. Ther.* **121**, 1–98.

Darzynkiewicz, Z. (1990). *In* "Flow Cytometry and Sorting" (M. R. Melamed, T. Lindmo, and M. L. Mendelsohn, eds.), 2nd ed., pp. 315–340. Wiley-Liss, New York.

Darzynkiewicz, Z., and Carter, S. P. (1989). *Exp. Cell Res.* **180**, 551–556.

Darzynkiewicz, Z., and Kapuscinski, J. (1990). *In* "Flow Cytometry and Sorting" (M. R. Melamed, T. Lindmo, and M. L. Mendelsohn, eds.), 2nd ed., pp. 291–314. Wiley-Liss, New York.

Darzynkiewicz, Z., Traganos, F., Sharpless, T., and Melamed, M. R. (1975). *Exp. Cell Res.* **90**, 411–428.

Darzynkiewicz, Z., Traganos, F., Sharpless, T., and Melamed, M. R. (1977). *Cancer Res.* **37**, 4635–4640.

Darzynkiewicz, Z., Traganos, F., Andreeff, M., Sharpless, T., and Melamed, M. R. (1979). *J. Histochem. Cytochem.* **27**, 478–485.

Darzynkiewicz, Z., Traganos, F., and Melamed, M. R. (1980). *Cytometry* **1**, 98–108.

Darzynkiewicz, Z., Traganos, F., Xue, S., Staiano-Coico, L., and Melamed, M. R. (1981). *Cytometry* **1**, 279–186.

Darzynkiewicz, Z., Williamson, B., Carswell, E. A., and Old, L. J. (1984). *Cancer Res.* **44**, 83–90.

Darzynkiewicz, Z., Traganos, F., and Kimmel, M. (1986). *In* "Techniques in Cell Cycle Analysis" (J. W. Gray and Z. Darzynkiewicz, eds.), pp. 291–336. Humana Press, Clifton, NJ.

Del Bino, G., Bruno, S., Yi, P. N., and Darzynkiewicz, Z. (1992). *Cell Proliferation* **25**, 537–548.

Evenson, D. P., Darzynkiewicz, Z., and Melamed, M. R. (1980). *Science* **210**, 1131–1134.

Gorczyca, W., Traganos, F., Jesionowska, H., and Darzynkiewicz, Z. (1993). *Exp. Cell Res.* **207**, 202–205.

Hotz, M. A., Traganos, F., and Darzynkiewicz, Z. (1992). *Exp. Cell Res.* **201**, 184–191.

Ichimura, S., Zama, M., and Fujita, H. (1971). *Biochim. Biophys. Acta* **240**, 485–489.

Kunicka, J. E., Darzynkiewicz, Z., and Melamed, M. R. (1987). *Cancer Res.* **47**, 3942–3947.

Kunicka, J. E., Olszewski, W., Rosen, P. P., Kimmel, M., Melamed, M. R., and Darzynkiewicz, Z. (1989). *Cancer Res.* **49**, 6347–6351.

Kunicka, J., Myc, A., Melamed, M. R., and Darzynkiewicz, Z. (1990). *Cell Tissue Kinet.* **23**, 31–39.

Subirana, J. A. (1973). *J. Mol. Biol.* **74**, 363–385.

Touchette, N. A., Anton, E., and Cole, R. D. (1986). *J. Biol. Chem.* **261**, 2185–2191.

Traganos, F., and Kimmel, M. (1990). *In* "Methods in Cell Biology" (Z. Darzynkiewicz and H. Crissman, eds.), vol. 33, pp. 249–270. Academic Press, San Diego.

Traganos, F., Darzynkiewicz, Z., Sharpless, T., and Melamed, M. R. (1979). *J. Histochem. Cytochem.* **27**, 382–389.

CHAPTER 34

Detection of Intracellular Virus and Viral Products

Judith Laffin and John M. Lehman

Department of Microbiology, Immunology and Molecular Genetics
Albany Medical College
Albany, New York 12208

I. Introduction

The utilization of flow cytometry (FCM) in the analysis of viral infection has grown to significant proportions. These analyses have included the appearance and quantitation of viral antigens, changes in cellular macromolecular synthesis, and events related to immunological changes following viral infection. Flow cytometry has provided an opportunity to assay single cells and the ability to correlate multiple parameters in a cell population. Further, it enables an analysis of the immunological system in response to the viral infection. All of these analyses have been performed with the flow cytometer utilizing a number of staining reactions. In this review we provide the protocol that our laboratory has utilized to study the permissive and nonpermissive infection of cells with Simian virus 40 (SV40). SV40 belongs to the papovavirus group and has a closed circular double-stranded DNA genome of 3×10^6 Da (~5200 bp) which codes for three virion proteins, VP_1, VP_2, and VP_3, and three nonstructural proteins, large (T) tumor antigen (94 kDa), small (t) antigen (17 kDa), and the agnoprotein (Brady and Salzman, 1986). SV40 is capable of a lytic infection (permissive) in monkey kidney cells and a transforming infection (nonpermissive) in numerous other cells (mouse, rat, hamster, human). In a permissive infection the early viral proteins are synthesized (T and t antigen), viral DNA is replicated, late viral proteins (VP_1, VP_2, and VP_3) are expressed, and complete infectious virions are assembled. In a nonpermissive infection only the early viral proteins are expressed with viral DNA integration and transformation of a few cells. The large T antigen has received considerable attention since this protein is known to initiate viral DNA synthesis in permissive cells and has multiple functions in the transformation of nonpermissive cells. The following events have been of particular interest in the studies of the role of T antigen in permissive and nonpermissive infection. The T antigen binds to the origin of replication of the viral DNA with several cellular proteins which leads to viral DNA replication. A number of activities have been associated with T antigen that are necessary for viral DNA replication; these include a helicase activity, an ATPase activity, and host proteins associated with DNA synthesis, including DNA polymerase α. Some of these host proteins have been demonstrated in cell-free systems such as PCNA (proliferating cell nuclear antigen), and topoisomerase I and II and the p53 and pRB gene products have been shown to bind to the large T antigen. While a definitive relationship to viral DNA synthesis has not yet been defined for p53 and pRB, these proteins are thought to play a role in viral DNA replication and cellular DNA synthesis stimulation in permissive and nonpermissive cells. With the synthesis of the viral DNA, the late proteins are synthesized (VP_1, VP_2, and VP_3) and can be detected in the nucleus as the virions begin to assemble. Along with the synthesis of these new viral proteins are associated certain cellular changes. The SV40 virus stimulates cellular DNA synthesis (Lehman and Defendi, 1970; Lehman et al., 1988). In the nonpermissive infection the early events occur (T and t antigen); however, viral DNA and the late proteins are not synthesized with a small percentage of cells acquiring a neoplastic phenotype. The large T antigen has received much attention since it has a role in the permissive infection but may be a major player in the transformation of cells.

The development of monoclonal antibody (mAb) to different epitopes of the T antigen has permitted an immunological analysis of the appearance and quantity of T antigen and its epitopes during the permissive and nonpermissive infections. The utilization of FCM with mAb to T antigen and the assay of other parameters, such as DNA, RNA, and other proteins, allow the quantity, distribution on single cells, and populations dynamics of T antigen to be correlated to the appearance of these macromolecules.

FCM analysis has provided information in defining the relationship of the initiation of cellular DNA synthesis with the appearance and quantity of T antigen during the lytic and nonpermissive infection (Lehman and Defendi, 1970; Lehman et al., 1988; Laffin et al., 1989). Of particular interest has been the initiation of multiple rounds of cellular DNA synthesis with SV40 virus in both permissive and nonpermissive cells. Without FCM we would have been unable to detect the stimulation of multiple rounds of DNA synthesis in permissive cells infected with SV40 virus. In studies with both wild-type and tsA30 SV40, a temperature-sensitive step has been defined utilizing the two-color fluorescent FCM analysis described below. The failure of the tsA30-infected CV-1 to be induced into a tetraploid S at 40.5°C identified a T antigen function which is required for the initiation of a second round of cellular DNA synthesis without a mitosis (Friedrich et al., 1992). When various monkey kidney cell lines were infected with differing multiplicities of SV40, cells differed in the amount of T antigen expressed per cell and the levels increased only 2-fold when 100-fold more virus was used. Further, while levels of T antigen differed, the permissive cells had higher G_2 phase levels of T antigen when compared to nonpermissive cells. This G_2 level was approximately a 2-fold increase over the G_1 levels (Lehman et al., 1993). In another series of studies with human diploid fibroblasts in pre- and postcrisis following infection with SV40, the DNA content of the infected cells shifted to tetraploidy and the levels of T antigen were higher in pre-vs postcrisis cells (Laffin et al., 1989). Further, the level of p53 increased in precrisis cells with increasing age and the levels of p53 were higher in SV40 postcrisis cells (Kuhar and Lehman, 1991). These investigations have provided information regarding the regulation of cellular DNA synthesis and how regulation may be modified by the SV40 viral T antigen. Future studies with FCM and other technologies will further define these viral events.

The following procedure outlines the staining protocol for SV40 T antigen in cells cultured in monolayer and stained with propidium iodide for DNA content (Laffin and Lehman, 1990). The mAb, PAb 101, was employed; however, antibody preparations once concentrated, purified, and characterized can be utilized. We have utilized the following monoclonals to T antigen: PAb 108, PAb 416, PAb 100; PAb 122 to p53; and polyclonal antibodies to the SV40 V antigens (Lehman et al., 1988; Laffin et al., 1989; Kuhar and Lehman, 1991). We have also applied this technology to other viral model systems to obtain information on various parameters of pathogenesis at the cellular level which might apply to the clinical disease. Results have indicated that the technique could be used as both a rapid diagnostic test and to study the pathogenesis of human infections such as herpes simplex virus (HSV-1), cytomegalovirus (CMV), and human immunodeficiency virus (HIV) (Elmendorf et al., 1986; McSharry et al., 1990a,b).

II. Materials and Methods

Procedures are summarized as follows:

1. Fixation
2. Reaction with primary antibody
3. Washes to remove primary antibody
4. Reaction with FITC-labeled secondary antibody
5. Washes to remove unbound secondary antibody
6. Incubation with RNase and propidium iodide (PI) (DNA staining)
7. Filtration of samples (prior to run on flow cytometer).

Following fixation of cells, which requires 30 min for 12 samples, the staining procedure is simple but long. It may take from 6 to 8 hr to complete, but only a small fraction of the effort is labor intensive. Thus, several runs of 12 samples can be processed in succession. Furthermore, the last incubation with RNase and PI can be combined and stained overnight at 4°C. Filtration is performed prior to flow analysis and can be accomplished in 10 min (12 samples). Time for data analysis varies:

A. Reagents for Fixation and Staining

1. Phosphate-buffered saline without Ca^{2+} and Mg^{2+} (PBS):
   NaCl, 8.0 g
   KCl, 0.2 g
   Na_2HPO_4, 1.2 g
   KH_2PO_4, 0.2 g.

Dissolve in 1 liter of distilled water (dH_2O), adjust pH to 7.4, filter sterilize, and store at room temperature (RT).

2. Trypsin-EDTA
   2.5% Trypsin (GIBCO, Grand Island, NY), 10.0 ml
   EDTA (tetrasodium ethylenediaminetetraacetate), 0.1 g
   PBS, 90.0 ml.

Filter sterilize, and store at 4°C.

3. Wash Solution (WS)
   Normal goat serum, 100.0 ml
   PBS, 900.0 ml
   Triton X-100, 20.0 μl
   Sodium azide, 1.0 g.

Heat-inactivate goat serum for 60 min at 56°C, filter sterilize, and store at 4°C.

4. Propidium iodide

> PI (Calbiochem-Behring, La Jolla, CA), 1.0 mg
>
> PBS, 100,0 ml
>
> Triton X-100, 20.0 μl
>
> Sodium azide, 0.1 g.

Store at 4°C in the dark.

5. RNase

> RNase A (Sigma Chemical Co., St. Louis, MO), 100.0 mg
>
> PBS, 100.0 ml
>
> Triton X-100, 0.1 g.

Boil for 1 hr and store at 4°C.

6. Antibodies

> *Primary Antibody.* A mAb or a polyclonal antiserum with high specificity is diluted in WS as determined by titration (Section III,D,1).
>
> *Secondary Antibody.* Commercially prepared affinity-purified goat anti-mouse

IgG F(ab')$_2$ fragments purchased from Boehringer-Mannheim (Indianapolis, IN) as a fluorescein conjugate (fluorescein isothiocyanate, FITC) are diluted in WS. The F(ab') is used to minimize nonspecific background of the Fc portion of the antibody. For a number of studies the whole antibody FITC-conjugated goat anti-mouse IgG has been used successfully and has been obtained from Antibodies, Inc. (Davis, CA).

B. Protocol

The following example demonstrates the processing of monolayers of SV40-infected human diploid fibroblasts (HDF) in 60-mm dishes.

Fixation procedure

1. Remove culture media and rinse once with PBS.
2. Add 1.0 ml trypsin-EDTA (warmed to 37°C) to each plate. Remove trypsin-EDTA after 15-30 sec.
3. Place at 37°C for 5 min or until cells detach. (Caution: Do not leave in trypsin-EDTA any longer than necessary.)
4. Add 1.0 ml of WS and resuspend cells using a 1000-μl pipetman.
5. Transfer to a 1.5-ml microfuge tube. Centrifuge in a variable speed microfuge (Fisher Model 59A) at 5000 rpm for 15 sec.

6. Remove supernatant; resuspend cells in 1.0 ml of cold PBS by vortexing gently.

7. Centrifuge as in step 5 above and remove supernatant.

8. Add 100 μl of cold PBS to cell pellet and thoroughly resuspend. Immediately add 900 μl of methanol at $-20°C$.

9. Count cells and adjust concentration to 1×10^6 cells/ml.

10. Samples can be immediately stained or stored at $-20°C$ for future use.

Staining with primary antibody

1. Centrifuge as in step 5 of above fixation procedure, remove fixative, and rinse once with 1.0 ml cold PBS.

2. Add 0.5 ml of primary antibody to each tube, and then vortex gently.

3. Incubate in a 37°C water bath for 2.0 hr.

Washes to remove unbound antibody

1. Centrifuge (fixation procedure, step 5), and then remove supernatant.

2. Add 0.5 ml of WS and vortex gently.

3. Place on ice for 15 min.

4. Repeat wash (steps 1–3 above).

Staining with FITC-labeled antibody

1. Add 0.5 ml of secondary antibody to each tube, and then vortex gently.

2. Incubate in a 37°C water bath for 2.0 hr.

Washes to remove excess antibody

Repeat above washing protocol (steps 1–4).

DNA staining

1. Add 0.5 ml of RNase solution and gently vortex.

2. Incubate at 37°C for 30 min.

3. Add 0.5 ml of PI solution to each tube and vortex (final volume 1.0 ml).

4. Place at 4°C in the dark for 30 min. (As an alternative, the RNase can be added to the samples after the final wash, mixed, and placed in dark at 4°C overnight; PI is then added approximately 1 hr before the samples are run.)

C. Filtration of Samples

Filter each sample prior to processing on the flow cytometer. This is accomplished by passing the sample through nylon mesh (50 μm) to remove cell clumps. The filtration should be repeated if samples have to be rerun at another time. Some preparations have been reanalyzed after a week at 4°C with little loss in signal, but generally samples should be processed within 48 hr. This is an important step since the cells may be prone to clumping which may clog the flow cytometer. Most mammalian cells that we have used are 25–30 μm in diameter; however, some cells are larger and therefore a larger nylon mesh and cytometer oriface may be needed. This can be determined by observing the cells with a microscope and measuring their diameter with an ocular micrometer.

Note that the antibody incubation times are optimal for the SV40 cell culture system. It may be the starting point for other antigen antibody measurements. For example, shorter incubations were adequate for assaying CMV-infected cells for early antigen (Elmendorf et al., 1988).

III. Critical Aspects of the Procedure

A. Criteria for Detection of Viral Antigen

1. Sample Processing

Sample choice is directed by the question being addressed but the initial processing of the cells may be handled by one of the following methods.

1. Experimental cell culture models, such as the SV40-HDF system, require removal of the cell monolayers with an enzyme (i.e., trypsin) without damaging the cells (Jacobberger et al., 1986; Lehman et al., 1988; Laffin et al., 1989).

2. Clinical samples including blood, bladder washings, and lung lavages must be prepared by a purification and concentration procedure (Elmendorf et al., 1988). Cells can then be fixed and stained (Section II,B).

3. Archival specimens (paraffin sections) may be used if the viral antigen is nuclear and stable. Techniques for the removal of paraffin and staining for DNA and a nuclear antigen have been described (Anastasi et al., 1987). Since cells are fixed, the sample may be processed immediately (see Hedley, Chapter 16, this volume).

4. Solid tissue must first be processed into a single-cell suspension using a method that neither damages the cells nor destroys the antigen. Cells are then fixed and stained (Section II,B).

2. Background

The amount of background fluorescence detected may vary with cell type. In studies with HDF, background fluorescence was high in the untransformed cell. When the cells were transformed by the SV40 T antigen, there was a lower background signal as a result of reduced autofluorescence and smaller cell size.

This made it imperative to establish standards of known negative and positive populations as controls for background fluorescence. In the SV40 studies we used an SV40-transformed Chinese hamster cell (A58) and as a control, B1, a normal Chinese hamster cell. The fluorescence of the negative cells (B1) and the positive cells (A58) was approximately 40–60 channels different on the y-axis (Laffin and Lehman, 1990). This provided a significant signal above the background.

3. Discriminating Factors

In addition to assaying for the antigen in question, it is desirable to use a second parameter relevant to the study to assist in identification of the positive population. For example at early postinfection time points of SV40-infected HDF, between 5 and 10% of the total cell population was positive for the PAb 101 epitope. When distribution of T antigen was examined in these cells using a second marker (DNA) to help discriminate the positive population, the histogram revealed that the majority of positive cells were in the G_2 peak (40–50%).

4. Cell Size

The size of the cell may influence the fluorescence measurements in several ways. First, it is more difficult to remove unbound antibody and PI from larger cells, and second, antigen concentrated in the nucleus of a small cell could have a brighter signal (peak value) than the same quantity of antigen diffused in the cytoplasm of a larger cell. The second difficulty can be eliminated by acquiring antigen signal using area (pulse integration) mode. Comparison of cell size can be monitored during flow analysis with forward angle light scatter. Additionally, the nuclear/cytoplasmic ratio may be evaluated and the location of the antigen identified by examination under the fluorescent microscope.

5. Cell Number

Cell number should always be kept constant. Ideal sample size is between 1.0 and 1.5×10^6 cells/ml. This allows for cell loss but retains the ability to collect between 10^4 and 10^5 cells per histogram. Lower cell counts are acceptable, but flow rate is adversely affected unless sample volume is adjusted.

B. Technical Difficulties in Fixation

Two factors that influence the selection of a fixative are (1) that the antigen is not lost or changed during processing and storage and (2) that the background fluorescence is not increased. A number of fixation techniques are available for analysis of intracellular antigens using FCM. To stain nuclear antigen and DNA, a method of lysis by detergent or osmotic pressure can be used, but results in the loss of cytoplasm. A second method utilizes detergent to permeablize the cell, followed by fixation with formaldehyde. A third method, alcohol fixation (methanol is used in our laboratory), both permeabilizes and fixes the cells (Jacobberger *et al.*, 1986). Methanol-fixed samples can be stored at $-20°C$ for >1 year with neither significant loss in quantity or quality of antigen nor change in background fluorescence, but repeated warming of a fixed sample appeared to be deterimental. Many potential problems can be identified by a preliminary evaluation under the fluorescent microscope. To this end, the cells should be grown on coverslips and then fixed and stained with the same reagents used for FCM.

In subsequent steps, fixation may result in a nonselective loss of cells. This is a consequence of cells adhering to surfaces and to each other. To maximize cell recovery, contact with new surfaces should be reduced by dispensing reagents into the sample tube and then resuspending by gentle vortexing. The pellet is dispersed easily when care is taken not to compact the cells. Aggregation is also reduced by the presence of serum and Triton X-100 in the staining solutions.

It is important to establish that the populations analyzed by FCM are a true representation of the original culture and that fixation did not result in modification of these populations. Changes in both the internal structure and the surface of an infected cell necessitate special attention during fixation and staining to prevent this type of alteration. During the lytic cycle of SV40 in monkey kidney (CV-1) cells, a concomitant increase in DNA content and cell mass with eventual cell lysis is observed. At late time points postinfection, there may be a selective loss of these large fragile cells by either lysis or removal during filtration. Additionally, the clumping caused by cell debris can bias the results. Careful monitoring of the sample condition at the time of collection allows adjustments to be made (i.e., shorter time in trypsin-EDTA).

C. Antibodies

One of the principal advantages of the FCM analysis of a population is the quantitation of viral and/or host proteins on a per cell basis. This demands that the antibody be specific and in saturation. Constant cell number per sample is required when the total population has a high level of epitope expression. A small percentage of very positive cells may not deplete the antibody, but if the sample contains a high proportion of cells with numerous epitopes, the results suggest an erroneous lower value of the antigen per cell (average FITC signal). This is a consequence of insufficient content of primary or secondary antibodies, with the result that not all epitopes are labeled.

Specificity and affinity of the antibody must be established. This was accomplished by employing a well-characterized antibody or determining reactivity of an undefined antibody using Western blot analysis and/or immunoprecipitation.

Again, examination of known positive and negative cells on coverslips may help determine whether staining was specific.

D. Standards

The sensitivity of this method necessitates including controls that will identify the problems in each of three areas: (1) antibody, both primary and secondary, (2) DNA content, and (3) sample. Some aspects have been covered in previous papers (Jacobberger et al., 1986; Laffin et al., 1989).

1. Antibody Controls

Primary antibodies are generally mAbs prepared from either cell culture or mouse ascites. After a 50% cut by ammonium sulfate, the antibodies are desalted and/or purified on a protein A-Sepharose column and stored at $-20°C$ in the presence of 0.1% sodium azide. Care should be taken not to contaminate the supply with proteases and DNases, which would lower the antibody titer and destroy DNA staining. The secondary antibody is a commercial anti-mouse IgG F(ab')$_2$ which proved to be very consistent. When needed, a dilution of each is made in WS and may be kept at 4°C for several weeks. There are numerous commercially available mAbs which may be used without purification but must be analyzed for background and dilution.

Each primary antibody is titrated using a standard positive and negative cell for that epitope (e.g., A58 and B1 were positive and negative, respectively, for T antigen epitope reactive with PAb101) (see Laffin and Lehman, 1990). The FITC-labeled secondary antibody is diluted 1:40, 1:60, and 1:80 and then tested with each of the primary antibody dilutions. The highest concentration that provides a >90% specific staining and does not alter other parameters should be selected. Examples of dilutions used were a 1:40 from a 50% cut of PAb 101 supernatant and a 1:80 for the secondary antibody. As a check, the positive and negative controls for that epitope are stained with the selected antibody concentration and processed with each experiment. Periodic microscopic examination of the antibody control samples is performed to confirm staining distribution. The technique described is an indirect immunofluorescent assay with a primary and second antibody. If the antigen detected is in low concentration multiple antibodies may be used to amplify the fluorescence in a multiple sandwich technique.

2. DNA Content

Included in each experiment described in this chapter, A58 and B1 cells (see Fig. 1) were used as standards for DNA content. These were checked for their x and y coordinates at the beginning and end of the instrument run and compared to previous assays. In addition, a methanol-fixed mouse lymphocyte control, treated with the RNase solution and stained with PI, was analyzed as a DNA content indicator and to confirm the red alignment. Each experiment also contained cells with a predicted DNA content for that cell type, stained in the same manner as the experimentals. Samples with questionable DNA content should be karyotyped.

3. Sample

Autofluorescence and nonspecific binding are determined by staining both infected and uninfected cells as follows: (1) one set with primary antibody and PI and no secondary, (2) one set with secondary and PI and no primary, (3) one set with PI only, and (4) one set with unrelated antibody of the same isotype as the primary, secondary, and PI. These samples are stained as described, substituting WS for the omitted reagent. The data obtained then allowed the selection of the proper control. An example of a normal cell with a high background is the uninfected HDF used in the human transformation experiments (Laffin *et al.*, 1989). These cells were then included as a control in each experiment with HDF.

IV. Flow Analysis

A. Instrumentation

Fluorescence measurements described in this chapter were performed with a Cytofluorograf IIs Model H.H. and a 2151 data analysis system (Ortho Diagnositcs, Inc., Westwood, MA), using a 5-W argon ion laser (Coherent, Palo Alto, CA) at a wavelength of 488 nm and 50 mW. For the past 2 years we have used the Omnichrome (Chino, CA) air-cooled argon laser Model 532 at a wavelength of 488 nm and 20 mW of power for excitation. This has provided a stable light source for our studies. Similar results have been obtained using a FACS IV (Becton–Dickinson, Sunnyvale, CA) and an EPICS V (Coulter Electronics, Hialeah, FL). The laser was aligned immediately before use with fluorescein-labeled microspheres (Polysciences No. 9847). For alignment, a yellow filter (40 nm width) was used for red fluorescence. The coefficient of variation (CV) was typically 1.0 for low-angle light scatter and 0.8 for both green and red. For analysis, a 535-nm band-pass filter was used for green and a 640-nm long pass filter was used for red fluorescence. No filter was used for light scatter. Data were collected and stored in a two-parameter cytogram as follows. First a gated population using light scatter versus red fluorescence (DNA) eliminated noncellular material. Cells passing through this gate were then analyzed using red area versus red peak for selection of single cells. Collected cells were then displayed in a third dual-cytogram of red area versus green area. These data were temporarily maintained in the Ortho data analysis system and then transferred to an IBM compatible PC for analysis and storage. Between 10,000 and 50,000 cells were collected on each control and sample.

B. Data Analysis

Direct evaluation of the data presented in this chapter was performed from the flow cytometer computer display. Estimation of the percentage positive and the relative DNA distribution was obtained when all controls were determined adequate. This lengthens the time between samples and increases the chance for variation. Each sample takes an average of 2–5 min to process. So, when possible, analysis using flow computer programs was delayed until after all samples were measured.

For storage and further analysis, the preliminary information was expanded by transfer of data to an IBM compatible PC. The broad range of FITC staining in many SV40 experiments necessitated collection of green fluorescence in log mode, which was then converted on the PC to linear as follows: $y' = 10[(y-40)/30]$. Red fluorescence had been collected in linear with the control diploid cells (B1 and A58) used to establish consistent location (channel number). For analysis, the cell number was correlated to both the x-axis (red, DNA) and the y-axis (green, T antigen). The mean y value was defined as the average FITC value after fluorescence compensation, which was necessary because of the overlap of the red and green signals. This can be accomplished either by a program available from the manufacturer of the flow cytometer or a computer subtraction program (Fogleman et al., 1994). The mean y value was used as the level of expression of the viral protein. The negative control was established by gating a known non-antigen expressing population of the same cell type so that <1% of this population was above the gate. For example, in SV40-HDF studies, uninfected HDF were used as the negative control. The percentage of positive was defined as the percentage of cells in a sample above this negative population. This allowed for correlation of cell-cycle compartment with quantity and percentage of cells expressing the viral protein.

V. Applications

The technique as outlined in this chapter has been used to study basic virus–cell interactions and the clinical aspects of virus infections. In our studies, the two-color fluorescence application described has been used to characterize a viral protein (T antigen) and cell-cycle changes of infected populations. These data have provided an opportunity to investigate how a virus alters the host cell cycle which produces either a lytic infection in permissive cells or abnormal growth in nonpermissive cells. Figure 1 demonstrates an analysis of the SV40 T antigen following infection of CV-1 cells with SV40. Figure 1A represents the background fluorescence of uninfected CV-1 cells with gate 1 defining the

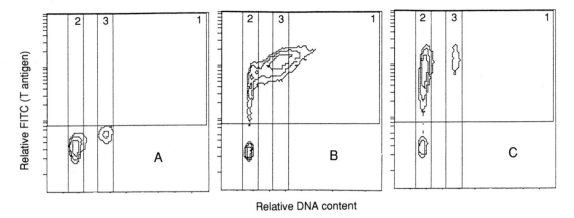

Fig. 1 Cell-cycle and T antigen expression. Mulitparameter contour plots of SV40-infected CV-1 cells stained with propidium iodide (PI) for DNA content and by indirect immunofluorescence (FITC) for the presence of T antigen using monoclonal PAb 101. Confluent CV-1 cells were infected with 100 pfu/cell of wild-type SV40 and incubated at 37°C. After 14 hr 400 μM of mimosine was added to half the cultures to arrest cells in late G_1. At various times postinfection, cultures were fixed and processed for FCM as described in the text. Histograms B and C represent the 33-hr infected cultures with and without the inhibitor, respectively. Gates were set using histogram A to identify the cells having a FITC value greater than the uninfected population (gate 1) and a DNA content of G_1 (gate 2) of G_2 (gate 3) phase of the cell cycle. These gates are shown in B and C to illustrate the changes in T antigen expression and cell-cycle progression.

T antigen-positive population (<1%). At 33 hr postinfection (Fig. 1B) cells are expressing T antigen (gate 1) with positive cells in both G_1 and G_2 phases of the cell cycle. Gates 2 and 3 represent all G_1 and G_2 cells, so additional gates are used to assess the cell-cycle phases of the positive population versus the negative population. When the cells are blocked by the inhibitor, minosine (Fig. 1C), the majority of the cells remain in the G_1 phase of the cell cycle (gate 2) and express T antigen as efficiently as unblocked cultures. By monitoring these data, information on the quantity and distribution of T antigen in relation to cell cycle will permit further studies on these populations using immunoprecipitation, Western blot analysis, and biochemical assays. This approach would provide an opportunity to investigate the T antigen's interaction with host proteins p53 and pRB, cyclins, and their associated kinases (Friedrich *et al.*, 1993). T antigen is a multifunctional protein with various regions identified as having specific functions. Monoclonal antibodies recognizing epitopes in these regions are available. Figure 2 provides an example of altered epitope expression following viral but not host DNA synthesis. FCM analysis revealed a loss of the carboxy-end epitope when viral replication occurred in tsA30-infected cultures at 37°C (Fig. 2A) but not at 40.5°C where viral replication is prevented. Since temperature might alter T antigen expression, cells were examined under the same conditions used in Fig. 2A but blocked before DNA synthesis occurred.

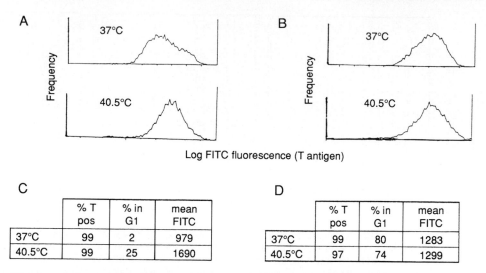

<table>
<tbody>
<tr><td>C</td></tr>
</tbody>
</table>

C

	% T pos	% in G1	mean FITC
37°C	99	2	979
40.5°C	99	25	1690

D

	% T pos	% in G1	mean FITC
37°C	99	80	1283
40.5°C	97	74	1299

Fig. 2 Decrease in T antigen epitope following viral replication. Temperature-sensitive T antigen stimulates both host and viral DNA synthesis in G_1-arrested cells at 37°C, but at 40.5°C, only host DNA is synthesized. To identify changes in T antigen associated with viral and not host replication, confluent CV-1 cultures were infected with tsA30 at 200 pfu/cell and placed at 37 and 40.5°C. At 45 hr postinfection, cultures were fixed and processed for FCM analysis and data collected as described in the text. The monoclonal PAb 101, specific for the carboxy end of T antigen, was used as primary antibody. A decrease in expression at 37°C compared to 40.5°C is presented both as (A) single-parameter histograms representing the distribution of total FITC fluorescence and (C) the mean FITC values for the entire population. To distinguish between the effect of temperature and replication on PAb 101 expression, cells were treated with 400 μM of mimosine at 12 hr postinfection and fixed at 45 hr postinfection. The distribution for the PAb101 epitope was similar at both temperatures (B) and the mean FITC values (D) were comparable. Further analysis of the multiparameter data is presented in (C) and (D).

As seen in Fig. 2B temperature did not alter T antigen expression when DNA synthesis did not occur. Multiparameter analysis confirmed the G_1 block and the distribution of T antigen in the population (Figs. 2C and 2D). These studies will be helpful in defining the role of T antigen in the permissive and nonpermissive infection.

There are a number of medically important viral infections that have been adapted to multiparameter FCM analysis. CMV causes significant morbidity and mortality in numerous groups of patients. Identification of this viral infection may take as long as 8 weeks to confirm with a combination of techniques such as virus replication in tissue culture, a rise in antibody titer, and the appearance of inclusion bodies. A number of new techniques, DNA probes, and mAb to CMV have been developed. We have utilized a specific mAb to an early CMV protein which can be identified in viral-infected cells with FCM thereby providing an opportunity to identify the virus (Elmendorf *et al.*, 1988). This early diagnosis would allow an assessment of the presence of the virus and the potential for drug treatment. An important application in viral diagnosis would be the ability to detect a small population of infected cells and follow that population throughout infection with CMV or with the HIV. There are a number of monoclonal antibodies available to HIV to detect the presence of the virus in cells and in patients (McSharry *et al.*, 1990a and b). These studies may prove useful in defining the development and progression of HIV infection.

A three-color immunofluorescence technique (Rabinovitch *et al.*, 1986) is available for the simultaneous measurement of the mAb and DNA content using standard 488 nM excitation. This would increase the sensitivity and permit the detection of the rare positive cells. More recently Srivastava *et al.* (1992) have modified this procedure, described above, by using a streptavidin–biotin based protocol for quantitation staining for the SV40 T antigen. This technique resulted the in a 2.4 times greater signal, therefore allows a gain in resolution in detecting the SV40 T antigen.

Acknowledgments

The authors gratefully acknowledge the contributions of Mr. David Fogleman and Dr. James Jacobberger in development of the staining protocol and Ms. Jo Ann D'Annibale for typing the manuscript. These studies were supported by Grant CA41608 from the National Cancer Institute.

References

Anastasi, J. A., Bauer, K. D., and Variakojis, D. (1987). *Am. J. Clin. Pathol.* **128**, 573–582.

Brady, J. N., and Salzman, N. P. (1986). *In* "The Papovaviridae: The Polyomaviruses" (N. P. Salzman, ed.), Vol. 2, pp. 1–26. Plenum, New York.

Elmendorf, S., McSharry, J., Laffin, J., Fogleman, D., and Lehman, J. M. (1988). *Cytometry* **9**, 254–260.

Fogelman, D., Jacobberger, J. W., and Lehman, J. M. (1994). In preparation.

Friedrich, T. D., Laffin, J., and Lehman, J. M. (1992). *J. Virol.* **66**, 4576–4579.

Friedrich, T. D., Laffin, J., and Lehman, J. M. (1993). *Oncogene* **8**, 1673–1677.

Jacobberger, J. W., Fogleman, D., and Lehman, J. M. (1986). *Cytometry* **7**, 356–364.

Kuhar, S., and Lehman, J. M. (1991). *Oncogene* **6**, 1499–1506.

Laffin, J., and Lehman, J. M. (1990). *In* "Methods in Cell Biology" (Z. Darzynkiewicz and H. Crissman eds.), Vol. 33, pp. 271–284. Academic Press, San Diego.

Laffin, J., Fogleman, D., and Lehman, J. M. (1989). *Cytometry* **10**, 205–213.

Lehman, J. M., and Defendi, V. (1970). *J. Virol.* **6**, 738–749.

Lehman, J. M., and Jacobberger, J. W. (1990). *In* "Flow Cytometry and Cell Sorting" (M. R. Melamed, P. F. Lindmo, and M. L. Medelsohn, eds.), 2nd ed. pp. 623–631. Wiley-Liss, New York.

Lehman, J. M., Laffin, J., Jacobberger, J. W., and Fogleman, D. (1988). *Cytometry* **9**, 52–59.

Lehman, J. M., Friedrich, T., and Laffin, J. (1993). *Cytometry* **14**, 401–410.

McSharry, J., Constantino, R., Robbiano, E., Echols, R. M., Stevens, R., and Lehman, J. M. (1990a). *J. Clin. Microbiol.* **28**, 724–733.

McSharry, J., Constantino, R., McSharry, M., Venezia, R., and Lehman, J. M. (1990b). *J. Clin. Microbiol.* **28**, 1864–1866.

Rabinovitch, P. S., Torres, R. M., and Engel, D. (1986). *J. Immunol.* **136**, 2769–2775.

Srivastava, P., Sladek, T. L., Goodman, M. N., and Jacobberger, J. (1992). *Cytometry* **13**, 711–723.

INDEX

VOLUMES IN SERIES

Founding Series Editor
DAVID M. PRESCOTT

Volume 1 (1964)
Methods in Cell Physiology
Edited by David M. Prescott

Volume 2 (1966)
Methods in Cell Physiology
Edited by David M. Prescott

Volume 3 (1968)
Methods in Cell Physiology
Edited by David M. Prescott

Volume 4 (1970)
Methods in Cell Physiology
Edited by David M. Prescott

Volume 5 (1972)
Methods in Cell Physiology
Edited by David M. Prescott

Volume 6 (1973)
Methods in Cell Physiology
Edited by David M. Prescott

Volume 7 (1973)
Methods in Cell Biology
Edited by David M. Prescott

Volume 8 (1974)
Methods in Cell Biology
Edited by David M. Prescott

Volume 9 (1975)
Methods in Cell Biology
Edited by David M. Prescott

Volume 10 (1975)
Methods in Cell Biology
Edited by David M. Prescott

Volume 11 (1975)
Yeast Cells
Edited by David M. Prescott

Volume 12 (1975)
Yeast Cells
Edited by David M. Prescott

Volume 13 (1976)
Methods in Cell Biology
Edited by David M. Prescott

Volume 14 (1976)
Methods in Cell Biology
Edited by David M. Prescott

Volume 15 (1977)
Methods in Cell Biology
Edited by David M. Prescott

Volume 16 (1977)
Chromatin and Chromosomal Protein Research I
Edited by Gary Stein, Janet Stein, and Lewis J. Kleinsmith

Volume 17 (1978)
Chromatin and Chromosomal Protein Research II
Edited by Gary Stein, Janet Stein, and Lewis J. Kleinsmith

Volume 18 (1978)
Chromatin and Chromosomal Protein Research III
Edited by Gary Stein, Janet Stein, and Lewis J. Kleinsmith

Volume 19 (1978)
Chromatin and Chromosomal Protein Research IV
Edited by Gary Stein, Janet Stein, and Lewis J. Kleinsmith

Volume 20 (1978)
Methods in Cell Biology
Edited by David M. Prescott

Advisory Board Chairman
KEITH R. PORTER

Volume 21A (1980)
**Normal Human Tissue and Cell Culture, Part A: Respiratory,
    Cardiovascular, and Integumentary Systems**
Edited by Curtis C. Harris, Benjamin F. Trump, and Gary D. Stoner

Series Editor
LESLIE WILSON

Volume 31 (1989)
Vesicular Transport, Part A
Edited by Alan M. Tartakoff

Volume 32 (1989)
Vesicular Transport, Part B
Edited by Alan M. Tartakoff

Volume 33 (1990)
Flow Cytometry
Edited by Zbigniew Darzynkiewicz and Harry A. Crissman

Volume 34 (1991)
Vectorial Transport of Proteins into and across Membranes
Edited by Alan M. Tartakoff

Selected from Volumes 31, 32, and 34 (1991)
Laboratory Methods for Vesicular and Vectorial Transport
Edited by Alan M. Tartakoff

Volume 35 (1991)
Functional Organization of the Nucleus: A Laboratory Guide
Edited by Barbara A. Hamkalo and Sarah C. R. Elgin

Volume 36 (1991)
***Xenopus laevis:* Practical Uses in Cell and Molecular Biology**
Edited by Brian K. Kay and H. Benjamin Peng

Series Editors
LESLIE WILSON AND PAUL MATSUDAIRA

Volume 37 (1993)
Antibodies in Cell Biology
Edited by David J. Asai

Volume 38 (1993)
Cell Biological Applications of Confocal Microscopy
Edited by Brian Matsumoto

Volume 39 (1993)
Motility Assays for Motor Proteins
Edited by Jonathan M. Scholey

Volume 40 (1994)
A Practical Guide to the Study of Calcium in Living Cells
Edited by Richard Nuccitelli

Volume 41 (1994)
Flow Cytometry, Second Edition, Part A
Edited by Zbigniew Darzynkiewicz, J. Paul Robinson,
   and Harry A. Crissman

Volume 42 (1994)
Flow Cytometry, Second Edition, Part B
Edited by Zbigniew Darzynkiewicz, J. Paul Robinson,
   and Harry A. Crissman

Volume 43 (1994)
Protein Expression in Animal Cells
Edited by Michael G. Roth

Volume 44 (1994)
Drosophila melanogaster: **Practical Uses in Cell and Molecular Biology**
Edited by Lawrence S. B. Goldstein and Eric A. Fyrberg

Volume 45 (1994) (in preparation)
Microbes as Tools for Cell Biology
Edited by David G. Russell

Volume 46 (1995) (in preparation)
Cell Death
Edited by Lawrence M. Schwartz and Barbara A. Osborne